7th Global Conference on Materials Science and Engineering (CMSE 2018)

IOP Conference Series: Materials Science and Engineering
Volume 474

Xi'an, China
1 - 4 November 2018

Editor:

Alexander Khotsianovsky

ISBN: 978-1-5108-8117-4
ISSN: 1757-8981

Printed from e-media with permission by:

Curran Associates, Inc.
57 Morehouse Lane
Red Hook, NY 12571

Some format issues inherent in the e-media version may also appear in this print version.

Printed by Curran Associates, Inc. (2019)

For permission requests, please contact the Institute of Physics
at the address below.

Institute of Physics
Dirac House, Temple Back
Bristol BS1 6BE UK

Phone: 44 1 17 929 7481
Fax: 44 1 17 920 0979

techtracking@iop.org

Additional copies of this publication are available from:

Curran Associates, Inc.
57 Morehouse Lane
Red Hook, NY 12571 USA
Phone: 845-758-0400
Fax: 845-758-2633
Email: curran@proceedings.com
Web: www.proceedings.com

Table of contents

Volume 474

The 7th Global Conference on Materials Science and Engineering

1–4 November 2018, Xi'an, China

Accepted papers received: 19 December 2018
Published online: 13 February 2019

Preface

Papers

IOP Conf. Series: Materials Science and Engineering **474** (2019) 011001 doi:10.1088/1757-899X/474/1/011001

Preface

IOP Conference Series: Materials Science and Engineering is publishing a volume of conference proceedings that contains a selection of papers presented at the 7th Global Conference on Materials Science and Engineering (CMSE 2018), which is an annual event started from 2012.

The 7th Global Conference on Materials Science and Engineering (CMSE2018) was held on November 1st-4th, 2018, in Xi'an, China. This conference was organized by School of Materials and Chemical Engineering, Xi'an Technological University, China, and it was hosted by Xi'an Technological University, Shaanxi Key Laboratory of Optoelectronic Functional Materials and Devices, Corrosion and Protection of Shaanxi Province, as well as supported by Northwestern Polytechnical University, Shaanxi University of Science and Technology, Xi'an Aeronautical University, China.

CMSE2018 was attended by about 200 participants from leading universities and research institutes as well as technological innovative enterprises from 26 countries, including Austria, Belgium, Brazil, China, Croatia, Czech Republic, France, Hungary, India, Italy, Japan, Kazakhstan, Lithuania, Poland, Republic of Korea, Romania, Russia, Singapore, Slovak Republic, Slovenia, Spain, Switzerland, Thailand, UK, Ukraine, and US.

The conference has provided a high-level national and international research for scholars, scientists, researchers, and engineers to present their results in the field of Materials Science and Engineering. Over 100 participants shared their oral presentations, and 26 shared their poster presentations mainly on 7 topics: Advanced Material Technologies, Energy materials and Semiconductors, Mechanics and Mechanical technologies, Nanomaterials, Surface and Coatings Technology, Characterization & Testing and Structural materials. Eight best oral presenters were awarded for their profound and impressive speeches.

This volume contains 62 papers, which were selected by a strict two-round blind-peer-reviews by international experts from the Conference Technical Committee and were considered innovative and instrumental to the further scientific research in the field of Materials Science and Engineering.

Finally, it is my pleasure to thank all participants for their contribution to CMSE2018 and the reviewers for their professional assistance in reviewing these papers, as well as the Organizing Committee, especially the Conference Secretary Ms. Li Ling, for the impressive results of the conference.

Guest Editor:
Dr. Alexander Khotsianovsky,
Managing Editor of Strength of Materials

Peer review statement

All papers published in this volume of *IOP Conference Series: Materials Science and Engineering* have been peer reviewed through processes administered by the proceedings Editors. Reviews were conducted by expert referees to the professional and scientific standards expected of a proceedings journal published by IOP Publishing.

Preparation of tin oxide/graphene composite for lithium-ion batteries' application

H L Cao, X Jiang, B J Song and Y C Shi*

College of Materials Science and Engineering, Shandong University, Jinan, Shandong 250061, China

*E-mail: yuanchangshi@sdu.edu.cn

Abstract. Tin oxide/graphene (SnO_2/GO) composites were successfully fabricated by a facile hydrothermal method with the seeding crystal method. The results of SEM and TEM observations strongly suggest that the excellent dispersibility of the product is influenced by the solution pH value. If the solution is pre-treated by the seeding crystal method before applying the hydrothermal method, the number of nanoparticles on the GO surface will be larger and the distribution of the product will be better, as compared to the product without such pre-treatment. The as-prepared SnO_2/GO also exhibits a reversible capacity of 750 mAh g^{-1} at the current density of 100 mAg^{-1}. It is mainly because nanomaterials facilitate the transfer of Li^+ between the active materials and the electrolyte. Besides, GO can act as a buffer during the charge/discharge process, which can effectively reduce the electrode damage during the volume change and ensure the electrochemical performance stability of the materials.

1. Introduction

Lithium-ion batteries (LIBs), as advanced energy storage devices, are among the most preferred power sources for electronic devices by the advantages of high energy density, high voltage and good life cycling [1-2]. In the composition of the LIBs, the selection of anode materials is critical for the electrochemical performance of the devices. The common use of anode materials is based on the graphite, because of its price and high electrochemical stability. However, their theoretical capacity can reach only 372 mAh g^{-1}[3], which restricts their promotion in the energy market. To meet the application demands, many researchers switch their focus to the transition metal oxide (TMO) materials, such as Fe_2O_3 [4-5], Co_3O_4 [6-7], SnO_2 [8-9], which have high theoretical capacity and can replace graphene as anode electrode materials. Among these materials, SnO_2 who is environmental friendliness, owns low intercalation potential, and high theoretical capacity (792 mAh g^{-1}) [10], has become one of the most preferred materials. However, its volume expansion is significant, and the volume change of tin particles could reach 300% during the charge-discharge process, which is similar to other pure TMOs. Most of the bulk materials will be damaged during the reactions. What's worse, leading self-discharge is also a general problem for SnO_2 materials, which is caused by the fragmented nanoparticles travel through the separator. In addition, further degrading the anodic also can happen at the same time after cycling [11]. Therefore, many researches came to focus on combining TMOs with carbon materials, like carbon, nanotube, graphene oxide, or reducing the size to nanoparticles. For example, Li et al. [12] synthesized SnO_2 nanoparticles with carbon coated and composite with graphene nanoribbons and carbon nanotubes, which have significantly improved the conductivity and improved its performance. As shown in the articles, the reversible capacity can reach 1038 mAh g^{-1} ath the current density of 100 mAh g^{-1}. After cycling 1000 cycles, the capacity can still remain about 500 mAh g^{-1}. Zhang et al.[13] fabricated phosphorus bridging SnO_2 and graphene composite electrodes by covalent bonding for LIBs. At the current density of 100 mA g^{-1}, the specific capacity can reach about 550 mAh g^{-1} over200 cycles. The electrochemical experiments showed that carbon materials used as component of the composites are helpful for the improvement of the performance. The widely used graphene oxide serve as the support materials to buffer the volume changes. Its high conductivity improves the transition of the ions as well. Moreover, the nano-scaled materials also provide more active points on the surface to improve the usage of the materials.

In this work, SnO_2 nanoparticles/graphene oxides composites are selected as anodes materials. The morphology of SnO_2 was adjusted by adjusting the pH of hydrothermal system in seed crystal-hydrothermal method, and the alkalinity solution is benefit for the formation of the composite with good distribution. And the number of the nanoparticles on the surface of the GO is also improved by the seed crystal method. When tested electrochemical performance of SnO_2/GO composites, it shows that the first charge / discharge capacity is 1420 mAh g^{-1} at the current density of 100 mA g^{-1}. After 11 cycles, the capacity remains about 750 mAh g^{-1}.

In this work, SnO_2 nanoparticles/graphene oxides composites are selected as electrode materials for lithium-ion LIBs. The morphology of SnO_2 was adjusted by adjusting the pH of the hydrothermal system in seed crystal-hydrothermal method, and the alkalinity solution is beneficial for the formation of the composite with a proper distribution. Moreover, the number of nanoparticles on the surface of the GO is also improved by the seed crystal method. When the electrochemical performance of SnO_2/GO composites was tested, it was found that the first charge/discharge capacity was 1420 mAh g^{-1} at the current density of 100 mAg^{-1}. After 11 cycles, the capacity remained at the level of 750 mAh g^{-1}.

2. Experimental

2.1. Materials

Tin (IV) chloride pentahydrate ($SnCl_4·5H_2O$, 99%), sodium hydroxide (NaOH, 99%), sodium dodecyl sulfate ($C_{12}H_{25}SO_4Na$, SDS, 99%), ethanol, N-methyl-2-pyrrolidone (NMP, C_5H_9NO, 99.0%) PVDF (C_5H_9NO, 99.0%) were purchased from Sinopharm Chemical Reagent Co., Ltd, China. Carbon black was supplied by Hefei Kejing Material Technology Co., Ltd, China.

2.2. Synthesis of SnO_2/graphene composite

It is prepared by seed crystal method and hydrothermal method. 3 mg GO was dissolved in 0.05 M 100 ml $SnCl_4·5H_2O$ aqueous solution and then left it for 12 h at the room temperature. Due to the hydrolysis reaction of $SnCl_4·5H_2O$, there would be some SnO_2 nanocrystals on the surface of the graphene. After that, the precipitation was collected after centrifugation and dried at 200°C. The final product was prepared by the hydrothermal method. The above precipitation was added into 35 ml aqueous solution with 0.1 M $SnCl_4$ • $5H_2O$, 0.1 M NaOH (0.5 M NaOH or 0.1 M HCl) and 0.3 M SDS, stirring about 30 min. Then the 100 mL Teflon-lined stainless-steel autoclave is used as the reaction container, which was added with the above solution and kept at 220 ℃ for 20 h. The products were centrifuged and washed with deionized water and ethanol for several times, respectively. After that, dry at 60 ℃ overnight for characterization and further usage. The product was named sample 1 in the following passage.

Another synthesize method was mainly based on the hydrothermal method. 50 mg $SnCl_4·5H_2O$ was dissolved in 40 ml GO solution (1mg ml^{-1}), which was added 10 ml SDS (0.3 M) and 10 ml NaOH (0.1 M). And then the above solution treated the same as above at 220 ℃ for 20 h. After that, collect the product by centrifugation and dry in the oven. Hereinafter, the product is referred to as "sample 2".

2.3. Characterizations

Rigaku Dmax-ray X-ray diffractometer is used to collect X-ray powder diffraction (XRD) pattern at a scanning rate of 4° min^{-1}, And a JSM-6700F field emission scanning electron microscope (SEM) is used

IOP Conf. Series: Materials Science and Engineering **474** (2019) 012001 doi:10.1088/1757-899X/474/1/012001

for structure examination. The transmission electron microscope (TEM) images are obtained by a JEOL JEM-2100 high-resolution TEM. Raman spectrometer was recorded on a LabRAM HR800.

2.4. Electrochemical measurement

Working electrodes were prepared as follows: mix the active material, PVDF, and carbon black at a weight ratio of 8: 1: 1. Then add NMP as the solution to form a homogenous slurry. After that coat slurry to a copper foil substrate and dried in a vacuum oven at 120 °C for 12 h. Li foil was used as the counter and reference electrode and 1MLiPF6 was used as electrolyte, with Celgard 2300 as separator. The cells were assembled in a glove box filled with argon atmosphere. As the active material, the performance of the cell was evaluated galvanostatically in the voltage range from 0.02 to 3 V at various current densities on a LAND CT2001A battery test system.

3. Results and discussion

3.1. Characterizations of SnO_2/graphene composite

The morphology and microstructure of the SnO_2/GO have been studied by SEM and TEM. As shown in Figure 1(a), the SEM image shows there are many nanoparticles on the surface of the GO, and the size of the particle is nanoscale. But the particles are all agglomerated with an uneven dispersion. When improving the concentration of the NaOH to 0.5 M, the number of particles on the surface of the GO is growing, showing in Figure 1(b). And the size of the particles stays the same as the product prepared in 0.1 M NaOH. The dispersion is also better, and most of the particles are located on the surface of the GO. However, in the product synthesized only by the hydrothermal method (as shown in figure 1(c) and (d)), the distribution of particles on the GO surface is uneven. Worse still, there are few particles on some of the GO surface. While in another place, there are more particles, but they are bonded in agglomerations.

Figure 1. SEM images of sample 1 at concentration of (a) 0.1 M NaOH, (b) 0.1 M NaOH; and images of sample 2 (c), (d).

The PH of the solution during the solvothermal and hydrothermal method has a significant effect on the structure formation of the products, which is widely used to control the morphologies of nano-/micro-scale materials [14-15]. Therefore, when the solution pH is transformed into acidity, by adding HCl, the structure of the product will be different. As shown in figure 2, there are no particles on the surface of GO, and the surface is much smoother than a product prepared in alkalinity solution. At the edges of the GO, some wrinkles are appearing on the surface, which may be caused by the erosion of the acid and the sedimentation of the irregular SnO_2. It means the SnO_2 particles cannot be formed and

IOP Conf. Series: Materials Science and Engineering **474** (2019) 012001 doi:10.1088/1757-899X/474/1/012001

deposit on the GO surface in the acid environment. These results also indicate the combination of the seed crystal and hydrothermal method can fabricate product with good dispersibility, and the different concentration of the NaOH will affect the dispersibility of the particles on the GO surface. The 0.5 M of the NaOH can obtain the largest number of the nanoparticles on the GO surface.

Figure 2. SEM image of SnO_2/GO composite prepared in acid solution.

Figure 3(a) shows the TEM image of SnO_2/GO composites prepared by seed crystal and hydrothermal method at high magnification. It can be found that the size of the particles is about 10 nm, and the lattice fringe can be observed. The typical lattice fringe spacing is measured about 0.48 and 0.68 nm, which were corresponded to the (200) and (100) crystal plane of SnO_2. Under low magnification image (figure 2(b)), the dispersibility can be studied, and the number of the particles is large. The product synthesized only by the hydrothermal method is displayed in figure 3(c), and there are few nanoparticles on the surface of the GO. It indicates that the size of the SnO_2 nanoparticles obtained by two different methods is almost the same. It proved that the crystal-hydrothermal method can effectively improve the number of nanoparticles on the surface of GO and can still maintain the morphology of the particles.

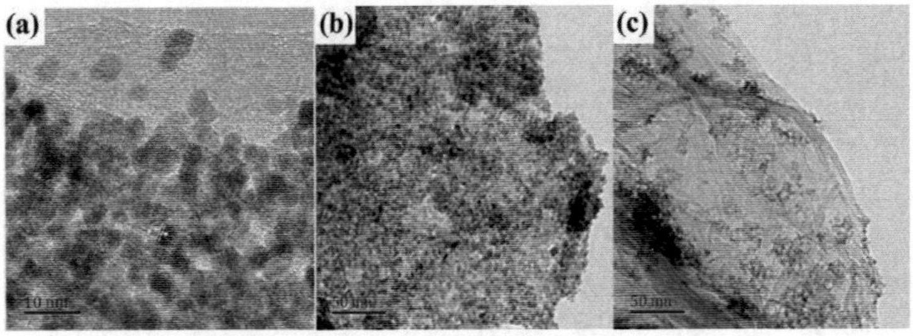

Figure 3. TEM images of (a), (b) sample 1 and (c) sample 2 SnO_2/GO composite.

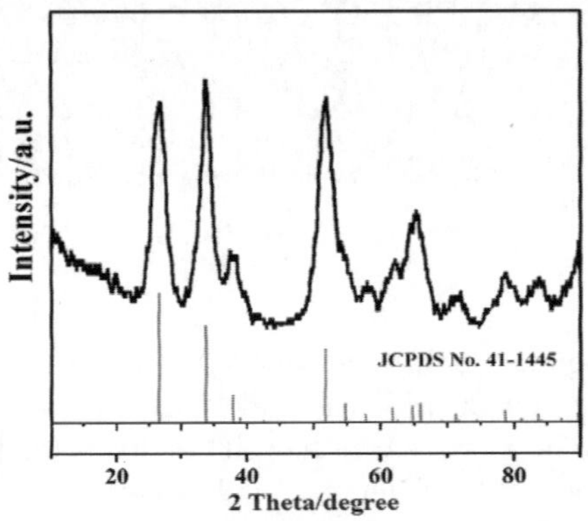

Figure 4. XRD patterns of the SnO_2/GO composite.

IOP Conf. Series: Materials Science and Engineering **474** (2019) 012001 doi:10.1088/1757-899X/474/1/012001

The crystalline phase and purity of products obtained by seed crystal and hydrothermal methods were studied by XRD. In figure 4, the diffraction patterns can be indexed to tetragonal SnO_2 (JCPDS No. 41-1445), and the peaks at 26.72, 33.92, 37.95, and 51.77° are indexed to the crystallographic planes of (110), (101), (200), (211), (112). There are still some small peaks that are fitted well with the JCPDS card, which means the prepared product is not purity, and efforts still need to be made to improve the purity. It may be caused during the washing process. The time for purifying should be prolonged.

Raman spectra are shown in figure 5. There are two peaks at about 1350 cm^{-1} (D band) and 1590 cm^{-1} (G band) observed in the composites, which are ascribed to the graphene substrate [16]. Besides, there is broad band from 360 to 960 cm^{-1} corresponding to the vibration modes of SnO_2 nanoparticles. Therefore, the Raman results are consistent with the TEM and SEM results, indicating the formation of the SnO_2/GO composite [17].

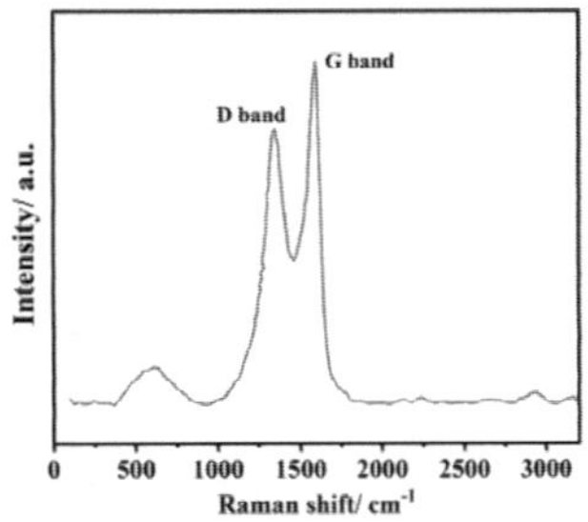

Figure 5. Raman spectra of SnO2/GO composite.

3.2. Electrochemical performance of the SnO_2/GO composite

In figure 6, the cycling performance of the SnO_2/GO is presented, which is obtained by seed crystal and hydrothermal methods, at the current density of 100 mA g^{-1}. The first discharge capacity can reach 1420 mAh g^{-1}, which is higher than theoretical capacity of 782 mAh g^{-1}. But the first Li^+ insertion capacity can only be reach 800 mAh g^{-1}. There is an obvious capacity loss, which is caused by the irreversible reaction and the formation of solid electrolytes interface. After 11 cycles, the specific capacity still maintains at 750 mAh g^{-1}. The coulombic efficiency of the product remains stable after 8 cycles, even though the number of the cycles is small. The reactions during charge and discharge process should be as below:[18]

$$SnO_2 + 4Li^+ + 4e- \leftrightarrow Sn + 2Li_2O \quad (1)$$

$$Sn + xLi^+ + xe- \leftrightarrow Li_xSn \quad (2)$$

It is assumed that nanomaterials facilitate the reactions between the electrolyte and the electrode, and the addition of the GO can buffer the volume change during the lithiation/delithiation process.

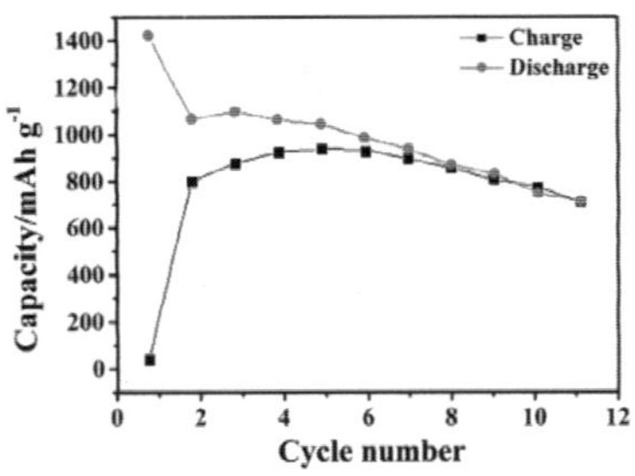

Figure 6. Cyclic performance of SnO_2/GO composite at the current density of 100 mAg^{-1}.

4. Conclusions

In conclusion, a simple method has been demonstrated to prepare the SnO_2/GO composite, and the SnO_2 is presented in the form of nanoparticles on the GO surface. In case of pre-treatment of the solution of $SnCl_4$ and GO before hydrothermal method, the dispersibility of nanoparticles on the GO surface will be better, and the number of the particles will be larger. Besides, the alkalinity solution is beneficial for the formation of the structure. When used as the anode materials, SnO_2/GO shows an excellent performance. At the current density of 100 mAg^{-1}, specific capacity can reach 750 mAh g^{-1} after 11 cycles, due to the nanostructure and addition of GO buffering the volume change during the reactions.

Acknowledgments
This work was supported by the National Natural Science Foundation of Shandong Province (ZR2014EMM017).

References

[1] Qu Z, Ren W, Xu L, Li F and Cheng H 2011 *ACS Nano* **5** 5463-71
[2] Li X, Gu H, Liu J, Wei H, Qiu S, Fu Y, Lv H, Lu G, Wang Y and Guo Z 2015 *RSC Advances* **5** 7237-44
[3] Lee J, Hwang T, Oh J, Kim J M, Jeon Y and Piao Y 2018 *J Alloy Compd* **736** 42-50
[4] Pan X, Duan X, Lin X, Zong F, Tong X, Li Q and Wang T 2018 *J Alloy Compd* **732** 270-9
[5] Ma T, Xu Y, Sun L, Liu X and Zhang J 2018 *Ceram Int* **44** 364-8
[6] Zhai X, Xu X, Zhu X, Zhao Y, Li J and Jin H 2018 *J Mater Sci* **53** 1356-64
[7] Li L, Jiang G, Sun R and Cao B 2017 *New J Chem* **41** 15283-8
[8] Nguyen T L, Park D, Hur J, Son H B, Park M S, Lee S G, Kim J H and Kim I T 2018 *Appl Surf Sci* **429** 218-24
[9] Sahoo M and Ramaprabhu S 2018 *Carbon* **127** 627-35
[10] Narsimulu D, Vinoth S, Srinadhu E S and Satyanarayana N 2018 *Ceram Int* **44** 201-7
[11] Dursun B, Topac E, Alibeyli R, Ata A, Ozturk O and Cakan R D 2017 *J Alloy Compd* **728** 1305-14
[12] Li X, Zhang X, Li T, Zhao X and Zhang X 2017 *J Alloy Compd* **729** 1064-71
[13] Zhang L, Zhao K, Yu R, Yan M, Xu W, Dong Y, Ren W, Xu X, Tang C and Mai L 2017 *Small* **13** 1603973-80
[14] Long H, Li Y and Zeng W 2017 *Mater Lett* **209** 342-4
[15] Hu C, Guo S, Lu G, Fu Y, Liu J, Wei H, Yan X, Wang Y and Guo Z 2014 *Electrochim Acta* **148** 118-26
[16] Liang J, Zhao Y, Guo L and Li L 2012 *ACS Appl Mater Inter* **4** 5742-8
[17] Li F, Song J, Yang H, Gan S, Zhang Q, Han D, Ivaska A and Niu L 2009 *Nanotechnology* **20** 455602-7
[18] Zhu X, Jan S. S, Zan F, Wang Y and Xia H 2017 *Mater Res Bull* **96** 405-12

Strain-controlled fatigue testing of TRIP800 high-strength quenching and partitioning steel

Q Y Jiang[1]
School of Mechanical & Electrical Engineering, Changchun Institute of Technology, Changchun 130012, China
E-mail: 472699610@qq.com

Abstract. In this paper, TRIP800 steel plates of thickness of 1.2 mm were subjected to the quenching and partitioning (Q&P) process and underwent strain-controlled fatigue tests. The respective experimental fatigue curves were plotted, and the influencing factors of fatigue resistance were identified and analysed, while the specimen fractographies were examined. The test results show that the existence of residual austenite inhibition of crack extension and strain induced the martensite phase transformation in the residual austenite. The residual compressive stresses improved the fatigue strength of Q&P steel plates under study. The fatigue fracture nucleation occurred in surface inclusions, the crack propagation zone exhibited the ductile fracture mode, while the transient area possessed the brittle fracture features.

1. Introduction

With the continuous expansion of production and marketing of automotive industry in China, the issues of energy saving, environmental protection, and safety become very topical. The efforts should also be focused on the formulation of relevant laws and regulations concerning the automobile material saving. High-strength steel application can ensure the light weight of automobiles, which is aimed by automobile and related industries [1-2].

In recent years, the researchers from Colorado University, US [3] developed a phase transformation-induced plasticity (TRIP) steel. After quenching and isothermal treatment above a certain temperature, the carbon and carbon became repeatedly included into martensite. The carbon by residual austenite to martensite distribution, thereby is gained by martensite and retained austenite consisting of two phases at room temperature of the high strength, plasticity and toughness. This process is called the quenching and partitioning process (Q&P) of martensite steel heat treatment [4-5]. Compared with the general high- and ultrahigh-strength steels, the proposed TRIP stee, which composition includes Fe, C, Mn, and Si, is cheap. Due to a lucrative combination of high strength and toughness achieved by the Q&P process, it draws the attention of international researchers, who seek to produce the Q&P steel at room temperature with low-carbon lath martensite and carbon-rich residual austenite.

However, the Q&P steel, as a lightweight substitute material for automobiles, must also have good fatigue properties, in order to have wider application prospects [6]. In this paper, through the Q&P processing of low-carbon and-low silicon TRIP800 steel plate, through the constant-strain fatigue test, the cyclic stress response of Q&P steel under the condition of constant-strain curve and strain fatigue life curve are established, and its fatigue resistance with the identification of influencing factors is analyzed. The macro- and microfracture of specimens are observed, the fatigue crack propagation and failure modes are discussed, and the basic data on fatigue design and reliability analysis of structural parts are provided.

2. Experimental

2.1. Test materials

The microstructure of TRIP steel is mainly composed of ferrite (F), superbainite (B + F), and residual austenite (γ), as shown in figure 1. Table 1 is the main component of TRIP800 steel plates for the test, while the main mechanical properties of steel plates are listed in table 2.

Table 1. Chemical composition of TRIP800 steel (wt %).

C	S	P	Mn	Si	Al
0.22	<0.005	0.021	1.87	1.54	0.050

Table 2. Mechanical properties of TRIP800 steel.

Tensile strength, MPa	Yield strength, MPa	Elongation, %
915.73	585.8	17.5

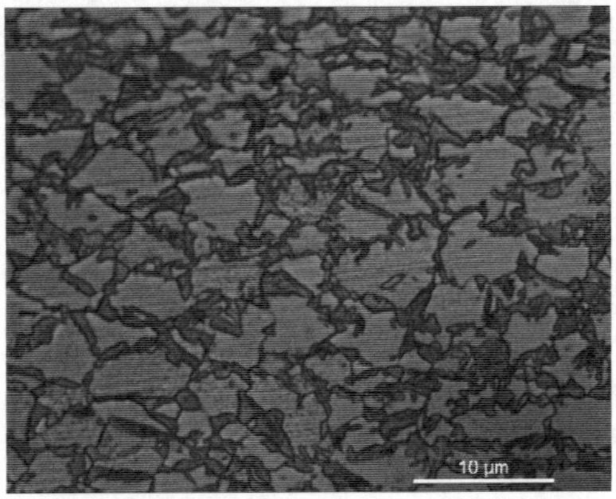

Figure 1. Microstructure of TRIP800.

2.2. TRIP800 steel plate Q&P processing

TRIP800 steel is in quenching after heating to 760 and 350°C, and then carbon distribution under the temperature of 400°C is to be heated for 1 min after cooling to the room temperature.

2.3. Test process

According to the national standard (GB -T 3075-2008) of China's metal axial fatigue test method, fatigue test is carried out on the electro - hydraulic servo fatigue test machine. The model of fatigue testing machine is the maximum test force of EHF-UM100k (SHIMADZU): dynamic plus or minus 100KN, static plus or minus 150KN, stress loading error plus or minus 0.5%, and strain measurement error plus or minus 0.5%. Load frequency is 0-50Hz, power is 380V. The microstructure of spot welding joint has been analyzed with the NiKonEPIPHOT300 optical microscope. The tensile fracture morphology has been observed by JSM-5600lv scanning electron microscope.

3 Results and Disscusion

3.1. Cyclic stress response curve

The cyclic stress response curve of the maximum stress of the corresponding cyclic number is shown in figure 2.

It can be seen in figure 2 that the curve shows a rising trend in the early stage of Q&P steel constant-strain fatigue experiment, which indicates that the phenomenon of cyclic hysteresis is presented. In the process of cyclic loading, while the average stress is less than the yield strength, but it still exists within the local microstructure of individual single crystal weak zone. Recycling early will appear when the slip. The slippage can cause the dislocation of the adjacent grain to be moved, and the dislocation will accumulate in the front of the obstacle, thus the deformation will be strengthened. When the dislocation reaches a certain critical value, the stress concentration reaches a certain critical value, and the dislocation will extend to other neighboring grains, because of the existence of the slip system, which is different from the production slip system. And, therefore, the strain-hardening occurs.

In the Q&P steel constant-strain fatigue test, the plastic deformation occurs in the maximum stress concentration zone. Because the existence of TRIP effect makes the residual austenite to martensite transformation, inhibition of the further plastic deformation occurs. If fatigue crack formation in the tip, the residual austenite to martensite transformation can produce significant work hardening, further set up the movement of the dislocation. Making the metal surface is not only difficult to form resides slip band, extrusion ridge of, or intrusion into a ditch, and delays the fatigue crack extension, but also possible strain aging, and can synthetically cause hardening phenomenon in this cycle until the number of cycles $N = 70000$ times or so is reached, the curve of a downward trend, stress gradually decreases with the cyclic time, and this stage for Q&P steel constant-strain fatigue test is of strain-softening stage. This stage is the initial boundary damage stage, which is accompanied by the formation of plastic deformation and a series of cavitation processes. Under the action of alternating load of reciprocating, the cyclic hardening reaches a turning point. Some fatigues reverse the process, such as dislocation reverses the movement caused by different number of dislocation interaction and offset, resulting in dislocation density decreased or cellular dislocation substructure change; or the interaction of dislocation can form micropores, which can be increased with the process development. In a region where the strain is aging, it dissolves and forms a weak zone with no aging. These processes are gradually playing an important role, which will lead to the strain-softening. The core of microcrack is formed when the microplastic deformation depletion and cavitation accumulation develop to a certain extent. The duration of this stage is the longest, and the evaluation of fatigue resistance of materials mainly considers the performance of this stage.

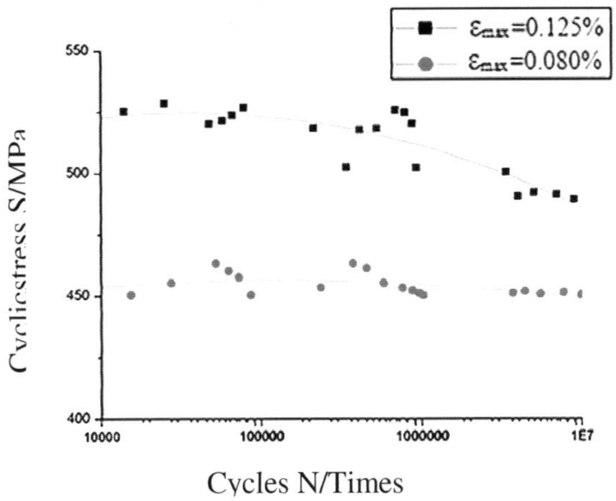

Figure 2. Cyclic stress response curve.

It can be clearly seen from both curves in figure 3 that the cyclic stress reduction trend corresponding to the two curves changes significantly during at N=1x10^5. The strain level is ε_{max}=0.125% at the final phase of the strain-softening, and as the number of cycles continues to increase, the stress shows the change of accelerating decline. The macroscopic crack occurs during this period and is visible to the naked eye, and the corresponding plastic deformation is increased rapidly, then the crack is rapidly expanded until the crack is broken. ε_{max}=0.080% strain levels for the curve, at the end of the strain-softening stage, with the continue increase of cycle number, small variations in stress, is gradually stabilized. The specimen did not break. In general, the process of cyclic hysteresis and strain-softening is finally terminated.

3.2. Fatigue crack source and crack fracture mechanism

Figure 3 depicts of the fatigue fracture of the specimen, which can be seen in figures 1 and 2. Figures 1 and 2 depict two fatigue sources are produced in different plane. The crack location in figure 1 is herringbone stripe extended to the right, and the fan is scaled up. The crack location in figure 2 in the final crack extension plane. The cracks in three places are connected by the coalescence, and the final fracture occurs by tear. The crack expansion plane produced by crack sources in figure 2 is very rough on the right side, indicating cracks originating from the crack source in figure 2 have propagated and led to the final fracture in this point.

Figure 4 is a scanning electron microstructure of specimen fatigue crack source. The fatigue of Q&P steel plate can be seen from the plate surface or the surface inclusion. The crack source is in the tear of the plate and there are many inclusions. The plastic flow is blocked here, causing a large stress concentration. The first initiation of the crack is the source of the crack.

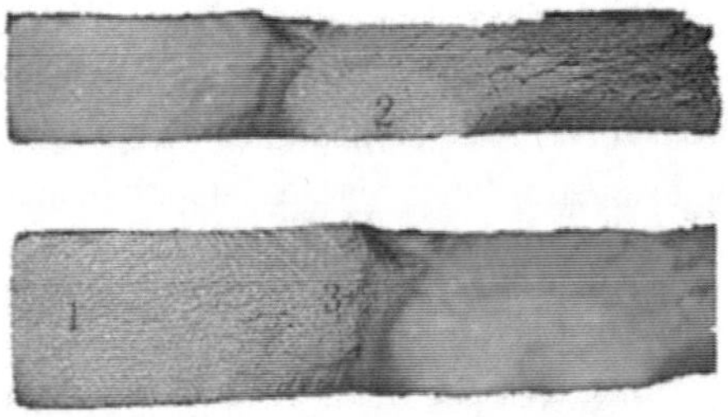

Figure 3. Fatigue fracture of specimen.

Figure 4. Scanning electron microstructure of specimen fatigue crack source.

Figures 5 and 6 are the SEM images of the fatigue crack propagation zone and the transient fracture zone. It is not difficult to see from figure 6 that there are two cracks in the crack propagation zone of the fatigue fracture. Because of the great stress at the tip of the crack, the Q&P steel can induce martensite around the crack propagation direction under the TRIP effect. The occurrence of martensite affects the stress field distribution at the crack tip, delays the crack propagation, and leads to the two cracks. Under the action of cyclic stress, the two cracks and the main crack expand at the same time. After a certain period of extension, the propagation rate of the main crack gradually slows down, and finally stops, while the new initiation of the two cracks replace the main crack and continues to expand. In the two cracks formation, a large amount of energy is absorbed. Therefore, the formation of the two cracks can well hinder the propagation of fatigue crack, which reflects the excellent strain fatigue property of fine grained structure of Q&P steel.

Figure 5. SEM of fatigue fracture crack propagation region.

Figure 6. SEM of fatigue crack in specimen.

As seen in figure 6, according to tearing edges and typical fracture, the plastic deformation occurred in the microregion with tearing edges. It is equivalent to the tensile fracture of plastic materials but here the necking zone, fracture zone fracture, and fatigue crack propagation area are more rough.

4. Conclusions

- ✓ The Q&P steel was subjected to strain-controlled cyclic tests with the loading frequency of 8Hz and zero strain ratio (R= 0). The fatigue strain level corresponding to the base number of cycles (10^7) was $\varepsilon_f = \varepsilon_{10^7} = 0.080\%$.
- ✓ The existence of residual austenite inhibition of crack extension and strain induced the martensite phase transformation in the residual austenite. The respective residual compressive stress improved the fatigue strength of Q&P steel plate.
- ✓ The fatigue fracture nucleation in the Q&P steel occurred in surface inclusions, the crack propagation zone exhibited the ductile fracture mode, while the transient area possessed the brittle fracture features.

Acknowldgement
This work was financially supported by the project of Jilin Provincial Science and Technology Department "The study on ultra-high gradient parts preparation process and organizational performance based on zoning stamping" (No. 20150060) and the project of Jilin Province Science and Technology Department "The research on 22MnB5 ultra-high strength steel hot stamping process". (No. 20140078)

References

[1] Melado A C, Nishikawa A S , Goldenstein H, Giles M A and Reed P A S. 2017. *IJF,* 40（2）104-109.
[2] Cheng G, Choi K S, Hu X and Sun X 2017 *Mater Design* 13（3）44-52.
[3] Chen L, Hu B, Xu J, Tian Y, Zheng X and Song J et al. 2017. *J Wuhan Univ Technol Mater. Sci. Ed.,* 32(5), 1179-1185.
[4] Dong J, Li C, Liu C, Huang Y, Yu L and Li H et al. 2017 *Mater Sci Eng A*, 24（2） 705-71
[5] Parthiban R, Chaudhury S G, Harikumar K C, et al. 2017 *Mater Sci Eng A* 10（5）340-347
[6] Melado A C, Nishikawa A S, Goldenstein H, et al. 2017 *IJ F,***12**（3）112-123

Reduction of stress concentration by polymer flexible joints in seismic protection of masonry infill walls in RC frames

A.Kwiecień*

Department of Civil Engineering, Cracow University of Technology
Warszawska 24, 31-155 Cracow, Poland

*Email:akwiecie@pk.edu.pl

Abstract. In the paper, the phenomenon of stress concentration occurring in brittle structural materials of infill structures is discussed. Four innovative repair methods based on flexible joints made of polyurethane PM are presented. They provide a reduction of the stress concentration and improvethe structural ductility. The effectiveness of newly proposed timber-polyurethane dissipative struts is proved by laboratory shear tests, which results are discussed in detail. The calculated dissipation energy and stiffness of the proposed system ensures a simultaneous transfer of loads and large deformations occurring during an earthquake to the introduced dissipative struts and improves the system safety. The recommended repair methods can be used efficiently by emergency teams in seismic areas.

1. Introduction

1.1. Masonry infills' behavior during an earthquake

Reinforced concrete (RC) frames' masonry infills are very popular structures in civil engineering, as constructions resistant to earthquakes. Anyway, they are vulnerable to large deformations, because stiff infills are rigidly connected to the RC frames. The infills contribute to lateral stiffness as well as the resistance of such construction systems,but are vulnerable to cyclic loads and crack easily during seismic excitation under stress concentrations generated by moving frames. Tensile stress initiates diagonal cracks and compression stress in corners causes crushing of masonry infills (figure1a) or even destroying of RC columns, what was presented in detail in [1-4]. Such cracked masonry panels may be very dangerous during aftershocks, because they may not be very well fixed to the framed structure as well as cohesion between defragmented masonry parts may be very low. This implies high costs and risks forinhabitants, state, and society because people are threatened by out-of-plane failure modes (figure 2). To protect such weakened structural elements, new innovative repair methods using flexible joints are proposed [5], which use flexible polyurethanes (PM polymer) – figure 1a.

1.2. Stiffness related to stress concentration

Currently, most common building materials transferring loads as concrete and masonry are characterized by relatively high stiffness (with the elastic modulusexceeding 1 GPa) and brittle behavior, in comparison to polyurethane joints, which elasticmodulus ranges from2 to 40 MPa. Taking into consideration large-scale structural elements carrying loads, like walls, beams, slabs or columns, high stiffness is an advantageous property because it allows constraining deflections (infill walls). On the other hand, high rigidity is accompanied by brittleness, manifested by a sudden failure of structural material (without warning). Results of large-scale compressive stress concentration are presented in figure 3, where crushing of corners in concrete pavement is caused by thermal elongation and "soft story" effect [6] initiated by infill strut. If the safety of users is required, this kind of behavior is preferably avoided in structural elements, especially in seismic areas, where ductile behavior is needed.

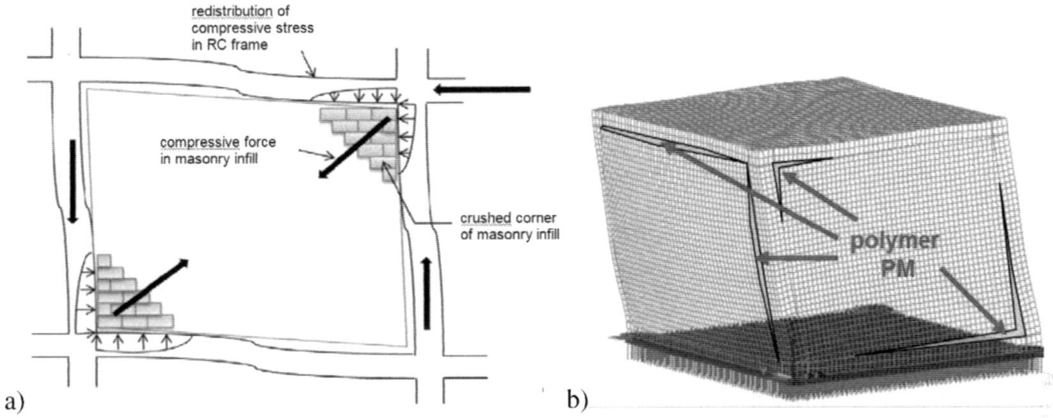

Figure 1. Infills behavior under large deformation: stiffly connected to RC frame (a), bonded to RC frame using flexible joint made of PM polyurethane(b).

Figure 2. Examples of infill failure: cracking and corner crushing (a), out-of-plane movement (b).

Figure 3. Examples of infill failure: concrete corner crushing in the pavement (a), RC column damaged by infill strut generating compressive stress concentration (b), "soft story" effect (c).

Focusing on the microscale behavior of concrete and masonry, the matrices of these materials include structural discontinuities (e.g., at an interface between gravel and mortar in concrete), being triggers of strain and stress concentrations when the material is loaded. Peaks of stress concentrations overcome the material strength with the increased load, forming microcracks, some of which get coalesced and propagated into the main crack. The rest of microcracks make the brittle material weaker in the vicinity ofnewly-formed cracks. In brittle materials, cracks develop and propagate, causing failure of structural elements. This is the reason why materials have to be protected against crack propagation and repaired [5].

1.3. Stiff and flexible repair injection
One of the repair methods is crack-filling using bonding materials (injection), which join cracks' sides, stopping their development. Various stiff bonding agents such as epoxy resins or cement-based fluids (with elastic modulus of 4-30 GPa and elastic properties) are used. Stiff inject bonds the crack, but microcracks in its surrounding are developing under stress concentration with reloading of the structure, creating a new rapidly propagating main crack even at the load level lower than the primary one. In such cases, the repair using stiff bonding agents failsto improve or even recover the initial strength of the cracked structure, because it fails to reduce peaks of stress concentration [7].

An innovative solution tothe external repair is usage of flexible bonding agents for crack-filling, which stiffness is much lower than that of the bonded structural material. Moreover, these flexible materials can withstand high deformations and are non-elastic, resulting in a reduction of stress concentration factor and stress redistribution. Such properties of the flexible repair bond allow for increasing the load-bearing capacity of cracked structural components made of brittle materials [5, 7]. The application conditions are quite strict (e.g., no dust, temperature over 5°C, etc.) and should be satisfied to ensure the above strength improvement.

2. Repair intervention in infill systems using flexible joints

2.1. Repair injection using polyurethane PM
As deformable injection material filling cracks in damaged masonry infills, flexible polyurethane Sika PM (with patented additives) is used. It is characterized by the following properties [1]: elastic modulus of 4.5MPa, tensile strength of 0.85÷1.95MPa, ultimate strain of 50÷150% (depending on the loading rate). This material was tested in many practical applications for full-scale structural elements and manifested itshigh efficiency [8].

Masonry infills are cracked in the in-and out-of-plane modes (figure 4a,b,c), causing the loss of stiffness resistance. In the location of cracks, stress concentrations are generated by earthquake-induced large deformation, causing further degradation. Repairs using stiff injection (cementitious mortar, epoxy resin) do not increase the ductility, which is required in seismic areas. On the contrary, the injection with polyurethane PM (figure 4c) allows to recover the in- and out-of-plane resistances, as well as significantly contribute to the ductility.

Flexible polyurethane can also be used as protection of infills and RC frames against stress concentration at the interface [1] in existing buildings. The space at the interface can be cut (1-3 cm-thick) and filled with Sika PM (figure 4d). In case o seismic excitation, local stress concentration (generated by the infill strut - figure 1a) is reduced by redistribution over a larger area. Moreover, ore ductile behavior of the whole system is introduced (figure 1b). When this kind of wall is damaged by a strong excitation, cracks can be repaired by polyurethane PM injection (figure 4e, f).

In newly-constructed buildings, a flexible joint can be made as a prefabricated layer of Sika PM. This layer is bonded at the inner surface of the RC frame, and next, masonry infill is erected (figure 5a). Properties of such system are very similar to the presented one, dedicated to existing buildings (figure 4d). Similarly to the previous case, a cracked wall damaged by a shock can be repaired by polyurethane PM injection (figure 5b,c). The efficiency of such a flexible interface solution was manifested in full-scale cyclic laboratory tests [1].

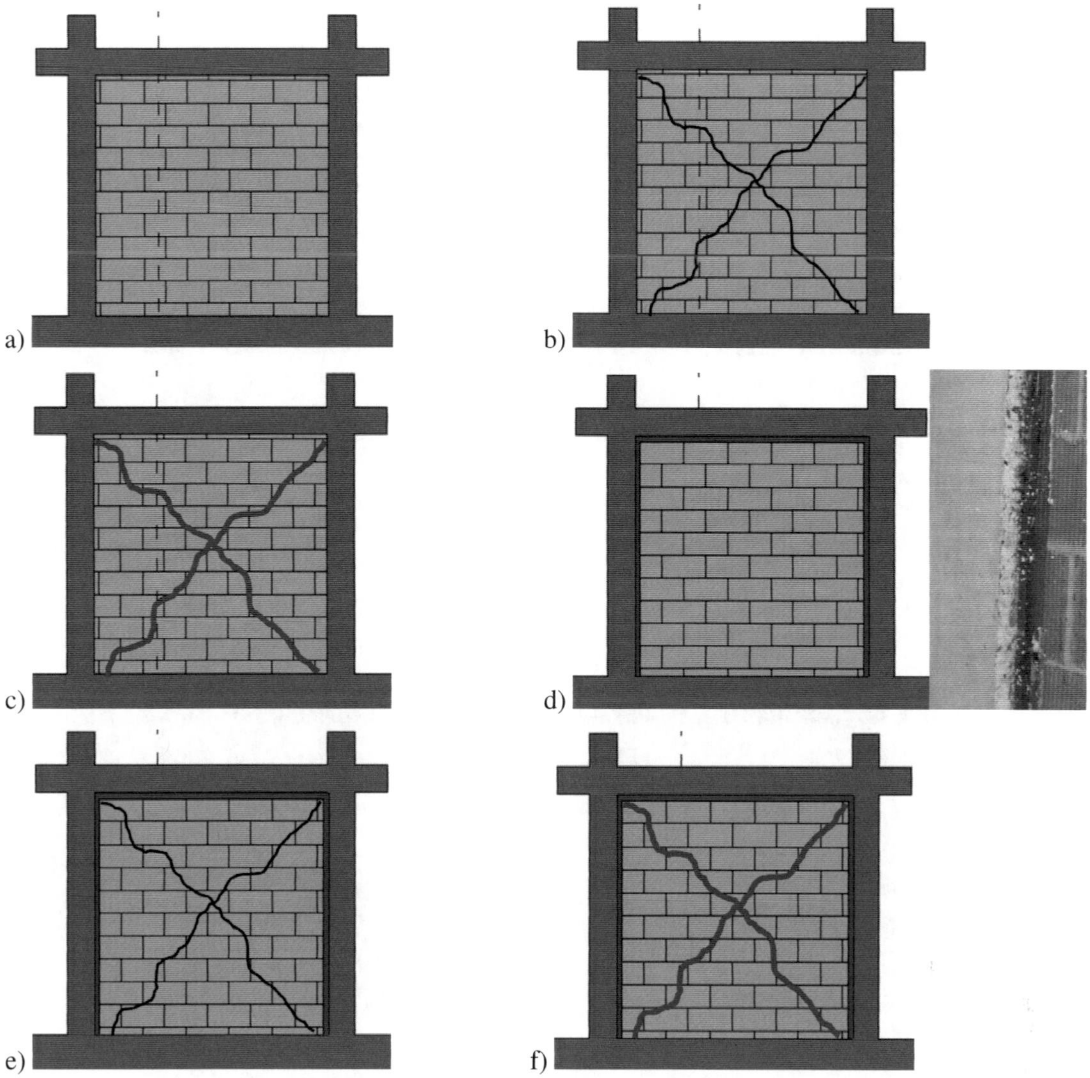

Figure 4. Infill walls with a stiff classical interface (a): cracked (b) and repaired by PM (c), and with flexible PM interface(photo of joint) in the existing building (d): cracked (e) and repaired by PM (f).

2.2. Emergency protection of "soft story."

Stress concentration causes cracks of infills and damages of RC columns in the shear mode (figure 3b). When infills collapse in out-of-plane mode (figure 2b), hinges are generated on both ends of damaged columns by inertial forces and initiate dangerous "soft story" effect (figure 3c). To protect the whole structure immediately against loss of stability and collapse, because of lack of shear resistance in empty RC frame (figure 5d), a new timber strut resistor with dissipative flexible joints is proposed. It consists of timber elements (with wood grainsparallel to the load) bonded together with polyurethane PM joints, working in two variants (figure 5e, f). Generally, the dissipative timber strut prevents horizontal movement of the upper frame part, blocking deformation at the compressive diagonal of the framein a visco-elasto-plastic manner. Energy is dissipated in both variants by shear deformation of the Sika PM joints. Extra dissipation is given by locking elements in the frame corners, fixed to the strut ends. In the first variant (figure 5e), extra dissipation assures deformation of the locking elements made of soft timber with wood grains perpendicular to the compressive load. In the second variant (figure 5f), the locking elements are made of polyurethane PM.

Figure 5. Newly-constructed infill with flexible PM interface (a): cracked (b) and repaired by PM (c); and "soft story" (d), protected by dissipative struts with ends made of soft wood (e) and PM (f).

2.3. Dissipation energy by timber elements and flexible joints

Innovative dissipation properties of the timber-polyurethane system were discovered during research in a project carried out by PalettenWerk on new wooden houses with flexible polymer joints, replacing steel dowels [9]. During shear tests of timber-polyurethane PM connections (figure 6), visco-elastoplastic deformation of PM joints and timber elements made of spruce class C24 (joint thickness 3 mm, bonding area 2x100x136 mm) was observed (figure 6b). Even after the damage of the polyurethane joint, residual resistance of the joint was observed (figure 6c and 7a,b,c). The representative element (one from 36 tested) dissipated the energy of 456 J (figure7a) during deformation, whereas the PM joint consumed only 130 J (figure 7b). The shear modulus of the connection (measured in the range of elastic behavior – figure 7c) was G = 1.15 MPa, which corresponded to the stiffness K = 88 kN/mm per 1 running meter of the flexible joint presented in figure 7d, when a horizontal load was applied tothe house designed by PalettenWerk. This example confirms significant dissipation and load-carrying abilities of the system.

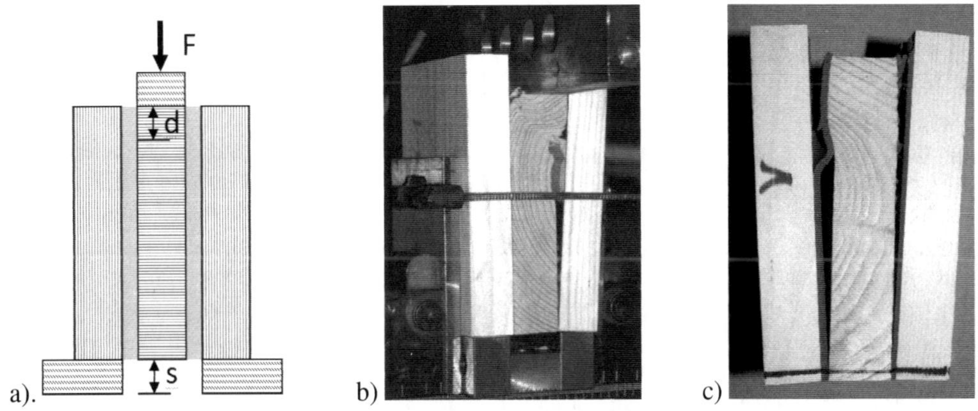

Figure 6. Shear test of PM joint between timber elements: setup (a), deformation of the specimen during the test (b), failure mode with visible permanent deformation of the inner element (c).

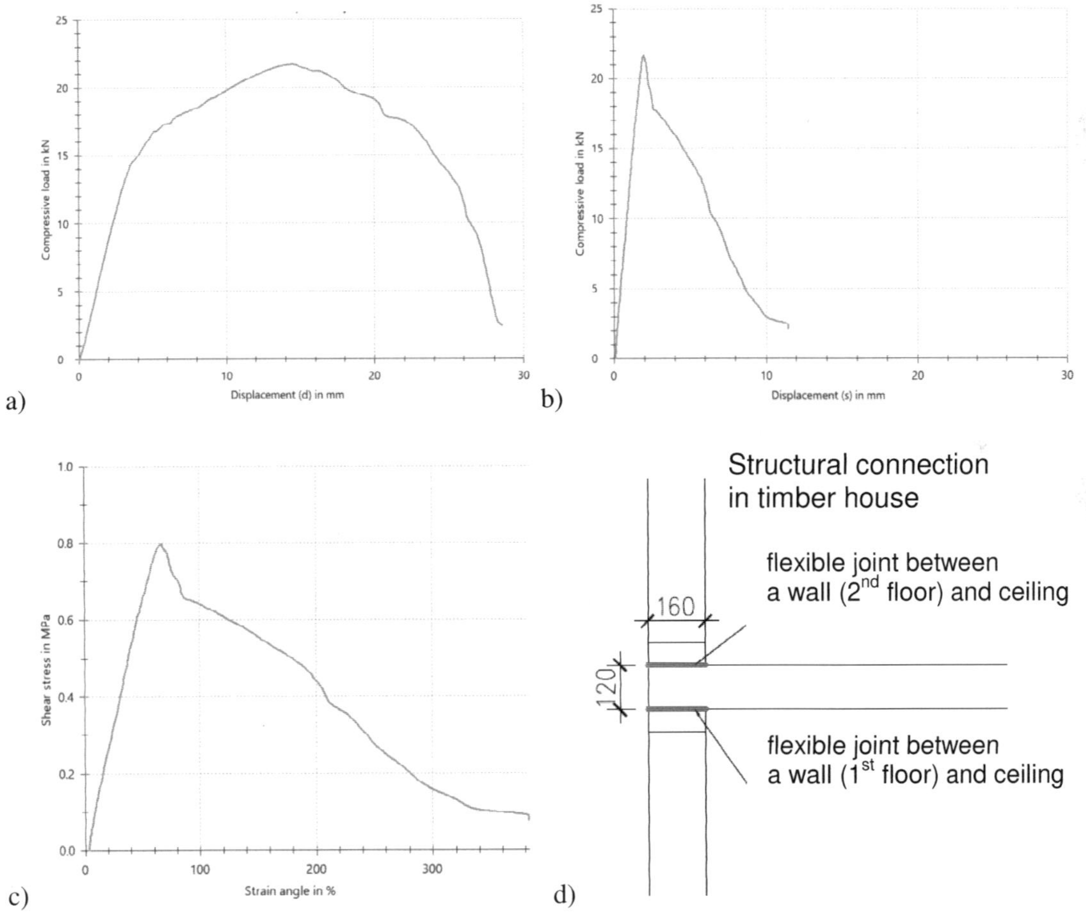

Figure 7. Test results: load vs. displacement of top surface "d" (a), load vs. displacement of bottom surface "s" (b), shear stress vs. strain angle (c); and structural connection using flexible joints (d).

3. Conclusions

The reduction of stress concentration by polymer flexible joints in seismic protection of masonry infill walls in RC frames was presented in 4ways, using polyurethane Sika PM, which reduced stress concentration and provided a significant stress redistributionin brittle materials (concrete, masonry), due to its high deformability and load-carrying capacity. The first method is based on PM injection, applied to cracks with damaged infills, recovering their load-bearing capacity and introducing higher ductility in them. The second one protects the existing infills against stress concentration and damage by creating a flexible interface in the space cut between infills and RC frames. The third one is a modification of the second method, as applied to newly-constructed buildings. The last proposal protects RC frames of "soft story" by using dissipative timber struts equipped with flexible joints made of polyurethane PM.

The first three methods use the innovative approach in civil engineering, in which highly deformable polyurethanes improve the load-carrying capacity and ductility ofrepaired structures, which were damaged during earthquakes, and assure a safe operation of infillstructures during aftershocks and further earthquakes. The fourth emergency method uses visco-elastoplastic properties of a new timber-polyurethane system for dissipating the deformation energy and protecting the "soft story" against collapse, by bearing the compressive load when inertial forces act during an earthquake. Additionally, locking elements of the strut ends with flexible timber elements (which grains are normal to the compressive load direction) or with PM ones, introduce extra dissipation abilities. Such a strut system is rapidly erectedand, thus, can be used by emergency teams in post-earthquake areas for the fast protection of damaged buildings against collapse in the expected aftershock. Advantageous properties of this system were confirmed by laboratory tests in the frame of the PalettenWerk project.

Acknowledgments

The authorwish to acknowledge the financial support from the project No POIR.01.01.01-00-0828/16, realized in Poland by the company PalettenWerk Kozik Sp. J.

References

[1] Kwiecień A, Gams M, Viskovic A, Kisiel P, Korelc J and Rousakis T 2017*Conference SMAR'2017, ZURICH (13-15.09.2017).*http://www.smar-conferences.org/smar/
SMAR_2017_Proceedings/papers/179.pdf

[2] Murty C V R, Brzev S, Faison H, Comartin C D, Irfanoglu A 2006*At Risk:A tutorial developed by a committee of the World Housing Encyclopedia.* Second Printing. Earthquake Engineering Research Institute, Oakland, California.

[3] Semnani SJ, Rodgers JE and Burton HV 2014*Earthquake Engineering Research Institute, Thornton Tomasetti Foundation.*

[4] Shah S A A, Shahzada K, Afzal S, Shabab M Eand Khattak N 2013.*International Journal of Advanced Structures and Geotechnical Engineering*, Vol. 02, No. 03, 106-109.

[5] Kwiecień A 2013*GSTF International Journal of Engineering Technology (JET),* **2**(1),pp 182-196.

[6] Guevara-Perez T L 2012 "SoftStory"and"WeakStory"inEarthquakeResistantDesign:AMultidisciplinaryApproach. *15th WCEE Conference*, Lisboa.

[7] Zdanowicz Ł, Kwiecień A and Seręga S 2017*Procedia Engineering* **193**, pp 517-524.

[8] Kisiel P 2015 *Procedia Engineering***108** pp 496 – 503.

[9] Kozik T and Kwiecień A 2018*2nd Forum Holzbau Polska 18.* (Warsaw 6-7.03.2018), ISBN: 978-83-937493-7-9, pp 57-71.

A generalized saturated capacitor model for hysteresis in piezoelectric materials

Y Liu*, M Huo and M Tang

Department of Aerospace Engineering, Harbin Institute of Technology, Harbin, 150001, China

*E-mail: liu-yanfang@hotmail.com

Abstract. A generalized saturated capacitor (GSC) model is proposed for hysteresis in piezoelectric materials. It introduces the Weiss domains (WDs) with a negative capacitance switch (NCS) into the saturated capacitor model. The WD with NCS is modeled as a parallel network of a charge-limited capacitor and perfect one. It is further made equivalent to a series network of a voltage-limited capacitor (VLC) with negative elastance values and a perfect capacitor, which induces the GSC model to become a series network of several VLCs with positive or negative elastances. The GSC model can capture non-convex hysteresis loops and is self-invertible, namely the inverse GSC model has the same structure as the original (forward) one but differs by its parameters. The proposed model effectiveness is corroborated by the experimental data, with the maximal discrepancy not exceeding 1%.

1. Introduction

Piezoelectric materials are characterized by many advantages such as fast response and high stiffness. They are widely used as sensors or actuators in many applications such as nano-manipulation, active vibration control, and nano-positioning [1]. However, the unwanted hysteresis limits their performance. It can lead to errors up to 15\% of the full range of a piezoelectric actuator (PEA) or even instability of the system [1, 2].

Modeling and compensating for hysteresis is reliable and straightforward. Thus, it becomes one of the most popular approaches to handle it [3, 4]. The hysteresis is described by models such as the Preisach model [5, 6], Prandtl-Ishlinskii model [7-11], and Bouc-Wen model [12, 13]. Then, the corresponding inverse model is constructed and connected in series with a PEA to eliminate hysteresis [9, 14]. Unfortunately, these models are a purely mathematical description of the observed hysteresis phenomenon and provide little physical interpretation. Hence, they hardly give any insight into the generation mechanism of the hysteresis.

A hysteresis model that is based on the physical principle is formulated by James C. Maxwell [15]. Its electric interpretation is referred to as the saturated capacitor (SC) model [16, 17]. The SC model represents the Weiss domain (WD) by a series network of a voltage-limited capacitor (VLC) and a perfect capacitor. Since a PEA contains many WDs, it is modeled by a series network of several VLCs. Unfortunately, as the SC model only considers WDs with positive capacitance switch (PCS), it can just capture convex hysteresis loops. A modified model that is capable of capturing non-convex hysteresis is proposed in [18]. However, it is too complicated to generate the inverse model.

In this paper, the WD with a negative capacitance switch (NCS) is introduced into the SC model. The WD with NCS is modeled by a parallel network of a charge-limited capacitor (CLC) and a perfect capacitor, which is further approximated by a serial network of a VLC with negative elastance and a perfect capacitor. Then a generalized SC (GSC) model is constructed, which has the same structure as the SC model but allows some VLCs to have negative elastances. If a GSC model is connected in series with a PEA, it can pre-shape the control input of the PEA and generate the inverse hysteresis to compensate for the nonlinearity.

2. Modeling and identification of hysteresis in PEAs

2.1. Saturated capacitor model

A PEA consists of many WDs, each of which has the same dipole orientation. When an electric field is applied to a WD, it performs like a capacitor. However, its capacitance is not constant. Without loss of generality, the continuous change of capacitance is simplified to be a switch between two constants.

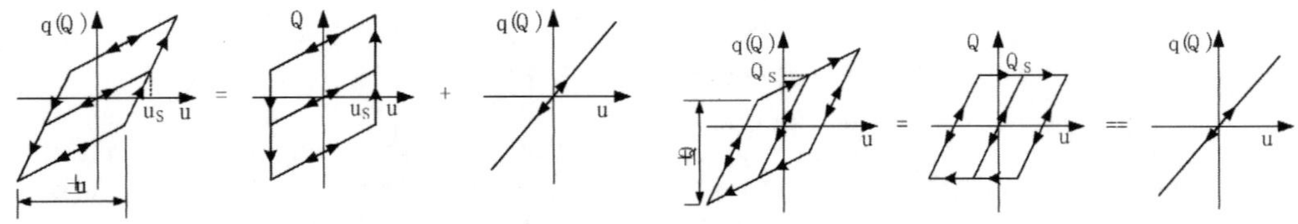

Figure 1. Charge-voltage property for a WD with (*a*) PCS and (*b*) NCS. Symbol u is the excitation voltage of the WD. Symbols Q and q are, respectively, the charges applied to and stored in the WD. Symbols u_S and Q_S represent the saturation voltage and the saturation charge, respectively. Symbols δu and δQ represent the changes of the applied voltage and charge, respectively. Symbols + and // represent series connection and parallel connection, respectively.

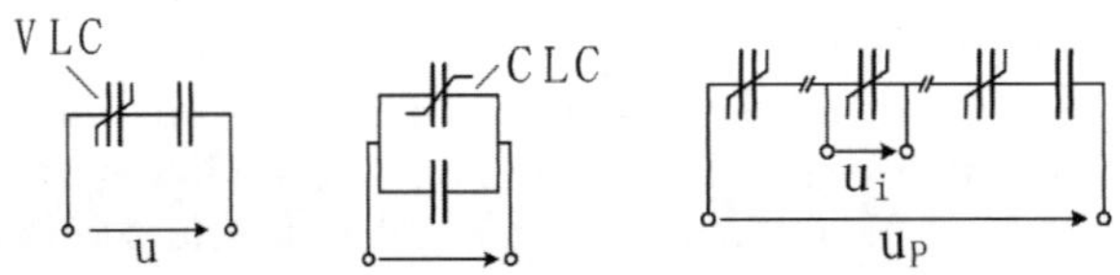

Figure 2. Equivalent circuits for a WD with (*a*) PCS, (*b*) NCS, and (*c*) for the SC model.

The charge-voltage property of a WD with PCS [16, 17] is shown in figure 1(*a*). The increase/decrease of the applied voltage δu has a threshold value. Before δu reaches this value, the WD performs like a perfect capacitor, namely the change rate of the charges stored and applied to it, i.e. $\dot{q}$, and $\dot{Q}$, are proportional to the change rate of the voltage applied to it, i.e., $\dot{u}$. Once the increase/decrease of the voltage reaches the threshold value, the capacitance switches to a more significant value. The capacitance always turns back to the smaller value if the change of the voltage switches direction. The WD with PCS can be equivalent to a series network of a VLC and a perfect capacitor, as shown in figure 2(*a*). The charge-voltage property of a VLC is also given in figure 1(*a*). The voltage across a VLC has a saturation value u_S . If $\dot{Q}$ does not change direction and u does not reach u_S, $\dot{q}=\dot{Q}=\dot{u}/S$. Once u_S is reached, charges cross the VLC without any resistance thus $u=u_S$ and $Q=Q_S$. Thus, the VLC is governed by

$$\dot{q}=\begin{cases}\dot{Q} & |q|<Q_S\\ 0 & |q|\geq Q_S\end{cases} \tag{1}$$

$$u=Sq \tag{2}$$

where S is the elastance (i.e., the inverse of capacitance) and $S \in \mathbb{R}^+$.

Insofar as PEA consists of many WDs, the SC model represents it by a series network of several VLCs and one ideal capacitor, as shown in figure 2(*c*). Herein, the ideal capacitor is equivalent to the series network of ideal capacitors of all the WDs. It can be seen as a VLC with infinite saturation charge. Meanwhile, it can avoid computation singularity in the simulation—the charge becomes infinite if the voltage increases slightly after all the VLCs are saturated, which leads to singularity in the simulation. Then, the governing equation of the SC model is

$$\dot{q}_i = \begin{cases} \dot{Q} & |q_i| < Q_{S,i} \\ 0 & |q_i| \geq Q_{S,i} \end{cases} \tag{3}$$

$$u_P = \sum_{i=1}^{n} S_i q_i \tag{4}$$

Its discrete form is

$$q_{i,k+1} = \begin{cases} \hat{q}_{i,k+1} & |\hat{q}_{i,k+1}| < Q_{S,i} \\ Q_{S,i}\mathrm{sign}\hat{q}_{i,k+1} & |\hat{q}_{i,k+1}| \geq Q_{S,i} \end{cases} \tag{5}$$

$$\hat{q}_{i,k+1} = u_{i,k} + Q_{k+1} - Q_k \tag{6}$$

$$u_{P,k+1} = \sum_{i=1}^{n} S_i q_{i,k+1} \tag{7}$$

where i and k are the serial numbers of VLCs and samples, respectively. Equation (5) is the state equation and updates q_i. Equation (6) calculates virtual states $\hat{q}_i$. Equation (7) gives the output voltage u_P.

As this model considers WDs with PCS only, it can hardly describe the non-convex hysteresis. Some WDs may have NCS. In the next subsection, the SC model is generalized by introducing WDs with NCS.

2.2. Generalized saturated capacitor model

The charge-voltage property of a WD with NCS is shown in figure 1(*b*). The increase/decrease of the applied charge δQ has a threshold value Q_S. Before δQ reaches this value, q and Q are proportional to u. The capacitance becomes smaller once δQ reaches Q_S. The capacitance always switches back to the more significant value if the change of u switches its direction to the opposite one. The WD with NCS can be represented by a parallel network of a CLC and a perfect capacitor, as shown in figure 1(*b*). The charge-voltage property of a CLC is also given in figure 2(*b*). The charge applied to a CLC has a saturation value Q_S. Before Q reaches to Q_S or the change of u switches direction, $\dot{q} = \dot{Q} = C\dot{u}$. The CLC has similar behavior as a perfect insulator once Q reaches Q_S, namely, charges cannot cross the CLC and $q = Q = Q_S$. Its governing equation is

$$\dot{Q} = \begin{cases} C\dot{u} & |Q| < Q_S \\ 0 & |Q| \geq Q_S \end{cases} \tag{8}$$

where C is the capacitance.

Considering WDs with NCS, the equivalent circuit of a PEA becomes quite complex. Fortunately, the parallel network of a CLC and a perfect capacitor can be represented by a series network of a VLC and a perfect capacitor. Let

$$C_I + C_C = \frac{C_V C_I}{C_V + C_I} \tag{9}$$

where C_I, C_C, and C_V are capacitances of the ideal capacitor, CLC, and VLC, respectively. Then

$$C_V = -\frac{C_I^2}{C_C} - C_I \tag{10}$$

Since C_I and C_C are positive, then C_V is negative. Thus, equation (10) indicates that a WD with NCS can be represented by a WD with PCS. However, the elastance $S = 1/C \in \mathbb{R}$. Thus, the GSC can be represented by the circuit given in figure 2(*c*) and the formula in equation (3).

2.3. Parameters' identification

The initial parameters can be calculated from the initial ascending curve [16, 17] or the major ascending/descending curve. The main points are separating the curve into n segments, calculating the average slope of each segment, and then obtaining $Q_{S,i}$ and S_i. Further, the parameters are optimized by solving the minimization problem

$$\min_{\substack{S_1,\dots,S_n\\ Q_{S,1},\dots,Q_{S,n}}} \sqrt{\frac{1}{N}\sum\nolimits_{k=1}^{N}\left[Q(t_k)-\mathrm{GSC}[u](t_k)\right]^2} \tag{11}$$

where GSC denotes the GSC model operator and N is the number of samples. For simplification, $Q_{S,i}$ is treated as fixed, i.e., evenly distributed in the desired range, and only S_i are optimized as follows

$$\min_{S_1,\dots,S_n} \sqrt{\frac{1}{N}\sum\nolimits_{k=1}^{N}\left[Q(t_k)-\mathrm{GSC}[u](t_k)\right]^2} \tag{12}$$

3. Properties of the GSC model and experimental validation

According to the above analysis, the generalization of the SC model is quite simple by replacing the constraint $S_i \in \mathbb{R}^+$ by $S_i \in \mathbb{R}$. However, this simple replacement yields some remarkable properties, namely (i) self-invertible property and (ii) non-convex hysteresis property.

3.1. Self-invertible property

Property 1 (Self-invertible property). The GSC model is self-invertible, which means that the inverse of a GSC model has the same structure as the GSC model but with different parameters.

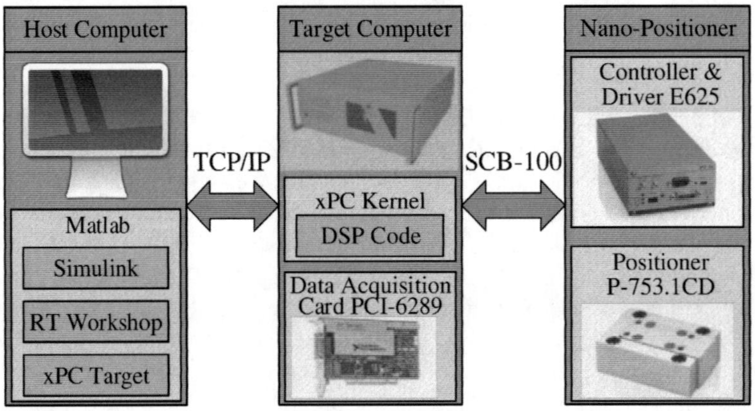

Figure 3. Experiment setup of the PEA-based nanopositioning system.

A PEA-based nano-positioning system is shown in figure 3. It is employed to collect data to demonstrate the self-invertible property. This system consists of a positioner P-753.1CD integrated with a capacitive sensor, a target computer equipped with a data acquisition card PCI-6289, and a host computer working under Matlab environment.

As shown in figure 4(*a*), in identifying the forward GSC model, the normalized input and output of the nanopositioning system are served as the input and output of the model, respectively. However, in identifying the inverse GSC model, the input and output are inverted, as shown in figure (*b*). The initial ascending curve is used to calculate the initial parameters, and the major hysteresis loop is used to optimize these parameters. Then the experiment system is excited by the triangle signal with decreasing amplitudes. The output displacement and the input voltage are collected and normalized. The normalized input u is applied to the identified forward GSC model and generated the model output y_F. Then, y_F is applied to the identified inverse GSC model and produced the model output u_I.

The parameters for the forward and the inverse models are given in table 1. As shown in figure 5, the hysteresis loops generated from the forward model are quite similar to that is obtained from the experiment system. These hysteresis loops are all counterclockwise. The area of the hysteresis loop represents the energy consumption of the nonlinearity. The normalized root mean square (NRMS) error is 0.58%, which demonstrates that the GSC model can capture the hysteresis loop directly and efficiently. The hysteresis loops generated from the inverse model are clockwise. They are symmetrical with the hysteresis loops that are generated from the forward model about the line $y=u$. The NRMS error is 0.26%, which illustrates that the GSC model is capable of describing the inverse hysteresis efficiently, too. From u to u_I, signals pass through the forward model and the inverse model. In other words, the inverse model is compensated by the forward model. The curve u - u_I shown in figure (*d*) represents the compensation results. The uncompensated error is 0.26%, which verifies that the forward model can linearize the inverse model effectively.

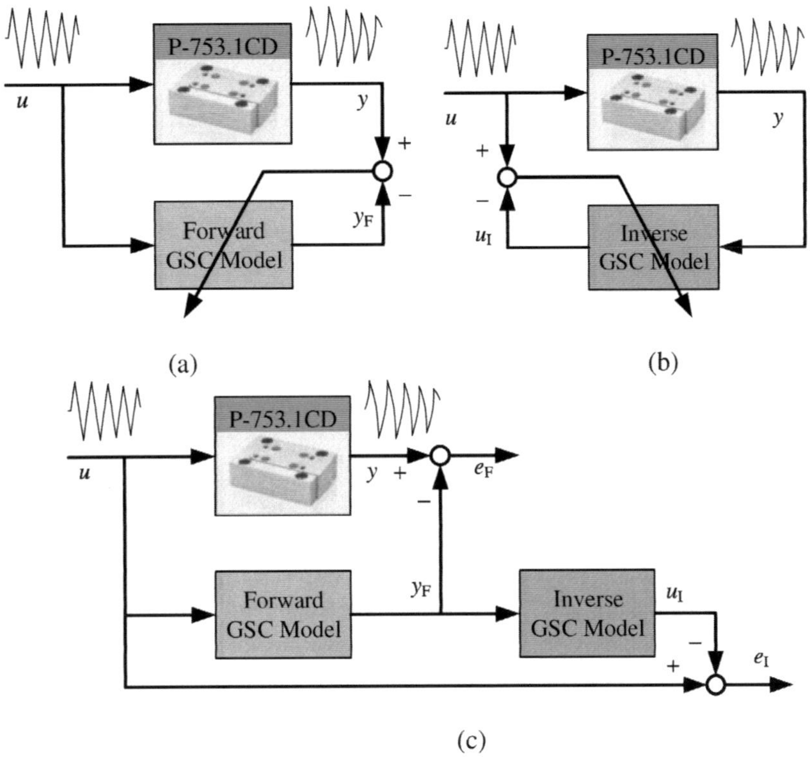

Figure 4. Identification of (*a*) the forward GSC model and (*b*) the inverse GSC mode and (*c*) model evaluation. Symbols u and y represent the normalized input voltage and output displacement of the experiment system. Symbol y_F represents the output of the forward GSC model. Symbol u_I represents the output of the inverse GSC model. Symbols e_F and e_I represent the modeling errors of the forward and the inverse GSC model, respectively.

Table 1. Identified parameters of the forward and the inverse GSC models.

Model	S_1	S_2	S_3	S_4	S_5
Forward GSC model	-0.1484	-0.1056	-0.0816	-0.0714	1.1628
Inverse GSC model	0.2448	0.1083	0.0586	0.0560	0.8544

If $S \in \mathbb{R}^+$, the ascending curve is convex. However, as shown in figure 5(*a*), the experimental ascending curves are concave. Thus, the original SC model can hardly capture the forward hysteresis loops. It is an inverse hysteresis model. The GSC model can forward-capture the hysteresis loops by utilizing negative elastances, which is also indicated by the identified parameters in Table 1.

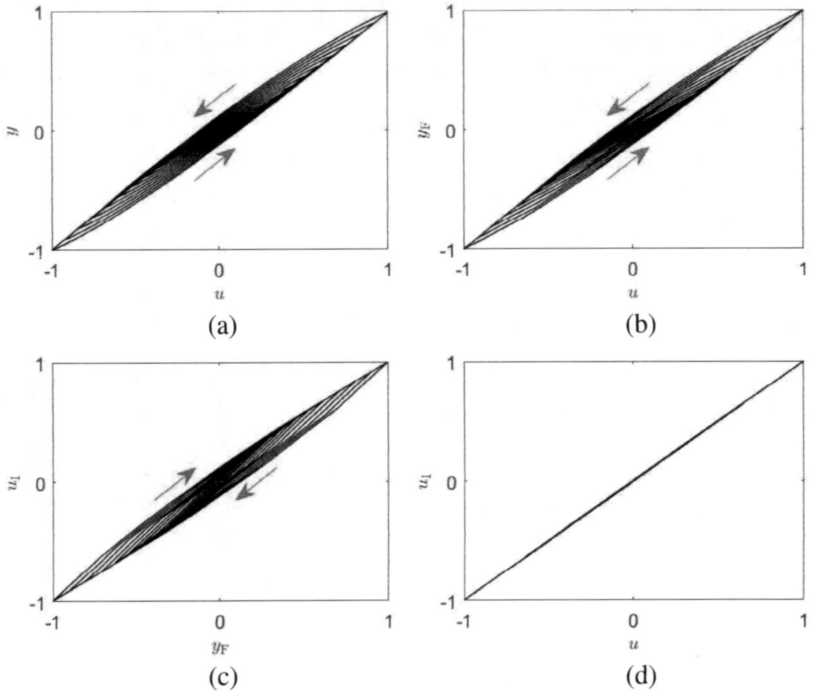

Figure 5. Hysteresis nonlinearities: (*a*) experiment, (*b*) forward model, and (*c*) inverse model and (*d*) compensated results.

3.2. Non-convex hysteresis property

Actually, in the above subsection, it has been demonstrated that the GSC model is capable of describing concave ascending curves. However, it is not sufficient to prove that the GSC model can define the non-convex hysteresis loops. In this subsection, experimental data will be employed to show that the GSC's capability of modeling the non-convex loops.

Property 2 (Non-convex hysteresis property). The GSC model is capable of generating non-convex hysteresis loops.

Table 2. Parameters of the GSC model for non-convex hysteresis loops.

i	S_i	i	S_i	i	S_i	i	S_i	i	S_i
1	-0.0688	4	-0.3334	7	0.4400	10	2.4334	13	0.1425
2	-0.1612	5	0.3420	8	-1.6728	11	2.4938	14	0.2370
3	0.2496	6	-0.5462	9	-4.3660	12	0.6859	15	0.1693

The data provided in [19] are depicted in figure 6(*a*). The experimental hysteresis loops are non-convex. The identified parameters are given in table 2. It is seen from the table that, to capture the concave part of the hysteresis loop, some VLCs have negative elastances. With these parameters, the hysteresis loops generated from the GSC model are also presented in figure 6(*a*). They follow the experimental ones closely. As seen in figure 6(*b*), the maximum error is at the position where the curve u - y intersects the u axis and is less than 5%. The NRMS error is about 0.94%. The u - y curve has a large slope when it crosses $y = 0$ the axis. Thus, for this position, more VLCs are needed to capture the change of the slope. In other words, an even distribution of $Q_{S,i}$ is not quite reasonable. Properly distributing $Q_{S,i}$ or increasing the number of VLCs may reduce the modeling error. On the other hand, instead of a piecewise linear function, employing a continuous function to capture the slope change of the ascending curves may also obtain improved performances. These efforts will be made in future studies.

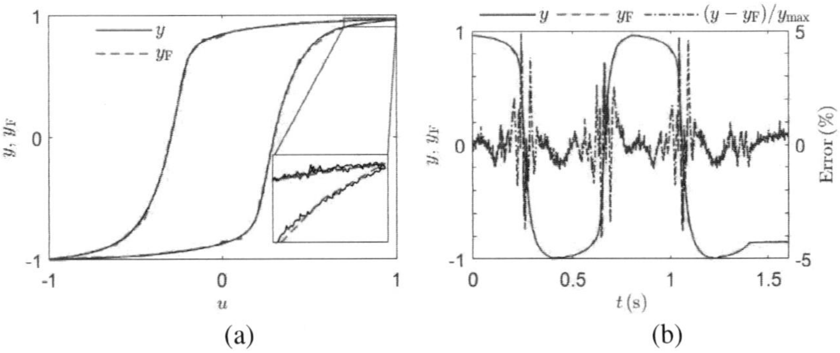

Figure 6. Performance of GSC model on modeling the non-convex hysteresis: (*a*) the hysteresis loops generated from experiment and model, i.e., *u*-*y* and *u*-y_F, and (*b*) modeling errors.

4. Conclusions

This paper proposes a generalized saturated capacitor (GSC) model for the hysteresis in piezoelectric materials. It introduces the Weiss domains (WDs) with negative capacitance switch (NCS) into the saturated capacitor model. The WD with NCS is represented by a parallel network of a charge-limited capacitor and a perfect capacitor and further described by a series network of a voltage-limited capacitor (VLC) with negative elastance and a perfect capacitor. Then, a piezoelectric actuator is modeled as a series network of several VLCs, which can have positive or negative elastances. The GSC model is capable of describing non-convex hysteresis and is self-invertible. Its effectiveness and self-invertible property are verified experimentally. Results show that the modeling error is less than 1% for both forward and inverse models. The maximum error occurs at the position where the hysteresis loop has a sharp slope. Increasing the number of VLCs distributed at this position may reduce the modeling error. Employing a continuous function to describe the hysteresis loop may also improve the model performance.

Acknowledgment

The authors express their gratitude to Prof. Christopher S. Lynch of the University of California for kindly providing the experimental data [19]. This work is partially supported by National Natural Science Foundation of China (11672093, 51705109, and U1737207), Special Foundation of Heilongjiang Postdoctoral Science (LBHTZ1609), the Innovation Fund of the Shanghai Academy of Spaceflight Technology under Grant SAST-2018-109, and Fundamental Research Funds for the Central Universities of China (HIT.NSRIF.201622).

References

[1] Devasia S, Eleftheriou E and Moheimani S R 2007 *IEEE Trans Control Syst. Technol* **15** 802-23
[2] Aguirre G, Janssens T, van Brussel H and Al-Bendera F 2012 *Mech Syst Signal Proc* b 218-31
[3] Wang Z, Zhang Z and Mao J 2012 *Electron Lett* **48** 1459-60
[4] Liu L, Tan K K, Chen S, Teo C S and Lee T H 2013 *IEEE Trans Indust Inform* **9** 859-68
[5] Zhang J, Merced E, Sepùlveda N and Tan X 2014 *IEEE-ASME Trans Mechatron* **19** 579-88
[6] Mackenzie I and Trumper D L 2016 *IEEE-ASME Trans. Mechatron* **21** 4-16
[7] Sun Z, Song B, Xi N, Yang R, Hao L, Yang Y and Chen L 2017 *IEEE Trans Ind Inform* **64** 5513-

23
[8] Chen X, Su C, Li Z and Yang F 2016 *IEEE Trans Ind Inform* **63** 6471-81
[9] Gu G, Zhu L and Su C 2014 *IEEE Trans Ind Inform* **61** 1583-95
[10] Li Z and Shan J 2017 *IEEE-ASME Trans Mechatron* **22** 1903-13
[11] Li Z and Shan J 2017 *IEEE Trans Ind Inform* **64** 8588-97
[12] Aschemann H and Schindele D 2014 *IEEE Trans Ind Inform* **61** 3620-29
[13] Xu Q 2013 *IEEE Trans Ind Inform* **60** 3927-37
[14] Janaideh M A and Krejči P 2013 *IEEE-ASME Trans Mechatron* **18** 1498-507
[15] Goldfarb M and Celanovic N 1997 *IEEE Contr Syst* **17** 69-79
[16] Liu Y, Shan J, Gabbert U and Qi N 2013 *Smart Mater Struct* **22** 5020
[17] Liu Y, Liu H, Wu H and Dan Z 2016 *Electron Lett* **52** 188-90
[18] Liu Y, Shan J, Meng Y and Zhu D 2016 *IEEE-ASME T Mech* **21** 38-43
[19] Rauls M B, Dong W, Huber J E and Lynch C L 2011 *Acta Mater* **59** 2713-22

IOP Conf. Series: Materials Science and Engineering **474** (2019) 012005 doi:10.1088/1757-899X/474/1/012005

Stable and environment-hard vacuum nanoelectronic devices for aerospace applications

M Nakamoto* **and J Moon**

Graduate School of Integrated Science and Technology, Shizuoka University, 3-5-1 Johoku, Naka-ku, Hamamatsu, Shizuoka 432-8011, Japan

*E-mail: nakamoto.masayuki@shizuoka.ac.jp

Abstract. Extremely stable field-emitter arrays (FEAs) having low work function as well as resisting harsh environments have been prepared by the transfer mold emitter fabrication method to make remarkably resistant and low operation-voltage vacuum nanoelectronic devices in harsh environments such as a strong radical atmosphere. Amorphous carbon and titanium nitride, which have a low work function as well as an environmental hardness, were used as the emitter material. The current emission fluctuations of amorphous carbon FEAs having no resistive layers, were as low as ±1.7%, (the lowest ever reported value), as compared to 5–100% of general FEAs fabricated by the conventional method and containing resistive layers. The amorphous carbon FEAs show the most stable emission characteristics, as compared to transfer mold FEAs coated with titanium nitride or platinum (Pt), because amorphous carbon FEAs have the most uniform tip radii. The amorphous carbon FEAs exhibit stable fluctuations of less than ±5.1% under in situ oxygen radical treatment. Panel and pixel electric thrusters have been proposed that use transfer mold FEAs. These electric thrusters make it easy to produce large and small thrusters. transfer mold amorphous carbon FEAs can be used to create highly efficient and reliable vacuum electronic devices such as electric propulsion engines and field-emission displays.

1. Introduction

Low operation-voltage field emitter arrays (FEAs) for extremely stable vacuum nanoelectronic devices, can be realized by using low-work function emitter materials and unique emitter fabrication method, that is also referred to as the transfer mold emitter fabrication method (TMEFM)[1–3]. This method produces sharpened emitter arrays uniformly and reproducibly.

However, FEAs are often exposed to high temperature environments and highly oxidizing atmospheres during the vacuum sealing or fabrication process of field emission displays (FEDs).

Besides, electrodynamic tethers (EDT) in a low-earth orbit (LEO) above 80–500 km has to undergo an exceedingly reactive oxygen-radical atmosphere. An EDT system, which is one kind of a spacecraft or a satellite, is comprised of 5 km–long conductive wire having electron emitters and collectors at both ends. It can be used for propulsion, stabilization or maintaining the trajectory of space spacecraft in orbit. Electrons were emitted from electron emitters at a spacecraft to space, while electrons were gathered by electron collector from space. Space in LEO has electrons and ion species having densities of about 1×10^{6} cm^{-3} [4]. If currents along the tether flow because of electric circuit formed by electron emitters, tether and electron collectors, the spacecraft or the satellite exert a Lorentz force against a magnetic field of Earth and can move. FEAs have the advantage that they require no fuel and warm-up time. Therefore, FEAs are suitable for an EDT.

However, one of the major issues is the degradation due to the harsh environments of FEAs in EDTs, because FEAs may be directly exposed to highly reactive environments such as oxygen radicals. Thus, evaluation of field emission characteristics during the exposure of highly reactive atmosphere such as in-situ oxygen radical treatment is necessary to realize reliable and stable electric propulsive system even in a highly reactive atmosphere. An electrical propulsion system, e.g., ion engines having advantages of very high fuel efficiency, can preserve impellent forces for long periods.

New ion engines as space propulsion systems by using an assembly of transfer mold field emitter arrays, were proposed in our previous report, can be useful to spacecraft applications [2].

2. Experimental methods

Figure 1 shows the fabrication process for FEAs by using the transfer mold method array.

FEAs were fabricated by using the transfer mold emitter fabrication method, which has various types such as Si mold-type method and coated type method [1–3]. Substrates of silicon mold were prepared by anisotropic etching of silicon using potassium hydroxide (KOH) aqueous solutions to form pyramidal holes (mold) having a very sharp vertex in figure 1(a).

Silicon dioxide layer was made by thermal oxidization method in Si substrate mold to obtain the intrinsic sharpening of emitter-tips in figure 1(a). The 20 nm-thickness amorphous carbon thin film as an emitter material, was deposited inside a Si mold formed by a thermal SiO_2 layer in figure 1(b).

The deposition of amorphous carbon thin film was carried out by using a plasma-enhanced chemical vapor deposition method with mixed gases of acetylene and argon at room temperature. Next, Ni as the supporting layer was filled in the remaining volume of the mold after the deposition of amorphous carbon thin film by using the sputtering method in figure 1(c). Then, mold and glass substrate were bonded by the anodic bonding method in figure 1(d). The elimination of Si Mold substrate and silicon dioxide film by using a tetramethylammonium hydroxide solution and a buffered hydrofluoric acid solution, respectively, is shown in figure 1(e). Then, amorphous carbon FEAs by TMEFM can be obtained.

The coated-type transfer mold emitter-fabrication method was described as follows. Ni was firstly filled by electroplating method in Si mold as shown in figure 1(b*). Mold and glass substrates were bonded by using the anodic bonding method in figure 1(c*). After the elimination of Si Mold substrate and silicon dioxide film, the coated-type TiN-FEAs or Pt-FEAs have been fabricated by titanium nitride or platinum coating on Ni FEAs using radio-frequency sputtering of TiN or Pt thin film in figures 1d and 1e [3].

For the evaluation of emitter-tip morphology for transfer mold, 30 images of FEAs on the random positions of all FEA segment area were taken by using high-resolution FE-SEM with the magnification of 650,000. Then, the average values of tip radius and their standard deviations were statistically estimated. The morphology and emission characteristics amorphous carbon were evaluated among those of TiN coated type FEAs and Pt-coated type FEAs.

According to the UPS results, titanium nitride and amorphous carbon thin film can withstand harsh environments and have as low as work functions of 3.2 eV and 3.6 eV, compared with those of 5.0 eV for multiwalled carbon nanotubes (CNTs). The results were compared with those of platinum-FEAs (5.1 eV); such FEAs do not have a low work function but can withstand harsh environments.

The morphology such as tip radius and base length of field emitter arrays was evaluated by using a field-emission scanning electron microscope (FE-SEM). transfer mold field emitter arrays have a pyramidal shape with the average base length of 1.6 μm, the pitch of 3.2 μm and the emitter height of 1.1 μm. The transfer mold FEAs had been fabricated by using the 6-inch Si wafer and been patterned by the FEAs having a segment of 4×4 pieces. The FEA area of one segment was 1.5×1.5 cm^2. The uniform morphology of all FEAs for all segments on the 6–inch size substrate had been evaluated by using FE-SEM. Then, one segment FEAs of 1.5×1.5 cm^2 was cut and was carried out emission current-electric field measurements. Anode electrode having the hemispherical shape with the diameter of 1 mm was located on the emitter array at a spacing of 10 μm. Field emission measurements were carried out at the random positions of the emitter area.

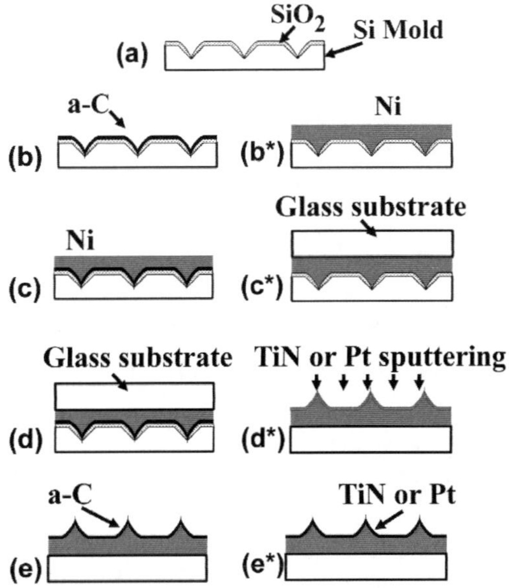

Figure 1. The fabrication process for FEAs by using transfer mold method.

The uniformity of field emission has been confirmed by measurement of all positions and segments FEAs on the 6–inch size substrate. I–V measurements were performed under vacuum conditions of 3.8×10^{-9} Torr without oxygen radical treatment and 7.5×10^{-7} Torr during oxygen radical treatments, respectively. Values of turn-on fields were defined with bias voltages at the condition of 10-μm-separation between anode and cathode and the 10 nA emission current.

The field-emission characteristics during in-situ oxygen treatment were evaluated by using the I–V measuring vacuum system with the oxygen radical source [2].

The flux of oxygen radical was 10^{15} atoms/cm^2·sec, which is as high as 10^7–10^8 times ordinary fluxes in the LEO (about 10^7–10^8 atoms/cm^2·sec) [4]. Work functions and chemical properties were estimated with ultraviolet photoelectron spectroscopy (UPS) and x-ray photoelectron spectroscopy (XPS).

3. Results and Discussion

Figure 2 shows FE-SEM images of amorphous carbon FEAs, TiN-FEAs, and Pt-FEAs fabricated by TMEFM, as well as their tips with and without treatment. Figures 2(a), 2(b), and 2(c) depict the amorphous carbon FEAs, TiN-FEAs, and Pt-FEAs, without treatment.

Tip radii of TiN-FEAs, amorphous carbon FEAs and Pt-FEAs were equal 6.8, 7.8, and 10.3 nm, respectively. Values of 6.8-10.3 nm are approximately equal or somewhat smaller than the diameter of multiwalled CNTs (5–10 nm). Moreover, amorphous carbon FEAs, TiN-FEAs, and Pt-FEAs had tip radii's standard deviations of 1.5, 1.9, and 2.0 nm, respectively.

Therefore, the FEAs fabricated by the ransfer mold method, are very uniform, unlike nonuniform tip radii, and are different in height and pitch from typical FEAs, e.g., carbon nanotube FEAs (CNT FEAs). Moreover, the amorphous carbon FEAs have the highest uniformity compared with those of TiN-FEAs and Pt-FEAs. RF sputtering method was used to prepare TiN-FEAs or Pt-FEAs by TiN thin film coating or Pt thin film coating on Ni FEAs. transfer mold amorphous carbon FEAs was prepared by the Si mold-type TMEFM having very sharp corners due to the intrinsic tip sharpening. Therefore, amorphous carbon FEAs have uniform tip-morphology.

The average values of tip radii for TiN-FEAs, amorphous carbon FEAs and Pt-FEAs, were 7.0, 8.0, and 10.5 nm, respectively, after 12-min radical treatment. Tip radii were approximately equal regardless of oxygen treatment.

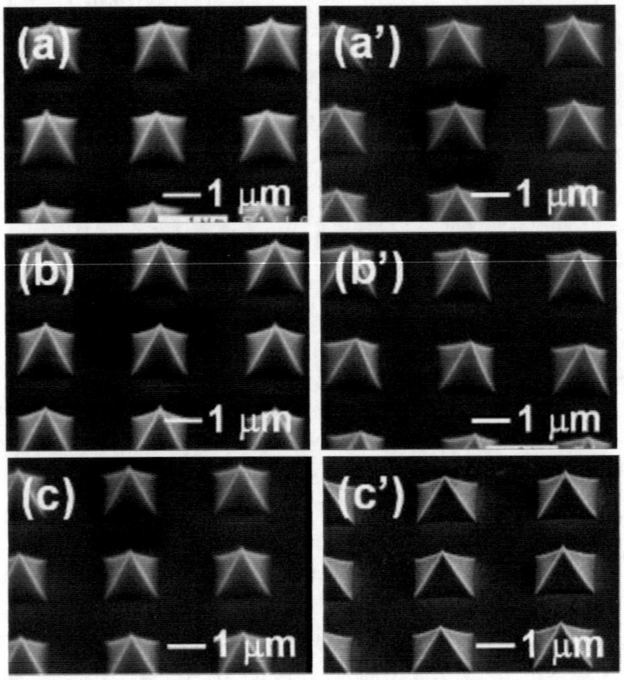

Figure 2. FE-SEM images about transfer mold FEAs and their tips. (a) Amorphous carbon FEAs without treatment; (a') Amorphous carbon FEAs with treatment for 12 min; (b) TiN-FEAs without treatment; (b') TiN-FEAs with treatment for 12 min; (c) Pt-FEAs without treatment, (c') Pt-FEAs with treatment for 12 min.

No damages of emitter tips for amorphous carbon FEAs, TiN-FEAs and Pt-FEAs occurred due to the radical treatment.

The amorphous carbon thin film has anticorrosive and solid properties, as well as low work-function compared with 4.3–4.5 eV of the common emitter materials such as molybdenum [5]. In this work, the work function (3.6 eV) of the amorphous carbon film was estimated by using UPS. According to our experimental XPS results, the sp^3 content of amorphous carbon on the emitter surface of amorphous carbon FEAs by TMEFM was analyzed to be 85.6%, estimated from the $sp^3/[sp^3+sp^2]$ ratio by using deconvolution of the XPS spectra concerning sp^3 and sp^2 peak. Amorphous carbon films have high elastic moduli of 55-143 GPa under condition of sp^3 contents higher than 60% [6]. Therefore, chemically inert amorphous carbon film having low hydrogen content and defect states can be achieved with high sp^3 contents of more than 60%.

Figures 3(a) and 3(b) show the relation between emission currents and electric fields, and Fowler-Nordheim (F-N) plots. Turn-on fields of TiN-FEAs, amorphous carbon FEAs, and Pt-FEAs were as low as 15.4, 17.5, and 27.1 V/μm, respectively.

Some research results were reported on turn-on field values of 2.8–3.6 V/μm for ultra-nanocrystalline diamond FEAs with no description of anode-to-emitter distance [7], 1.1–1.2 V/μm for CNT FEAs with 2-mm anode-to-distance and work function of 5 eV, 3.5 V/μm for CNT FEAs with 1.8 mm anode-to-distance [8-9]. However, in agreement with the quantum mechanics, even under the nanoscale tip radii and electrode separation, to leave electrons from Fermi energy level to the vacuum energy level, one must assume that their energy level is between the Fermi level and the vacuum level, which implies the work function of emitter materials.

IOP Conf. Series: Materials Science and Engineering **474** (2019) 012005 doi:10.1088/1757-899X/474/1/012005

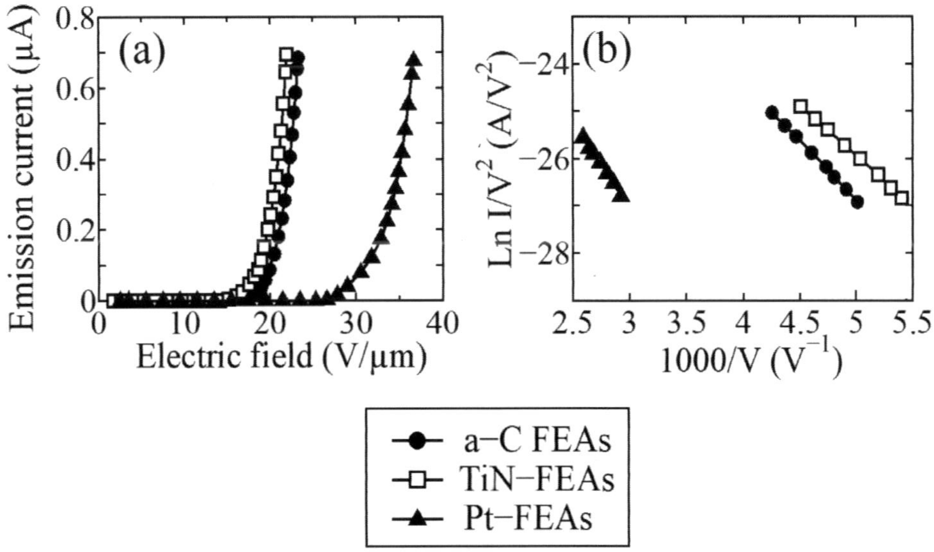

Figure 3. (a) Relation between emission current and electric fields, and (b) F-N plots of amorphous carbon FEAs (a-C FEAs), titanium nitride-FEAs (TiN-FEAs), and platinum-FEAs (Pt-FEAs).

The turn-on E-field with a larger work function is required. Even though the turn-on E-field of field emitters with a substance having work function of 5 eV, is as low as 1 V/μm in the condition of 1 mm anode-emitter separation (also referred to as electrode separation) and the biased voltage of 1 kV, the turn-on E-field should increase from 1 V/μm at 1 mm electrode separation. Then, the biased voltage has to be higher than 5 V so that the turn-on E-field is required with higher than 50 V/μm in the condition of 0.1 μm electrode separation. Usually, the turn-on E-fields under condition of 1–2 mm electrode separation, can be reduced to 100–300 μm.

However, the gradual increase of turn-on E-fields occurs in the range of less than 100–300 μm electrode separation. Turn-on E-fields enormously increase in the range of less than 10–30 μm electrode separation.

Thus, this electrode separation is one of the most critical parameters to compare and evaluate the performance of FEAs and must be mentioned. However, many reports indicate that electric fields of less than work function values per micrometer such as 1–4 V/μm, do not clarify the electrode separation such as anode-to-cathode distance or circumscribe the range of relatively long electrode separation, e.g., 100 μm to 2 mm. If the gated transfer mold FEAs for vacuum nanoelectronic devices are considered, the separation between emitter cathode and the gate electrode of fewer than 10 μm is significant for the I-V measurement. Considering 4.5 eV of Mo work function, the turn-on voltage of molybdenum FEAs by TMEFM is 7 V, which is nearest to the theoretically lower limit turn-on voltage of 4.5 V [1]. The turn-on field is 17.5 V/μm, which is approximately estimated regarding the gate-to-emitter separation of 0.4 μm. Therefore, the values of 15.4–27.1 V/μm are as low as those of typical FEAs like Spindt FEAs and CNT FEAs, which have 50−600 V/μm, approximately estimated as the condition of a short anode-to-emitter separation of less than 30 μm [10–15].

TiN-FEAs showed lower turn-on fields than those of amorphous carbon FEAs and Pt-FEAs, because TiN-FEAs with smaller tip radius, exhibited higher geometric factors than those of amorphous carbon FEAs and Pt-FEAs. The relevant parameters in the Fowler–Nordheim (FN) field-emission theory are the geometric factor (β) which can be estimated by the emitter's microstructure geometry, and the work function (φ) [2, 3].

Martin et al. [16] showed that the β factor (β_1) in the range of $D \gg r$, can be approximately calculated as (1). D is anode-to-emitter tip separation, and r is the emitter-tip radius.

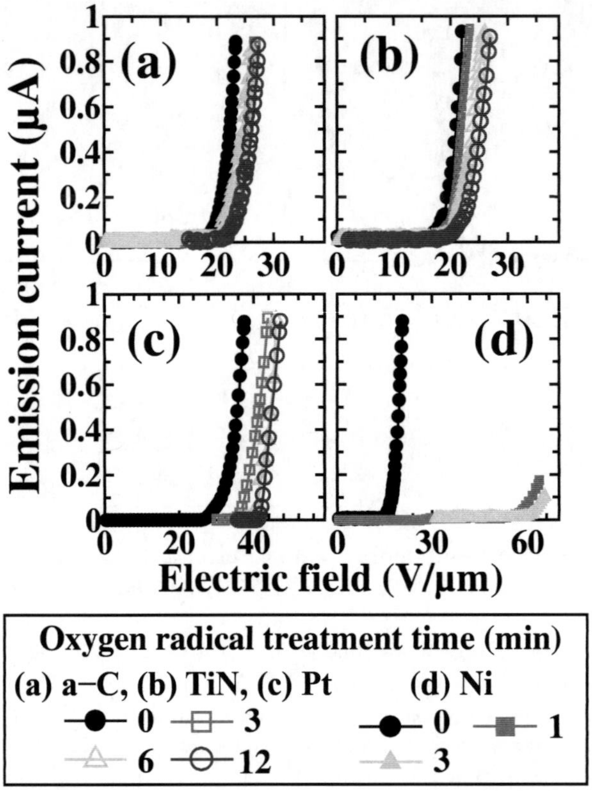

Figure 4. The relation between emission current and electric field for (a) amorphous carbon FEAs, (b) TiN-FEAs, (c) Pt-FEAs and (d) Ni-FEAs depending on the radical treatment time.

The β_1 can be approximated by

$$\beta_1 = 2\left[r \ln(4D \cdot r^{-1})\right]^{-1}. \tag{1}$$

In our experiment, D is 10 μm and r = 6.8–10.5 nm. The value of D is 950–1470 times larger than the values of r. Therefore, (1) can be applied for evaluation of geometric factor.

Charbonnier et al. [17] approximated the β factor (β_2) as

$$\beta_2 = 1.4 \cdot \left[r^{2/3} D^{1/3} (\tan\theta)^{0.2} [(D+h)\cdot h^{-1}]^{1/2}\right]^{-1}, \tag{2}$$

where θ is a half curvature-angle at an apex of emitter tip, r is an emitter-tip radius, D is an anode electrode-to-emitter separation, and h is a height of emitter. The experimental values of the geometric factor from F-N plots slope are given by

$$\beta_{ex} = 6.5 \times 10^7 \cdot m^{-1} \cdot \varphi^{3/2}, \tag{3}$$

where φ is the emitter materials' work function, and m is the F-N plot slope.

Using (1), β_1 was found to be 3.04×10^7 cm^{-1}, 3.44×10^7 cm^{-1} and 2.38×10^7 cm^{-1} for amorphous carbon FEAs, TiN-FEAs, and Pt-FEAs, respectively. Using (2), β_2 was 0.62×10^7 cm^{-1}, 0.68×10^7 cm^{-1} and 0.52×10^7 cm^{-1} for amorphous carbon FEAs, TiN-FEAs, and Pt-FEAs, respectively. Using (3) and the work function found with UPS, β_{ex} were 1.51×10^7 cm^{-1}, 2.31×10^7 cm^{-1} and 1.50×10^7 cm^{-1}for amorphous carbon FEAs, TiN-FEAs, and Pt-FEAs, respectively. The 3.6 eV, 3.2 eV and 5.1 eV of work functions for amorphous carbon, titanium nitride, and Pt, were obtained by the UPS measurement.

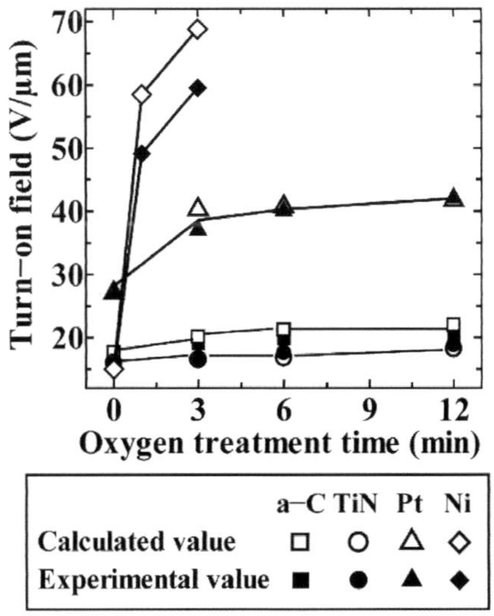

Figure 5. Calculated values and experimental values of turn-on fields depending on radical treatment time for amorphous carbon FEAs, TiN-FEAs, Pt-FEAs, and Ni FEAs.

Figure 4 shows emission current-electric fields characteristics of amorphous carbon FEAs, TiN-FEAs, and Pt-FEAs, depending on the radical treatment time. Turn-on field of 14.9 V/μm for Ni FEAs before in-situ radical treatment, increased approximately fourfold to 59.4 V/μm after 3-min oxygen radical treatment and even arcing was happened after a 3-min-radical treatment.

Turn-on fields for the amorphous carbon FEAs, TiN-FEAs, and Pt-FEAs were varied as small as 17.5 to 21.8 V/μm, 15.4 to 18.8 V/μm, and 27.1 to 42.1 V/μm, respectively. It means that FEAs with having amorphous carbon, TiN and Pt as emitter materials, possess more stable field emission characteristics even in highly oxidizing environments.

Besides, turn-on fields for Pt-FEAs were higher than those of TiN-FEAs and amorphous carbon FEAs because the tip radii of Pt FEAs were larger. Pt-FEAs had a smaller geometric factor (2.38×10^7 cm^{-1}) than those of TiN-FEAs (3.04×10^7 cm^{-1}) and amorphous carbon FEAs (3.44×10^7 cm^{-1}).

The amorphous carbon FEAs and TiN-FEAs have as low turn-on fields as 17.5 V/μm and 15.4 V/μm, which were estimated with 10-μm anode-to-emitter separation. This is compared with 50–600 V/μm for conventional FEAs. Thus, FEAs fabricated by TMEFMcan be used in vacuum nanodevices. Moreover, the amorphous carbon FEAs have the most stable turn-on fields even though highly oxidizing environments.

Figure 5 shows turn-on field, which was estimated by the FN equation with geometric factors, where we have used (1), and work functions, where we measured by UPS results, concerning radical treatment time.

The experimental values of turn-on fields almost coincide with the theoretical values of turn-on fields about amorphous carbon FEAs, TiN-FEAs and Pt-FEAs, because geometric factors of these FEAs do not change with oxygen radical treatment.

According to our experimental XPS results, amorphous carbon FEAs exhibited as small as the peak-area ratio of 6.3%, which was estimated with oxidized portions at the XPS spectra, compared with those of 11.8% of TiN-FEAs. Besides, C_{1s} peak at 285.3 eV does not shift for amorphous carbon FEAs with 12 min-oxygen radical treatment [18]. However, TiN $2p_{1/2}$ peak of TiN-FEAs, shift from 464.6 eV to lower energy level of 464.6 eV with 12 min-treatment [19]. Ti–O bonds at the surface of TiN-FEAs have increased by oxygen radical treatment so that Ti-N bonding peak shifts due to increasing amount of oxidized matters. Thus, it can be explained that amorphous carbon FEAs can maintain more stable field-emission characteristics in highly oxidizing environments than those of TiN-FEAs.

However, nickel (Ni) FEAs have a very intense Ni–O peak in the range of 853.5–857 eV binding energy with the radical treatment. NiO peak-area ratio was increased by 7.1 times (5.5% to 39.0%) by a 3-min radical treatment. Ni $2p_{1/2}$ peak shifted with 0.9 eV from 870.3 eV to 871.2 eV after 12-min radical treatment. Ni–O bonds at the surface of Ni-FEAs were significantly increased by the oxygen radical treatment so that Ni $2p_{1/2}$ peak shift occurred due to the increasing amount of oxidized matters. Thus, it can be explained that Ni-FEAs cannot maintain stable field emission characteristics in highly oxidizing environments. Then, turn-on fields of Ni-FEAs, as shown in figure 5, were rapidly changed.

According to work function, experimentally measured by UPS spectra, the work function of 4.5 eV for Ni FEAs were changed to 6.0 eV after 3-min treatment. Measured φ by UPS for amorphous carbon FEAs of 3.6 eV, TiN-FEAs of 3.2 eV and Pt-FEAs of 5.1 eV, varied to 3.9 eV, 3.5 eV, and 5.2 eV, respectively, after 12-min treatment. The amorphous carbon FEAs, TiN-FEAs, and Pt-FEAs have the endurance to harsh environments. Moreover, FEAs fabricated by TMEFMare very uniform in the morphology so that the calculated values of work function, coincide with the measured values.

Usually, conventional field emitter arrays have severe geometrical nonuniformity so that those FEAs with resistive layer, exhibit as high emission-current fluctuations as 5–100%. Moreover, those FEAs with no resistive layer, exhibit and several hundred % emission-current fluctuations. According to the Fowler–Nordheim theory, the calculated work function is originated from the slope of the *FN* plot derived from the I–V characteristics. The slope of *FN* plot itself is ambiguous because emissions current has significant fluctuations.

Besides, geometric factors, which can be obtained by *FN* plots, usually, cannot express correct values, because of the high fluctuation of emission currents. The high fluctuations are originated from nonuniform geometrical morphology. Therefore, the calculated work function for the conventional FEAs cannot be expressed by a reliable value.

Besides, in the case of FEAs fabricated by transfer mold method, the geometric factor can express the true values because of stable emission currents having low fluctuation. These small fluctuations are originated from remarkably uniform geometrical morphology so that work functions obtained by the true slope of *FN* plots, can express the reliable value. In conclusion, for transfer mold FEAs, the value of theoretically calculated work function coincides with those of the measured work function.

However, in the case of Ni FEAs, the calculated work functions did not correspond with the measured ones because of unreliable values due to unstable emission characteristics and arcing formation.

According to the surface analysis with XPS, in case of amorphous carbon FEAs and TiN-FEAs, C–O peak at 285.5 eV and Ti–O peak at 455 eV, respectively, were not shown with oxygen radical treatment. Besides, Pt-FEAs have no oxide peaks after radical treatment. No oxidized matters at the surface of emitters can explain keeping work functions or stable emission characteristics in highly oxidizing environments. However, in the case of Ni FEAs, a dominant peak of NiO was shown at 857.2 eV. Thus, it can be explained that Ni-FEAs cannot maintain stable field emission characteristics in highly oxidizing environments. Then, turn-on fields of Ni-FEAs in figure 5, were rapidly changed.

Figure 6 shows the emission-current fluctuations of (a) amorphous carbon FEAs, (b) TiN-FEAs, and (c) Pt-FEAs, and (d) Ni FEAs depending on in-situ oxygen radical treatment. Usually, FEAs fabricated by using typical methods, e.g., Spindt-type FEAs and CNT-FEAs, exhibit as high as 5-100% fluctuations with the resistive layer, and more than 100% fluctuations with no resistive layer.

Table 1 shows the emission fluctuations of amorphous carbon FEAs, TiN-FEAs, Pt-FEAs, and Ni FEAs depending on radical treatment. Those FEAs have no resistive layers. Those emission fluctuations were as low as values of ±1.7, ±2.5, and ±2.5%. Moreover, ±1.7% is one of the weakest fluctuations being ever reported.

No deformation of emitter tips for amorphous carbon FEAs, TiN-FEAs, and Pt-FEAs happened by the oxygen radical treatment. After field emission measurements during 30 min, those of amorphous carbon FEAs, TiN-FEAs, and Pt-FEAs were 8.0±2.0, 7.2±2.2, and 10.4±2.2 nm, respectively.

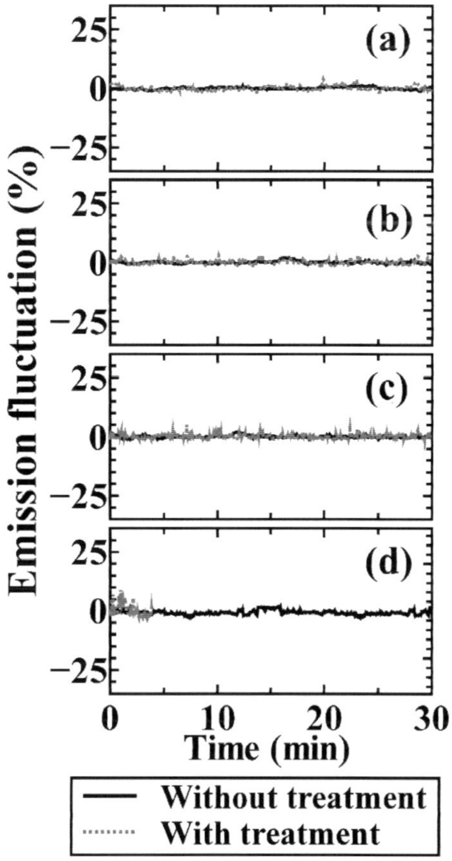

Figure 6. Emission fluctuations of (a) amorphous carbon FEAs, (b) titanium nitride-FEAs, and (c) platinum-FEAs, and (d) Ni FEAs, depending on radical treatment.

Table 1. Emission current fluctuations of amorphous carbon FEAs, TiN-FEAs, Pt-FEAs, and Ni FEAs with and without the radical treatment.

	Without radical treatment (%)	With radical treatment (%)
Amorphous carbon FEAs	±1.7	±5.1
TiN-FEAs	±2.5	±5.1
Pt-FEAs	±2.5	±8.1
Ni FEAs	±4.0	±10.1

However, the Ni FEAs were damaged and evaporated due to the arcing within 3 min, so that there was no pyramidal shape of FEAs. Fluctuations of Ni FEAs having no resistive layer increased from ±4.0% to ±10.0% concerning radical treatment in figure 6(d).

Moreover, in the case of our experiment system, the flux of oxygen radicals was 1×10^{15} atoms/cm^2·sec, 10^7–10^8 times higher than typical fluxes experienced by orbital satellites (about 10^7–10^8 atoms/cm^2·sec). The 30 minutes under in-situ oxygen radical treatment is equal to the corresponding time of the actual oxidation reaction for 0.3–3 years in LEO. Therefore, experimental time of 30 minutes under in-situ oxygen radical treatment is quite enough to evaluate harsh environment characteristics.

Furthermore, amorphous carbon FEAs and TiN-FEAs, having a low work function as well as resistance to harsh environments, exhibit as stable fluctuations as ±5.1% even though FEAs were suffered by highly reactive oxygen radicals.

However, ±4% fluctuations for Ni FEAs increased ±10% in the condition of no resistive layer. Ni FEAs using a typical metal emitter material were destroyed and melted away. Work functions strongly influence the flicker noise concerning emission-current fluctuations [20]. It is regarded that work function at the surface of Ni FEAs is dependent on the adsorption (Ni–O) of active radicals such as O and the desorption processes. The work function increases due to adsorbates covering the Ni FEAs' surface so that current fluctuations, which closely depend on the work function, also increase. Therefore, amorphous carbon FEAs and TiN-FEAs with a low work function as well as resistance to highly oxidizing atmospheres exhibit comparatively stable field-emission characteristics. Especially, amorphous carbon FEAs exhibit more endurable field-emission characteristics than do FEAs coated with TiN because the amorphous carbon FEAs have the most uniform tip radii due to the lowest standard deviation of tip radii.

Nakamoto has come up with both of panel type and pixel type devices by using vacuum nanodevices for aerospace applications [3]. The electric propulsion system, e.g., hall thrusters, ion engines, pulse plasma thrusters (PPT), field-emission electric propulsion (FEEP) and EDTs, has good advantages in aerospace applications. Typical ion engines' designs such as are cylindrical structures; e.g., hollow-cathode for electron confinement and microwave discharge electrodes as waveguide tube. Those cylindrical structures cannot be fabricated on flat substrates, nor can their size or shape be modified.

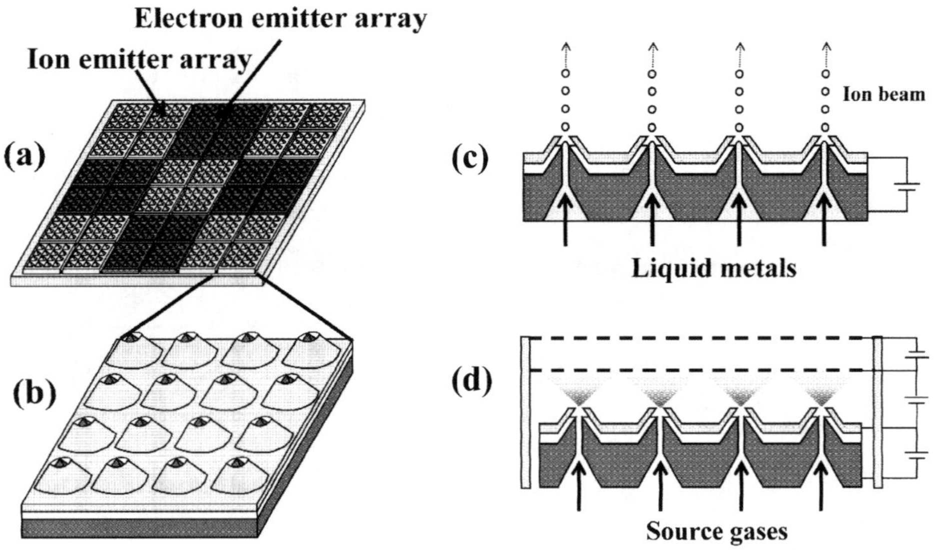

Figure 7. Proposed (a) panel-type and (b) pixel-type thrusters. Proposed schematic for thrusters using transfer mold FEAs for (c) field emission electric propulsion (FEEP) devices and (d) ion engines.

Figure 7 provides illustrations of ion-emitter arrays and electron-emitter arrays. Figures 7(a) and 7(b) show the proposed panel- and pixel-type electric thrusters, respectively. The (flat) panel type electric thrusters are assembled with nanoscale electron-emitter arrays, ion-emitter arrays and power supply units like an on-chip system. Electron-emitter arrays can compensate positive-charged ones as a neutralizer. Figures 7(c) and 7(d) exhibit the new pixel-type ion emitter arrays, which are made of either FEEP or ion engines having field emitter arrays. In the case of FEEP, electrons emit from FEAs so that metal cations are generated by alkali liquid metals such as Cs, which have a low ionization potential (3.87 eV). In the ion engines, electrons emit from FEAs and are accelerated. Then, collisions between electrons and gas molecules in plasma can generate cations such as Xe^+ ions. The panel- and pixel-type electric thrusters can be assembled in a combination with electron emitters and ion emitters. Each electron and ion emitter is individually and digitally operated by variable numbers of emitters. Such a combination of pixel- and panel-type electric thrusters can be readily applicable to satellites regardless of their size.

4. Conclusion

Amorphous carbon FEAs produced by TMEFM can serve as remarkably stable vacuum nanodevices in highly oxidized atmospheres.

The evaluation results for field-emission stability were compared with those for TiN-, Pt-, and Ni-based FEAs. Amorphous carbon FEAs having no resistive layers had as low emission fluctuations as ±1.7%, which is one of the lowest ever reported. The amorphous carbon FEAs exhibit the most stable emissions, in comparison with emission fluctuations (±2.5%) of TiN-FEAs, because amorphous carbon FEAs have the most uniform tip radii. Also, FEAs fabricated by TMEFM exhibited both high environmental resistance and low work function as emitter materials, as well as possessed low fluctuations of about ±5.1% in highly oxidized atmospheres.

It is also recommended that the new ion engines and other aerospace devices that use the transfer mold FEAs should be assembled as combinations of pixel- and panel-type electric thrusters with ion-emitter arrays and electron-emitter arrays. Such combinations can be readily applicable to satellites of any size.

In summary, the remarkably uniform and low operation voltage amorphous carbon FEAs fabricated by the TMEFM show the most stable field-emission characteristics and the highest endurable field-emission characteristics in highly oxidized atmospheres. Thus, such FEAs are lucrative for making highly reliable and efficient electric thruster devices for satellites operating in various aggressive and highly oxidizing environments.

References

[1] Nakamoto M and Fukuda K 2002 *Appl Surf Sci* **202** 289–94
[2] Nakamoto M and Moon J 2011 *J Vac Sci Technol* B **29** 02B112-1–02B112-5
[3] Nakamoto M, Moon J and Shiratori K 2010 *J Vac Sci Technol* B **28** C2B1–C2B5
[4] Abe T, Knudsen D J, Yau A W, Watanabe S and Sagawa E 2001 *Adv Space Res* **27** 1403–12
[5] Schafer J, Ristein J, Ley L 1997 *J Vac Sci Technol* A **15** 408–14
[6] Robertson J 1992 *Phys Rev Lett* **68** 220–3
[7] Joseph P T, Chen H C, Chen C H, Niu H, Tai N H and Lin I N 2011 *Mater Express* **1** 68–73
[8] Lahiri I, Wong J, Zhou Z and Choi W 2012 *Appl Phys Lett* **101** 063110-1–063110-5
[9] Nam J W, Alegaonkar P S, Park J H, Yoo J B, Choe D H, Kim J M and Kim W S 2007 *J Vac Sci Technol* B **25** 306–11
[10] Spindt C A, Brodie I, Humphrey L and Westerberg E R 1976 *J Appl Phys* **47** 5248–62
[11] Spindt C A, Holland C E, Schwoebel P R and Brodie I 1996 *J Vac Sci Technol* B **14** 1986–9
[12] Itoh J, Tohma Y, Morikawa K, Kanemaru S and Shimizu K 1995 *J Vac Sci Technol* B **13** 1968–72
[13] Lee N S, Chung D S, Han I T, Kang J H, Choi Y S, Kim H Y, Park S H, Jin Y W, Yi W K, Yun M J, Jung J E, Lee C J, You J H, Jo S H, Lee C G and Kim J M 2001 *Diam Relat Mater* **10** 265–70
[14] Han I T, Kim H J, Park Y J, Lee N, Jang J E, Kim J W, Jung J E and Kim J M 2002 *Appl Phys Lett* **81** 2070–2
[15] Pflug D G, Schattenburg M, Smith H I, Akinwande A I 2001 Tech Dig of the 47th IEEE IEDM, Washington, DC, USA, December 2–5, 2001, 179–82
[16] Martin E E, Charbonnier F M, Dolan W, Dyke W P, Pitman H W and Trolan J K 1960 Technical Report of Wright Air Development Division **5**, 9–20

[17] Charbonnier F M, Mackie W A, Hartman R L and Xie T 2001 J Vac Sci Technol B **19** 1064–72
[18] Filik J, May P W, Pearce S R J, Wild R K, Hallam K R 2003 *Diam Relat Mater* **12** 974
[19] Matsuo P J, Standaert T E, Allen S D, Oehrlein G S, Dalton T J 1999 *J Vac Sci Technol* B **17** 1435–47
[20] Gotoh Y, Nagao M, Matsubara M, Inoue K, Tsui H and Ishikawa J 1996 *Jpn J Appl Phys* **35** L1297–L1300

IOP Conf. Series: Materials Science and Engineering **474** (2019) 012006 doi:10.1088/1757-899X/474/1/012006

Hydraulic system elaboration and simulation for single-drive light-load monorail locomotive in fully mechanized coal mining applications

Z A Song*, F Jiang

College of Mechanical and Electronic Engineering, Shandong University of Science and Technology, Qingdao, China

*E-mail: 292323718@qq.com

Abstract. To solve the problem of light-load material transportation at the top and bottom of fully mechanized mining face, a hydraulic control system for single-drive monorail light-duty locomotive was designed. The importance of proper clamping force is mainly introduced, and the clamping pressure and braking time are determined. The relationships between the average travelling speed of the locomotive under different loads, the load and slope path of the locomotive are obtained. The simulation path was designed according to the field operation, and the AMESIM software was used for modeling and simulation. The simulation results meet the technical and engineering requirements and briefly introduce the use of the rotary valve in the traction circuit and the clamping circuit.

1. Introduction

In fully mechanized coal mining applications, the transportation of materials and equipment, at present, is generally ensured by pneumatic or electric monorail cranes with explosion-proof motor and explosion-proof switch power supply. The wire travels along the guide rail along the locomotive, and the distance of the excavation is the driving distance of the locomotive. It is symmetrically arranged on the guide rail + reducer + drive wheel by the explosion-proof motor, and the driving mode of the explosion-proof component such as the brake power failure is realized by the disc motor. The heavy-duty Scharf Monorail System (Germany) uses a low-speed, high-torque motor controlled by a variable displacement pump, which applies an electro-hydraulic valve to achieve braking [1-2].

The hydraulic single-drive monorail locomotive introduced in this paper is illustrated in figure 1. It is aimed to solve the problem of light-duty transportation in underground coal mines. Its braking and clamping are controlled by the hydraulic pressure. Only the motor needs the protection from explosion hazard, insofar as it has to run in the inflammable gas environment. It is especially suitable for mines with a severe bottom heave and poor floor conditions. The maximum lifting weight of the locomotive is 5 tons. It can move forward and backward along the suspension guide and needs only one operator to fulfil the task that previously required 3 or 4 ones. The alleviation of labor intensity of operators improves the production efficiency.

This paper starts from the light-load transportation law of a single-drive monorail crane. The hydraulic system of the hydraulic control system has the advantages of the existing heavy-duty transportation monorail transportation main circuit and safety auxiliary circuit, travelling, clamping, braking, traction, etc. Full hydraulic controls the manual and electric hoist lifts. At present, the design requirements for such technical indicators as virtual and physical prototypes have been satisfied and verified via the underground industrial tests,.

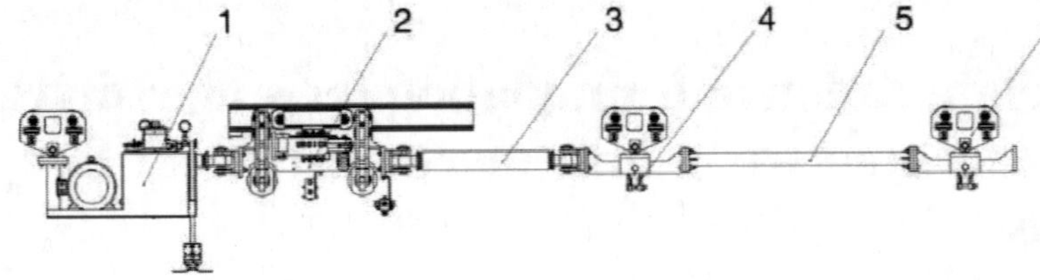

Figure 1. The single-drive monorail crane hydraulic virtual prototype.

2. Transport main circuit

2.1. Clamping circuit

The hydraulic clamping circuit consists of an inlet check valve, a pressure reducing valve, a rotary clamping valve, and a clamping cylinder. As shown in figure 2, the function of the check valve is to complete the first clamping. The check valve and the clamping cylinder chamber will close the pressure of the hydraulic medium between the two. Even if the machine is shutdown, it can connect the drive wheel to the guide rail. The control of the board, so that whenever the start, the brake cylinder is supplied with pressure oil to improve the brake release time and ensure the safety of the locomotive. When the driving portion is detached from the guide rail, the rotary valve can be placed in the position of the switch, and the liquid in the clamping cylinder enters the oil tank to release the clamping [3].

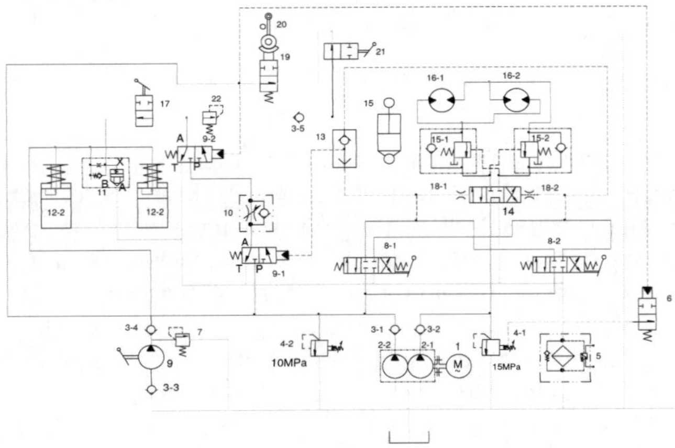

Figure 2. The single driving monorail crane control system.

According to the design requirements, the traction value is 5kN. The periphery of the drive wheel is made of rubber. The friction coefficient with the steel guide rail is 0.25. The positive pressure of the drive wheel and the guide rail is FJ=5/0.25=20kN.

The force is generated by the rear pump pressure liquid of the double gear pump acting on the rod cavity of the clamping cylinder to produce a pulling force on the driving wheel. The set pressure of the pressure reducing valve is:

$$p_J \geq \frac{4F_J}{\pi(D_J{}^2 - d_J{}^2)} \tag{1}$$

where N_J is driving wheel clamping force (N); D_j is clamping cylinder piston diameter (designed $D_J = 60mm$); d_J is piston rod diameter of clamping cylinder (designed $d_J = 36mm$); p_J is setting pressure after the pump overflow valve (N/m^2). The calculation of setting pressure of the pressure relief valve yields: $p_J \geq 1.98$ MPa.

According to reference GB2346-80 on the outlet pressure of the pressure-reducing valve circle, the value $p_J = 2.5MPa$. Therefore, the size of the clamping force to meet the design requirements is:

$$F_J = \frac{\pi}{4}(60^2 - 36^2) \times 2.5 = 252.8(\text{kN}) > 200\text{kN}$$

2.2. Travelling circuit.

The travelling circuit is composed of a manual hydraulic control reversing valve (manual pilot valve and main hydraulic valve), a two-way balancing valve and a hydraulic motor, which operate the manual pilot valves located on both sides of the locomotive, The control fluid from the rear pump acts on one side of the main pilot valve. Pushing the main spool to move [4-6], the oil generated by the front pump controls the direction of rotation of the hydraulic motor. The actual flow output of the pump before the double pump is:

$$q_B = V_B n \eta_v \tag{2}$$

where q_B is the actual flow of the output of the hydraulic pump (m^3/s); n is motor speed (n = 1450 rev/min); V_B is displacement of the front pump ($V_B = 10 \times 10^{-6} m^3 / r$); η_v is volumetric efficiency ($\eta_v = 92\%$).

The output of the hydraulic pump in the front part of the double gear pump drives a pair of motors. Assuming that the flow of both engines is equal, the speed of each motor can be calculated as follows:

$$n_M = \frac{V_B n \eta_v}{2V_M} \tag{3}$$

where V_M is motor displacement. In the case under study, $V_M = 470 \times 10^{-6}$ m^3/r.

The average running speed is related to the motor axial driving wheel diameter D_Q as follows:

$$v = D_Q \pi n_M \tag{4}$$

From Eq.4, we get $D_Q = 355 \times 10^{-3}$ m.

The data input into formula (4) yields the average running speed of the locomotive:

$$v = 0.2638(\text{m/s})$$

Assuming that the friction coefficient of bearing trolley's travelling wheel and rail is μ, the traction force required for a smooth road is:

$$F_p = mg\mu \tag{5}$$

where m is the total weight, including of the dead weight (kg); F_p is traction force of flat road surface ($F_p = 10kN$); μ is the friction coefficient of the travelling wheel and track. In our case, $\mu = 0.2$.

In the upward motion, the necessary traction for the total weight m is:

$$F_u = mg(\sin\theta + \mu\cos\theta) \tag{6}$$

where F_u is maximum traction load required for upward motion (N); θ is ramp inclination angle (the

IOP Conf. Series: Materials Science and Engineering **474** (2019) 012006 doi:10.1088/1757-899X/474/1/012006

maximum value is 25°).

The single driving monorail crane load can be measured by the car weight testing tools: its value plus locomotive weight 1218 kg yields the total weight of the single driving monorail crane m [7-9].

When θ changes from 0° to 25°, the locomotive moves uphill, lifting the weight m under different slope values, the relationship is shown in figure 3: the higher the slope, the smaller the lifting weight.

When the locomotive moves downhill, the driving force is as follows:

$$F_d = mg(\sin\theta - \mu\cos\theta) \,, \quad (7)$$

where F_d is downhill required traction load, N.

Above each action on the driving wheel, the product of the traction force and the radius of the driving wheel is torque acting on the motor shaft.

$$T = F \times R \,, \quad (8)$$

where R is motor drive wheel radius. In our case, R=0.1775m.

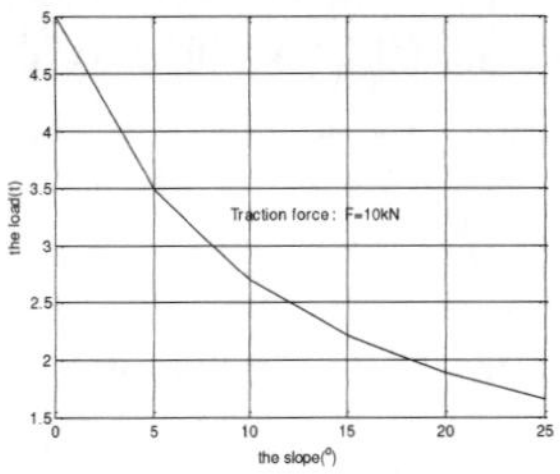

Figure 3. The load operation limit diagram.

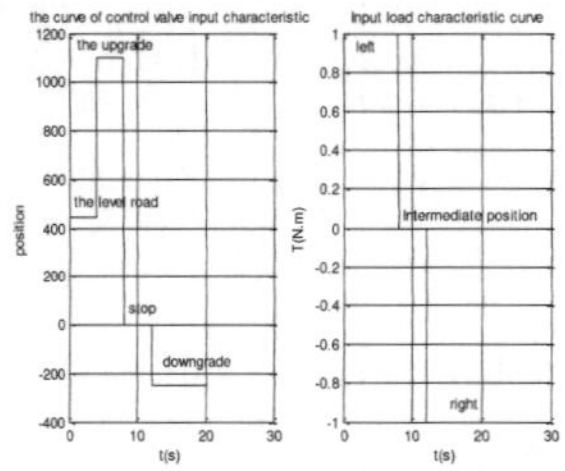

Figure 4. The pilot main valve position and the input load input signal.

2.3. Travelling circuit simulation.

As an example of operation condition, we used the Anju Coalmine (China) with a slope of 10° . The load was set at 2.50t. For the design of a round-trip transportation mission, AMESIM system was used to carry out simulation tests.

2.3.1. Path planning.

Forward motion. The required steps are as follows. Use the hand hoist to load the transported materials - pull roadway side manual pilot valve handle to the left position - smooth road - 10° upward slope – put the manual pilot valve handle to the middle position - ready to unload the transported materials [10].

For the smooth-road torque acting on a single motor shaft, formulas (5) and (8) yield:

$$T_p = 2500 \times 10 \times 0.2 \times 0.1775/2 = 443.75(N.m)$$

For the upward slope of 10 degrees, the torque acting on a single motor shaft is also derived via equations (6) and (8):

$$T_u = 2500 \times 10 \times 0.1775 \times (\sin(10^\circ) + 0.2 \times \cos(10^\circ))/2$$
$$= 2518.725(N.m)$$

Return trip. The required steps are as follows. Use the hand hoist to unload the transported materials, push roadway side manual pilot valve handle to the right position - smooth road - 10° downward slope - put the manual pilot valve handle to the middle position - ready to load the transported materials.
For the downward slope of 10 degrees, the torque acting on a single motor shaft can be derived via formulas (6) and (8) as follows:

$$T_{dh} = 1218 \times 10 \times 0.1775 \times (\sin(10^{o}) - 0.2 \times \cos(10^{o}))/2$$
$$= -248.5(N.m)$$

For the time period of 0~4s, the load on the flat surface of the hydraulic motor is shown in figure 4 (b). According to formula (9), during the time period of 4~8s, the hydraulic motor is subjected to the uphill load; at 8~12s, the hydraulic motor does not rotate. According to formula (10), at 12~16 s, the hydraulic motor is subjected to downhill load; at 16~20s, the hydraulic motor does not rotate.

2.3.2. Configuration scheme.

As shown in figure 5, the hydraulic circuit of AMESIM is simplified to scheme, in which the motor control is used to convert the mechanical energy of the main circulation pump into a hydraulic energy device. The motor rotation speed is set to 1450 rpm, and the displacement of the hydraulic pump is configured as follows. The pressure is controlled by a relief valve with a pressure set at 16 MPa. A three-position four-way valve with an O-type neutral function is selected as the hydraulic control main valve. There are two balancing valves between the hydraulic motor and the hydraulic control main valve, which are used to prevent the "landslide" of the slope during travelling and ensure a soft start, smooth running, soft and reliable stop at the required position on the slope. The displacement of the hydraulic motor is $470 \times 10^{-6} m^3 / s$, the rotary inertia is: $34153 kg \cdot mm^2$. The reversing valve input signal I is set as shown in figure 4 (a), and the extra torque load input signals II and III are arranged as shown in figure 4 (b). After running AMESIM, the output signal provides the torque, angular velocity, differential pressure, and flow to form a resulting data file.

2.3.3. Simulation analysis.

Using MATLAB software to process the data of the simulation result file, after the data processing of the simulation results file, the program constructed characteristic curve shown in figure 6. From the simulation characteristic curve, one can see the following patterns.

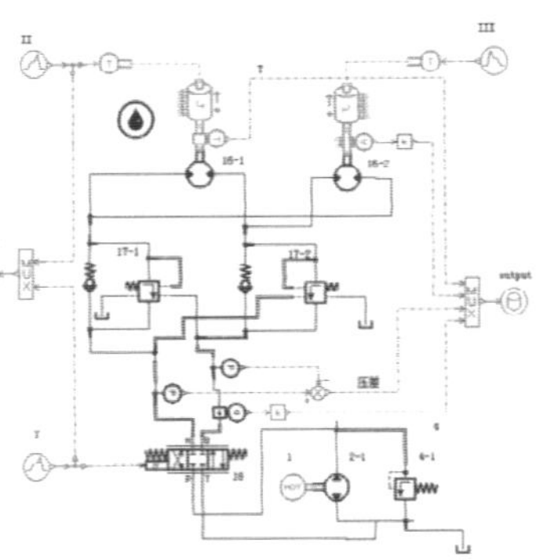

Figure 5. Scheme of the AMESIM transportation main circuit.

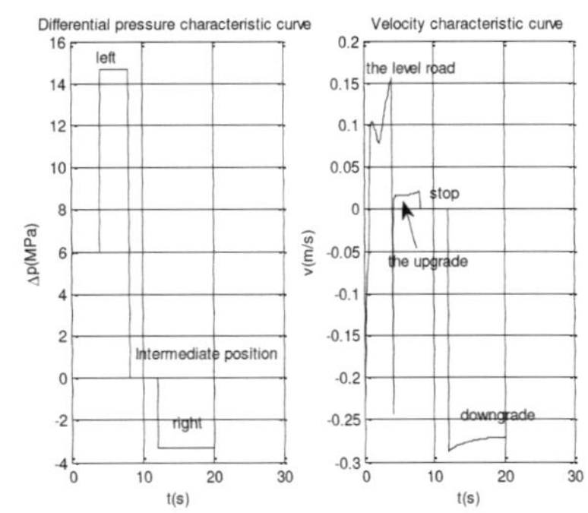

Figure 6. The characteristic curves of differential pressure and travel velocity.

1) The differential pressure between the two ends of the hydraulic motor is related to the load by the following formula:

$$\Delta p = \frac{2\pi T}{V_M \eta_M},$$

where Δp is motor differential pressure, N/m^2; η_M is motor mechanical efficiency. In our case, $\eta_M = 0.9$

The comparison of figures 4 (b) and 6 (a) shows that both characteristic curves are consistent, that is, the pressure is proportional to the load applied to the motor.

2) The locomotive running speed is inversely proportional to the torque load, so the speed and traction force satisfy the following formula:

$$v = \frac{\Delta p q_B \eta_V \eta_M}{F}$$

Using formula (8), we get:

$$v = \frac{\Delta p q_B \eta_v \eta_M R}{T}$$

Comparing Fig. 4 (b) with Fig. 6 (b), the rotational speed is inversely proportional to the torque load. At the time period of 4~8s, the torque load applied to the motor is the largest, and the rotational speed is the lowest; at 12~16s we get the downhill portion of the pass, with a high speed and easy stall.

The above findings prove that the simulation results are consistent with the theory of hydraulic transmission.

3. Safety auxiliary circuit

3.1. Working brake

The condition for the locomotive working brake is that the cartridge valve of the external emergency braking module in figure 2 is closed [11].

Regardless of the throttling effect and drag loss of the throttle valve, the brake action of the locomotive is equivalent to the hydraulic resistance of the orifice and the two throttle valves connected in series, according to the principle of flow continuity:

$$C_d A_T \sqrt{\frac{2p_X}{\rho}} = \frac{\pi d_0^{\ 4}(p_B - p_X)}{128\mu l_0}$$

where A_T is flow area of the throttle valve, m^2; l_0 is the length of the damping hole, m; d_0 is the damping hole diameter, m; p_X is the cartridge valve control cavity pressure, MPa; p_B is effect of pressure on the surface of the ring cavity B, at the beginning of the brake $p_B = 10MPa$; ρ is the oil liquid density, kg/m^3; μ is the dynamic viscosity of the oil, N.s/m^2

The above formula can be reduced to the following form:

$$A_T = \frac{\pi d_0^{\ 4}(p_B - p_X)}{128\mu l_0 C_d} \sqrt{\frac{\rho}{2p_X}}$$

The last equation analysis yields several patterns:

(1) $p_X = p_B$, $A_T = 0$; and at the same time w $p_X \to \infty$, p_X is adjusted by the overflow valve, indicating that the throttle valve at this time is closed. Thus, both ends of the damping hole are closed to the liquid flow. For a stationary fluid, the pressure value is constant, port A is not connected to B port, and the cartridge valve closes.

(2) $p_X = 0$, $A_T \to \infty$, the throttle valve and overflow channel are open. In the hydraulic control valve spring position, the cartridge valve is opened, ports A and B are connected, the locomotive implements the emergency brake. In the hydraulic control valve liquid control position, the cartridge

valve is closed.

3.2. Emergency brake

The centrifugal brake works mechanically and is integrated into the drive section of the monorail crane, consisting of a drive wheel and a mechanical structural unit with a centrifugal counterweight and a trigger pin. When the load is moved downward, it can be seen from figure 6(b) that when the empty car runs downward, the speed is high. When it is loaded is extremely prone to stalling, when the speed reaches the speed of the trigger of the factory settings (such as v=1m/s), the trigger pin will hit the trigger control rocker, shifting the machine control valve (see figure 2), from the motor position to spring position, the oil under pressure moves back to the tank. The hydraulic control valve is shifted to the spring position, the liquid flow occurs by A>T, $p_X = 0$.

When there is no liquid flow in the damping hole of the external emergency brake module, ignore the spool weight, and viscous friction, fluid statics analysis spool force of the two-way cartridge valve is:

$$F_B = F_X + F_y + F_k \tag{9}$$

where F_B is the force on the acting face of the ring cavity B, $F_B = \frac{\pi}{4}(D^2 - d^2)p_B$, *N;* D is the diameter of the cartridge valve spool, m; d is the control chamber of A cavity hole diameter, m; p_B is pressure on the acting face of the ring cavity B, N/m^2; F_X is the downward force on the acting face of the ring cavity, $F_X = \frac{\pi}{4}D^2 p_X$, *N;* F_k is cartridge valve reset stretch, $F_k = k(x_0 + x)$, *N;* k is the reset spring stiffness, N/m; x_0 is initial compression of the reset spring, m.

Steady-state fluid dynamics can be interpreted as the reaction force of the cartridge valve spool pointing in the direction, in which the valve port closes as liquid enters and exits the valve cavity due to changes in the direction of fluid flow.

The calculation formula for the steady-state flow force [12] is as follows:

$$F_y = \rho q v \cos\alpha \tag{10}$$

where ρ is the oil liquid density, kg/m^3; α is the liquid jet angle, $\alpha = 69^\circ$; q is flow through the valve port of the cartridge valve, m^3/s;

$$q = C_d \pi (D-d) x \sqrt{\frac{2(p_B - p_X)}{\rho}} \quad (11) \quad v = C_v \sqrt{\frac{2(p_B - p_X)}{\rho}} \tag{11}$$

After substitution of equation (11) into (10), we get:

$$F_y = 2C_d C_v \pi (D-d)\cos 69^\circ (p_B - p_X)x \tag{12}$$

After re-arrangement, formula (9) can be reduced to the following form:

$$p_B = \frac{(\pi D^2 - 42C_d C_v \pi (D-d)\cos 69^\circ x)p_X + 4k(x_0 + x)}{\pi(D^2 - d^2) - 42C_d C_v \pi (D-d)\cos 69^\circ x} \tag{13}$$

The results obtained show that the opening and closing of the two-way cartridge valve depends on the X chamber pressure p_X and the reset spring stiffness of the cartridge valve, the size of the pressure p_X between the damping hole and the throttle valve related to the setting of the flow area of the throttle valve. During the centrifugal trigger action, the upper type can be reduced to the following form:

$$p_B = \frac{4k(x_0 + x)}{\pi(D^2 - d^2) - 42C_d C_v \pi(D-d)\cos 69^\circ x} \tag{14}$$

The force acting on the annular area of the cartridge valve is much larger than the return spring force of the cartridge valve. Therefore, the force balance is broken, the brake cylinder is subjected to high- and low-pressure variations, and the cylinder chamber is filled due to the expansion of the cylinder volume of the brake cylinder. In the piston chamber cycle, the oil return route is short, and the braking time is also shortened accordingly.

4. Conclusions

The main contents of this paper are summarized as follows:

✓ The relationships between the average running speed of the locomotive, slope, and the load under different working conditions are obtained. The simulation path is based on the engineering design requirements. The simulation results using the AMESIM software show that the operation of the hydraulic motor between the two end points is determined by the load. The running speed is inversely proportional to the traction force, which shows a high consistency of simulation results and the hydraulic drive theory.

✓ This paper describes the conditions for realizing the brake operation, deriving the expression of the flow passage area of the throttle valve, and analyzing the practical significance of adjusting the throttle valve. The clamp circuit of the check valve ensures the priority of the brake release after the motor rotation.

✓ The traction and clamping circuit is controlled by a rotary valve, closed traction rotary valve, manual operating pump, and brake release control. The opening of the rotary clamping valve can reduce the clamping force of the drive wheel and the guide rail.

✓ The hydraulic system analysis performed in this paper can be extended to the multi-drive or scheduling monorail cranes.

References

[1] Zhang J, Yang H and Lu Y 2000. *Chin J Mech Eng-En*, **36**(7).
[2] Dai J X 2009 *Coal Mine Machinery,* **2**(184).
[3] Wang Q, Zhu A D and Xu X H 2014 *Agricultural Equipment & Vehicle Engineering,* 19-21.
[4] Situ R, Grech S, Santarossa L G and Bruce R C 2014. *Conf. of Australian Society of Sugar Cane Technologists Held at Gold Coast.*
[5] Couchot F 2016 *Commun Algebra*, **1**:381-389.
[6] Meng Y, Fang Y and Liu C 2010 *Int. Conf. on Advanced Computer Theory & Engineering*. IEEE.
[7] Chang K H 2013 *Motion Simulation and Mechanism Design with SolidWorks Motion 2011* Beijing: Mechanical Industry Press, **9**:20-24.
[8] Song Z A, Zhang X, Song Y F 2015 Beijing: National Defense Industry Press,184-209.
[9] Liang Q, Su Q Y 2014. Beijing: Mechanical Industry Press, 256-297.
[10] Gao F, Xiao L J, Guo S and Ma H G 2013 *Applied Mechanics & Materials, 278-280*, 350-353.
[11] Zhao W, Guo C and Wang F 2016 *Journal of Brazilian Society of Mechanical Sciences & Engineering,* **38**(2), 525-533.
[12] Han M, Song Y, Zhao W, Cheng Y and Ji X 2015 *Journal of Fusion Energy,* **34**(3), 571-571.

IOP Conf. Series: Materials Science and Engineering **474** (2019) 012007 doi:10.1088/1757-899X/474/1/012007

Layup structure effect on the composite laminate impact resistance: experimental and finite element studies

M Gong[1,2], S G Sun[1], L T Teng[3,*] and Q S Tu[3]

[1]School of Mechanical, Electronic and Control Engineering, Beijing Jiaotong University, Beijing, 100044, China
[2]CRRC Institute, Beijing, 100070, China
[3]CRRC Qingdao Sifang Co., LTD, Qingdao, 266111, China

*E-mail: Tengletian@cqsf.com

Abstract. Four kinds of sandwich laminates with different layup structures and same thickness are designed and fabricated by autoclave and heat pressing process respectively. The impact tests of four composite laminates were carried out to compare the impact resistance of composite laminates with different ply structures. The impact finite element model is established, and the impact process is simulated by the finite element method. To some extent, the impact failure mechanism of composite laminates and their impact failure results have been obtained.

1. Introduction

Composite materials have a number of advantages, such as high static and cyclic strength, substantial rigidity, light weight, vibration damping, high-temperature resistance, and high performance. These advantages imply their wide application in aviation, aerospace, energy, transportation, construction, and other industries [1-7].

In recent years, the high-speed impact resistance of composite laminates has been paid more and more attention, and the corresponding theoretical and experimental studies have been carried out [8-10]. Zhu [8] put forward the failure modes of Kevlar fiber-reinforced laminates, such as deformation, fiber delamination, and fiber failure, in the penetration of projectiles. From the observation of 3.3g cube projectile impact test of UHMWPE (Ultra-high molecular weight polyethylene) fiber laminated plates, Wang [9] also revealed the failure modes of UHMWPE in shear, impact delamination and tensile deformation, etc. By simulating the penetration behavior of fiber-reinforced composite laminates, Qin [10] found that the penetration mode of the flat projectile was a plugging mode, and the penetration of the ball projectile and cone projectile was a stabbing pattern. Besides, Naik [11] and Gu [12] have established an analytical model for the ballistic performance of composite laminates with orthotropic laminates. Several typical energy-absorbing forms of target plates were considered. Since the object of these models was orthotropic laminated composite laminates, the in-plane anisotropy was not apparent, and the results were acceptable. However, for the composite laminates with arbitrary layers, the analysis of high-velocity impact damage and numerous test results revealed that the anisotropy significantly influenced the damage shape and, thus, had to be considered in the further analysis [13-14].

To explore the impact resistance of the composite laminates with different layer structures, we designed and fabricated, in accordance with the UIC651 standard, four different layups with the same thickness and foam core, based on the driver's cabin of the last generation subway train. The experimental and simulation works have been carried on. In this paper, upon the impact tests, the composite lamination layup and fiber arrangement angle were analyzed to assess the effect of layer structure on the impact resistance performance of composite laminates. The impact failure mechanism of composite laminates was simulated by the Hyperworks13.0 finite element analysis software.

2. Materials

The test materials were prepreg and foam. The prepreg was made of carbon fiber, glass fiber, and aramid fiber. The specifications of the material are shown in table 1. The prepreg was provided by SCN (Shanghai CEDAR &NISSEI Co., Ltd., China), while the foam material (model PMI, density of 52 kg/m^3, thickness of 59 mm) was provided by Evonik Industries (Germany).

Table 1. Specifications for prepreg.

Code	Material	FAW(g/m^2)	RC(%)	Thickness(mm)
3kC	3K Twill(carbon)	200.0 g/m^2	37%	0.20 mm
K	Plain(Aramid)	235.0 g/m^2	40%	0.25 mm
G	Broken twill(glass)	290.0 g/m^2	40%	0.24 mm
C	Unidirectional(carbon)	150.0 g/m^2	40%	0.15 mm

3. Specimen preparation

The specimens are sandwich structures with inner and outer skins. They had different ply structures, as shown in table 2. The skin size was 1100*800 mm, fabricated with autoclave molding, at 130°C and pressure of 0.6 MPa; the curing time was no less than one hour. The cured sandwich composites were cut into 1000*700 mm specimens with CNC machine.

Step 1. Cut the clothes:

Cut and prepare the clothes according to the size of specimens.

Step 2. Cut the forms:

Cut and prepare the forms according to the size of specimens.

Step 3. Lay the skins:

Lay the skins according to the lay-up sequence of specimens. Mark the direction of the fiber at the same time. The tolerance of angle is ±2°.

Step 4. Form the skin:

Skins are formed in an autoclave; the curing program is shown below:

Air heating rate: 3 Ai;

Curing temperature: 80±3° C, maintain 30mins, 80±3° C, maintain 15mins, 130±3° C, maintain 60mins;

Pressurization time: when the temperature of 80±3° C is achieved, apply 0.3±0.01MPa, maintain 30mins, and then hoist the pressure to 0.6±0.01 MPa;

Pressure rate: 0.05MPa/min;

The specimens should be in the state vacuum of -0.9 atmosphere. Make sure that specimens were in the state vacuum until they are cured;

Air cooling rate: 3° C/min;

Pressurization time: when the temperature of 60±3° C is attained, drop the pressure;

Pressure drop rate: 0.05MPa/min;

Step 5. Bond the skin and form:

The composite specimen is made by the hot pressing process. The structure of the specimen consists of outer skin, epoxy bond film, PMI foam, epoxy bond film, inner skin. The production process condition: pressure 0.2 MPa, temperature 130° C, 1.5h.

Step 6. Machine the specimens:

The cured sandwich composites are cut into 1000*700 mm specimens with CNC machine.

IOP Conf. Series: Materials Science and Engineering **474** (2019) 012007 doi:10.1088/1757-899X/474/1/012007

Table 2. Ply structure of composite laminates.

Specimen No.		Ply structure	Thickness (mm)
1	Outer	The outer structure of composite laminate cut into 1000	2.30
	Inner	±Inner structure of composite laminates	1.80
2	Outer	3KC/(0/90/0)G/±45C/±45C/(0/90/0)G	2.24
	Inner	(0/90)G/±45C/±45C/±45C/(90/0)G	1.86
3	Outer	3KC/(0/90)G/±45C/K*2/±45C/(90/0)G	2.26
	Inner	(0/90)G/±45C/K/±45C/(90/0)G	1.81
4	Outer	3KC/(0/90)G/±45C/K/+45K/±45C/(90/0)G	2.26
	Inner	(0/90)G/±45C/K/±45C/(90/0)G	1.81

4. Impact test

According to the UIC651 standard, the specimen size was 1000*700 mm, and the cylindrical projectile with a hemispherical tip was used to impact the specimen. The total weight of the projectile was1kg, and its structure is shown in Figure 1. The aluminum alloy cylindrical projectile hit four specimens at a speed of 300.67, 309.60, 300.57, and 304.93 km/h, respectively. During the testing, the temperature of the test board was kept at 15-35°C. The impact velocity of the projectile was assessed as follows:

$$V_P=V_{max}+160 \quad (1)$$

where V_P is projectile velocity at impact (km/h) and V_{max} is the maximum speed for a towing device (km/h).

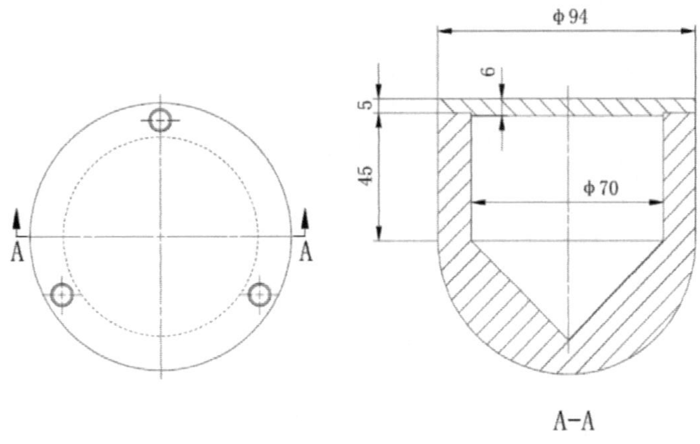

Figure 1. Structure and size of the projectile for the impact test.

Table 3. Results of impact tests.

Specimen number	Thickness (mm)	FAW(Inner + outer skin) (g/m²)	Impact velocity (km/h)	Test result	Conclusion
1	63.10	4100	300.67	Penetrated	Not pass
2	63.10	4600	309.60	Penetrated	Not pass
3	63.07	4425	300.57	Penetrated	Not pass
4	63.07	4425	304.93	Not penetrated	Passed

In the actual test, the input impact angle was 90° to the composite lamination with the impact velocity of no less than 300 km/h. The test temperature was 20° C, and the test results are shown in table 3.

IOP Conf. Series: Materials Science and Engineering **474** (2019) 012007 doi:10.1088/1757-899X/474/1/012007

5. Experimental data and analysis

The impact test results are shown in table 3: specimen No.4 passed the impact test, while specimens 1-3 failed. Figure 2(a)-(d) show the front and back of the specimen after impact. In figure 2(d), the projectile did not penetrate the test specimen, while the rest of specimens were penetrated.

It was found that the inner skin of one side of the specimen was severely damaged, and the main forms of damage were matrix cracking and delamination.

Part of the project interpenetrated the out skin of the test specimen 4, and was embedded between the fiber layer and PMI foam. A portion of the inner skin was bent and delaminated from the PMI foam core. Although some fibers were broken, the inner skin was not penetrated.

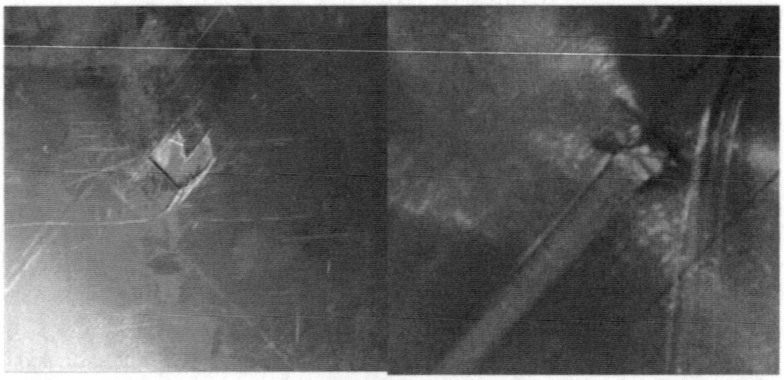

(a) Specimen 1

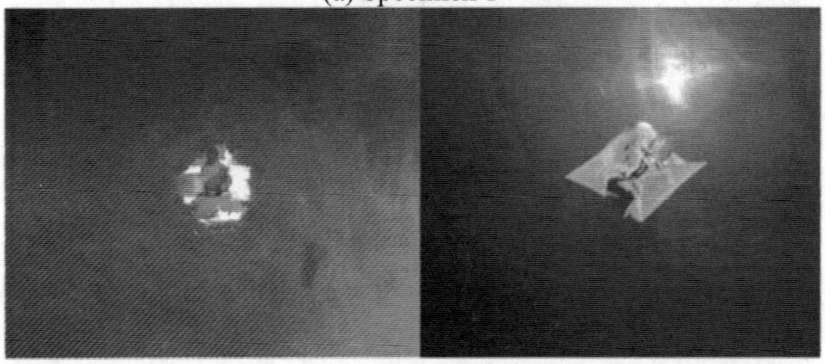

(b) Specimen 2

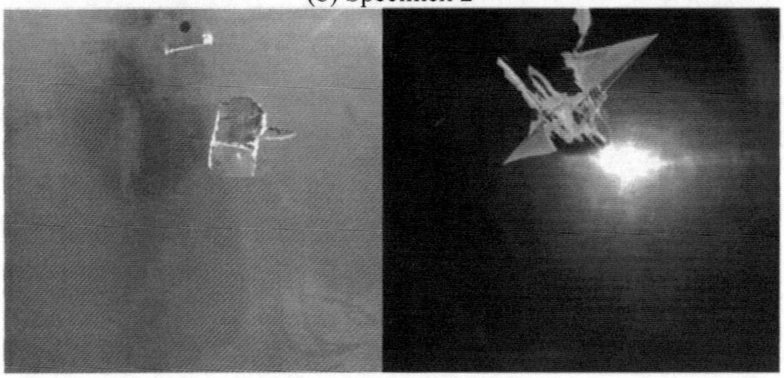

(c) Specimen 3

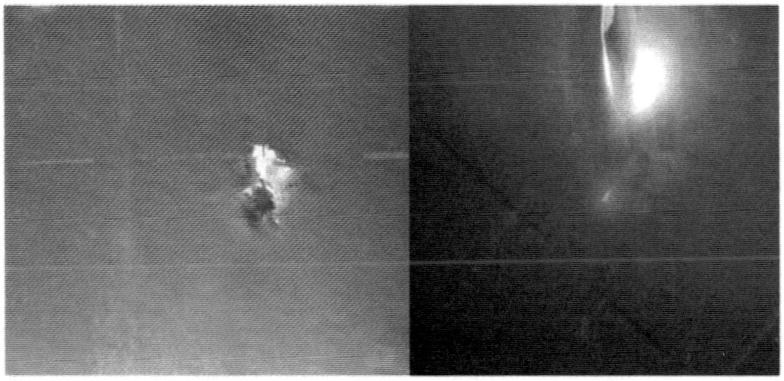

(d) Specimen 4

Figure 2. The front and back of the specimen after impact.

The layer structure of specimen 4 is orthotropic, and the mechanical properties in the normal direction to the fiber length are quite high. Besides, the anti-impact performance of aramid layer was observed: so it had excellent impact resistance.

As compared to specimen 3, the number of aramid fibers was increased, the impact resistance was strengthened, and the impact resistance of test specimen 4 was improved.

The skin of specimen 1 was a pure carbon fiber structure, and specimen 2 was a hybrid structure, which consisted of carbon and glass fibers. As compared to specimen 4, it lacked aramid fibers, so it had poor impact resistance.

6. Impact simulation

6.1. The finite element analysis model

The finite element method was used to analyze the impact strength of specimens under study. The finite element mesh was prepared by the pre-treatment Hyper-Mesh, and RADIOSS was applied to calculate the specimen strength of the specimen. Loads and constraints were set according to the standard UIC651 "Layout of driver's cabs in locomotives, railcars, multiple unit trains and driving trailers".

The simulation model of high-speed impact on composite sandwich panels adopts the form of skin + core material + skin, and the skin layer is subdivided into "laminate- interface-laminate" form. The single layer thickness of the material was shown in table 1. Each layer is simulated with interface properties in the thickness of zero. The dimension requirements of the impact head are shown in figure 1, and the weight is 1 kg. The boundary condition of the model is that the sandwich plate is fixed around it, the initial velocity of 300 km/h is applied to the warhead, and the contact type is integrated interface contact.

In the calculation, the carbon fiber layer was separated by finite element method with 2D shell element SHELL4N. The 3D solid element HEXA8N was used in the finite element dispersion of foam. The finite element model is shown in figure 3. The whole simulation model was constituted by 16352 shell elements, 56000 solid elements, and 57269 nodes.

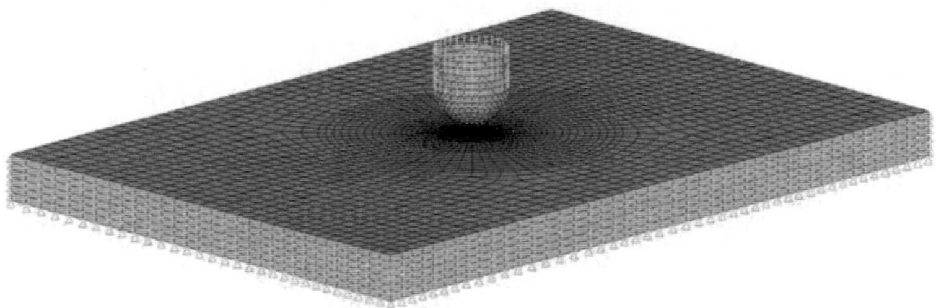

Figure 3. Finite element analysis model of the high-speed impact of the composite.

IOP Conf. Series: Materials Science and Engineering **474** (2019) 012007 doi:10.1088/1757-899X/474/1/012007

Carbon fiber/epoxy resin, glass fiber broken twill fabric/epoxy resin, aramid fiber plain fabric/epoxy resin with the volume fraction of 60%, its performance parameters are shown in tables 4~6, respectively. The interlayer interface uses the material properties of the matrix, and its performance parameters are shown in table 7.

Table 4. Carbon fiber/epoxy composite property parameters.

E1 (GPa)	E2=E3 (GPa)	n12	n23	G12=G13 (GPa)	G23 (GPa)
135	8.5	0.29	0.3	4	3.5
XT (MPa)	XC (MPa)	YT (MPa)	YC (MPa)	S12=S13 (MPa)	S23 (MPa)
2750	911	35	130	63	32

Note: Subscript 1 corresponds to the fiber main direction, Subscript 2 to normal to the main direction, Subscript 3 to thickness direction, XT is tensile strength in the main direction, XC is main direction compression strength, YT is tensile strength 90°-direction, YC is compression strength in 90°-direction, S is shear strength.

Table 5. Aramid fiber plain fabric/epoxy composite property parameters.

E1 (GPa)	E2 (GPa)	n12	n23	G12=G13 (GPa)	G23 (GPa)
37.3	15.9	/	0.3	4	3.5
XT (MPa)	XC (MPa)	YT (MPa)	YC (MPa)	S12=S13 (MPa)	S23 (MPa)
517	200	67.9	130	77.2	80

Table 6. Glass fiber broken twill fabric/epoxy composite property parameters.

E1 (GPa)	E2 (GPa)	n12	n23	G12=G13 (GPa)	G23 (GPa)
28.1	24.6	/	0.3	4	3.5
XT (MPa)	XC (MPa)	YT (MPa)	YC (MPa)	S12=S13 (MPa)	S23 (MPa)
346	140	252	130	20	18.5

Table 7. Performance parameters of matrix materials.

E(GPa)	G(GPa)	T(MPa)	S(MPa)
8.3	4.5	55	60

6.2. Failure criteria

1) Damage of single layer composite

IOP Conf. Series: Materials Science and Engineering **474** (2019) 012007 doi:10.1088/1757-899X/474/1/012007

The composite unit strength is defined with the Tsai-Wu criterion. Before the limit of the Tsai-Wu criterion is reached, it is assumed that the material is an orthotropic linear elastic material.

$$F_1\sigma_1+F_2\sigma_2+F_{11}\sigma_1^2+F_{22}\sigma_2^2+F_{44}\sigma_{12}^2+2F_{12}\sigma_1\sigma_2\leqslant F(W_p,\dot{\varepsilon}) \quad (2)$$

where σ_1, σ_2, σ_{12} are stresses in the material coordinate system

Beyond the limit of the Tsai-Wu criterion, the material is transformed into nonlinear one, and the maximum strain rate and the maximum strain energy failure theory of brittle materials are adopted. The material hardening is simulated by the plastic strain energy and strain rate to set the limit of the Tsai-Wu criterion. The delamination failure is reduced to delamination induced by the shear strain.

The coefficients in the Tsai-Wu criterion are determined by ultimate stresses.

$$F_1=-\frac{1}{\sigma_1^c}+\frac{1}{\sigma_1^t}F_{22}=\frac{1}{\sigma_2^c\sigma_2^t} \quad (3)$$

$$F_2=-\frac{1}{\sigma_2^c}+\frac{1}{\sigma_2^t}F_{44}=\frac{1}{\sigma_{12}^c\sigma_{12}^t} \quad (4)$$

$$F_{11}=\frac{1}{\sigma_1^c\sigma_1^t}F_{12}=-\frac{\alpha}{2}\sqrt{F_{11}F_{22}} \quad (5)$$

Here superscripts c and t represent compression and tension, respectively.

In the above formula, F is a variable Tsai-Wu criterion limit function defined by the strain energy (W_p) and true strain rate ($\dot{\varepsilon}$).

$$F(W_p,\dot{\varepsilon})=\left(1+b\left(\frac{w_p}{w_p^{ref}}\right)^n\right)*\left(1+c\ln\left(\frac{\dot{\varepsilon}}{\dot{\varepsilon}_0}\right)\right) \quad (8)$$

where w_p^{ref} is the reference strain energy, B is the strain-hardening coefficient, n is the strain-hardening index, $\dot{\varepsilon}_0$ s the true strain rate, and C is the strain rate factor.

The damage between the initial and maximum fracture elongation rates ε_{ti} and ε_{mi} , respectively, is controlled by the damage constant d_i.

$$d_i=max\left(\frac{\varepsilon_i-\varepsilon_{ti}}{\varepsilon_{mi}-\varepsilon_{ti}},d_{max}\right),\ i=1,2 \quad (7)$$

According to the damage constant d_i, stresses in the damaged layer in the laminates are reduced as follows:

$$\sigma_i^{reduced}=\sigma_i(1-d_i) \quad (8)$$

The maximum value of damage constant di is d_{max}, strain energy per unit volume is $w_{p*}=\frac{w_p}{w_{ref}^p}$, while the maximum strain energy per unit volume is $w_{p*max}=\frac{w_{p_{max}}}{w_{ref}^p}$.

In case of $w_{p*}>w_{p*max}$, the fiber fracture occurs; when the strain in direction 1 is higher than ε_{m1}, the fiber fracture occurs; when the strain in direction 2 exceeds ε_{m2}, the matrix fracture occurs; the stress in the unit drops to zero, and the unit is removed.

2) Delamination damage

The simplified delamination failure theory is based on out-of-plane shear strain (γ_{31}, γ_{23}). When the shear stress $\gamma=\sqrt{(\gamma_{13})^2+(\gamma_{23})^2}$meets the condition $\gamma_{max}>\gamma>\gamma_{ini}$, stress σ_{31}, σ_{23} in the element decreases gradually. If the value of $\frac{\gamma-\gamma_{ini}}{\gamma_{max}-\gamma_{ini}}$ in a layer of the laminate is higher than d_{max}, the delamination occurs, and the unit is removed.

3) Foam damage

The material is linear elastic until the yield stress is reached. When the stress in the material is exceeds the yield stress, we get

$$\sigma=(a+b\varepsilon_p^n)\left(1+c\ln\left(\frac{\dot{\varepsilon}}{\dot{\varepsilon}_0}\right)\right) \quad (9)$$

where a is yield stress, b is strain-hardening coefficient, c is strain rate coefficient, ε_p^n is plastic strain, and $\dot{\varepsilon}$ is strain rate.

When the plastic strain of the material is higher than the maximum allowable plastic strain, failure occurs, and the unit is removed.

IOP Conf. Series: Materials Science and Engineering **474** (2019) 012007 doi:10.1088/1757-899X/474/1/012007

6.3. Simulation results

Specimen 1 is the representative specimen, which has not passed the impact test (failed). Specimen 4 is the specimen, which passed the impact test. Simulations were performed separately. The impact simulation is shown in figures 4 and 5.

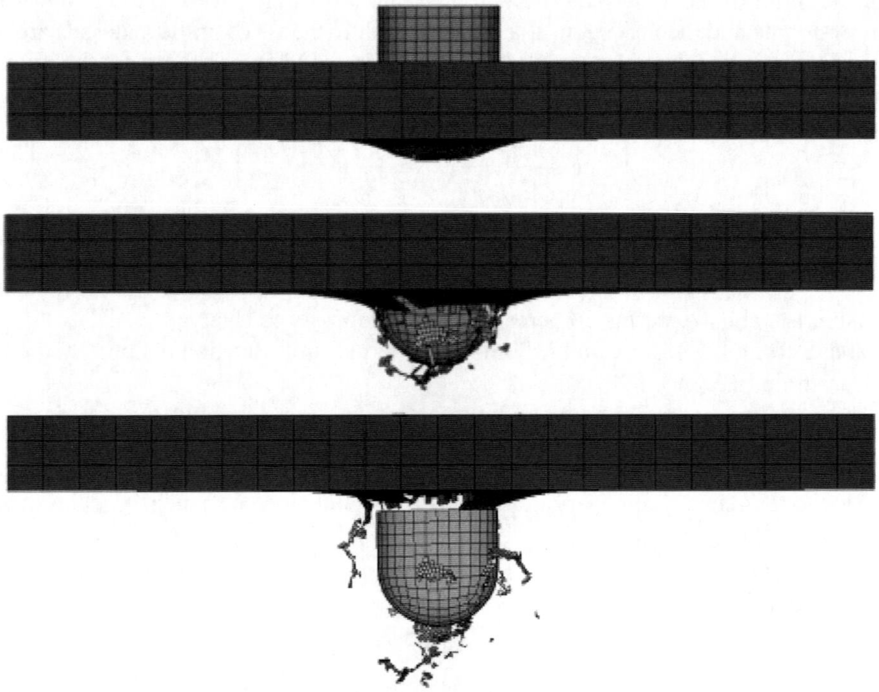

Figure 4. Simulation diagram of the shock process of specimen 1.

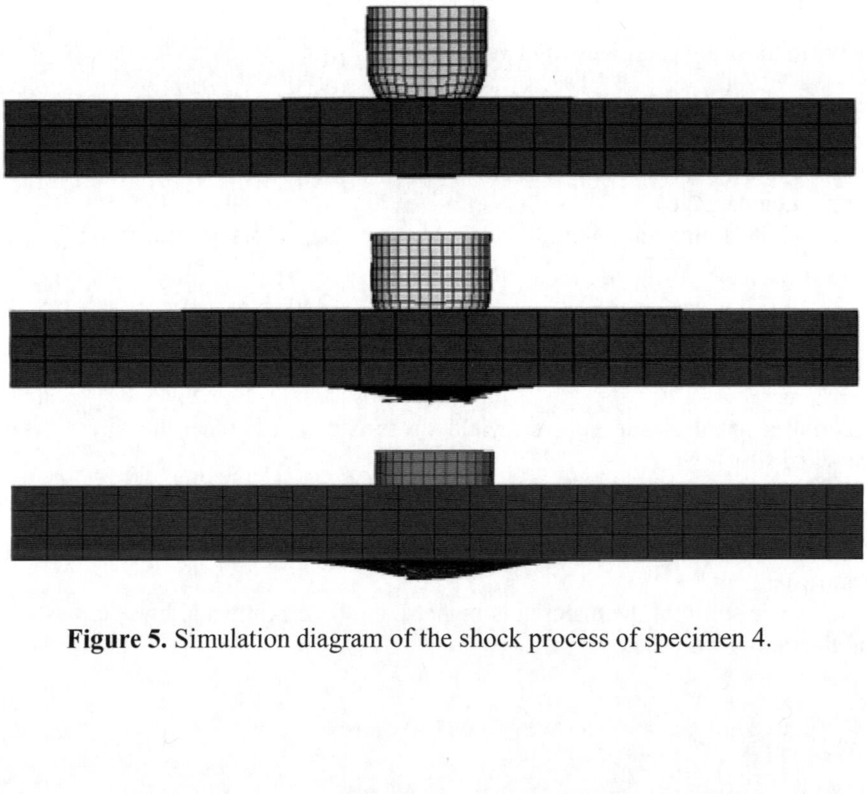

Figure 5. Simulation diagram of the shock process of specimen 4.

From the finite element analysis results depicted in figure 4, the composite core plate had a tapered projection in the thickness direction after being penetrated by the projectile. The damage area increased gradually from the point near the projectile to the direction away from the projectile. Matrix collapse damage occurs mainly in contact with projectiles. In cases of matrix cracking or matrix collapse damage, the distribution of any damage in the thickness direction had a funnel shape and was very similar to the form of delamination damage. This implies that the matrix damage is related to the delamination damage.

Through the finite element simulation calculation, the velocity change curve in the impact process can be obtained. As shown in figure 6, different specimens are impacted by the impact head and change in velocity during the impact. And figure 7 also shows that the final velocity and kinetic energy of specimen 4 are reduced to zero, while the remaining specimens 1~4 eventually had some residual velocities, indicating that the plate was eventually broken down.

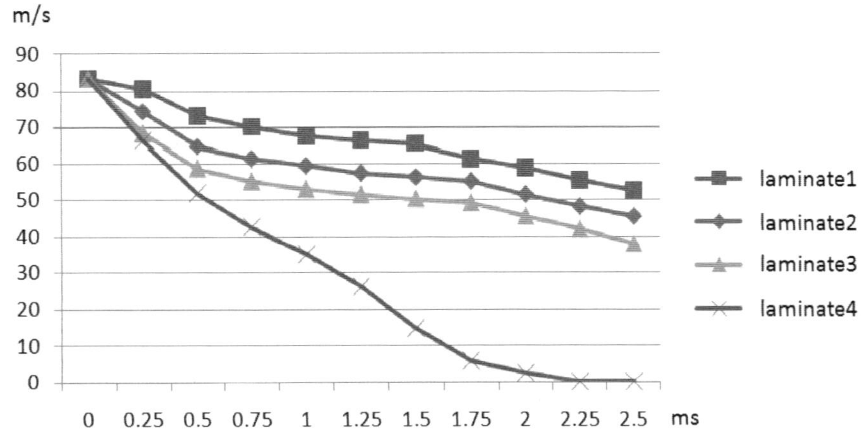

Figure 6. Velocity curves during impact.

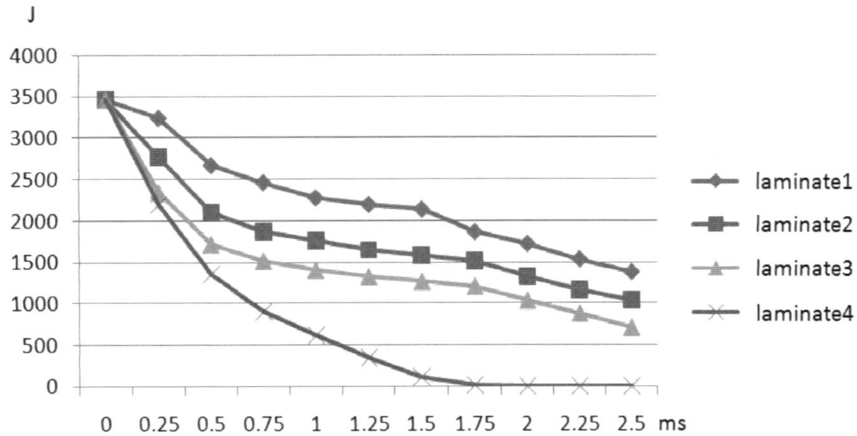

Figure 7. Kinetic energy curves during impact.

7. Conclusions

The test results show that some specimens of carbon/glass/aramid fiber structure with alternative layups passed the impact test, while other specimens failed.

The thickness of specimens 1~4 was the same, as well as thickness and density of the foam. The test specimens 1~3 with (i) anisotropic layer carbon fiber, (ii) carbon fiber/glass fiber, and (iii) non-anisotropic layer carbon/glass/ aramid fiber hybrid composite layers, respectively, failed during the impact test. In specimen 4, the aramid material (available in specimen 3) was replaced by the alternately anisotropic ply. As compared to specimens 1 and 2, aramid material with better impact resistance was added, which improved its impact resistance. As compared to specimen 3, specimen 4 is was alternately anisotropic. As a result, specimen 4 had the most excellent impact resistance and passed the UIC651 test standard.

Through the finite element simulation, the impact process and penetration results of specimens with different ply structures can be effectively predicted. The finite element numerical model elaborated in this study is considered instrumental in the impact failure prediction of composite laminates.

References

[1] Shen G L and Hu G K 2006 *Mecha. Compos. Mater.*
[2] Moura M F S F and Marque A T 1993 *J Compos.: Part A* **33** 361-8
[3] Moura M F S F and Gonçalves J P M 2004 *J Compos. Sci. Technol.* **64** 1021-7
[4] Choi H Y, Downs R J and Chang F K 1991 *J Compos. Mater.: Part I* **25** 992-1011
[5] Hitchen S A and Kem P M J 1995 *J Compos.* **26** 207-14
[6] Rong M Z, Zhang M Q, Y Liu, et al. 2002 *J Compos. Mater.* **12** 105-26
[7] Chang F K and Lessad L B 1991 *J Compos. Mater.* **25** 2-43
[8] Zhu G, Goldsmith W and Dharan C K H 1992 *Int. J. Solids Structure* **4** 399-420
[9] Wang X Q, Zhu X, Mei Z Y and Chen X 2009 *J Explo. Shocks* **29** 29-34
[10] Qin J B, Han Z J, Liu Y Y and Lu G Y 2013 *J Vibra Shock* **32** 122-6
[11] Naik N K and Shrirao P 2004 *J Compos. Struct.* **66** 579-90
[12] Gu B H 2002 *ACTA Mater Compos.* **19** 92-6
[13] Ching H Y and Rodney B K 1987 *J Inter. Impact Eng.* **5** 729-38
[14] Nunes L M, Paciornik S D and Almeida J R M 2004 *J Compos. Sci. Technol.* **64** 945-54

Design optimization and experimental study of coaxial powder-feeding nozzle in the laser cladding process

H Ju, Z J Zhang, C X Lin*, Z J Liu and H L Jiang

College of Naval Architecture and Ocean Engineering, Dalian Maritime University, Dalian 116026, China

* E-mail: lchxin@dlmu.edu.cn

Abstract. To improve the powder convergence and cladding layer quality, the three-dimensional axisymmetrical powder flow through a coaxial feeding nozzle was simulated by the Eulerian-Eulerian two-fluid model. With the composite design via the response surface method, the significance between the main geometric dimensions of nozzle and fluidization characteristics of powder flow was analyzed and regression-fitted. Moreover, a mathematical model was constructed to optimize the nozzle structure, and the optimal nozzle was used in laser cladding experiments. The results show that appropriate powder flow can be obtained in the condition of the carrier gas more than 9 l/min, and the powder sending rate less than 7.5 g/min. The structural parameters of the optimized nozzle are as follows: nozzle angle of 62.94°, outlet width of 1.0mm, exit radius of 5.83mm, and flow separation angle of 3.00°. With the optimized nozzle, the waist diameter and focus depth of the original powder flow decreased by 39.4 and 46.4%, respectively, and the discrepancy between the experimental and simulation results was less than 9%. With the appropriate technological parameters, a layer with good geometry morphology and appropriate dilution rate can be prepared in the laser cladding process with a coaxial feeding nozzle.

1. Introduction

Laser cladding, as a rapidly developing surface modification technology, has unique advantages of high energy, rapid solidification velocity, and quick cooling rate. Besides, the laser alloyed coatings displayed fine microstructure, excellent solid metallurgical bonding and small intermixing with the substrate [1-2]. It has been widely used in many practical applications such as surface metallization, component-repairing and 3D rapid prototyping [3-5]. The powder-feeding method is the crucial technological parameter in the laser cladding process and the coaxial feeding nozzles which allow the powder flow to flow coaxially with the laser beam provides the capability of precise deposition. Thus, the coaxial laser cladding has gained extensive application and much more feasibility in automatic manufacturing [6]. Considering the morphology and quality of the cladding is directly influenced by the powder flow, the feeding powder flow and the powder nozzle are always studied.

The research on the powder flow and structural design of the powder-feeding nozzle have been performed by domestic and foreign scholars [7-9]. However, since most studies were related to the powder flow rule of the two-dimensional nozzle model, no obvious improvement of the powder-feeding effect was obtained. Meanwhile, most nozzles have such deficiencies as poor powder convergence, deep focal position, large focus diameter, and low powder-usage efficiency.

In this paper, a three-dimensional coaxial feeding nozzle model was established to obtain the three-dimensional flow field by the Eulerian-Eulerian two-fluid models. And the influencing factor of nozzle structure on powder flow is multiple-square regressive analyzed by the response surface analysis. And the response model of nozzle structure and powder-feeding performance is obtained to study the influence of structural parameter on the powder flow. Moreover, the layer with good geometry morphology and appropriate dilution rate can be prepared in the laser cladding process with the coaxial feeding nozzle after a single-factor experiment of processing technologies.

2. Simulation of the powder flow stream

2.1. Simulation of the powder flow stream

The powder flow of coaxial feeding belongs to gas-solid two-phase flow, and the Eulerian-Eulerian two-fluid models are chosen to simulate the powder-gas two-phase flows. The ANSYS Fluent software is used to simulate the flow stream of the coaxial nozzle, and the simulation conditions are set as follows:

(1) The governing equation:

Based on the law of mass conservation, energy conservation, and momentum conservation, the Euler-type conservation equations are obtained.

(2) Three-dimensional model and mesh generation:

Figures 1 (a) and 1 (b) depict the 3D structure and 2D segmentation images, respectively, while the main dimensions of the nozzle model refer to the existing nozzle structure, as shown in table 1. There, the main influence factors on the nozzle flow stream are nozzle angle (θ), outlet width (w), exit radius (*r*). To increase the simulation speed, a quarter of three-dimensional nozzle model with main structure parameters are adopted, and the different flow field (I, II, III) is respectively meshed with different meshing size (2, 0.4, and 1 mm) as is shown in figure 1 (c).

Table 1. The values of nozzle structural parameters.

Parameter	θ(°)	w(mm)	r(mm)	λ(°)	α(°)	e(mm)	g(mm)	a(mm)	h(mm)	b(mm)
Size	67	2.2	5.00	5.0	37	60	7	25	50	30

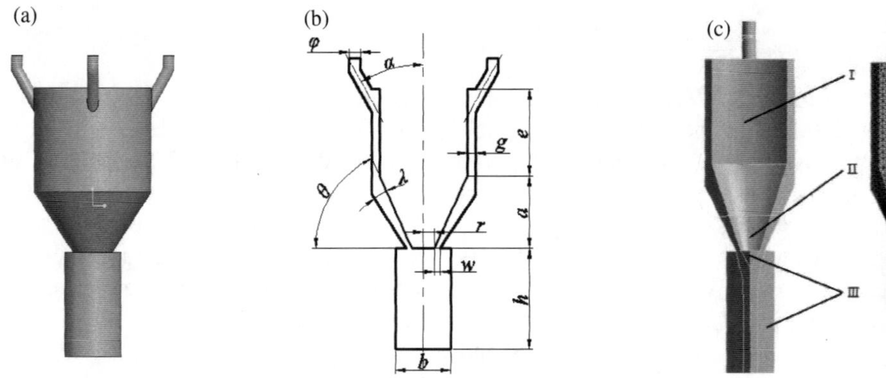

(a) 3D model; (b) sketch of a cross-section; (c) regions and their meshing

Figure 1. Three-dimensional model and meshing of the nozzle.

(3) Material properties

The carrier gas is argon, whose viscosity and density are 2.125×10^{-5} kg/(m·s) and 1.623 kg/m^3, respectively. The powder density and particle sizes are 7.5×10^3 kg/m^3 and 35 μm, respectively.

(4) Boundary conditions

The inlet of carrier gas stream and powder flow is set as speed type, and the outlet of powder flow is configured as outflow type. The gravitational acceleration is 9.8 m/s^2, and the ambient gas pressure is 1.013×10^5 Pa. The wall and two sides of the model are set as standard boundary and periodicity boundary conditions, respectively.

2.2. Simulation results of powder flow

The carrier gas flow and powder-feeding rate possess great influence on the characteristic and converge morphology of the powder flow field, and they are the significant parameters in the process of coaxial powder-feeding. To confirm proper feeding process parameters, the effect of a single factor (carrier gas flow and powder-feeding rate) on the volume concentration distribution is researched by the post-processing of the Fluent software.

Under the specific powder-feeding rate (5 g/min), figure 2 shows the volume concentration and flow velocity of powder with different carrier gas flow. As shown in figure 2 (a), when the carrier gas flow is minimum, the gathered position of powder is nearest to the nozzle bottom, and its powder volume concentration is maximum (about 8%). With the increment of carrier gas rate, the focal position of the powder flow gradually declines. It can be seen in figure 2 (b) that the axisymmetric powder flow rate increases with the carrier gas flow.

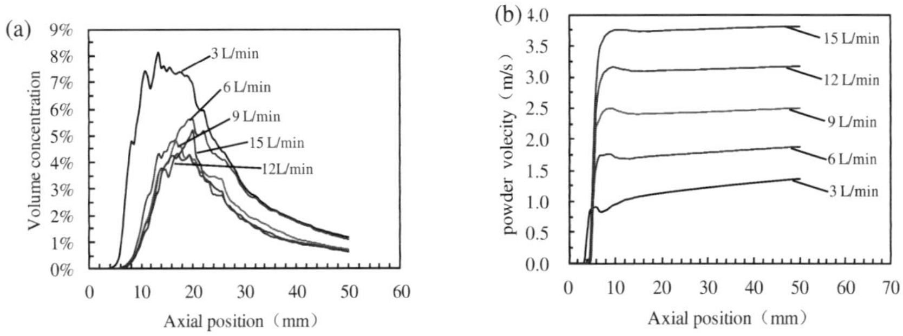

Figure 2. Powder flow characteristic under different carrier gas flow: (a) Concentration; (b) Velocity.

To study the effect of powder-feeding rate on the powder flow, the distribution of powder flow under different powder-feeding rate and specific gas carrier rate (9 l/min) is shown in figure 3. It can be seen in figure 3 (a) that the powder-feeding rate has a prominent influence on powder volume concentration. That is the powder concentration of the entire nozzle model increases with the powder-feeding rate, and the maximum powder concentration can reach 20%. Figure 3 (b) shows that the velocity gradually reduces when the powder-feeding rate arises from 2.5 to 12.5 g/min, and the value remains unchanged until the level of 7.5 g/min is reached.

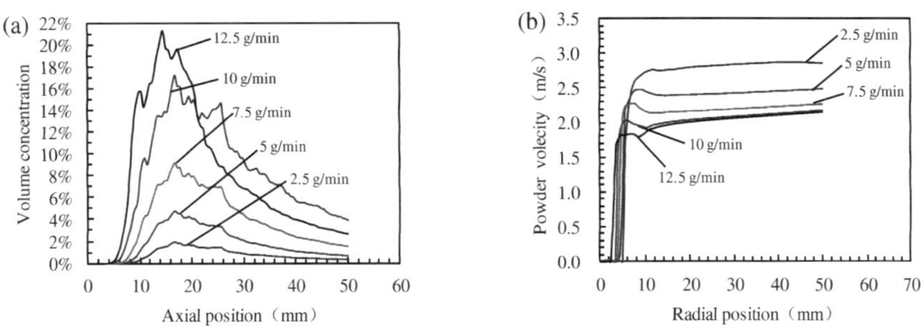

Figure 3. Powder flow characteristics at different feeding rates (a) volume concentration; (b) powder velocity

Based on the above analysis, to avoid blocking the outlet by melting powder spatters, the focal position of powder flow should not be too close to the nozzle outlet. Meanwhile, to enhance the utilization efficiency of the powder and decrease the laser attenuation by the powder flow, the focal position and volume concentration of powder flow could not be too large. Thus, the carrier gas rate should not be lower than 9 l/min, and the powder-feeding rate should not exceed 7.5 g/min for a better feeding effect.

3. Structure optimization of coaxial nozzle

3.1. Regression model between nozzle structure and stream characteristic

The response surface method [10] was used to establish the regression model between feeding nozzle structure and powder flow characteristics. The center composite design [11] was adopted to study the effect of nozzle major structural parameters (nozzle angle (θ), outlet width (w), exit radius (r) and shrinkage angle (λ)) on the powder-feeding characteristics. The five levels (high-level, low-level, zero-level and two alpha levels) set in the center composite design, and the value ranges of nozzle angle, outlet width, exit radius, shrinkage angle were chosen as 55~65°, 1~2mm, 4~6mm, and 3~9°, respectively, according to available literature and nozzles. The design scheme included 30 points (16 comprehensive test points, 8 axial test points, and 6 central test points). The simulated tests and response values are obtained under the conditions that carrier gas flow of 9 l/min and the powder-feeding rate of 5 g/min. The Design Expert software was used to process the experimental data by the variance analysis and regression fitting, and the non-significant factor is removed by the stepwise selecting method to obtain more precise response surface regression model.

It can be seen from the variance analysis of the flow waist diameter that its F-value in the regression model is 33.82, which proves the model efficiency. The P-value is less than 0.0001, thus the difference is almost impossibly caused by the sampling error. And the phenomenon that F-value of unfitting degree is greater than 0.05 shows the fitting degree of regression model is better. Through the above analysis, three test values indicate that the regression model is prominent, and the significant items include first-order terms θ, w, r, interaction items θ × w, θ × r and second-order term θ^2. By removing the non-significant item, the regression model of flow waist diameter is described by equation (1):

$$d=19.151-0.570\theta-0.935w+0.193r-0.543\lambda+0.030\theta w+0.009\theta\lambda+0.004\theta^2 \quad (1)$$

In the same way, the regression model of the focal distance and depth also can be obtained. By the regression fit and significance analysis, the models always possess significance and reliability. The regression models about powder flow focal distance and depth are shown as equations. (2) and (3), respectively.

$$f=-7.478+0.099\theta+2.442w-2.438r+1.381\lambda+0.076\theta r-0.020\theta\lambda-0.026\lambda^2 \quad (2)$$

$$l=-33.105+1.244\theta-0.179w-2.835r+0.268\lambda+0.046\theta r+0.538wr-0.011\theta^2-0.027\lambda^2 \quad (3)$$

3.2. Structural optimization of coaxial feeding nozzle

Depending on the response-surface model, the multiple-response desirability function approach is used to optimize the nozzle structure. The optimal objective is to select the minimum of waist diameter and focal depth. Meanwhile, the focal distance should not be lower than 15 mm. And the structure parameters of the optimized nozzle are shown in table 2.

Table 2. The nozzle structural parameters.

Parameter		θ(°)	w(mm)	r(mm)	λ(°)	α(°)	φ(mm)	e (mm)	g (mm)	a (mm)
Size	non-optimized	67.0	2.2	5.00	5.0	37.0	5.0	60	7.0	40
	optimized	62.9	1.0	5.83	3.0	37.0	5.0	30	4.0	25

The flow field characteristics of non-optimized and optimized nozzles are shown in table 3, which indicates that the primary powder flow characteristics are improved to a certain extent. That is, the waist diameter and focal depth are reduced by 39.4 and 46.4%, respectively. Although the focal length of powder flow is decreased by 27.6%, it is still larger than 15 mm, to avoid the nozzle bearing the excessive temperature in the laser beam vicinity.

Moreover, the actual and simulative powder flows were comparatively analyzed, and the morphologies and data obtained are presented in figure 4 and table 4.

Table 3. The comparison of experimental data between before and after optimization.

	Nozzle angle (°)	Width (mm)	Radius (mm)	Shrinkage angle (°)	Diameter (mm)	Focal distance (mm)	Focal depth (mm)
un-optimized	67.00	2.2	5.00	5.0	5.02	21.0	12.5
optimized	62.94	1.0	5.83	3.0	3.04	15.2	6.7
variation ratio	-6.1%	-54.2%	16.7%	-40.0%	-39.4%	-27.6%	-46.4%

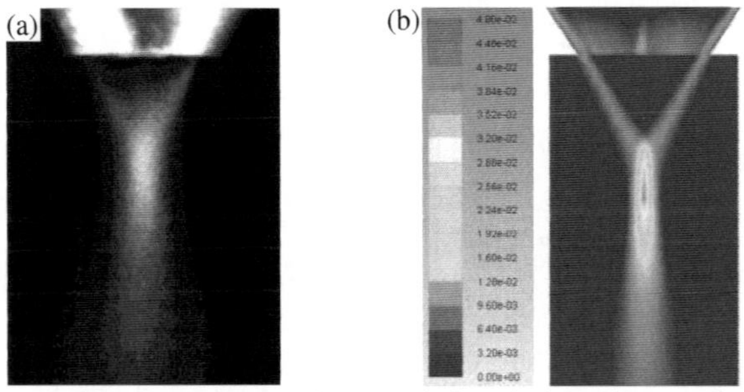

Figure 4. The experimental (a) and simulation (b) powder flows.

Table 4. The comparison between experimental and simulated data.

	Nozzle angle (°)	Width (mm)	Radius (mm)	Shrinkage angle (°)	Diameter (mm)	Focal distance(mm)	Focal depth (mm)
Simulation	62.94	1.0	5.83	3.0	2.80	14.8	6.4
Experimental	62.94	1.0	5.83	3.0	3.04	15.2	6.7
Error	0	0	0	0	8.6%	2.7%	4.7%

The discrepancies between simulated and experimental data on the powder flow waist diameter, focal distance, and focal depth are 8.6, 2.7, and 4.7%, respectively. Since these values are less than 9%, the proposed simulation model possesses high reliability.

4. Laser cladding process with the coaxial feeding nozzle

To evaluate the performance of optimized nozzle and select the appropriate process parameters, the cladding powder is 304 stainless steel powders: powder size (-400 mesh), powder density (2.6-3.5 g/cm3), fluidity (25-35S/ 50g) and compressibility (greater than 6.55 g/cm3) in this experiment, and the cladding substrate is also 304 stainless steel plates. A DL-LPM-V CO_2 laser with a wavelength of 10.6 μm and maximum power of 5 kW was used for the material processing. According to the practical tests, the main influencing factors of the cladding layer quality are carrier gas flow rate, scanning speed, powder flow defocusing distance, and powder-feeding rate. Meanwhile, the values of the laser power and laser focal distance were fixed at 2.4 kW and 30 mm, respectively.

4.1 Effect of process parameters on cladding layer quality

To study the influence of carrier gas flow rate on the cladding quality, the scanning speed, powder flow defocusing distance and powder-feeding rate were fixed at 600 mm/min, 0 mm and 3.6 g/min, respectively. By measuring the morphology features of a single layer with the metallographic, the effect of carrier gas flow on the cladding layer quality is illustrated by figure 5.

Similarly, to study the influence of scanning speed, powder flow defocusing distance, powder-feeding rate on the cladding layer quality, each factor's effect on the cladding layer quality was plotted in figure 6.

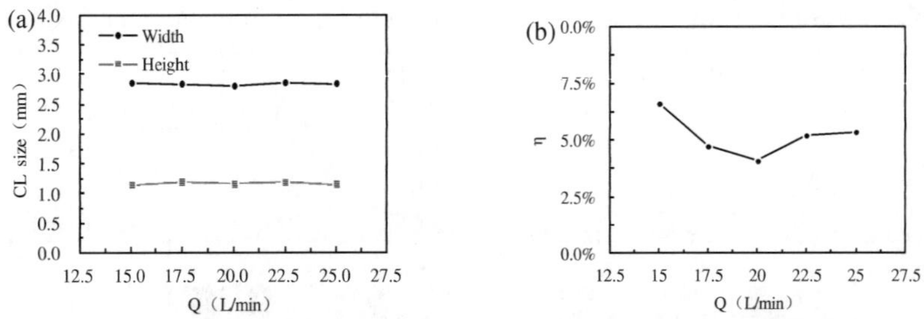

Figure 5. Effect of carrier gas flow on cladding layer: (a) size; (b) dilution rate

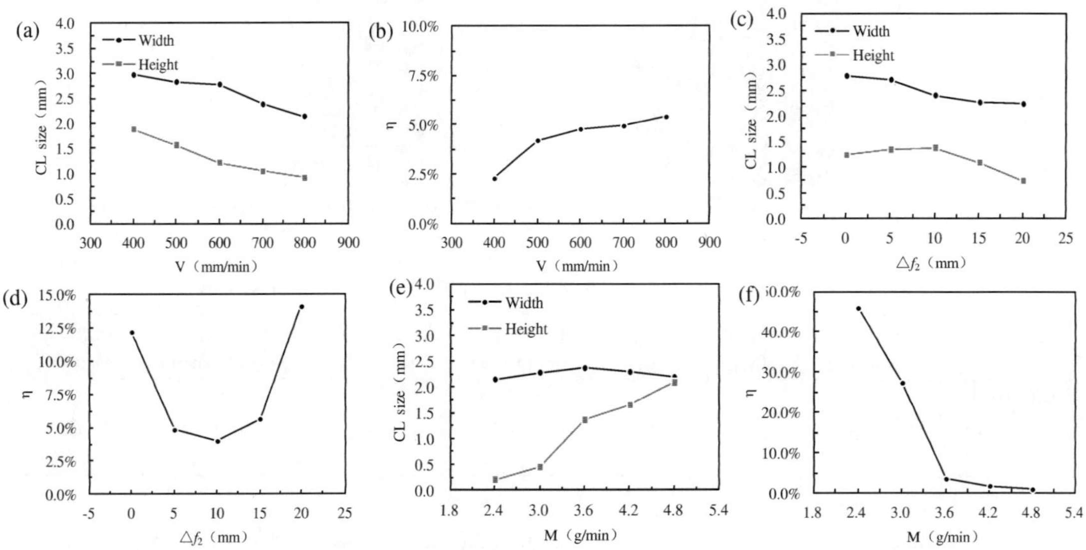

Figure 6. Revealed relationships: (a) scanning speed vs. cladding layer (CL) size; (b) scanning speed vs. dilution rate (η); (c) powder flow defocusing distance (Δf_2) vs. CL size; (d) distance (Δf_2) vs. dilution rate (η); (e) powder-feeding rate (*M*) vs. CL size; (f) powder-feeding rate (*M*) vs. dilution rate (η).

4.2. Optimal technological parameters in signal laser cladding

Considering the well geometrical morphology and the appropriate dilution rate, the technological parameters are selected as follows: 2.4 kW (laser power), 30 mm (laser defocusing distance), 15 l/min (carrier gas flow), 800 mm/min (scanning speed), 0 mm (powder flow defocusing distance) and 3.6 g/min (powder-feeding rate).

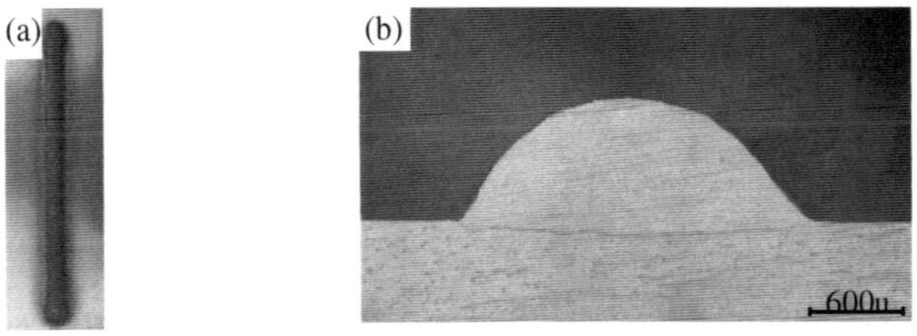

Figure 7. The macro- (a) and cross-sectional morphology (b) of single pass laser cladding layer.

The macro-morphology of a single-pass laser cladding layer is shown in figure 7, which proves that the layer surface is smooth and contains no beads, bubbles, cracks or any other macroscopic defects. Meanwhile, to analyze the layer quality in more detail, the metallographic microscope was used to observe the cross-sectional morphology (figure 7 (b)). The calculated dilution rate (η) was 11.1%. So, the layer possesses a proper geometrical morphology and the appropriate dilution rate under the optimal technology. In general, the optimized coaxial feeding nozzle can be used for the laser cladding layer preparation with excellent cladding quality.

5. Conclusions

Based on the Fluent analysis software and the response surface method, the following main conclusions can be drawn:

- ✓ The powder flow of three-dimensional coaxial feeding nozzle is simulated by Eulerian-Eulerian two-fluid models, and the reasonable range of powder-feeding technology is obtained. They are shown as follows: carrier gas flow should exceed 9 l/min, and the powder-feeding rate should be less than 7.5 g/min.
- ✓ Based on the central composite design of the response surface method, the regressive mathematical model between nozzle geometric parameters and powder flow fluidization characteristics is established. The optimal parameters of the nozzle are: nozzle angle of 62.95°, outlet width of 1mm, exit radius of 5.83 mm, and shrinkage angle of 3.00°.
- ✓ Comparing the powder convergence effect of the un-optimized feeding nozzle with the optimized feeding nozzle, the latter waist diameter and focal depth are reduced by 39.4 and 46.4%, respectively. The error between the simulative and experimental powder flow characteristics is small, and the results indicate that the simulation model and optimized result possess reliability.
- ✓ The effect of the technological parameter on the quality and dilution rate of the laser cladding layer was explored by a single-factor experiment. Then, the appropriate technological parameters were selected as follows: 2.4 kW (laser power), 30 mm (laser defocusing distance), 15 l/min (carrier gas flow), 800 mm/min (scanning speed), 0 mm (powder flow defocusing distance) and 3.6 g/min (powder-feeding rate). These parameters can guarantee a proper geometrical morphology and appropriate dilution rate of the laser cladding layer.

Acknowledgments

This work was support by the Fundamental Research Funds for the Central Universities of China (No. 3232016354).

References

[1] Ye X Y, Ma M X, Cao Y X, *et al.* 2011 *Phys. Procedia* **12** 303-12
[2] Zeng C, Tian W, Liao W H, *et al.* 2016 s *Surf. Coat.Tech.* **294** 122-30
[3] Smurov I, Doubenskaia M and Zaitsev A 2013 *Surf. Coat. Tech.* **220** 112-21
[4] Baldridge T, Poling G, Foroozmehr E, *et al.* 2013 *Opt. Laser Eng.* **50** 180-4
[5] Calleja A, Tabernero I, Fernandez A *et al.* 2014 *Opt. Laser Eng.* **56** 113-20
[6] Chew Y, Pang J H L, Bi G, *et al.* 2015 *J. Mater. Process. Tech.* **224** 89-101
[7] Lu Q P, Zhang A F, Li D C, *et al.* 2010 *Chinese J. Lasers* **37** 3162-6

[8] Liu H, He X L, Yu G, *et al.* 2015 *Sci. China Phys. Mech.* **58** 1-10
[9] Balu P, Leggett P and Kovacevic R 2012 *J. Mater. Process. Tech.***212** 1598-610
[10] Wang H X, Shi C, Wang C S, *et al.* 2010 *Trans. China Welding Inst.* **31** 69-72
[11] Box G E P and Wilson K B 1951 *J. Royal Stat. Soci.* **13** 1-45

Dynamical Casimir effect meets material science

V V Dodonov

Institute of Physics and International Center for Physics, University of Brasilia, P.O. Box 04455, Brasilia 70919-970, Federal District, Brazil

E-mail: vdodonov@fis.unb.br

Abstract.
The aim of this study is to analyze attempts to observe the so called Dynamical Casimir effect in cavities, changing their optical lengths by means of fast time variations of material properties (dielectric permeability or conductivity) of thin slabs attached to the cavity walls. The emphasis is made on the case of semiconductor slabs excited by short laser pulses. Considering the evolution of the classical electromagnetic field in this case, an approximate analytical solution for an infinite set of coupled ordinary differential equations for the mode amplitudes is derived under certain simplifying assumptions. According to this solution, an amplification of the initial field can be made, provided the induced dielectric permeability can become negative with a large absolute value. Evaluations of the feasibility of such a scenario are given.

1. Introduction

In 1970, Moore [1] showed that motions of ideal boundaries of a one-dimensional cavity can result in a generation of quanta of the electromagnetic field from the initial vacuum quantum state. Since that time, such a remarkable effect (called after [2,3] as *Dynamical Casimir Effect* – DCE) attracted attention of many researchers. Reviews of numerous publications on this subject can be found in [4,5]. One of the main theoretical results obtained in the middle of 1990s was a simple formula for the mean number of quanta that could be generated during time t inside an ideal three-dimensional non-degenerate cavity whose walls perform a periodic motion with the *double* eigenfrequency $2\omega_0$ of the fundamental cavity mode (the parametric amplification of the vacuum fluctuations) [6] $\langle n\rangle(t) = \sinh^2(\epsilon\kappa\omega_0 t)$. Here ϵ is the maximal relative displacement of the boundary with respect to the wavelength $\lambda = 2\pi c/\omega_0$ and $\kappa < 1$ is a numerical coefficient which depends on the cavity geometry. For example, $\kappa = [\lambda/(2L_0)]^3$ for a rectangular cavity, where L_0 is the average distance between vibrating walls.

Scholars seeked for a possible experimental verification of the DCE since the early 1990s. One of the main obstacles is the extreme smallness of parameter ϵ that could be achieved for real cavities with moving walls. Indeed, consider cavities with dimensions of the order of a few centimeters, possessing the fundamental eigenfrequencies $\omega_0/2\pi$ of the order of a few GHz. An idea of [6] was not to move the wall as a whole (since it seems practically impossible at high frequencies [4]), but to excite oscillations of the *surface* of the cavity wall. Note that the amplitude a of a standing acoustic wave at frequency $\omega_w = 2\omega_0$ is related to the relative deformation amplitude δ as $a = v_s\delta/\omega_w$, where v_s is the sound velocity. Since the maximal deformation of usual materials cannot exceed $\delta_{max} \sim 10^{-2}$, the maximal possible velocity of the boundary is $v_{max} \sim \delta_{max} v_s \sim 50$ m/s for $v_s \sim 5$ km/s. Therefore, the maximal relative displacement $\epsilon = a/\lambda$ equals $\epsilon_{max} \sim v_s\delta_{max}/(4\pi c) \sim 10^{-8}$. Using this value, one could get $\sinh^2(5) \sim 5000$ photons in the initially empty ideal cavity with $\omega_0/(2\pi) = 3\,\mathrm{GHz}$ and $\kappa = 1/2$ after $t \sim 50\,\mathrm{ms}$. The required cavity quality factor Q is about 10^8. If the value of δ_{max} could be increased by one order of magnitude (and this is the material science problem), then the necessary time and the cavity quality factor could be diminished by the same one order of magnitude. The excitation of the surface oscillations with the maximal amplitude at such a high frequency is a challenge. Probably, this can be achieved with the aid of the so-called 'film bulk acoustic resonators' (FBARs) [7], as was proposed in connection with the DCE in [8,9].

In fact, the fundamental mechanism of the DCE is the parametric amplification of initial vacuum fluctuations due to temporal changes of the cavity eigenfrequencies. But such changes can be achieved not only by means of changing the cavity geometry, but also by changing the material properties of the cavity. Hence the idea of *simulating* the DCE and other quantum effects was suggested by Yablonovitch about three decades ago [2]. Namely, he proposed to use a medium with a rapidly decreasing in time refractive index ('plasma window'). Similar ideas were put forward later in [10,11]. A possibility to use *quantum circuits* instead of cavities to simulate the DCE was discussed in [12–16], and it was realized in [17,18]. The dielectric permeability can be changed in nonlinear optical materials illuminated by strong laser pulses. This way of modelling effects predicted by the quantum field theory in a laboratory was considered by several authors, started from [19–21] (see [22] for a short review). However, no real experiments were performed till now due to the smallness of the nonlinear optical parameters. This is one of challenges for the material science: to find (construct) materials with high nonlinear optical coefficient $\chi^{(3)}$ in the microwave domain.

On the other hand, fast big variations of electric properties can be achieved in semiconductors illuminated by laser pulses, as was pointed out by Yablonovitch [2]. Following this idea, attempts to observe the analogue of 'true' DCE in *microwave cavities* were performed [23,24]. In that experiment, a semiconductor slab was periodically illuminated by chains of short (a few picosecond) laser pulses, so that its conductivity changed from almost zero to almost metallic values. This resulted in periodical changes of the fundamental mode frequency of the order of 10^{-3} (or bigger). However, in spite of many efforts, the amplification of the microwave field was not observed. The main difficulty consists in the inevitable losses inside the semiconductor slab with a finite electric conductivity. Theoretical models accounting for these losses at the quantum level were elaborated in [4,25]. However, it seems that some important factors were not taken into account there (although experimental results showed a good agreement with some theoretical predictions). In order to find a possible drawback, the preliminary analysis of the problem in the *classical* regime was performed in [26]. The main question was as follows: is it possible in principle to amplify the classical EM field by means of strong temporal changes of the electrical properties of a thin conducting slab inside the cavity? What kinds of materials should be used? Some answers (generalizing results of [26]) are given in the next sections.

2. The classical EM field dynamics: approximate solutions for a thin slab with time-dependent parameters

Firstly, consider the Maxwell equations for the EM field inside an ideal cavity, filled in with a stationary, isotropic, non-dispersive and non-magnetic dielectric material:

$$c\,\mathrm{rot}\mathbf{B} = \partial\mathbf{D}/\partial t, \qquad c\,\mathrm{rot}\mathbf{E} = -\partial\mathbf{B}/\partial t, \qquad \mathbf{D}(\mathbf{r},t) = \varepsilon(\mathbf{r})\mathbf{E}(\mathbf{r},t), \quad \varepsilon(\mathbf{r}) \geq 1. \qquad (1)$$

These equations admit the complete set of solutions in the form $\mathbf{E}(\mathbf{r},t) = \mathbf{E_n}(\mathbf{r})\exp(-i\omega_{\mathbf{n}}t)$, $\mathbf{B}(\mathbf{r},t) = -i\mathbf{B_n}(\mathbf{r})\exp(-i\omega_{\mathbf{n}}t)$, where real time independent functions $\mathbf{E_n}(\mathbf{r})$ and $\mathbf{B_n}(\mathbf{r})$ satisfy the equations (where the 'vector' index $\mathbf{n}$ combines three integers (l,m,n) and $k_{\mathbf{n}} = \omega_{\mathbf{n}}/c$)

$$\mathrm{rot}\mathbf{B_n} = \varepsilon(\mathbf{r})k_{\mathbf{n}}\mathbf{E_n}, \quad \mathrm{rot}\mathbf{E_n} = k_{\mathbf{n}}\mathbf{B_n}. \quad \int\varepsilon(\mathbf{r})\mathbf{E_n}\mathbf{E_m}dV = \int\mathbf{B_n}\mathbf{B_m}dV = 8\pi\delta_{\mathbf{nm}}. \qquad (2)$$

In the general case, $\mathbf{E}(\mathbf{r},t)=\sum_{\mathbf{n}}\mathbf{E}_{\mathbf{n}}(\mathbf{r})f_{\mathbf{n}}(t)$ and $\mathbf{B}(\mathbf{r},t)=\sum_{\mathbf{n}}\mathbf{B}_{\mathbf{n}}(\mathbf{r})g_{\mathbf{n}}(t)$, the total energy can be written as

$$W=\int dV\left(\varepsilon\mathbf{E}^2+\mathbf{B}^2\right)/(8\pi)=\sum_{\mathbf{n}}\left(f_{\mathbf{n}}^2+g_{\mathbf{n}}^2\right). \tag{3}$$

For the conducting and nonstationary medium, described by means of functions $\sigma(\mathbf{r},t)$ and $\varepsilon(\mathbf{r})+\delta\varepsilon(\mathbf{r},t)$, the first equation in (1) is replaced with $c\,\mathrm{rot}\mathbf{B}=4\pi\sigma\mathbf{E}+\partial\left[(\varepsilon+\delta\varepsilon)\mathbf{E}\right]/\partial t$. This form implies the neglect of temporal dispersion, which seems reasonable for microwave fields. Using (2), one can write $\sum_{\mathbf{n}}\mathbf{E}_{\mathbf{n}}\left[\varepsilon\left(\dot{f}_{\mathbf{n}}-g_{\mathbf{n}}\omega_{\mathbf{n}}\right)+\delta\varepsilon\dot{f}_{\mathbf{n}}+f_{\mathbf{n}}\left(4\pi\sigma+\partial\delta\varepsilon/\partial t\right)\right]=0$. Then, following the standard procedure with account of (2), one can replace the partial differential equations with the ordinary ones for the time-dependent amplitudes

$$\dot{g}_{\mathbf{m}}=-\omega_{\mathbf{m}}f_{\mathbf{m}},\qquad \dot{f}_{\mathbf{m}}=\omega_{\mathbf{m}}g_{\mathbf{m}}-\sum_{\mathbf{n}}\alpha_{\mathbf{mn}}f_{\mathbf{n}}-\frac{d}{dt}\sum_{\mathbf{n}}\beta_{\mathbf{mn}}f_{\mathbf{n}}, \tag{4}$$

$$\alpha_{\mathbf{mn}}(t)=\int dV\sigma(\mathbf{r},t)\mathbf{E}_{\mathbf{m}}(\mathbf{r})\mathbf{E}_{\mathbf{n}}(\mathbf{r})/2,\quad \beta_{\mathbf{mn}}(t)=\int dV\delta\varepsilon(\mathbf{r},t)\mathbf{E}_{\mathbf{m}}(\mathbf{r})\mathbf{E}_{\mathbf{n}}(\mathbf{r})/(8\pi). \tag{5}$$

These equations can be significantly simplified for rectangular cavities with $0<z<L$, $|x|<L_x/2$, $|y|<L_y/2$, provided functions $\varepsilon(\mathbf{r})$, $\sigma(\mathbf{r},t)$ and $\delta\varepsilon(\mathbf{r},t)$ depend on the single longitudinal variable z only. Suppose that $\varepsilon=const>1$ for $0<z<L_s<L$ and $\varepsilon\equiv1$ for $L_s<z<L$. Then, the consequence of equations (2) is the Helmholtz equation $\Delta\mathbf{E}_{\mathbf{n}}+\mathbf{k}_{\mathbf{n}}^2\varepsilon(z)\mathbf{E}_{\mathbf{n}}=0$. It is known [25, 27, 28] that the TM configuration of the electromagnetic field is much more advantageous than the TE one for the field excitation due to the DCE. In this case,

$$B_z\equiv0,\quad E_z=\Phi(\mathbf{r}_\perp)\psi(z),\quad E_x=\mathbf{k}_\perp^{-2}\frac{\partial\Phi}{\partial x}\frac{d\psi}{dz},\quad E_y=\mathbf{k}_\perp^{-2}\frac{\partial\Phi}{\partial y}\frac{d\psi}{dz}.$$

$$\Delta_\perp\Phi+\mathbf{k}_\perp^2\Phi=0,\qquad \psi''+\left[\mathbf{k}_n^2\varepsilon(z)-k_\perp^2\right]\psi=0.$$

Since the tangential components of the electric field must disappear near all surfaces of the ideal cavity, the boundary conditions for function $\psi(z)$ are $\psi'\equiv d\psi/dz=0$ at $z=0$ and $z=L$, whereas $\Phi\left(\mathbf{r}_\perp\right)=\cos\left(k_lx\right)\cos\left(k_my\right)$, $k_l=(1+2l)\pi/L_x$, $k_m=(1+2m)\pi/L_y$. As a consequence, modes with different indexes l and m are independent. We assume that $l=m=0$. Then, equations (4) can be written in the vector form

$$\dot{\mathbf{g}}=-\Omega\mathbf{f},\qquad \dot{\mathbf{f}}=\Omega\mathbf{g}-\mathcal{A}\mathbf{f}-d(\mathcal{B}\mathbf{f})/dt, \tag{6}$$

where $\mathbf{f}\equiv(f_0,f_1,f_2,\ldots)$, $\mathbf{g}\equiv(g_0,g_1,g_2,\ldots)$, $f_m\equiv f_{00m}$, $g_m\equiv g_{00m}$. The diagonal matrix Ω has elements $\omega_m\equiv\omega_{00m}$. The elements of symmetrical matrices $\mathcal{A}(t)$ and $\mathcal{B}(t)$ are as follows:

$$a_{mn}(t)=\int_0^{L_s}dz\sigma(z,t)\chi_{mn}(z),\quad b_{mn}(t)=\int_0^{L_s}dz\delta\varepsilon(z,t)\chi_{mn}(z)/(4\pi), \tag{7}$$

$$\chi_{mn}(z)=(L_xL_y/8)\left[\psi_m(z)\psi_n(z)+\mathbf{k}_\perp^{-2}\psi_m'(z)\psi_n'(z)\right]. \tag{8}$$

An immediate consequence of (6) is the equation $dW/dt=-2\left[\mathbf{f}\mathcal{A}\mathbf{f}+\mathbf{f}d(\mathcal{B}\mathbf{f})/dt\right]$, where $W=\mathbf{f}^2+\mathbf{g}^2$. Since $\sigma(z,t)\geq0$, matrix $\mathcal{A}$ is *non-negative*: $\mathbf{f}\mathcal{A}\mathbf{f}\geq0$ for any vector $\mathbf{f}$. Therefore, $dW/dt\leq0$ if $\sigma>0$ and $\mathcal{B}=0$. This means that the total energy W always decreases if $\delta\varepsilon=0$.

Since matrices Ω, $\mathcal{A}$, and $\mathcal{B}$ do not commute, equations (6) can be solved only numerically in the general case. However, approximate analytical solutions can be found under certain assumptions for thin slabs with $L_s\ll L$. For the totally empty cavity, one has

$$\psi_n(z)=N_n\cos\left(\pi nz/L\right),\quad N_n^{-2}=L_xL_yL\left(1+\mu n^2\right)\left(1+\delta_{n0}\right)/(64\pi), \tag{9}$$

$$\omega_n^2 = \omega_0^2\left(1+\mu n^2\right), \quad \omega_0 = c\pi/(L\sqrt{\mu}), \quad \mu = L_x^2 L_y^2\left[L^2\left(L_x^2+L_y^2\right)\right]^{-1}. \tag{10}$$

If $L_s \ll L$, then the electric field outside the slab remains practically the same as in the stationary case, as well as the eigenfrequencies and normalization constants. But E_z component of the electric field inside the slab is smaller: $\psi_n(z < L_s) \approx N_n \cos\left(\zeta_n \pi n z/L\right)/\varepsilon$, $\zeta_n \approx \sqrt{\varepsilon + (\varepsilon-1)/(\mu n^2)}$. Therefore, the ratio of the second term (containing the product of derivatives) to the first one in the function $\chi_{mn}(z)$ (8) inside the slab equals approximately $\mu(\pi m n \varepsilon L_s/L)^2$, so that this term can be neglected for not very big values of mode numbers m and n and the dielectric constant ε. Using one more approximation $\cos\left(\zeta_n \pi n z/L\right) \approx 1$, one can simplify matrices $\mathcal{A}(t)$ and $\mathcal{B}(t)$ as follows:

$$\mathcal{A}(t) = a(t)\Psi, \quad \mathcal{B}(t) = b(t)\Psi, \quad a(t) = \int_0^{L_s} \sigma(z,t)dz/L, \quad b(t) = \int_0^{L_s} \delta\varepsilon(z,t)dz/(4\pi L). \tag{11}$$

Here Ψ is the symmetric matrix with the factorized elements $\Psi_{mn} = \Psi_m \Psi_n$, $\Psi_0 = \sqrt{4\pi}/\varepsilon$, $\Psi_{n\neq 0} = \left[\varepsilon^2\left(1+\mu n^2\right)/(8\pi)\right]^{-1/2}$. Moreover, $\Psi^n = \rho^{n-1}\Psi$ for $n = 1, 2, 3, \ldots$, where [26]

$$\rho = \mathrm{Tr}(\Psi) = \sum_{k=0}^{\infty} \Psi_k^2 = \left(4\pi/\varepsilon^2\right)\left[1 + 2\sum_{n=1}^{\infty}\left(1+\mu n^2\right)^{-1}\right] = 4\pi^2\left(\varepsilon^2\sqrt{\mu}\right)^{-1}\coth\left(\pi/\sqrt{\mu}\right) \tag{12}$$

Consequently, any function of matrix Ψ is proportional to Ψ itself. In view of this finding, let us make one more approximation, assuming that matrix Ω is proportional to the unity matrix I: $\Omega = \varpi I$. Of course, this assumption is rather artificial, but it enables us to solve equations (6) analytically:

$$\mathbf{g}(t) = \left[u_g(t)I + v_g(t)\Psi\right]\mathbf{g}(0) + \left[z_g(t)I + w_g(t)\Psi\right]\mathbf{f}(0), \tag{13}$$

$$\mathbf{f}(t) = \left[u_f(t)I + v_f(t)\Psi\right]\mathbf{g}(0) + \left[z_f(t)I + w_f(t)\Psi\right]\mathbf{f}(0). \tag{14}$$

The time-dependent scalar functions in (13) and (14) must obey the following set of linear ordinary differential equations and initial conditions:

$$\dot{u}_f = \varpi u_g, \quad \dot{u}_g = -\varpi u_f, \qquad \dot{z}_f = \varpi z_g, \quad \dot{z}_g = -\varpi z_f, \tag{15}$$

$$\dot{v}_f = \varpi v_g - a(t)\left(u_f + \rho v_f\right) - \frac{d}{dt}\left[b(t)\left(u_f + \rho v_f\right)\right], \qquad \dot{v}_g = -\varpi v_f, \tag{16}$$

$$\dot{w}_f = \varpi w_g - a(t)\left(z_f + \rho w_f\right) - \frac{d}{dt}\left[b(t)\left(z_f + \rho w_f\right)\right], \qquad \dot{w}_g = -\varpi w_f, \tag{17}$$

$$u_g(0) = z_f(0) = 1, \qquad u_f(0) = z_g(0) = v_f(0) = v_g(0) = w_f(0) = w_g(0) = 0. \tag{18}$$

Solutions to (15) are obvious: $u_g(t) = z_f(t) = \cos(\varpi t)$, $u_f(t) = -z_g(t) = \sin(\varpi t)$. Equations (16) and (17) are more complicated, because their coefficients can be arbitrary functions of time satisfying the conditions $a(0) = b(0) = 0$. To simplify the problem, let us suppose that functions $a(t)$ and $b(t)$ grow from zero to some constant values a and b during a very short time interval $\delta \to 0$, then remain at constant levels until the instant $t_* - \delta$, and finally return very quickly to the initial zero values at $t = t_*$. Then (16) and (17) turn into linear inhomogeneous equations with constant coefficients a and b for $\delta < t < t_* - \delta$. To find the new initial conditions at $t = \delta$ we integrate equations (16) and (17) from 0 to δ, assuming that all functions remain bounded during this short interval, and then take the limit $\delta \to 0$. Thus we arrive at the conditions at $t = 0+$: $v_f(0+) = v_g(0+) = w_g(0+) = 0$, $w_f(0+) = -b/(1+b\rho)$. Using the standard scheme, one can find solutions for any instant of time in the interval $\delta < t < t_* - \delta$. To find the values at $t = t_*$, one has to integrate equations (16) and (17) from $t_* - \delta$ to t_* and again take the limit

IOP Conf. Series: Materials Science and Engineering **474** (2019) 012009 doi:10.1088/1757-899X/474/1/012009

$\delta \to 0$. Then $v_g(t_*) = v_g(t_* - \delta)$, $w_g(t_*) = w_g(t_* - \delta)$, whereas $v_f(t_*) = v_f(t_* - \delta)(1 + b\rho) + bu_f(t_*)$ and $w_f(t_*) = w_f(t_* - \delta)(1 + b\rho) + bz_f(t_*)$. The final results are

$$v_f(t_*) = (r\rho)^{-1}\left[2\varpi\beta e^{-\lambda t_*}\sinh(\kappa t_*) - r\sin(\varpi t_*)\right] = -w_g(t_*),$$

$$v_g(t_*) = (r\rho)^{-1}\left\{r\left[e^{-\lambda t_*}\cosh(\kappa t_*) - \cos(\varpi t_*\right] + a\rho e^{-\lambda t_*}\sinh(\kappa t_*)\right\},$$

$$w_f(t_*) = (r\rho)^{-1}\left\{r\left[e^{-\lambda t_*}\cosh(\kappa t_*) - \cos(\varpi t_*\right] - a\rho e^{-\lambda t_*}\sinh(\kappa t_*)\right\},$$

where

$$\lambda = a\rho/(2\beta), \quad \kappa = r/(2\beta), \quad r = \sqrt{(a\rho)^2 - 4\varpi^2\beta}, \quad \beta = 1 + b\rho. \tag{19}$$

The total energy variation equals

$$\Delta W \equiv \mathbf{f}^2(t_*) + \mathbf{g}^2(t_*) - \mathbf{f}^2(0) - \mathbf{g}^2(0) = h_{ff}Y_f^2 + h_{gg}Y_g^2 + 2h_{fg}Y_fY_g, \tag{20}$$

$$Y_f = \sum_{n=0}^{\infty}\Psi_n f_n(0), \quad Y_g = \sum_{n=0}^{\infty}\Psi_n g_n(0), \tag{21}$$

$$\rho h_{ff} = e^{-2\lambda t_*}\left\{[a\rho\sinh(\kappa t_*) - r\cosh(\kappa t_*)]^2 + 4\varpi^2\beta^2\sinh^2(\kappa t_*)\right\}/r^2 - 1 \tag{22}$$

$$\rho h_{gg} = e^{-2\lambda t_*}\left\{[a\rho\sinh(\kappa t_*) + r\cosh(\kappa t_*)]^2 + 4\varpi^2\beta^2\sinh^2(\kappa t_*)\right\}/r^2 - 1 \tag{23}$$

$$h_{fg} = -4a\varpi\beta e^{-2\lambda t_*}\sinh^2(\kappa t_*)/r^2. \tag{24}$$

3. Discussion and conclusions

Formulas (20)-(24) show that the possibility of the EM field amplification depends crucially on the sign of coefficient $\beta = 1 + b\rho$. Indeed, the energy variation in the short time limit ($|\kappa|t_* \ll 1$ and $|\lambda|t_* \ll 1$) equals $\Delta W = -2at_*Y_f^2/\beta + \mathcal{O}(t_*^2)$. Consequently, for short pulses of variations of material properties, the total energy always decreases if $\beta > 0$, but it can increase if $\beta < 0$. A similar behavior of ΔW can be observed for long time as well. If $\beta > 0$, then $\lambda > 0$. If κ is real, then $\kappa < \lambda$, so that $h_{ff} = h_{gg} = -1/\rho$ and $h_{fg} = 0$ for $(\lambda - \kappa)t_* \gg 1$. The same asymptotical values of these coefficients arise for $\lambda t_* \gg 1$ if κ is an imaginary number. Therefore, the total energy always decreases for long excitation pulses if $\beta > 0$. On the contrary, if $\beta < 0$, then κ is real and $\lambda < 0$, so that the total energy can grow as $\exp\left[(|\lambda| + |\kappa|)\,t_*\right]$ for $|\kappa|t_* \gg 1$ due to some kind of parametric instability.

The fundamental conclusion is that the possibility of the DCE simulation in cavities by means of temporal variations of conductivity inside a thin slab depends on the sign and maximal absolute value of the accompanying dielectric permeability variation $\delta\varepsilon$. If $\delta\varepsilon > 0$, then $b(t) > 0$, so that no field amplification can be expected. However $\delta\varepsilon$ can be negative, e.g., according to the Drude model: $\delta\varepsilon(\mathbf{r}, t) = -4\pi n(\mathbf{r}, t)e^2/\left[m_{ef}\left(\omega^2 + \gamma^2\right)\right]$. Here $n(\mathbf{r}, t)$ is the concentration of free carriers created by the laser pulse, e the electron charge, m_{ef} the effective mass, and γ is the collision frequency. Therefore, the intriguing question is: whether condition $|b|_{max}\rho > 1$ can be fulfilled in realistic situations? For realistic values $\mu < 1$ one can replace the coth function in the right-hand side of (12) by unity. In particular, for the cubical cavity, $\mu = 1/2$ and $\rho \approx 56/\varepsilon^2$. Since we consider the non-dispersive case, we have to assume that $\varpi \ll \gamma$. Then, using equations (10)-(12), one can arrive at the requirement $\gamma^2 < 8\pi^2 n_{max}e^2L_s/(\lambda_0 m_{ef}\varepsilon^2)$, where n_{max} is the maximal concentration of free carriers inside the slab and $\lambda_0 = 2\pi c/\omega_0$ is the wave length corresponding to the fundamental cavity eigenfrequency. Certainly, n_{max} cannot be higher than the electron concentration in good metals $n_{met} \sim 10^{23}\,\text{cm}^{-3}$. Using this value together with the free electron mass for m_{ef}, $\varepsilon \sim 10$, and taking $L_s/\lambda_0 \sim 10^{-3}$, one obtains the upper limit for the collision frequency $\gamma < 10^{14}\,\mathrm{s}^{-1}$. For $n_{max} \sim 10^{19}\,\mathrm{cm}^{-3}$ (as for inferior metals) the limitation is $\gamma < 10^{12}\,\mathrm{s}^{-1}$. In view of (11), the requirement $|b|_{max}\rho > 1$ means that it is necessary to achieve negative values of $\delta\varepsilon$ of the order of -10^4 (under the same assumptions). Variations of the dielectric function in the photo-excited GaAs to the negative values of the order of -10 were reported in [29] for the visible light (in a good agreement with the Drude approximation). For the microwave frequencies, one can expect bigger absolute values of these negative changes, but this seems to be a task for future experiments.

Note that coefficient b can be written as $b = \chi\mathcal{N}_s/L$, where $\mathcal{N}_s$ is the total number of carriers created by the laser pulse per unit area of the semiconductor surface and χ is the dielectric susceptibility per one carrier ($\delta\varepsilon = 4\pi\chi n$). In turn, $\mathcal{N}_s = \xi\Phi/E_g$, where Φ is the fluence of laser pulse, E_g the energy gap of the semiconductor and ξ the efficiency of free carriers generation. In addition we have $L\sqrt{\mu} \approx L_\perp$ - the transverse cavity dimension. Therefore, the condition $|b_{max}|\rho > 1$ results in the following requirement for the fluence: $\Phi > \varepsilon^2 L_\perp E_g/\left(4\pi^2\chi f\right)$. For $L_\perp \sim 10\,\mathrm{cm}$, $E_g \sim 2\,\mathrm{eV}$, $f = 1$, $\gamma = 10^{12}\,\mathrm{s}^{-1}$ and other parameters used above, one obtains $\Phi > 400\,\mathrm{J/m}^2$. This value is close to the threshold $1\,\mathrm{kJ/m}^2$ of the permanent damage for the GaAs material [29]. Note the strongly negative role of the factor $\varepsilon^2 \sim 100$ for the semiconductors like GaAs. Consequently, materials with smaller ε would be much better. Concluding, the dream to observe analogs of the DCE in cavities with periodical variations of the effective length is still alive, although the main challenges are nowadays in the realm of the material science.

Acknowledgments
A partial support of the Brazilian funding agency CNPq is acknowledged.

References

[1] Moore G T 1970 *J. Math. Phys.* **11** 2679–91
[2] Yablonovitch E 1989 *Phys. Rev. Lett.* **62** 1742–5
[3] Schwinger J 1992 *Proc. Nat. Acad. Sci. USA* **89** 4091–3
[4] Dodonov V V 2010 Current status of the dynamical Casimir effect *Phys. Scr.* **82** 038105
[5] Dalvit D A R, Maia Neto P A and Mazzitelli F D 2011 *Casimir Physics* (*Lecture Notes in Physics* vol 834) ed D Dalvit, P Milonni, D Roberts and F da Rosa (Berlin: Springer) p 419–57 (*Preprint* arXiv:1006.4790)
[6] Dodonov V V 1995 *Phys. Lett.* A **207** 126–32
[7] Tadigadapa S and Mateti K 2009 *Meas. Sci. Technol.* **20** 092001
[8] Kim W-J, Brownell J H and Onofrio R 2006 *Phys. Rev. Lett.* **96** 200402
[9] Sanz M, Wieczorek W, Gröblacher S and Solano E 2018 Electro-mechanical Casimir effect *Quantum* **2** 91 (*Preprint* arXiv:1712.08060v3)
[10] Okushima T and Shimizu A 1995 *Japan. J. Appl. Phys.* **34** 4508–10
[11] Lozovik Y E, Tsvetus V G and Vinogradov E A 1995 *Phys. Scr.* **52** 184–90
[12] Man'ko V I 1991 *J. Sov. Laser Res.* **12** 383–5
[13] Segev E, Abdo B, Shtempluck O, Buks E and Yurke B 2007 *Phys. Lett.* A **370** 202–6
[14] Dodonov A V 2009 *J. Phys.: Conf. Series* **161** 012029
[15] Fujii T, Matsuo S, Hatakenaka N, Kurihara S and Zeilinger A 2011 *Phys. Rev.* B **84** 174521
[16] Nation P D, Johansson J R, Blencowe M P and Nori F 2012 *Rev. Mod. Phys.* **84** 1–24
[17] Wilson C M, Johansson G, Pourkabirian A, Simoen M, Johansson J R, Duty T, Nori F and Delsing P 2011 *Nature* **479** 376–9
[18] Lähteenmäki P, Paraoanu G S, Hassel J and Hakonen P J 2013 *Proc. Nat. Acad. Sci. USA* **110** 4234–8
[19] Lobashov A A and Mostepanenko V M 1991 *Teor. Mat. Fiz.* **86** 438–47 [*Theor. & Math. Phys.* **86** 303–9]
[20] Grishchuk L P, Haus H A and Bergman K 1992 *Phys. Rev.* D **46** 1440–9
[21] Hizhnyakov V V 1992 *Quant. Opt.* **4** 277–80
[22] Dodonov V V 2015 *J. Phys.: Condens. Matter* **27** 214009
[23] Braggio C, Bressi G, Carugno G, Lombardi A, Palmieri A, Ruoso G and Zanello D 2004 *Rev. Sci. Instrum.* **75** 4967–70
[24] Agnesi A, Braggio C, Bressi G, Carugno G, Della Valle F, Galeazzi G, Messineo G, Pirzio F, Reali G, Ruoso G, Scarpa D and Zanello D 2009 *J. Phys.: Conf. Series* **161** 012028
[25] Dodonov V V and Dodonov A V 2006 *J. Phys.* B **39** S749–66
[26] Dodonov V V and Dodonov A V 2016 *J. Russ. Laser Res.* **37** 107–22

IOP Conf. Series: Materials Science and Engineering **474** (2019) 012009 doi:10.1088/1757-899X/474/1/012009

[27] Crocce M, Dalvit D A R and Mazzitelli F D 2002 Quantum electromagnetic field in a three-dimensional oscillating cavity *Phys. Rev.* A **66** 033811
[28] Uhlmann M, Plunien G, Schützhold R and Soff G 2004 *Phys. Rev. Lett.* **93** 193601
[29] Huang L, Callan J P, Glezer E N and Mazur E 1998 *Phys. Rev. Lett.* **80** 185–8

Structural analysis of hydrogenated diamond-like carbon coatings prepared using magnetron sputtering

Y L Su[1,*], S H Yao[2], Y H Mao[1], Z L You[1]

[1]Department of Mechanical Engineering, National Cheng Kung University, No.1, University Road, Tainan City 701, Taiwan
[2]Program in Technology Management, Chang Jung Christian University, No.1, Chang-Da Road, Tainan City 711, Taiwan

*E-mail: suyl@mail.ncku.edu.tw

Abstract. In this study, a series of hydrogenated diamond-like carbon (H-DLC) coatings were deposited using an unbalanced magnetron sputtering system. For hydrogenation, acetylene gas was used as the hydrogen source. The typical conditions for preparing DLC coatings were used, with additional acetylene gas being introduced into the depositing chamber. The acetylene flow rate was varied from 0 to 16 sccm at intervals of 4 sccm and finally set at 24 sccm. The chemical composition, microstructure, and phases of the resulting coatings were evaluated. Fourier transform infrared spectrometry, X-ray photoelectron spectroscopy, and Raman analysis were performed, and specific mechanical properties were examined. The results indicated that hydrogenation influenced the H-DLC coatings due to the formation of sp^3 bonding. The coating prepared at a flow rate of 8 sccm of acetylene exhibited the highest hardness of 15.4 GPa because it had the highest number of sp^3 bonds.

1. Introduction

Protective coatings have been commonly used in industrial applications for several decades [1-2]. Amorphous carbon (including a-C and a-C: H) coatings, better known as diamond-like carbon (DLC) coatings, exhibit mechanical properties comparable to diamonds, such as high hardness. DLC coatings have been widely applied in diverse industries, such as in the fields of aerospace, automobiles, oil and gas, construction, and power generation [3-5]. Aisenberg and Chabot (1971) were the first to employ a deposition technique to prepare DLC coatings [6]. For more than 30 years, various deposition techniques have been developed for obtaining DLC coatings. Among them, magnetron sputtering is one of the most commonly used methods [7-8]. DLC coatings deposited using magnetron sputtering exhibit lower residual stress, higher loading capacity, and higher hardness than coatings prepared using other techniques, such as radio frequency and plasma-activated processes [9]. A published study sought to transition from DLC to a-C:H coatings and to achieve high deposition rates by applying carbon-carrying gas to working atmospheres [10]. Jacob and Moller controlled the ratio of sp^2 bonds to sp^3 bonds to obtain various forms of C-H alloys in the ternary phase diagram of an amorphous C-H coating [11]. Liu et al. indicated that the acetylene flow rate and substrate temperature during the deposition determined the structures of DLC coatings [12]. Conditions such as the chamber atmosphere and preparation method considerably influence a coating's final performance, while the addition of H also affects DLC coatings. Tillmann and Hoffmann found that the hardness decreased with increases in H content when the H content was more than 20 at.% [13]. Rabbani and Vogelaar prepared hydrogenated

DLC (H-DLC) coatings by using CH_4 gas and found that H addition led to an increase in the I_D/I_G ratio and surface roughness (RMS) [14]. Piao et al. reported that increases in the H_2 flow rate did not lead to a remarkable increase in the deposition rate but caused a decrease in the I_D/I_G ratio and surface roughness (RMS) [15]. Toro et al. determined that the deposition rate decreased with increases in the H_2 flow rate [16]. In this research, a series of H-DLC coatings with high H content was prepared by increasing the flow rate of acetylene up to 24 sccm. The influence of H addition on the coating structures was systematically investigated.

2. Experiment

2.1. Coatings

The H-DLC coatings were deposited using an unbalanced magnetron sputtering system (KD-550U, Taiwan), featuring four targets positioned orthogonally around the rotating specimen holder. SKH51 steel (HRC 65) and p-type (100) Si wafers were used as substrate materials. The steel was used for evaluating the mechanical properties, whereas the Si wafers were used for structural analysis. Three graphite and one Ti targets were used (300 mm × 109 mm × 10 mm, 99.5% purity, Nutek Material Co.). The sample holder rotated at a speed of 3 rpm. The targets were oriented towards the coated samples, and the distance between the samples and the targets was 13 cm. Each coating comprised a Ti interlayer followed by a TiC transitional layer and finally the main hydrogenated carbon layer. The Ar gas flow was maintained at a constant rate of 45 sccm during the deposition process. The Ti interlayer was deposited using one Ti target under a Ti current of 1 A and a bias voltage of −80 V for 10 minutes. The TiC transitional layer was prepared under a Ti current of 1 A, a C current of 2 A, and a bias voltage of −80 V for 6 minutes. The two layers (Ti interlayer, and TiC transitional layer, had a total thickness of near 0.1 μm, aiming at enhancing adhesion between the coating and the substrate. To prepare a series of main layers with varying H content, the flow rate of acetylene gas was varied from 0 to 16 sccm at intervals of 4 sccm and finally set at 24 sscm. The other conditions were fixed, with a bias voltage of −40 V and a C current of 2.5 A. Under the aforementioned conditions, amorphous (a-) C coatings were obtained. The coatings were named according to the acetylene flow rates at which they were produced. The coatings were given the following names: a-C, a-C:H04, a-C:H08, a-C:H12, a-C:H16, and a-C:H24.

2.2. Analysis equipment

A Fourier transform infrared (FTIR) spectrometer (Thermo Nicolet Nexus 470, Japan) was used to detect the H content in the coatings. The FTIR measurements were performed in the scan range of 2800–3500 cm^{-1} with a resolution of 0.1 cm^{-1}. Raman measurements (HR800 Micro-PL, Horiba, Japan) were conducted at room temperature with the 633-nm line of a He–Ne laser in the working spectral range of 1000–2000 cm^{-1}. An X-ray photoelectron spectroscope (PHI 5000 VersaProbe, Japan) was employed with monochromatic Al K-alpha radiation ranging from 282 to 290 eV. The OriginLab software (OriginLab Corporation, USA, Massachusetts), which was used in FTIR spectroscopy and Raman analysis, was also used to deconvolute the X-ray spectroscopy (XPS) spectrum according to the Gaussian–Lorentzian method. The ratios between the deconvoluted bonds were calculated using the areas under each deconvoluted peak. The cross-sectional microstructure and thickness of the coatings were examined using a field emission scanning electron microscope (SEM) (HITACHI SU-5000, Japan). The Ra value was analyzed using a surface profilometer (Kosaka SE-300, Japan) with a cut-off of 0.25 mm. The nanohardness was measured using an LBI nanoindenter (UNAT-M, Germany) under an indentation load of 10 mN and depth of 100–130 nm. The adhesion was evaluated using a scratch tester (FM-POD-200NT, Taiwan) under a load of 0–100 N applied at a rate of 10 N/mm and increment of 1 N per second by a diamond head with a diameter of 30 μm. During scratching, a critical failure load (Lc) value was obtained, which was defined as the point when the friction coefficient rose abruptly.

3. Results and Discussion

Figure 1 displays the FTIR analysis results, which indicate the main peak at approximately 2920 cm^{-1}, two shoulder peaks at 2980 and 2850 cm^{-1}, and a small peak at 3010 cm^{-1}. The main peak corresponds to the sp^3 CH_2 asymmetric stretching mode, while the two shoulder peaks to the sp^3 CH_3 asymmetric and symmetric stretching modes, respectively [17–19]. The small peak at 3010 cm^{-1} was from the sp^2 CH_2 asymmetric stretching vibration [20]. Considering the effects of flow rate of acetylene, in spite of no remarkable difference between the a-C:H04 and a-C:H08 coatings, the strength of peaks corresponding to the C-H stretch increased when the flow rate increased, which indicated increased bonding between C and H. In addition, the increased sp^2 CH_2 can be found.

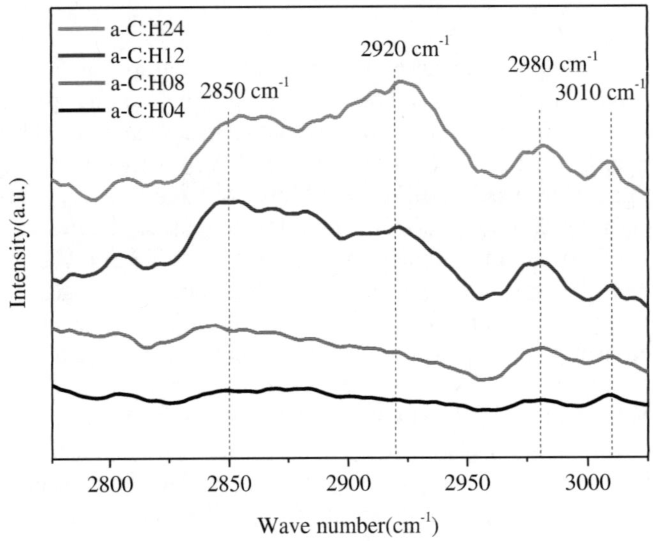

Figure 1. FTIR analysis results of the a-C:H coatings.

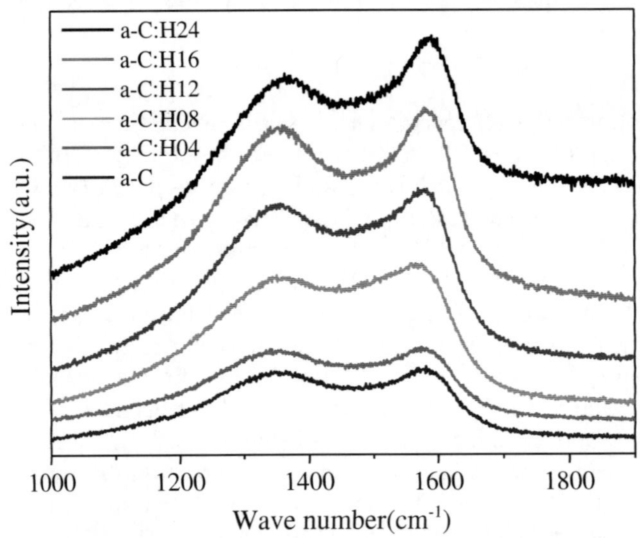

Figure 2. Raw Raman spectra of the a-C:H coatings.

Figure 2 illustrates the raw Raman spectra, and the Raman analysis results are listed in table 1. The data indicate that with increasing H content, the wave number of the G peaks first decreased and then increased. The a-C:H08 coating had the lowest wavenumber for its G peak. The D peaks exhibited an opposite trend. When the number of sp^3 bonds increased, the G peak shifted to a low position [21]. An increased fraction of sp^3 bonding caused an increase in the degree of bonding distortion because of the rise in the shear-to-bulk-modulus ratio of the sp^3 structure, which increased the defects in the C structure [22].

Table 1. Raman analysis results of the a-C:H coatings.

Coating	D peak position (cm^{-1})	G peak position (cm^{-1})	I_D/I_G
a-C	1364.1	1581.1	1.53
a-C:H04	1363.8	1579.9	1.52
a-C:H08	1372.4	1571.1	1.67
a-C:H12	1366.9	1575.9	1.18
a-C:H16	1362.1	1580.9	1.12
a-C:H24	1367.7	1583.8	1.07

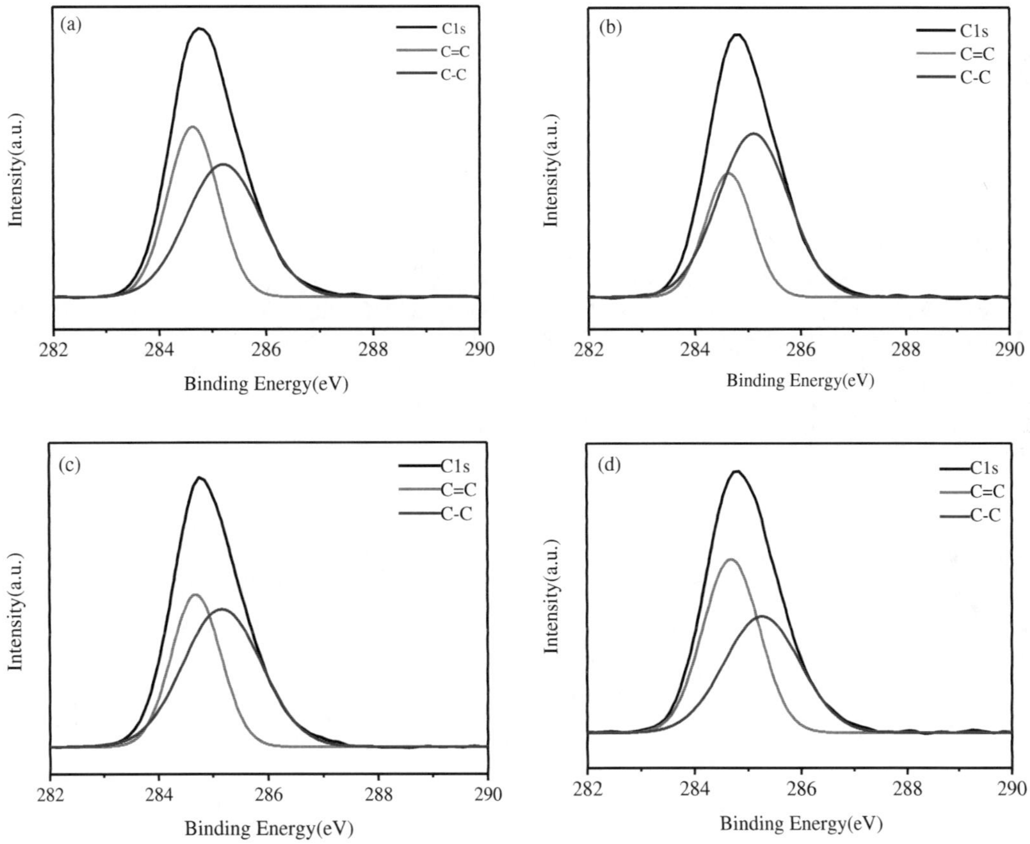

Figure 3. Typical XPS C1s spectra of coatings: (a) a-C, (b) a-C:H:08, (c) a-C:H16, and (d) a-C:H24.

It was learned from the FTIR results that the H addition led to the formation of C-H and C=C bonds. Figure 3 displays the typical XPS results. Because the C-H bonding energy is very close to the C-C bonding energy, deconvolution is not meaningful. Thus, the C-C peak comprised C-C and C-H bonds.

As seen in these figures in Figure 3, the height of the C-C peak increased and then decreased with increasing H_2 flow rate, with the a-C:H08 coating exhibiting the highest C-C peak. The fractions of C-C and C=C bonds were calculated using the XPS spectra (table 2). The a-C:H08 coating showed the highest fraction of sp^3 bonding. Diesselberg et al. indicated that sp^2 bonding changes to sp^3 bonding with a small amount of H addition [23]. When the H content continues to increase, the fraction of sp^3 bonding decreases and that of sp^2 bonding increases [24]. The results here are consistent with these findings from previous studies.

Table 2. The calculated bond fraction in the XPS C1s spectra.

Sample	C-C (%)	C=C (%)
a-C	53.1	46.9
a-C:H04	60.2	39.8
a-C:H08	66.7	33.3
a-C:H12	58.9	41.1
a-C:H16	58.7	41.3
a-C:H24	47.5	52.5

Figure 4 displays the SEM micrographs of the surface morphologies and cross sections of the coatings. The coating thickness increased from 1.4 to 3.1 μm with an increase in the acetylene flow rate. The a-C:H08 coating (figure 4c) exhibited the densest cross-sectional structure. The a-C:H24 coating (figure 4f) presented the least dense structure due to high H content. No columnar structure was observed on any of the coatings, which suggested the formation of amorphous coatings.

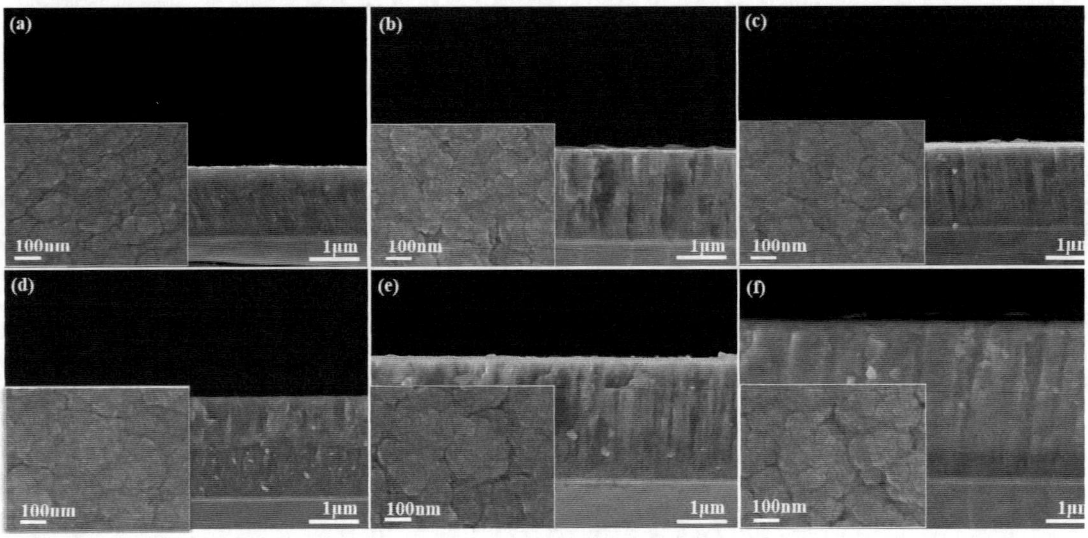

Figure 4. SEM micrographs of the surface morphologies and cross sections of the coatings: (a) a-C, (b) a-C:H04, (c) a-C:H08, (d) a-C:H12, (e) a-C:H16, and (f) a-C:H24.

Table 3 lists the characterization results of the coatings. All the coatings had very smooth surfaces with ultralow Ra values. The scratching test results indicated that the coatings had comparable adhesion strength except for the poor adhesion of the a-C:H24 coating [20, 25–26]. Table 4 provides the results of the nanohardness test and the calculated H^3/E^2 values. The three indices in 4able 4 exhibit an increasing trend with a small addition of H. The a-C:H08 coating showed the maximum value for all three indices. As the H content continued to increase, the nanohardness and modulus of elasticity decreased drastically.

Table 3. Thickness, surface roughness, and adhesion of the coatings.

Coating	Thickness (μm)	Ra (nm)	*Lc* (N)
a-C	1.4	37	83
a-C:H04	1.9	33	73
a-C:H08	2.0	34	75
a-C:H12	2.1	36	83
a-C:H16	2.6	38	80
a-C:H24	3.1	36	68

Figure 5 illustrates the relationship between the hardness and sp^3 bonding fraction of the coatings. As the sp^3 bonding fraction increased, the nanohardness increased. The a-C:H08 coating had the highest hardness. The a-C:H16 coating had a nanohardness of merely 7 GPa, which indicates the increased sp^2 C-H bonding.

Table 4. Nanohardness, modulus of elasticity, and H^3/E^2 values of the H-DLC coatings.

Coating	Nanohardness (GPa)	Modulus of elasticity (GPa)	H^3/E^2
a-C	13.3	191.8	0.064
a-C:H04	13.4	192.1	0.065
a-C:H08	15.4	198.8	0.092
a-C:H12	13.9	182.6	0.081
a-C:H16	7.0	99.8	0.034
a-C:H24	5.4	66.1	0.036

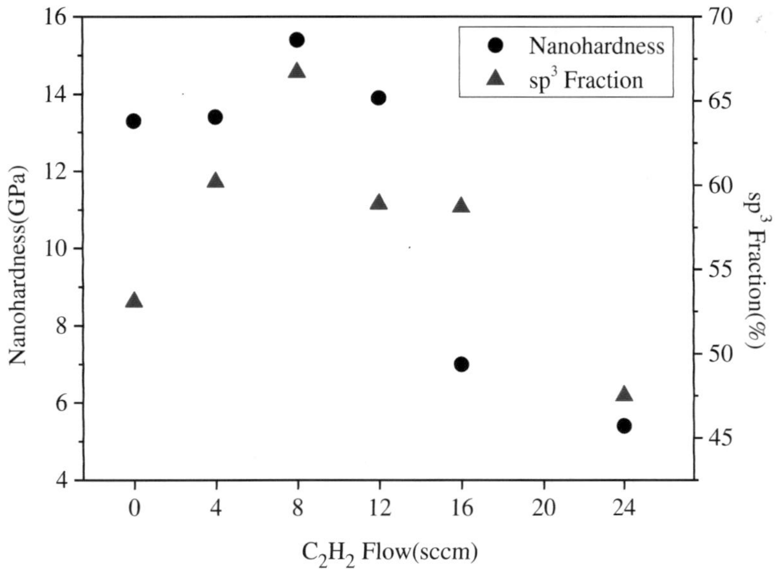

Figure 5. Correlation between the nanohardness and sp^3 fraction of the coatings.

The FTIR spectroscopy, Raman analysis, and XPS indicated that H addition led to the formation of sp^3 C-H bonding, and the sp^3/sp^2 bond fraction varied with H addition. sp^3 C-H bonding was less favorable than sp^3 C-C bonding. Moreover, due to the significant difference in the atomic sizes of C and H, sp^3 C-H bonding was considered as a defect, and the overall C structure was distorted. However, a small number of defects produced by the appropriate amount of H atoms may enhance the performance of H-DLC coatings. The a-C:H08 coating, which exhibited the highest hardness, had an optimal number of defects caused by H addition.

4. Conclusions

DLC coatings are widely applied in industries and require enhancement. In this study, acetylene gas with a flow rate up to 24 sccm was introduced during deposition to obtain a group of H-DLC coatings. The obtained H-DLC coatings exhibited an amorphous structure. The H content increased with increasing acetylene flow rate. Hydrogenation led to the formation of various types of bonds with hybridized sp^{in} states. The fraction of sp^3 bonding in the coatings varied with the H content. The a-C:H08 coating grown with 8 sccm acetylene had the highest fraction of sp^3 bonding. This coating also exhibited the highest nanohardness of 15.4 GPa. A certain small amount of sp^3 bonding in the H-DLC coatings enhanced their hardness.

Acknowledgment

The authors wish to express their gratitude to the Ministry of Science and Technology for financing this research under the grant MOST 105-2221-E-006 -071. The support from the Centre of Noble Equipment, NCKU, Taiwan is also acknowledged.

References

[1] Butter R S and Lettington A H 1995 *J Chem Vapor Depos* **3** 182-92
[2] Huang M S, Yang Y S, *et al.* 2012 *J Polym Eng Sci* **32** 389-99
[3] Loir A S, Garrelie F, *et al.* 2004 *J Thin Solid Films* **453–454** 531–6
[4] Veerasamy V S, Luten H A, *et al.* 2003 *J Thin Solid Films* **442** 1–10
[5] Dai H Y, Zhan C, *et al.* 2016 *J Optik* **127** 2512-5
[6] Aisenberg S and Chabot R 1971 *J Appl Phys* **42** 2953-8
[7] Grill A 2003 *J Diamond Relat Mater* **12** 166–70
[8] Smietana M, Szmidt J, *et al.* 2008 *J Diamond Relat Mater* **17** 1655–9
[9] Bordin D and Coelho P 2018 *J Dent Mater* **3123** e1-e10
[10] Rubio-Roy M, Corbella C, *et al.* 2009 *J Diamond Relat Mater* **18** 923-6
[11] Jacob W and Moller W 1993 *J Appl Phys Lett* **63** 1771-3
[12] Liu W J, Guo X J, *et al.* 2009 *J Thin Solid Films* **517** 4227-32
[13] Tillmann W and Hoffmann F 2011 *J Surf Coat Technol* **206** 1705-10
[14] Rabbani F and Vogelaar B M 2004 *J Diamond Relat Mater* **13** 170-9
[15] Piao J X, Kumar M, *et al.* 2016 *J Surf Coat Technol* **307** 1119-23
[16] Toro R G, Calandra P, *et al.* 2017 *J Surf Interface Anal* **6** 60-71
[17] Wang Y, Ye Y, *et al.* 2011 *J Surf Coat Technol* **205** 4577-81
[18] Dai H, Chen J, *et al.* 2013 *J Non-Cryst Solids* **363** 77-83
[19] Wu Y, Li H, *et al.* 2014 *J Tribol Int* **71** 82-7
[20] Tomasella E, Meunier C, *et al.* 2001 *J Surf Coat Technol* **141** 286-96
[21] Diesselberg M H, Stock -H, *et al.* 2004 *J Surf Coat Technol* **188** 612-6
[22] Robertson J 1992 *J Surf Coat Technol* **50** 185-203
[23] Neuville S and Matthews A 2007 *J Thin solid films* **515** 6619-53
[24] Ishihara M, Suzuki M, *et al.* 2005 *J Diamond Relat Mater* **14** 989-93
[25] Cheng Y H, Wu Y P, *et al.* 1999 *J Diamond Relat Mater* **8** 1214-9
[26] Grill A and Patel V 1993 *J Diamond Relat Mater* **2** 1519-24

IOP Conf. Series: Materials Science and Engineering **474** (2019) 012011 doi:10.1088/1757-899X/474/1/012011

On using photoconductivity decay to determine Si free carrier recombination lifetime: possibilities and challenges

I M Anfimov, M V Anfimov, D S Egorov, S P Kobeleva*, K V Pushkov, I V Schemerov and S Yu Yurchuk

National University of Science and Technology "MISiS", Moscow 119049, Russia

*E-mail: kob@misis.ru

Abstract. Free carrier recombination lifetime (τ) in Si single crystals is a parameter, that characterizes the quality of the material. The primary procedure for τ measurement in ingots is photoconductivity decay (PCD) analyses, including contactless μ-PCD and eddy current methods. One of the main problem in this methods is a surface recombination influence on the PCD curve. It is usually assumed that in ingots the influence of surface recombination can be neglected. Well-known dependence of effective lifetime (τ_{eff}) on τ, surface recombination velocity (S) and sample thickness (d) describes the maximum value of a lifetime. Based on the numerical calculations of the one dimensional continuity equation we have shown that if the sample thickness exceed $5L_D$, the maximum lifetime cannot be achieved before PCD signal decline to 5% of maximum, so the question arises what is the interval of the decay curve where the experimental value of the τ_{eff} has to be determined. We used the range from 45-5% of the maximum PCD signal following the recommended SEMI MF 1575 standard (τ_{SEMI}). The dependencies of τ_{SEMI} on τ, d and S were calculated for different versions of PCD measurements and measurement equipment. The recommended thickness for lifetime measurements of non-passivated samples is in the range $(1\text{-}5)L_D$. The hardware capabilities of contactless methods are compared.

1. Introduction

In semiconductors, unlike in metals, the charge carriers concentration can be widely changed at constant temperature due to doping level and generation/recombination processes. In direct-bandgap materials, basic recombination mechanism is radiative recombination-dependent on the concentration of free charge carriers. In indirect-bandgap semiconductors basic mechanism is a Shockley-Read-Hall (SRH) process: the electron passes from conductivity band to valence band through energy states created within the bandgap by an impurity. The free carrier recombination lifetime (τ) is inverse proportional to deep level impurities concentration (DL) [1, 2]. Therefore τ in the non-direct-bandgap material, like Si and Ge, is a main quality characteristic as well as the resistivity. From measurements of recombination lifetime in indirect-bandgap semiconductors the concentration of DL can be detected with high precision ($<10^{10}$ cm^{-3}) that is impossible with other methods.

The most common method to measure the recombination lifetime in a semiconductor without making any structures or contacts to sample is photoconductivity decay measurements. The are some contactless PCD measurement techniques that allow estimating the parameter on ingots and single-crystal wafers before the production process begins [2-10].

There is a fundamental problem of PCD curve analysis. Relaxation curve is strongly affected by the surface recombination velocity (*S*) [6-10]. In 1955, the work was published where analytical expressions for charge density distribution and time dependence in light illuminated samples were obtained [11]. It was proposed to take into account the influence of the surface recombination as the boundary conditions for solving the continuity equation and take into account the first term in the expansion of the carrier flux concerning the excess carrier concentration on the surface $\Delta n(0,d)$. In this case, *S* was the parameter determining the relation between the flux and $\Delta n(0,d)$. In [11] was shown that the relaxation curve is described as the sum of exponents. Contributions of each one depend on the initial charge density distribution $\Delta n(0,x)$. Characterization times depends on the diffusion coefficient *D*, sample thickness *d*, and *S*:

$$\tau_{eff}^{-1} = \tau^{-1} + \tau_S^{-1} = \tau^{-1} + 4\xi D/d^2 \quad (1)$$

Parameter ξ can be obtained from transcendental equation [11, 12]:

$$\cot\xi = 2\xi D/dS \quad (2)$$

The maximum value of ξ in (2) from the first quadrant define the maximum value $\tau_{eff} = \tau_{max}$. Exact solution for τ_{max} can be obtained in two extreme cases: for $dS \to \infty$, $\xi = \pi/2$ (infinity surface recombination velocity case):

$$\tau_{eff}^{-1} = \tau^{-1} + \frac{\pi^2 D}{d^2} \quad (3)$$

and for $d \to 0$ (thin sample case):

$$\tau_{eff}^{-1} = \tau^{-1} + \frac{2S}{d} \quad (4)$$

In [13], by numerical solution it was shown that, in general, τ_{max} can be defined with great accuracy as a sum of those two cases:

$$\tau_{eff}^{-1} = \tau^{-1} + \left(\frac{d}{2S} + \frac{d^2}{\pi^2 D}\right)^{-1} \quad (5)$$

Obviously, in ingots with non-passivated surfaces and large *d*, it can be expected that the measured effective lifetime will be equal to the bulk lifetime. Unfortunately, in quite many investigations concerning the influence of various factors on the determination of recombination lifetime, it is not discussed in which part of the relaxation curve the above maximum value of the effective lifetime is reached [2-10].

Based on a numerical solution of the continuity equation in papers [12-14] it was shown that when sample thickness rise, the relative contribution of the "fast" exponents increases. To calculate the τ_{eff}, it was proposed to use the recommendations of the SEMI MF1535 standard [15] and calculate τ_{eff} as the slope of the logarithm of the relaxation curve in the interval from 45-5% of the maximum signal intensity (τ_{SEMI}).

The work aims to analyze the influence of the main factors which affect the shape of the PCD curve for non-passivated samples and to determine the conditions of the applicability of Eqs. (3-5) to the calculation of the bulk carrier lifetime (τ) using its effective experimental value.

2. Thickness influence on the PCD curve

The one-dimensional continuity equation was solved by a finite difference method. The simulation was carried out for samples of n- and p-type silicon single-crystals. The samples thickness varies from 0.5 to 16 mm, τ varies from 10 μs to 10 ms, *S* varies from 10 to 10^4 cm/s, and the illuminating LED wavelength (λ) ranges from 0.8 to 1.107 μm. The limitation of the penetration depth of the microwave field in semiconductor takes into account by integration of excess concentration in the bounded area.

In [12] it was shown, that for a frequently used wavelength 1.06 μm for τ varying from 10 to 10^4 μs (this interval includes most popular τ values for microelectronics) correlation between τ_{SEMI} and *d* have a generalized form if the thickness is expressed in units of diffusion length ($L_D^2 = D\tau$). If the sample thickness is less than $5L_D$, expression (5) can be used. Moreover, if the thickness is in the range of $(1\text{-}5)L_D$, a more simple Eq. (3) can be used. In this case, it is not necessary to know the value of surface recombination velocity. Consider the value of differential lifetime, defined as the cotangent of the inclination angle of the dependence of the logarithm of the excess concentration Δn on time t ($\tau_d = -\left(\frac{\partial \ln \Delta n}{\partial t}\right)^{-1}$). Dependencies of τ_d on dimensionless time t/τ_{max} for n-type silicon samples with different thickness are shown in figure 1. Curves were calculated using the following parameters: $\tau = 700$ µs ($L_D = 1$ mm), $S = 10^4$ cm/s (unpassivated surface).

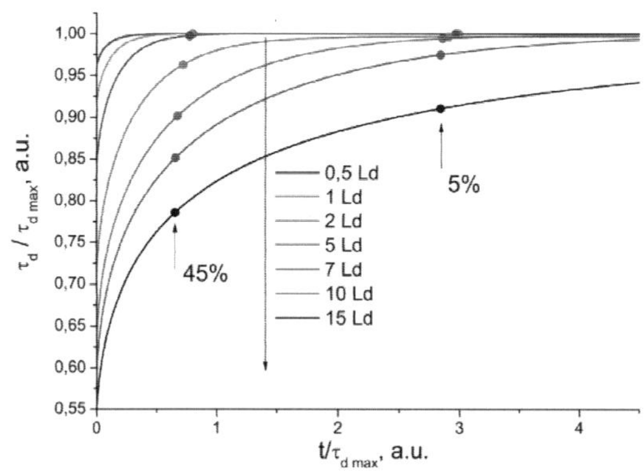

Figure 1. Dependence of the differential lifetime for samples with different thickness. The arrows indicate the time at which the excess concentration is 45% and 5% of the maximum.

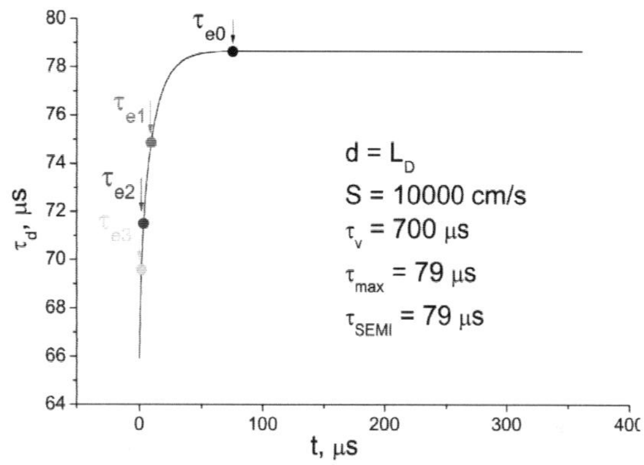

Figure 2. The variation of the differential lifetime for the sample with $d = L_D$.

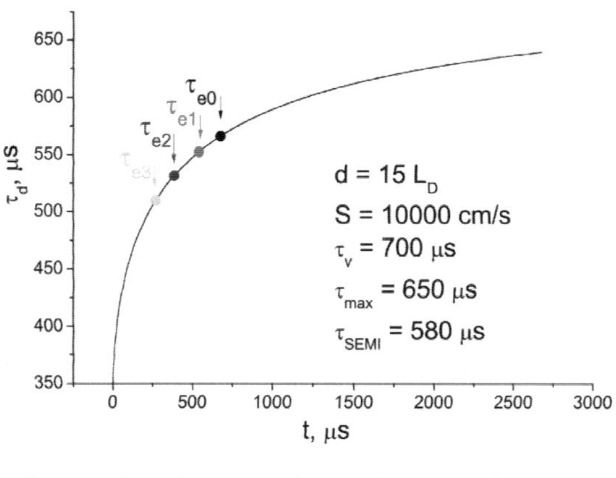

Figure 3. The variation of the differential lifetime for the sample with $d = 5L_D$.

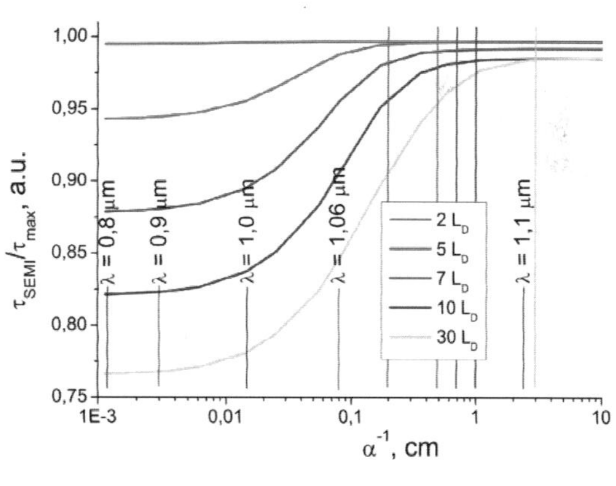

Figure 4. The variation of the differential lifetime for the sample with $d = 15L_D$.

It can be seen in figure 1 that for a small thickness saturation of the differential lifetime is established almost instantly. With a further increase of the thickness, the contribution of "fast" surface exponents increases, and for $d > 5L_D$, τ_{SEMI} will be noticeably lower than expected by (5). Two reasons may cause this decline. The first one is a magnification of fast surface exponents contribution to the decay curve. The second one relates to inhomogeneity of the generation of the excess charge carriers across the thickness. Increasing of the effective lifetime for several quadrants of Eq. (2) is shown in figure 2-4. The circles indicate the time on the relaxation curve, at which the exponents from different quadrants decrease by a factor of e (2.718).

As can be seen in figures 2-4, an increase in the sample thickness leads to an expansion of the characteristic times of the surface exponents. When the thickness of the sample is less than $5L_D$, relaxation curve becomes monoexponential starting at the level of 45% of the maximum.

3. Influence of light wavelenght on the PCD curve

The wavelength of illuminating LED influence on the homogeneity of generation of excess charge carriers because of different value of absorption coefficient α. In figure 5, it is shown how τ_{SEMI} depends on the inverse absorption coefficient. Also in the figure marked the typically LED wavelength used in modern measuring devices.

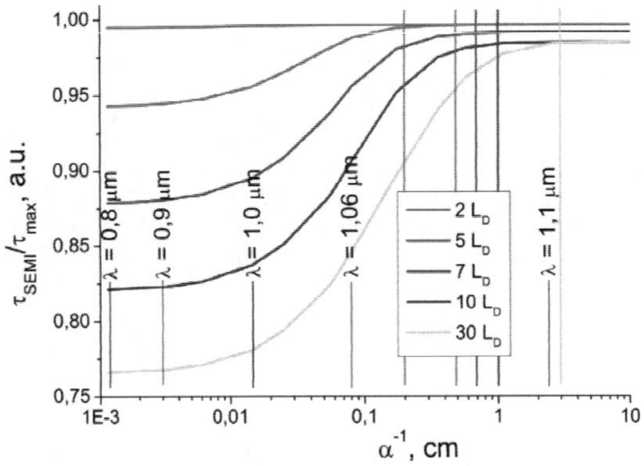

Figure 5. The relation between the effective lifetime and the inverse absorption coefficient α^{-1} of single-crystal silicon. Colored vertical lines indicate thicknesses of those samples. Simulations were carried out for n-type silicon with $\tau = 700$ µs ($L_D = 1$ mm).

As can be seen in figure 5, for thicknesses of $(1\text{-}3)L_D$, the effective lifetime does not depend on the illumination wavelength, and it can be calculated from (3). For $d > 5L_D$ and LEDs with $\lambda < 1$ µm τ_{SEMI} changes from (3) and goes to saturation when average photon mean free path until absorption is equal to the thickness of the sample.

From the viewpoint of the uniformity of the minority carrier distribution, the best light emitting diodes for the excitation is the 1.1 µm ones. However, the small absorption coefficient of silicon (α = 0.375 cm^{-1}) leads to a PC signal being too low for the registration. Thus, the LED with λ = 1.05 µm (α = 18.3 cm^{-1}) may be considered optimal because in this case the deviation from the values calculated using Eq. (3) does not exceed 5%, regardless of thickness. The best for the measurements is a sample thickness of $(1\text{-}5)L_D$. For such samples one can choose LEDs with high absorption coefficients (λ = 0.9 µm, α = 314 cm^{-1}).

4. Skin-layer limiting effect

For contactless methods of the photoconductivity decay detection, two methods can be used: absorption of a microwave signal by free charge carriers (µ-PCD method) and the change in the quality factor of the RC circuit (transient eddy current HF methods) [9, 10, 14-18].

The influence of the illumination wavelength is equally relevant for both µ-PCD and HF methods. But the µ-PCD has limitations that make the HF method more suitable. First of all, skin effect limits the depth of penetration of the electromagnetic field into the conducting material. The skin effect will be manifested if the thickness of the sample exceeds the thickness of the skin layer. In this case, a decrease in frequency, for example from 10 GHz, recommended in SEMI 1535 [16], to 2.5 GHz (all other things being equal) increases the depth of the skin layer by a factor of 2 (see table 1). The effect on the measurement results will be observed when the thickness of the skin layer is less than $5L_D$. For resistivity above 1 Ω·cm, the maximum lifetime is limited by recombination through impurity centers. For the concentration of impurity centers of 10^9 - 10^{10} cm^{-3}, the maximum lifetime is 10 - 1 ms ($5L_D$ > 5 - 26 mm),respectively. As can be seen from table 1, the use of the recommended frequency for measuring non-passivated samples with long lifetimes can lead τ_{SEMI} to values estimated by (3), even if the thickness of the sample lies in the optimal range from the viewpoint of use (4). From this point of view, the HF method is also advantageous, since for MHz frequencies the thickness of the skin layer is from one to hundreds of centimeters.

Table 1. Dependence of the thickness of the skin layer (mm) on the frequency of the electromagnetic field and resistivity.

			ρ, Ω·cm		
f, GHz	0.01	0.1	1	10	100
3.2 E-2	8.9	28.1	88.9	281.2	889.2
2.5	1.0	3.2	10.1	31.8	100.6
5.0	0.7	2.2	7.1	22.5	71.1
10	0.5	1.6	5.0	15.9	50.3
34	0.3	0.9	2.7	8.6	27.3

We make an analysis of PCD curve for a low level of injection of the excess free carrier. With a high level of injection, the differential lifetime in the material depends on the level of injection: it will decrease with the decrease of the signal. Taking into account this factor requires exact knowledge of the initial carrier concentration and maximum nonequilibrium concentration, e.g., leads to a significant complication of measurement devices.

5. Experimental investigation

To verify the simulation results FZ-grown n-type single-crystal silicon wafers were measured. The resistivity of silicon was 2500 Ω·cm. Samples were cut out from the same ingot, so the bulk lifetime was the same for samples of different thickness. A thickness of the samples varied from 1 to 32 mm. The measurements of the PCD curve were carried out by μ-PCD method [16]. Samples were illuminated with IR-LED (λ = 1.06 μm). Results of measurements (τ_{SEMI}) are shown in table 2.

Table 2. Measurement results for single-crystal silicon wafers.

Thickness, mm	32	5	2	1.2
Thickness/L_D, a.u.	21.3	3.3	1.3	0.8
τ_{SEMI}, μs	1350 ± 130	820 ± 80	230 ± 20	91 ± 8
τ (3), μs	1380 ± 150	1600 ± 200	1540 ± 600	1400 ± 800

The evaluation of τ based on $1/\tau_{eff}$ dependence on $1/d^2$ (3) gives D = 15 ± 1.5 and τ = 1550 ± 150 (L_D ≈ 1.5 mm).

Substituting these values of D and d into Eq. (3) τ can be estimated (see τ(3) in table 1). It can be seen, that for samples with the thicknesses of 2 mm and 5 mm τ(3) is close to the expected τ. For 1.2 mm sample τ(3) is less because the equation (5) must be used, but we don't know S exactly.

In [13], it was shown that for single-crystal silicon samples with thicknesses above $5L_D$, the following equation could be used for τ calculation:

$$\tau = A \exp\left(-\frac{d}{\delta}\right) + t_0 \qquad (6)$$

where d is the sample thickness [cm], and τ is the recombination lifetime [s].

The following dependences of parameters A, δ and t_0 on τ_{eff} [s] were obtained:

- $A = 600\ \tau_{eff}^{1.3}$, $\delta = 1.9\ \tau_{eff}^{0.41}$, $t_0 = 1.6\ \tau_{eff}^{1.038}$ for n-type single-crystal silicon;

- $A = 31000\ \tau_{eff}^{1.4}$, $\delta = 2.2\ \tau_{eff}^{0.43}$, $t_0 = 1.6\ \tau_{eff}^{1.033}$ for p-type single-crystal silicon.

For the 32 mm sample, it can be obtained $\tau = 1680 \pm 200$.

In figure 6, the logarithm of normalized PCD curves is plotted versus dimensionless time t/τ_{max}. PCD signal was normalized to the level of 5% of maximum value. The circles indicate the time at which the excess conductivity reached 45% of the maximum value.

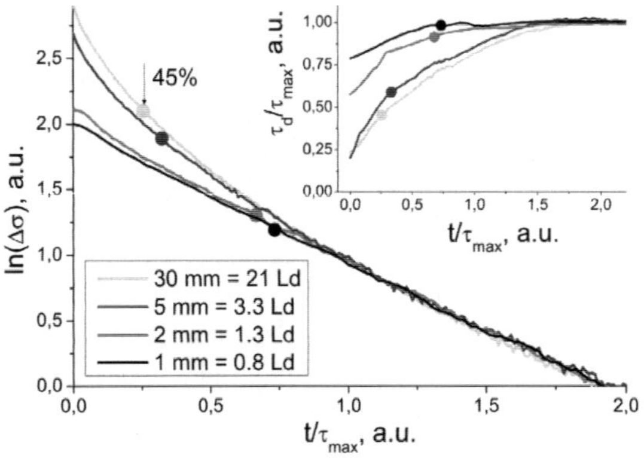

Figure 6. The normalized PCD signal and differential lifetime in silicon wafers.

From figure 6 it can be seen that the distortion of the initial part of the relaxation curve strongly depends on the sample thickness. The part of the relaxation curve for τ_{SEMI} calculation is almost linear if sample thickness is less than $3L_D$.

6. Conclusions

- ✓ Numerical simulation of the decay curve and differential lifetime shows that, for non-passivated samples, attaining the lifetime value described by the well-known Eq. (5) depends on the relative sample thickness. When calculating the effective lifetime using the interval between 45% and 5% of the maximum signal intensity, as recommended by SEMI MF 1535 (τ_{SEMI}), (5) may be used for the calculation of the bulk lifetime for thicknesses up to $5L_D$.
- ✓ At thicknesses above $5L_D$, the maximum lifetime value is not achieved because of the increasing contribution of faster components to the decay curve, so that a numerical simulation must be used for a correct calculation of the bulk lifetime.
- ✓ In the interval from $1L_D$ to $5L_D$, it is possible to use the convenient formula considering an infinite recombination rate (3).
- ✓ For thicknesses up to $5L_D$, the τ_{SEMI} value practically does not depend on the wavelength of the excitation light. For thicker samples, τ_{SEMI} depends on the excitation wavelength and achieves a maximum value when the sample thickness does not exceed the inverse light absorption coefficient.
- ✓ The lifetime calculation error will be smaller if the PCD curve is detected by the eddy current technique.

References

[1] Shockley W and Read W T 1952 *Phys Rev* **87** 835-42
[2] Schuler N, Hahn T, Schmerter S, Hahn S, Dornich K and Niklas J R 2010 *J Appl Phys* **107** 064901
[3] Runyanan W R and Shaffner T J 1998 *Semiconductor Measurements \& Instrumentation* (N.Y.: McGraw-Hill)
[4] Schroder D K 2006 *Cancun; Mexico: ECS Transactions* **3** 321-37
[5] Horanyi T S, Tutto P and Kovacsics Cs 1996 *Electrochem Soc* **143** 216-20
[6] Karazhanov S Zh 2001 *Semicond Sci Technol* **16** 276-87

[7] Brody J, Rohatgi A and Ristow A 2003 *Sol Energ Mat Sol C* **77** 293-301
[8] Lauer K, Laades A, Ubensee H, Metzner H and Lawerenz A 2008 *J Appl Phys* **104** 104503-1-9.
[9] Lagowski J and Wilson M D 2014 *U.S. Patent* US8912799 B2
[10] Schuler N, Anger S, Dornich K and Bothe K 2011 *Sol Energ Mat Sol C* **98** 245-52
[11] Stevenson D T and Keyes R G 1955 *J Appl Phys* **26** 190-5
[12] Anfimov I M, Kobeleva S P and Schemerov I V 2015 *Inorg Mater+* **51** 1447-51
[13] Anfimov I M, Kobeleva S P, Pylneva A V, Schemerov I V, Egorov D S and Yurchuk S Yu 2017 *Russian microelectronics* **46** 585-90
[14] Anfimov I M, Kobeleva S P, Schemerov I V and Orlova M N 2014 *Journal of Nano- and Electronic Physics* **6** 03018(3c)
[15] ASTM F1535-00, Standard Test Method for Carrier Recombination Lifetime in Silicon Wafers by Noncontact Measurement of Photoconductivity Decay by Microwave Reflectance (Withdrawn 2003), ASTM International, West Conshohocken, PA, 2000
[16] Anfimov I M, Kobeleva S P and Schemerov I V 2016 *Instrum Exp Tech+* **59** 420-4
[17] Borodovskii P A, Buldygin A F and Tokarev A S 2004 *Semiconductors* **38** 1005-11
[18] Melikov A, Halliop B, Mandelis A and Kherani N P 2012 *Thin Solid Films* **520** 5309-13

IOP Conf. Series: Materials Science and Engineering **474** (2019) 012012 doi:10.1088/1757-899X/474/1/012012

Effect of tool/material interface temperature on flow stress and the Zener-Holloman parameter in Al 6082 friction stir butt welds

M Krishna Kishore [1, 2, *], A Kumar[2] and N Ch Kaushik[3]

[1]Mechanical Engineering Department, Gayatri Vidya Parishad College of Engineering, Andhra Pradesh-530048, India
[2]Mechanical Engineering Department, National Institute of Technology-Warangal, Telangana-506004, India
[3]Mechanical Engineering Department, SOET, BML Munjal University, Haryana-122413, India

*E-mail: mugada.krishnakishore@gmail.com

Abstract. This paper outlines, the temperature around the tool/material interface during friction stir welding (FSW) of Al 6082-T6 alloy. The first section gives a brief overview on temperature measurement around the tool periphery, i.e. trailing edge (TE), leading edge (LE), retreating side (RS), advancing side (AS), and their corners associated with it. The second section was focused on the plasticity behavior of welded material as a function of temperature, governed by flow stress and Zener Holloman (Z)-parameter. Around the tool periphery in a clockwise direction, the critical observations were made from the obtained data. The peak temperature surges with increasing shoulder diameter and it was higher for the weld done with 24mm shoulder diameter tool. It was noted that during every cycle there was asymmetric temperature distribution at tool/material interface, the rate of heat generation was higher in TE to RS-LE zone and lower in RS-LE to TE zone. The calculated flow stress and log Z values were higher on RS and lower in AS to RS-LE zone of the tool shoulder periphery. Flow stress was almost uniform for all the shoulder diameters except on RS of 18mm shoulder tool. The optimum tool shoulder diameter was found to be 21mm, below which the flow stress was high and above which the plasticity at tool shoulder periphery was negligible.

1. Introduction

Al 6082 alloy is a relatively high-strength heat-treatable 6xxx series aluminum alloy and a replacement to conventional Al 6061 alloy. The welding of these alloys leads to coarsening or dissolution of the fine precipitates due to the developed thermal cycles [1]. The use of friction stir welding (FSW) for aluminum alloys, a promising solid-state welding method came into existence since the 1990s. It has numerous advantages over conventional fusion welding techniques [2]. The process occurs in three simple steps, namely: plunging, dwelling, and traversing. The frictional contact heat of shoulder and frictional as well as shear deformation heat of pin is responsible for the formation of bond [3]. The review on the FSW process in the aspects of heat generation was done by Nandan et al. [4]. It was proposed that the temperature, tool force, and torque depend on the process parameters, tool geometry, and workpiece material. The flow of the material and heat generation were governed by tool geometry, which consists of a pin and shoulder. The profiles of the pin and shoulder were responsible for frictional contact heat generated during the process, and for deforming the material respectively [2]. It was observed that the contribution of the pin is approximately 3% of shoulder heat generation [5].

The shoulder size influence on frictional heat generation and weld formation were studied by many researchers [6-15]. Zhang et al. [6] studied on the effect of the size of the shoulder on temperature distribution, they established a quadratic relation for the efficient power as a function of shoulder diameter and concluded that with increasing shoulder size, the temperature, and effective power increases. In the later attempts by Zhang et al. [7], the effect of shoulder diameter and taper angle on heat generation was analyzed. It was observed that the shoulder-workpiece contact area generates more heat compared to the pin-workpiece contact area. Elangovan and Balasubramanian [8] investigated on the effect of increasing shoulder diameter and concluded that shoulder diameter was proportional to heat generation, the maximum shoulder diameter in their case was 21mm which resulted in the wider thermo-mechanically affected zone (TMAZ). The work carried by Akinlabi and Akinlabi [9] at a constant tool tilt and depth of penetration shows with the increasing shoulder diameter, axial force and torque were increasing.

Kumar and Kailas [10] observed the shoulder interaction with the workpiece and stated that with their increase in interaction, the material outflows from the weld and diverted again towards the weld region. Thomas and Nicholas [11] also reported, increasing the shoulder region leads to the flash generation on sides, which confines to the certainty of critical shoulder diameter. The Zener Holloman (Z-parameter) relation, the influence of shoulder diameter on plastic deformation at peak temperature was reported by Ramanjaneyulu et al. [12]. It was concluded that with an increase in shoulder diameter, temperature and Z-parameter increases up to a certain diameter and beyond that the shoulder contribution is less for the quality of the joint as maximum friction heat was generated in the material. For the selection of optimum shoulder diameter, Arora et al. [13] developed a model based on the utilization of maximum torque. It was understood that the sticking torque signifies the resistance to the flow of plasticized material and it was maximum at critical shoulder diameter. Mehta et al. [14] reported on the influence of shoulder diameter on thermal cycles, forces and torque requirement with a three-dimensional heat transfer and material flow model. Further with the optimum tool shoulder diameter, Zhang et al. [15] observed an asymmetric temperature distribution around the tool shoulder during numerical investigations, and they reported that the temperature gradient was higher in leading edge that in trailing edge. With the existing literature, it was noted that significant research work was published on the effects of shoulder diameter in the FSW process. But for data relating to the temperature flow, material plasticity governed by flow stress and Zener-Holloman parameter was scarce for Al 6082 alloy. In the present study, the work was classified into two categories, i.e., one focusing on the temperature distribution and the other on the flow stress evaluation as well as the Zener-Holloman parameter around the tool shoulder.

2. Experimental procedure

2.1. Methods and materials

A taper cylindrical pin profiled tool made of H13 tool steel (Hardness: 50 HRC) with three different flat shoulder diameters were chosen for welding of Al 6082 alloy sheet of 6mm thickness in T6 heat treated condition. The chemical composition for Al 6082 alloy work material was indicated in table 1.

The experiments were conducted on the Friction Stir welding machine of Make: RV Machine Tools Pvt. Ltd., Coimbatore; Model: FSW-3T-NC. Based on the pilot experiments, the optimum process parameters and tool geometry (as indicated in table 2) were adopted from the previous studies given in references [17-21]. The schematic layout for data acquisition system to measure temperature and force measurements during FSW is shown in figure 1 The K-type (grounded) thermocouples were brought in to contact at various locations as indicated in figure 2, i.e., at every 45^0 steps around the tool shoulder outer boundary, $Layer^0$ (ϕ P mm). Similarly, from $Layer^0$, the temperatures were also recorded at $Layer^1$ (ϕ Q mm) and $Layer^2$ (ϕ R mm) as shown in figure 2, where P = 18, 21 and 24 mm, corresponding to Q = 28, 31 and 34 mm, & R = 38, 41 and 44 mm, respectively. By using the data acquisition system (DAQ) of Model: USB- TC, Make: Measurement Computing Corporation Pvt. Ltd. the temperature readings were recorded.

Table 1. The chemical composition of Al 6082-T6 work material.

Element,	Cu	Mg	Si	Fe	Mn	Zn	Ti	Cr	Al
% wt.	0.03	0.96	1.16	0.27	0.54	0.05	0.01	0.02	Remainder

Table 2. The process parameters and tool geometry adopted [20].

Process Parameter	Value	Process Parameter	Value
Tool rotational speed ω, (RPM)	800	Pin bottom radius R_{P_2} (mm)	1.5
Tool traverse speed V_f, (mm/min)	40	Pin length h_p, (mm)	5.5
Plate thickness, t_p (mm)	6	Taper angle, α (degree)	15.6^0
Shoulder radius R_s (mm)	9, 10.5, 12	Tilt angle (degree)	1
Pin top radius R_{P_1} (mm)	3	Plunge depth (mm)	5.7

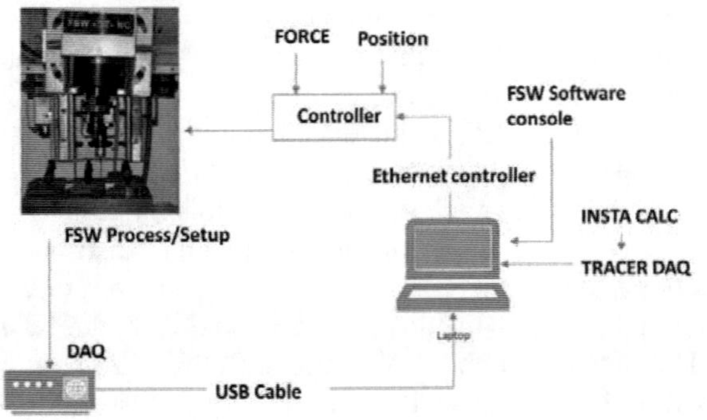

Figure 1. Schematic layout of the temperature & force measurements.

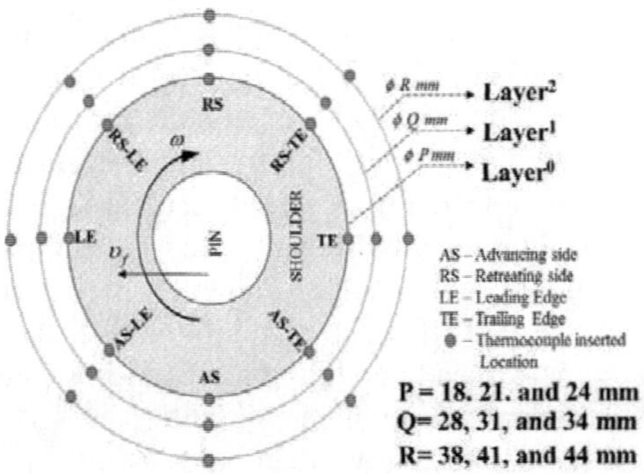

Figure 2. The arrangement of thermocouples around the shoulder.

2.2. Evaluation of the Zener Holloman parameter (Z)

The recorded values of forces and torque acting on the tool during the entire FSW process were adopted for modeling the Z-parameter and flow stresses around the tool shoulder. It is essential to understand the amount of heat generated when different shoulder diameters were adopted. The heat input as a function of torque and process parameters, for the different shoulder diameters according to Akinlabi et al. [9] and Ramanjaneyulu et al. [12] is given in equation (1).

$$q = \eta \frac{(2\pi\omega T_t)}{V_f} \quad (1)$$

where q is heat input (J/mm), η is efficiency factor (equal to 0.9 for aluminum alloys), ω is tool rotational speed, V_f is tool traverse speed, and T_t is tool torque (N-m).

According to Perzyna model [4, 18], the viscosity (μ) is a function of effective flow stress (σ_e) and effective strain rate ($\dot{\varepsilon}$), according to the following equation:

$$\mu = \frac{\sigma_e}{3\dot{\varepsilon}},\ where\ \dot{\varepsilon} = \left(\frac{2}{3}\dot{\varepsilon}_{ij}\dot{\varepsilon}_{ij}\right)^{1/2}\ and\ \dot{\varepsilon}_{ij} = \frac{1}{2}\left(\frac{\partial u_i}{\partial x_j} + \frac{\partial u_j}{\partial x_i}\right) \tag{2}$$

where $\dot{\varepsilon}$ is effective strain rate, which can also be expressed in terms of flow stress and activation energy as per Sellars and Tegart's viscoplastic model [20-22]. This model was used to solve large strain deformation processes it was therefore applicable for FSW also. Equation (3) shows the expression for the effective strain rate as a function of flow stress, activation energy, and temperature. In this model, the flow stress was obtained by performing hot deformation tests of a solid which were equivalent to its recovery or recrystallization steady state of stresses.

$$\dot{\varepsilon} = A(\sinh \alpha\sigma_e)^n \exp\left(-\frac{Q}{RT}\right) \tag{3}$$

where σ_e is effective flow stress; Q is activation energy, 153 kJ/mol; R is universal gas constant, 8.314 J/mol/K; T is temperature in ^{0}K; while the material constants n, A, and α have the following values for Al 6082 alloy [20]: n = 2.976, ln A=19.29 s^{-1}, α=0.045 MPa^{-1} By rearranging equation (3) we get equation (4):

$$\sigma_e = \frac{1}{\alpha}\sinh^{-1}\left[\left(\frac{Z}{A}\right)^{\frac{1}{n}}\right] \tag{4}$$

where Z is the Zener-Holloman (Z) parameter representing the strain rate, which is given in equation (5).

$$Z = \dot{\varepsilon}\exp(Q/RT) \tag{5}$$

The rate of strain means the strain w.r.t time $(\dot{\varepsilon} = \varepsilon/t)$ where time (t) in seconds. The natural strain for the deformed shape of material (ε) can be evaluated using equation (6) [20-22]:

$$\varepsilon = \ln\left(\frac{l}{APR}\right) + \left|\ln\left(\frac{APR}{l}\right)\right| \tag{6}$$

where APR is tool advance per revolution, l = length deformed = $2r\cos^{-1}((d-x)/d)$, d = diameter probe in mm, x = perpendicular distance to the weld direction i.e. from the retreating side (RS) to the streamline in contact to tool pin.

3. Results and Discussion

3.1. Force measurement

For various shoulder diameters, the axial force was measured, and the heat input values were calculated using equation (1), as shown in figure 3. It was noted that with the increase of shoulder diameter the axial force and heat input increased significantly up to 21 mm and later they remained constant. This behavior can be attributed to the material characteristics, where the frictional or shear deformation energy reaches to maximum, and even on the further increase of the size of shoulder the effect is insignificant. The total tool torque (T_t) and axial force (F_z) was higher for 21 mm shoulder diameter, i.e., 56.49 N-m and 6.28 kN respectively.

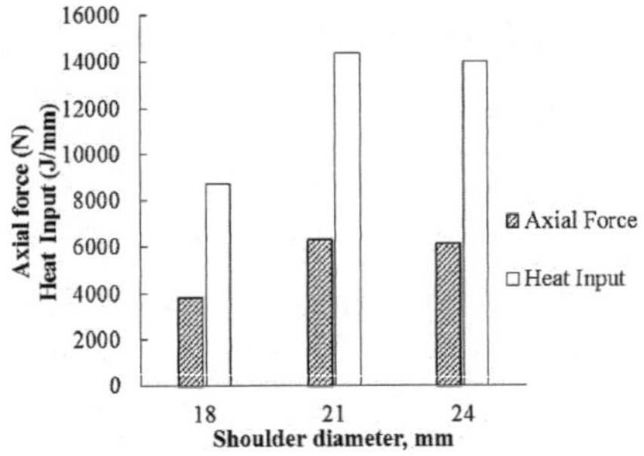

Figure 3. Axial force and Heat Input w.r.t shoulder diameter.

3.2. Macrostructural observation

Figure 4 shows the macrostructure of weld zone obtained by different shoulder diameters. The flow of weld line remnant was observed in all the welds. For weld with 24 mm shoulder diameter, the macrostructural observation revealed a crack, which was originated at shoulder tip and extended to the root of the weld. This could be the reason for lower values of axial force and tool torque. However, for welds with 18 and 21 mm shoulder diameters, the weld region was defect free. It was depicted that the weld region is wider with increasing the shoulder diameter. With increasing the shoulder diameter, the size of the weld zone increases as reported in the earlier literature [12].

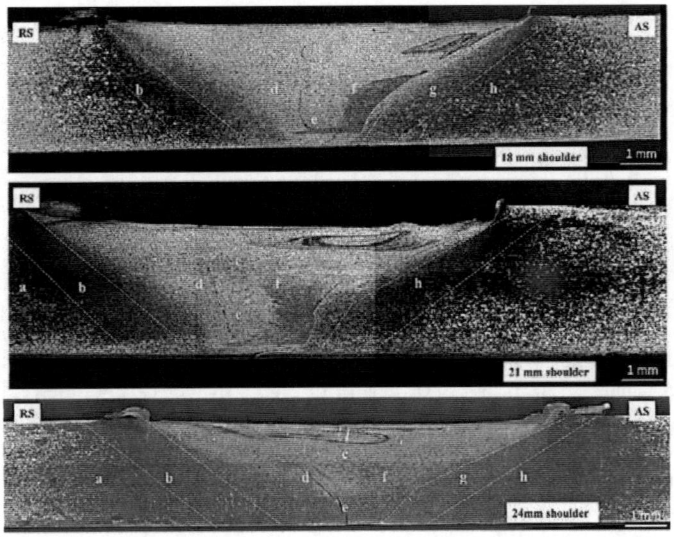

Figure 4. The macrostructure of weld cross-section for various shoulder diameters.

3.3. The Zener-Holloman parameter and flow stress at shoulder periphery

The Zener-Holloman parameter (Z) used to quantify the plasticity level for a given material. The flow stress at higher temperatures is the function of Z. The asymmetric behavior of temperature flow in the FSW region and their associated effects on material plasticity around the tool shoulder should be apparent. The measured temperature and computed log Z values for different shoulder diameters at different layers were shown in tables 3-5. The values of Z were found to be higher on RS, stable in RS-LE, and higher on LE in case of layer^0, layer^1, and layer^2 respectively. As reported by Ramanjaneyulu et al. [12], with increasing the shoulder diameter, the trend of Z- value at periphery indicated a negligible plasticity.

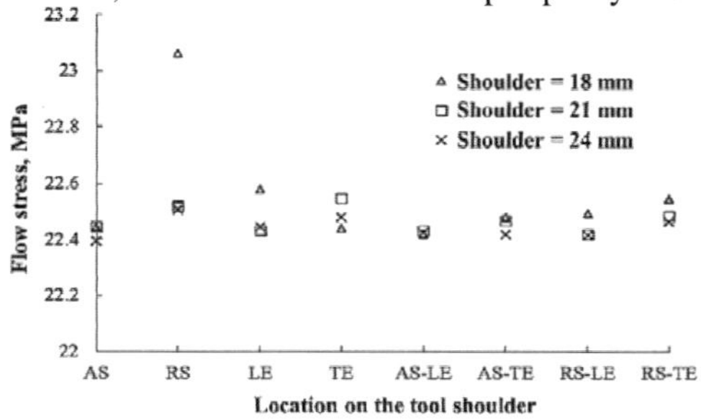

Figure 5. Variation of flow stress for Layer^0.

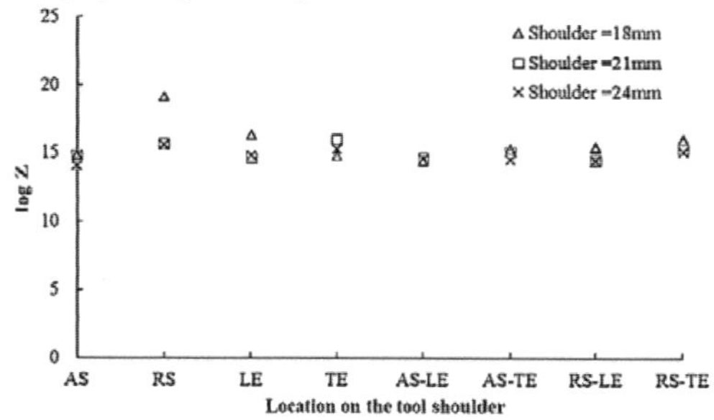

Figure 6. Variation of log Z for Layer^0.

Table 3. The variation of temperature, log Z for layer^0 around the shoulder.

Location → / Parameter ↓			AS	RS	LE	TE	AS-LE	AS-TE	RS-LE	RS-TE
Shoulder diameter (mm)	18	*T* (^{0}C)	366	202	342	367	383	342	335	311
		log Z	14.8	19.1	16.3	14.8	14.7	15.3	15.4	15.9
	21	*T* (^{0}C)	364	322	373	310	374	350	383	339
		log Z	14.8	15.7	14.7	16	14.7	15.1	14.5	15.4
	24	*T* (^{0}C)	406	328	364	342	379	383	386	352
		log Z	14.1	15.6	14.8	15.3	14.6	14.5	14.4	15.1

Table 4. The variation of temperature, log Z for layer^1 around the shoulder.

Location → / Parameter ↓			AS	RS	LE	TE	AS-LE	AS-TE	RS-LE	RS-TE
Shoulder diameter (mm)	18	*T* (^{0}C)	237	190	197	286	235	272	208	224
		log Z	18.0	19.6	19.3	16.6	18.0	16.9	18.9	18.4
	21	*T* (^{0}C)	260	270	207	270	245	270	231	264
		log Z	17.3	17.0	19.0	17	17.7	17.0	18.2	17.2
	24	*T* (^{0}C)	252	256	242	288	252	301	223	302
		log Z	17.5	17.4	17.8	16.5	17.5	16.22	18.4	16.2

The calculated flow stress was found to be uniform around the tool shoulder for all the shoulder diameters. It is seen in figure 5 that the flow stress is higher for smaller diameters in RS because the material is cooler on RS, as compared to AS, therefore more heat is required to warm and plasticize the material. The similar trends were observed in log Z plot as shown in figure 6, where the log Z values are higher on RS. Comparatively, the log Z values are lower on AS, the temperature is higher in the zone of AS to RS-LE in the clockwise direction. This shows that the material was plasticized in the zone of AS to RS-LE and transported around the tool pin thereby depositing with forging action in the TE, as reported in the earlier literature [16].

Table 5. The variation of temperature, log Z for layer^2 around the shoulder.

Location → / Parameter ↓			AS	RS	LE	TE	AS-LE	AS-TE	RS-LE	RS-TE
Shoulder diameter (mm)	18	$T\,(^0C)$	193	148	150	251	186	209	161	190
		$log\ Z$	19.5	21.3	21.2	17.6	19.7	18.9	20.7	19.8
	21	$T\,(^0C)$	215	193	150	212	208	226	194	202
		$log\ Z$	18.7	19.5	21.2	18.7	18.9	18.3	19.4	19.0
	24	$T\,(^0C)$	240	236	184	274	223	248	180	278
		$log\ Z$	17.9	18.0	19.8	16.9	18.4	17.6	19.9	16.8

4. Conclusions

The following conclusions were drawn from the present research work:

- ✓ In the domain of experiments conducted in the present study, the maximum heat input was observed at 21 mm shoulder diameter. The rate of heat generation was higher in TE to RS-LE and lower in RS-LE to TE. At 21 mm shoulder diameter, the higher values of total tool torque (T_t) and axial load (F_z) were noted, i.e., 56.49 N-m and 6.28 kN, respectively.
- ✓ Around the tool shoulder periphery in a clockwise direction, the calculated log Z and flow stress values were higher on RS and lower in the zone of AS to RS-LE. This was because of cool and warm material conditions respectively. With the increasing shoulder diameter, the peak temperature increases and was higher for 24mm shoulder diameter in AS to AS-LE zone, i.e., 406 ^{0}C.
- ✓ The log Z values were lower on AS, and the temperature was higher on the zone of AS to RS-LE clock-wise direction this shows that the material was plasticized in the zone of AS to RS-LE and transported around the tool pin thereby depositing with forging action in the TE.

References

[1] Threadgill P L, Leonard A J, Shercliff H R and Withers P J 2009 *Int Mater Rev* **54** 49–93
[2] Mishra R S and Ma Z Y 2005 *Mater Sci Eng* **50** 1–78
[3] Schmidt H and Hattel J 2005 *Sci Technol Weld Join* **10** 176–86
[4] Nandan R, Debroy T and Bhadeshia H K D H 2008 *Prog Mater Sci* **53** 980–1023
[5] Schmidt H, Hattel J and Wert J 2004 *Model Simul Mater Sci Eng* **12** 143–57
[6] Zhang Z, Liu Y L and Chen J T 2009 *Int J Adv Manuf Technol* **45** 889–95
[7] Zhang Z, Wu Q and Zhang H W 2014 *Trans Nonferrous Met Soc* (English Ed. **24)** 3293–301
[8] Elangovan K and Balasubramanian V 2008 *Mater Des* **29** 362–73
[9] Akinlabi E and Akinlabi S 2011 *Adv Mat Res* **5** 299–300
[10] Kumar K and Kailas S V 2008 *Mater Sci Eng A* **485** 367–74
[11] Thomas W and Nicholas E 1998 *Mater Des* **18** 269–73
[12] Ramanjaneyulu K, Reddy G M and Rao A V 2014 *Trans Indian Inst Met* **67** 769-780

[13] Arora A, De A and Debroy T 2011 *Scr Mater* **64** 9–12
[14] Mehta M, Arora A, De A and T Debroy 2011 *Metall Mater Trans A Phys Metall Mater Sci* **42** 2716–22
[15] Zhang J, Shen Y, Li B, Xu H, Yao X, Kuang B and Gao J 2014 *Mater Des* **60** 94–101
[16] Reynolds A P 2008 *Scr Mater* **58** 338–42
[17] Mugada K K and Adepu K 2018 *Int J Adv Manuf Technol* **99** 1553–66
[18] Mugada K K and Adepu K 2018 *J Inst Eng Ser C 1 -8*
[19] Mugada K K and Adepu K 2018 *J Manuf Proc* **32** 625–34
[20] Mugada K K and Adepu K 2018 *IOP Conf Ser Mater Sci Eng* **330** 1-7
[21] Zienkiewicz O C and Cormeau I.C. 1972 *Arch Mech* **24** 873-889
[22] Sheppard T and Jackson A 1997 *Mat Sci Tech* **13** 203-9

IOP Conf. Series: Materials Science and Engineering **474** (2019) 012013 doi:10.1088/1757-899X/474/1/012013

Effects of microfiller on the rheological behavior and structure of SBS modified bitumen

L M Zhou[1], T Xia[1,*], J H Xu[2,3,], Y P Qin[1], W Q Chen[1] and J F Dai[1,3]**

[1]College of Material Science and Engineering, Chongqing Key Laboratory of Mold Technology, Chongqing University of Technology, Chongqing 400054, China
[2]College of Polymer Science and Engineering, Sichuan University, Chengdu 610065, China
[3]Chongqing Zhixiang Paving Technology Engineering Co., Ltd., China Merchants Chongqing Communications Technology Research and Design Institute, Chongqing, 401336, China

*E-mail: xiatian@cqut.edu.cn, **E-mail: xujianhui@cmhk.com

Abstract. Rheological behavior of styrene-butadiene-styrene (SBS) modified bitumen with the presence of micro glass bead (GB) is investigated. It is found that either the GB fraction or size can influence the rheological property of bitumen/SBS/GB blends. Phase separation structures in all the modified bitumen systems keep a droplet-matrix type due to the sizeable compositional asymmetry between SBS-rich phase and bitumen-rich phase. The difference in the interfacial area fraction, obtained by Image J analysis reflects the effect of GB size on the rheological property. Further study on the rheology-structure correlation is put special attention on bee structure (catana phase) measured by AFM. It is found that the ratio of specific amplitude to the wavelength (A_0/λ_0) of a bee-like structure may be a valid indicator of stiffness for the modified bitumen concerning the effect of GB size difference on the rheological property. When the chemical composition is the same, the lower value of A_0/λ_0 indicates the modified bitumen with higher stiffness.

1. Introduction

Styrene-butadiene-styrene (SBS) modified bitumen has been widely investigated regarding their compatibility, microstructure, rheological property induced by phase separation [1-4]. Our previous work [5] found that with an increase of SBS addition, the phase separation mechanism would shift from nucleation-and-growth to viscoelastic phase separation, leading to a transition of phase structure from droplet-matrix structure to a viscoelastic network of SBS-rich phase. This change was determined by both compositional asymmetry and dynamic asymmetry between SBS and bitumen, which originated from the competition between SBS and asphaltenes to absorb light components [6].

Recently, the effect of silica nanoparticles, with different surface nature on the rheology and structure of bitumen/SBS blends has also been investigated [7]. It was found that the addition of SiO_2 nanoparticles improved the rutting resistance of bitumen/SBS blends, which was revealed by complex modulus enhancement and phase angle reduction. This improvement, in agreement with literature reports [8, 9], was strictly related to the distribution of nanoparticles and their surface nature. The blends modified by hydrophobic SiO_2 exhibit the better rheological property due to their more preferential attraction with bitumen, which was revealed by the apparent decrease in bee-like structure size formed in asphaltenes. Bee-like structures are characterized by elongated regions with alternating higher and lower bands in the surface topography of bitumen, which are surrounded by a flat area [10, 11]. The structure is also referred to a catana phase, which is immediately surrounded by a peri phase and is isolated from the para phase by the peri phase [12, 13]. The cause for the bee-like structure formation has been attracted much attention, and various possible mechanisms were proposed. For example, the bee structure was taken as a surface wrinkling, which was the result of phase separation and differential contraction due to differences in the elastic modulus of the material phases [11], or a chemically and mechanically distinct surface film that separated from the bulk bitumen [14]. In other reports, the formation of bee structure is due to the crystallization of wax or waxy molecules in asphaltenes [15, 16].

Compared with nanoparticles, the particles with microscale have lower surface free energy whose aggregation, mobility and affinity to other molecular chains will be relatively weaker. This difference would influence the rheology and structure of bitumen/SBS blends, accordingly. Besides, the effect of microparticles with a different size on the rheology and microstructure of bitumen/SBS blends is also worthy of study, which is the primary purpose of this work.

2. Materials and experiments

2.1 Materials and sample preparation

Bitumen (AH-70#) used in this paper was provided by the SK company (Korea). The fraction of saturates, aromatics, resins and asphaltenes in this bitumen were 5.9, 59.8, 19.1 and 15.2% respectively, measured by thin-layer chromatography with flame ionization detection (TLC-FID) [5, 17]. SBS (1401) was supplied by Yueyang Baling Company, Sinopec, which was a linear structure polymer containing 40% styrene by weight with M_w of 10^5 g/mol. The SBS/bitumen blends were prepared by continuous mixing of the components in toluene. The neat blend in this study is bitumen/SBS (100/4). The fraction of glass beads is 1 or 4 wt.% of bitumen. Three kinds of glass beads with a size of 2.5, 10 and 15 μm were used as microfillers without any surface modification. The samples of bitumen modified by SBS and glass beads are abbreviated as B-X%-Yμm, where X denotes the loading fraction of glass beads into bitumen/SBS blend, and Y indicates the size of the glass bead. Samples for morphological observation and rheological measurements were prepared by casting the solution on a glass slide and PTFE plate, respectively. After the removal of solvent in a desiccator and vacuum oven alternatively, these samples can be tested.

2.2 Experiments

The rheological test was performed on a rheometer (AR 1500ex, TA Instruments, USA) with 25 mm parallel plate with the fixed gap of 1 mm. A dynamic frequency sweep was carried out at 60 °C with a sweep frequency range of 0.01~100 rad/s. The strain in both kinds of experiments was kept as 1 % which is low enough to ensure the location of the linear viscoelastic region. It was necessary to note that all the rheological samples were annealed at 160 °C for 3 h, whose temperature and time was close to the usual mixing process for polymer modified bitumen in practical production [2,18].

The sample on a glass slide was annealed for 60 min under 160 °C before structure observation. The optical microscope (BK-POL-TR) was supplied by Optec Company in China. Atomic force microscopy (NX10, Park) was applied for the nano- and micro-structures observation. Height and Z-images were scanned using an etched silicon probe with a non-contact scanning mode. The resonance frequency of cantilever was 286 kHz. The sample was oscillated in the z-direction as the sample was scanned line by line at a rate of 0.5 Hz perpendicular to the cantilever. Every image was built up by 256*256 pixels each originating from the force curve evaluation.

3. Results and discussion

As shown in figure 1 (a), once the addition of glass beads into bitumen/SBS (100/4) blend, the complex modulus (G^*) is improved in the entire frequency sweep region, no matter how large the particle size is. The G^* of B-1%-15μm almost overlaps with that of B-1%-2.5μm, revealing their similar rutting resistance ability. The G^* at 10 rad/s of the neat blend, B-1%-2.5μm, B-1%-10μm, and B-1%-15μm systems are 29300 Pa, 46250 Pa, 52110 Pa, 44040 Pa, respectively. As a whole, the G^* of B-1%-10μm sample shifts slightly upward, as compared to B-1%-15μm, indicating the best rutting resistance brought by 10μm glass bead.

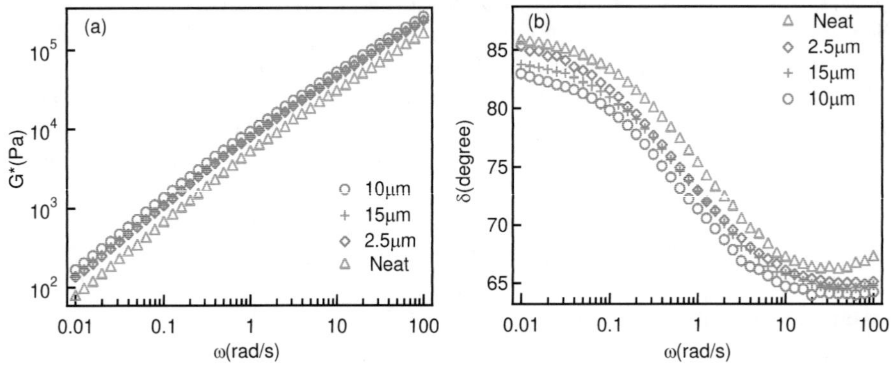

Figure 1. Variation of (a) complex modulus G* and (b) phase angle with the frequency of bitumen/SBS/GB (100/4/1) with different GB sizes under 60 °C. The strain is kept at 1%.

The phase angle δ comparison in figure 1(b) reflects the ability of resistance to permanent deformation of materials. The δ value displays a contrary order to G^* that δ (B-1%-10μm) < δ (B-1%-15μm) < δ (B-1%-2.5μm) < δ (Neat). As we know, the phase angle reveals the viscoelastic balance of the material behavior, which is defined as the phase difference between stress and strain in an oscillatory test. It is noticed that the δ of all the samples are located in the region of δ >45° within the whole frequency range. This result suggests the dominant viscous response of all the SBS modified bitumen systems since the predominant viscous bitumen matrix and only 4wt.% SBS elastomer addition. The relatively lowest δ value of B-1%-10μm reflects the best resistance to rutting or deformation, consistent with complex modulus data reveals.

Based on figure 1, the addition of only 1wt.% GB into SBS modified bitumen blends can enhance their rheological property slightly, in spite of the particle size of GB. To further improve this property, the amount of GB addition is increased to 4 wt.%.

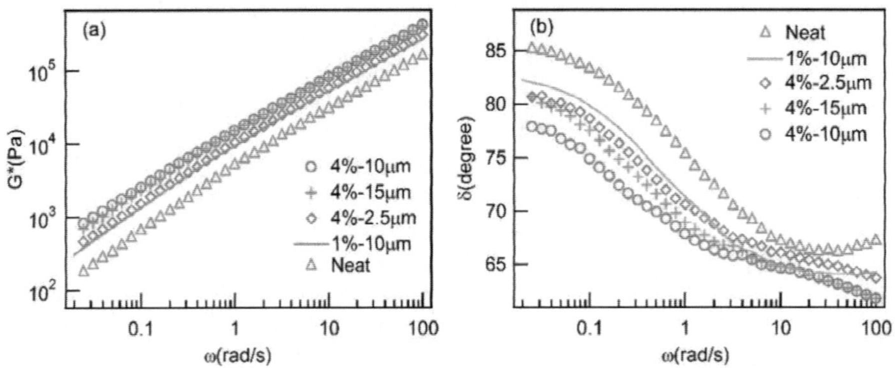

Figure 2. Variation of (a) complex modulus G* and (b) phase angle with a frequency of bitumen/SBS/GB (100/4/4) with different GB size under 60 °C. The strain is kept as 1%.

As can be seen in figure 2 (a), the G^* of B-4%-10μm has almost improved one magnitude compared with a neat blend. All the G^* curves of bitumen/SBS/GB (100/4/4) blends locate above the B-1%-10μm (solid line), whose G^* value is the highest among the bitumen/SBS/GB (100/4/1) blends, according to figure 1(a). This comparison implies that the GB fraction rather than GB size plays a dominant role in complex modulus, which reflects the strength of materials. The δ variation has a contrary tendency to G^*. Combined with figure 2 (a) and (b), the highest G^* and lowest δ of B-4%-10μm blend suggest that 10 μm has the best modification effect regarding resistance to rutting and permanent deformation.

It is known that rheological behavior depends on the structure of materials. Morphology evolution of bitumen/SBS/GB blends observed by optical microscopy after annealing under 160 °C for 60 min is shown in figure 3.

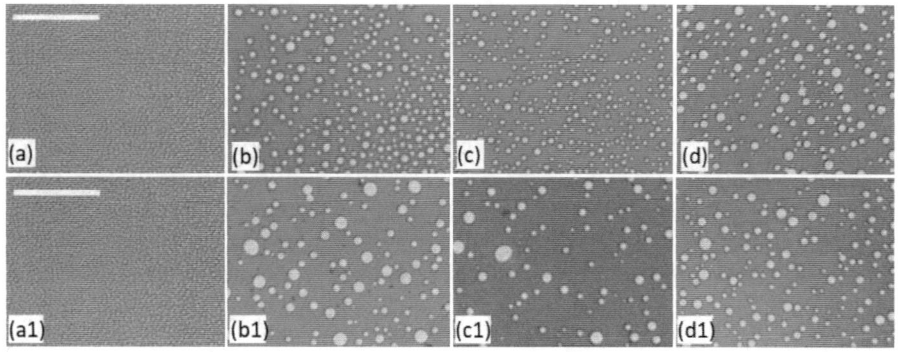

Figure 3. Morphologies of (a-a1) bitumen/SBS (100/4) blend, (b-d) bitumen/SBS/GB(100/4/1) blends, (b1-d1) bitumen/SBS/GB(100/4/4) blends. The GB size in (b-d) or (b1-d1) are 2.5μm, 10μm and15μm respectively. The scale bar denotes 50 μm.

All the blends form a typical droplet-matrix structure in which the white and dark regions represent SBS-rich dispersed phase and bitumen-rich matrix, respectively. Upon the addition of 1% GB into the neat blend, the droplet size increased but the number of droplets reduced. According to figure 3, the area fractions of SBS-rich phase for (a-d) analyzed by Image J via using the "Make Binary" option are 46%, 15%, 12%, and 16%, respectively. Thus, although the average size of droplets in bitumen/SBS/GB is larger than that of the neat blend, the total interfacial area of droplets after phase separation is reduced after the addition of GB with different particle size, especially for B-1%-10μm. From this point of view, the compatibility between bitumen and SBS gets better after the incorporation of 1% GB particles. Besides, the excellent distribution of GB fillers in bitumen can also make a remarkable contribution to the rheological behavior. Consequently, these synergistic effects lead to a gradual improvement on rutting resistance of bitumen/SBS blends by the addition of GB, which is embodied by G^* increase and δ decrease in figure 1.

The area fraction of SBS-rich phase for figure 3 (a1-d1) analyzed by Image J via the "Make Binary" option are 46, 12, 8, and 12%, respectively. This contrasting result is consistent with 1 wt.% GB filled bitumen/SBS (100/4) blends. That means the similar effect of GB particle size on the microstructure, and rheological behavior revealed by figure 1 and figure 2 either the fraction of GB is 1 wt.% or 4 wt.%. It should be noted that no visible GB particles are observed on the surface of the blends either in figure 3. This may be attributed to the relatively higher density of GB particle due to their microscale size, which leads them to sink to the bottom which is covered by the bitumen-rich and SBS-rich phases.

Besides the microstructure observed by optical microscopy, the smaller structures of bitumen/SBS/GB were also investigated by AFM. Regarding the same effect of GB size on rheological properties and microstructure, only bitumen/SBS/GB (100/4/1) systems are studied by AFM. As shown in figure 4 (a-d), lots of bee-like structures (catana phase) are observed. The bee structures in each sample are not completely symmetric. Some of the bees are relatively longer and bent, as denoted by the red circle. These bent bees look like composed of two neighbor bees whose arrangement directions has a minor angle. This appearance of bent bees is probably caused by the existence of droplet phase structure induced by phase separation between bitumen and SBS, which results in a large amplitude of surface at some region and accordingly influence the bee structure growth.

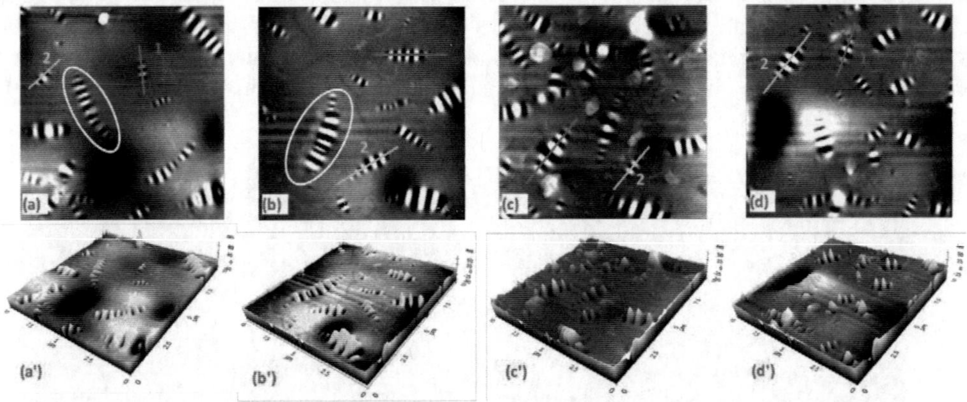

Figure 4. AFM topographic (10μm*10 μm) images of (a) bitumen/SBS (100/4) blend, (b) B-1%-2.5μm, (c) B-1%-10μm and (d) B-1%-15μm blends. The bottom images are the corresponding 3D AFM results.

The 3D structure in figure 4 (a'-d') implies that the bee structure consists of tubers [19] and chases, alternatively. It was reported that [20] the bees are caused by the surface wrinkling due to buckling of the bee laminate since this phase is stiffer and does not contract as much as the continuous matrix phase during cooling from melt temperatures. The bee laminate phase is phase separated from the bulk of the bitumen and blooms to the surface of the bitumen since the wax has a lower surface free energy.

Due to the size heterogeneity, it is difficult to compare the average size of the bee structure of bitumen/SBS blends before and after the addition of GB particles only based on figure 4. However, the number of bees for GB modified bitumen/SBS blends in figure 4(b)-(d) is more than that of the neat blend in figure 4 (a). This may be the reason for the higher complex modulus of bitumen/SBS/GB blends compared with the neat blend.

The bee structure shown in figure 4 accords with the hypothesis that bee structure is wrinkled, a stiff thin film of crystallized paraffin wax supported by a compliant substrate [11, 21]. Based on the relevant models proposed by Huang et *al.* [22-25], if a thin stiff film is subjected to a lateral compression strain $_{load}$ beyond a critical compressive strain ε_c on a soft substrate, it tends to buckle into sinusoidal wrinkles with specific wavelength λ_0 and amplitude A_0 to reduce the total elastic strain energy. The characteristic factors satisfy the following equations:

$$\lambda_0 = \frac{\pi h}{\sqrt{\varepsilon_c}} \quad (1)$$

$$A_0 = h\sqrt{\frac{\varepsilon_{bee}}{\varepsilon_c} - 1} \quad (2)$$

$$\varepsilon_c = \frac{1}{4}\left(\frac{3\overline{E_s}}{\overline{E_f}}\right)^{2/3} \quad (3)$$

where h represents the thickness of bee structure rather the whole thickness of blend film, ε_{bee} denotes the native compressive strain which could cause the wax thin film to wrinkle into bee structure, $\overline{E_f}$ and $\overline{E_s}$ are the plane-strain modulus of the top thin film with the existence of bee structure and that of the substrate, respectively. The substrate material is thought to be a para phase and probably is the major material component of bitumen, which controls its bulk properties.

Further quantitative analysis of the bee structure presented in figure 4 (a-d) is given in figure 5. Taking curve 1 in figure 5 (a) as an example, the specific amplitude A_0, wavelength λ_0 of sinusoidal wrinkles can be derived, and the values are listed in table 1.

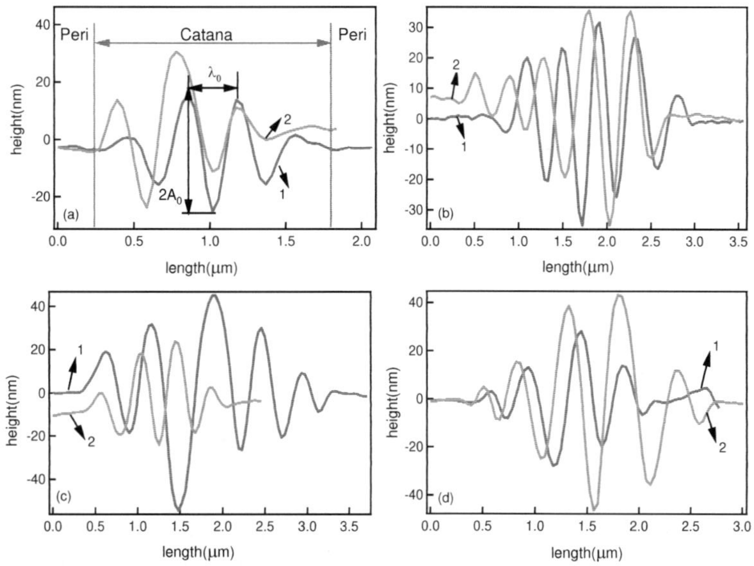

Figure 5. The characteristic height lines of two bees of (a) neat bitumen/SBS (100/4) blend, (b) B-1%-2.5μm, (c) B-1%-10μm and (d) B-1%-15μm corresponding to figure 4(a-d).

Table 1. Characteristic values of bee structure obtained from figure 5.

Samples	Lines	λ_0 (nm)	A_0 (nm)	A_0/λ_0
Neat	No.1	320	20	0.063
	No.2	420	27	0.064
B-1%-2.5μm	No.1	378	33	0.087
	No.2	444	35	0.078
B-1%-10μm	No.1	826	50	0.060
	No.2	394	24	0.060
B-1%-15μm	No.1	518	28	0.054
	No.2	482	45	0.092

Table 1 reveals that the height of selected bees ($2A_0$) is in the range of 40~100 nm. The previous study reported that the average height of bee structure in base bitumen is 22 ~ 85 nm [26]. This increased height of bees reveals the SBS/GB modification effect on bitumen. The characteristic size comparison has a strong dependence on the bee selection due to its size heterogeneity. To avoid this effect, it is better to compare the ratio of A_0/λ_0 instead of their value. According to table 1, the G-1%-10 μm blend has the minimum average value of A_0/λ_0. By combining equations (1) and (2), one can obtain the following equation:

$$\frac{A_0}{\lambda_0} = \frac{\sqrt{\varepsilon_{bee} - \varepsilon_c}}{\pi} \tag{4}$$

Based on equation (4), smaller A_0/λ_0 values correspond to a smaller difference between ε_{bee} and ε_c. The following discussion only emphasizes the comparison of B-1%- Yμm blend systems, to avoid the chemical component difference between the binary and ternary modified bitumen brought by GB. The ε_{bee} is thought to arise naturally from the curvature elastic strain of a paraffin single crystal [21,27]. Given that ε_{bee} for B-1%-Yμm systems have approximate value due to their same chemical components, there would be a positive correlation between ε_c and $\overline{E_s} / \overline{E_f}$ based on equation (3). Then it can be deduced that the A_0/λ_0 would decrease as $\overline{E_s}$ increases. The substrate is assumed to be a para phase and probably is the major material component of bitumen, which controls its bulk properties [24]. Thus, it can be deduced that smaller A_0/λ_0 values correspond to the SBS/GB modified bitumen with higher stiffness.

According to table 1, the average values of A_0/λ_0 are 0.060, 0.073, and 0.083 for B1%-10μm, B-1%-15μm, and B-1%-2.5μm, respectively. Accordingly, the stiffness values of modified bitumen $\overline{E_s}$ can be ranged in the decreasing order as follows: B-1%-10μm > B-1%-15μm > B-1%-2.5μm. This conclusion is consistent with the resistance to deformation of B-1%-Yμm materials revealed by phase angle δ as shown in figure 1(b). To exclude the randomness of this result, we also change the scanning region for each sample and measure the other bee structures. The measurements still reveal the same trend of A_0/λ_0 comparison for the three kinds of bitumen/SBS/GB blends. This comparison suggests that the GB size can influence the characteristic size of bee structures, which reflects a difference in stiffness of B-1%-Yμm systems. The origin of structure difference in SBS modified bitumen induced by GB particles with different size is still under study. This may be associated with components interaction, phase separation behavior, and distribution or aggregation of microfillers.

4. Conclusions

In this study, both effects of GB fraction and particle size on the rheological behavior of bitumen/SBS (100/4) blends were studied. It is found that either GB fraction or GB size can influence the rutting resistance reflected by G^* and δ. As for the GB with a particles size of 10 μm is the best candidate modifier for the bitumen/SBS blends regarding rheological property, the phase separation structure and bee-like structure observed by optical microscopy and AFM respectively are investigated. Further qualitative and quantitative investigations on the bee structure elucidate the correlation between the ratio of specific amplitude to the wavelength of (A_0/λ_0) the bee structure and the stiffness of bitumen materials ($\overline{E_s}$). Roughly speaking, a lower value of A_0/λ_0 indicates the modified bitumen with higher stiffness. This conclusion is limited in the modified bitumen with the same chemical component just only difference on the GB particle size. This stiffness reflected by bee structure information is consistent with the resistance to deformation revealed by the phase angle. Thus, the A_0/λ_0 may be an effective indicator of stiffness instead of individual A_0 or λ_0, especially for the modified bitumen with multicomponent, in which the bee structure is liable to be influenced by many factors, such as phase separation, modifier distribution and so on.

Acknowledgments
We are grateful for the financial support from the National Natural Science Foundation of China (51403026).

References

[1] De Carcer Í A, Masegosa R M, Viñas M T, Sanchez-Cabezudo M, Salom C, Prolongo M G, Contreras V, Barceló F and Páez A 2014 *Constr. Build. Mater.* **52** 245-52

[2] Fu H, Xie L, Dou D, Li L, Yu M and Yao S 2007 *Constr. Build. Mater.* **21** 1528-33
[3] Larsen D O, Alessandrini J L, Bosch A and Cortizo M S 2009 *Constr. Build. Mater.* **23** 2769-74
[4] Zhang H, Jia X, Yu J and Xue L 2013 *Constr. Build. Mater.* **40** 224-30
[5] Xia T, Xu J, Huang T, He J, Zhang Y, Guo J and Li Y 2016 *Constr. Build. Mater.* **105** 589-94
[6] Airey G D 2003 *Fuel* **82** 1709-19
[7] Xia T, Zhou L, Lan S and Huang X 2017 *Constr. Build. Mater.* **153** 957-64
[8] Fisher M L, Colic M, Rao M P and Lange F F 2001 *J. Am. Ceram. Soc.* **84** 713-8
[9] Blanco I, Bottino F A, Cicala G, Cozzo G, Latteri A and Recca A 2015 *Polym. Compos.* **36** 1394-400
[10] Loeber L, Sutton O, Morel J, Valleton J M and Muller G 1996 *J. Microsc.* **182** 32-9
[11] Lyne Å L, Wallqvist V, Rutland M W, Claesson P and Birgisson B 2013 *J. Mater. Sci.* **48** 6970-6
[12] Masson J F, Leblond V and Margeson J 2006 *J. Microsc.* **221** 17-29
[13] Masson J F, Leblond V, Margeson J and Bundalo-Perc S 2007 *J. Microsc.* **227** 191-202
[14] Hung A M and Fini E H 2015 *RSC Adv.* **5** 96972-82
[15] Zhang H, Wang H and Yu J 2011 *J. Microsc.* **242** 37-45
[16] Zhang H, Yu J, Wang H and Xue L 2011 *Mater. Chem. Phys.* **129** 769-76
[17] Feng Z, Yu J and Kuang D 2013 *Pet. Sci. Technol.* **31** 113-20
[18] Liang M, Xin X, Fan W, Luo H, Wang X and Xing B 2015 *Constr. Build. Mater.* **74** 235-40
[19] Wua S P, Pang L, Mo L T, Chen Y C and Zhu G J 2009 *Constr. Build. Mater.* **23** 1005-10
[20] Laurell A L, Wallqvist V, Rutland M W, Claesson P and Birgisson B 2013 *J. Mater. Sci.* **48** 6970-6
[21] Pauli T, Grimes W, Beiswenger J and Schmets A J 2014 *J. Mater. Civ. Eng.* **27** C4014001
[22] Huang R 2005 *J. Mech. Phys. Solids* **53** 63-89
[23] Im S and Huang R 2005 *J. Appl. Mech.* **72** 955-61
[24] Huang R and Im S H 2006 *Phys. Rev. E* **74** 026214
[25] Jiang H, Khang D Y, Song J, Sun Y, Huang Y and Rogers J A 2007 *Proc. Natl. Acad. Sci.* **104** 15607-12
[26] Masson J F, Leblond V and Margeson J 2006 *J. Microsc.* **221** 17-29
[27] Toda A, Okamura M, Hikosaka M and Nakagawa Y 2005 *Polymer* **46** 8708-16

IOP Conf. Series: Materials Science and Engineering **474** (2019) 012014 doi:10.1088/1757-899X/474/1/012014

Automatic device elaboration for grouting bag setting under free spanning of submarine pipelines

B Zhang[1], Z Wang[1,*] and T Wang[2]

[1]College of Mechanical and Electrical Engineering, Harbin Engineering University, Harbin 150001 China
[2]School of Mechanical Engineering, Hebei University of Technology, Tianjin, 300401 China

*E-mail: wangzhuo_heu@hrbeu.edu.cn

Abstract. A new grouting support method is proposed to solve the problem of divers. Those dispose of the free spanning submarine pipelines by the traditional grouting support method, which may have low efficiency and security. The equipment, which is designed to replace the divers to operate by setting grouting bags under the spanning pipes automatically, reduces the difficulty of the divers' operation and the working time in the water, and makes the working process more automatized. Based on the wave theory, the wave load on the grouting support and the frictional force of the seabed on the support are analyzed. The deformation of the grout bag is calculated by the finite element method, to ensure that the stiffness is sufficient to meet the strength requirements. After the mechanical and electrical system design of the equipment and the simulation experiment, it was proved that the equipment operation under work conditions is reliable, and the proposed modification is effective.

1. Introduction

Free span of pipelines can cause vortex-induced vibration of pipes, which may lead to their fracture [1]. This is one of the main factors threatening the safety of submarine pipelines. At present, methods of disposing of spanning pipes can be subdivided into four categories: landfill treatment, support management, blocking flow and smooth siltation technology, along with a pipe. For the sake of technology, the engineers used the falling stone method to support the spanning pipes [2] in the Ormen Lange project in Norway. The treatment used in this project and this innovative type method was introduced in [3-4].

The technology consists of many specific methods, such as sandbag support method, steel column support method, grouting bags support method, etc. The sandbag support method was utilized to deal with the spanning pipes in the oilfield of the Gulf of Mexico and the North Sea [5]. In the Canyon Express oilfield project in the Gulf of Mexico and the Ceiba Field project in West Africa, a steel column support method was used to manage the pipe-suspension [6]. Grouting support method is to set grouting bags under the spanning part of the pipe and then inject grout into the grouting bag through grouting equipment on the working ship. After the solidification of cement in the grouting bag, the bag will provide reliable support to the pipe [7].

In the United States, Britain, Mexico, and other countries, engineers have used the bionic water plants method in the pipe spanning project [8]. The engineers have used steel structure support method along with bionic water plants method to dispose of the spanning of pipes in the platform numbered CB35 in

Cheng island oilfield [9]. Guide plate method has been used in Green Canyon Block oilfield in the Gulf of Mexico and North Sea [10]. The floating curtain blocking method was invented by Shanghai Jiao Tong University, which has been proved to be feasible to manage the spanning pipes [11].

After a comparison of each method, support management technology is advantageous for being easy to implement and provide stability. The grouting support method has better adaptability to the seabed, and it can offer more extended support to the pipe. As a result, the method has the shortcoming of low operating efficiency and high risk to divers. The research is on to improve the traditional grouting method. The focal point of the new type of grouting method is to design equipment which can fulfill the task of setting grouting bags automatically.

The suspending grouting bag treatment device is designed for the submarine pipeline in this paper which is different from other equipment. In this method, 5-6 grouting bags are in water every time and processing 5-6 suspending points in turn. Moreover, the specially designed laying grouting bag mechanism can realize the automatic laying of the grouting bag underwater.

In this paper, the load of the grouting support in the working condition is calculated through the wave theory to analyze its stability. The deformation of the grouting bag is analyzed by the finite element method to check its safety. After the analysis of the theory, the authors have prepared the design of the equipment and built a prototype machine to do the experiment.

2. Brief introduction and feasibility for the analysis of the new type of grouting method

2.1. Introduction of the scheme

The purpose of this paper is to modify the traditional grouting support method. According to the statistics from the investigation of projects about spanning pipes, these are finished before the general height of the suspension, 0.6~2.0 m, and the maximum depth of water about 60 m. The equipment is the critical point of this work, and it needs to be reliable in the working conditions.

Considering the situation that the grouting support may sink with the seabed movement, so the engineers need to inject the cement into the grouting bag for the second time. To increase the contact area of the pipe and the support, the A-shaped structure is selected to the outline of the bag. And the sandwich layer at the bottom of the bag is designed for the second injection.

2.2. Reliability analysis of grouting support

The impact of seawater on the grouting support is composed of wave and current load. The wave load produces a periodic horizontal impact on the grouting support. The current load produces a constant effect on grouting support.

Based on the shape of the grouting bag, the load of the bag can be calculated by the Morison equation. To facilitate the analysis, the following coordinate system was established, as is shown in figure 1.

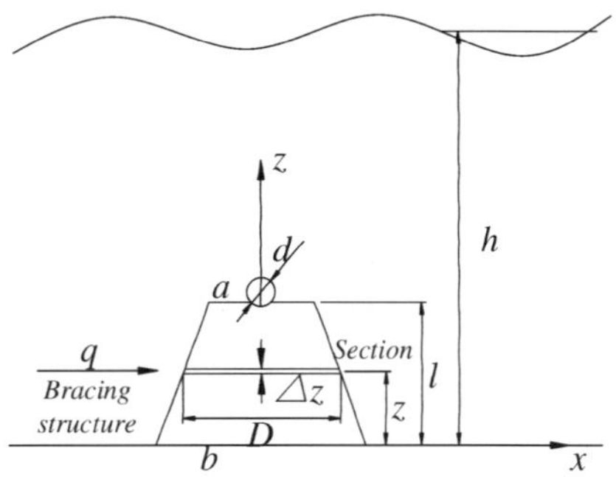

Figure 1. Schematic diagram of the impact of the wave to the support.

IOP Conf. Series: Materials Science and Engineering **474** (2019) 012014 doi:10.1088/1757-899X/474/1/012014

The horizontal wave force of single grouting support structure is calculated by the microelement method. It can be calculated via Eq. (1).

$$F = \int_0^l \frac{1}{2} C_D \rho u_x |u_x| \left[b + \frac{(a-b)}{l} z \right] dz + \int_0^l C_M \rho \left[b + \frac{(a-b)}{l} z \right]^2 \frac{\partial u_x}{\partial t} dz \quad (1)$$

where F is the horizontal force generated by the waves working on the grouting support; l is the height of the whole grouting support; C_D is the coefficient of the drag force generated by the waves; ρ is the density of seawater; u_x is the horizontal velocity of the particle of the wave at the position of microelement without the grouting support; a is the side length of the upper surface of the grouting support; b is the side length of the lower surface of the grouting support; z is the distance between the microelement and the surface of the seabed; dz is the height of the microelement; C_M is the coefficient of the inertial force; $\partial u_x / \partial t$ is the horizontal acceleration of the particle of the wave at the position of microelement without the grouting support.

Given the working conditions, the horizontal velocity and acceleration of the waves can be calculated by the Airy wave theory [12].

$$\begin{cases} u_x = \dfrac{\pi H}{T} \dfrac{\cosh kz}{\sinh kh} \cos(\theta) \\ a_x = \dfrac{2\pi^2 H}{T^2} \dfrac{\cosh kz}{\sinh kh} \sin(\theta) \\ \theta = kx - \omega t \end{cases} \quad (2)$$

where H is the height of the waves; T is the cycle time of the waves; h is the depth of seawater; θ is the phase angle of the waves; k is the number of the waves in certain time; x is the horizontal coordinate; ω is the angular frequency of the waves; t is the time of the motion of waves. Based on Eq. (2), the effect of each factor on the horizontal velocity and acceleration of waves is depicted in figure 2.

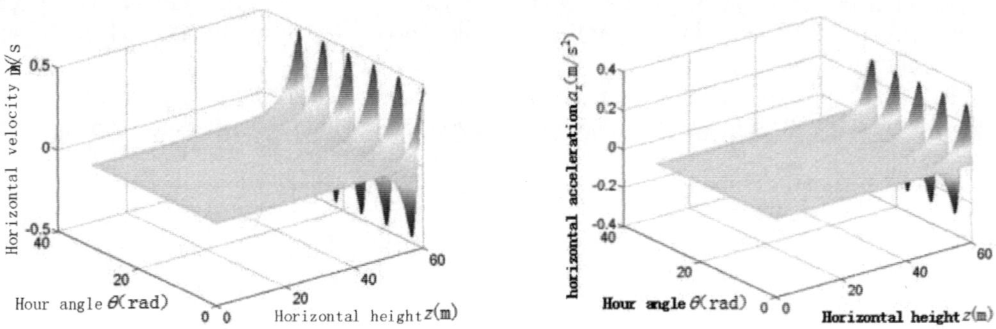

(a) **(b)**

Figure 2. Effect of phase angle and depth of water on the horizontal velocity and acceleration of the waves.

As shown in figure 2(a) and (b), the effect of phase angle and the depth of water on the horizontal velocity and acceleration of waves are increasing gradually with the increasing distance between the seabed and microelement. At the position of sea level, the speed and acceleration have the maximum value. In other words, with the increasing depth of water, the value of velocity and acceleration decreases gradually. When the depth of water reaches a certain value, the speed and acceleration have been close to zero, and then their impact can be neglected. From the analysis, the horizontal velocity and acceleration of waves are affected by many factors. In engineering projects, the depth of the working area is often known from pre-observation and engineers choose the maximum velocity and acceleration of the motion of waves to do designing calculation [13]. It is generally considered that the depth of water exceeds the length of waves by 0.5 times, the area can be regarded as a deep water area. Taking u_x and a_x into the Morison equation, the horizontal force generated by the waves working on the grouting support can be calculated by Eq. (3).

$$F = K_D A_D \cos\theta|\cos\theta| + K_M A_M \sin\theta \tag{3}$$

In Eq. (3), the coefficient can be calculated by the following equations from the combination of the Morison equation and the Airy wave theory.

$$\begin{cases} K_D = \dfrac{(b-a)l}{4} + \dfrac{a\sinh 2kl}{4k} + \dfrac{(b-a)\sinh^2 kl}{4k^2 l} \\ A_D = \dfrac{\pi^2 C_D \rho H^2}{2T^2 \sinh^2 kh} \\ A_M = \dfrac{2\pi^2 C_M \rho H}{T^2 \sinh kh} \\ K_M = \dfrac{a\sinh kl}{k} + \dfrac{(a-b)(1-\cosh kl)}{k^2 l} \end{cases} \tag{4}$$

Next, we calculate the derivation of time in the Eq. (3) and assume a zero value of the derivative, to find out the maximum horizontal load of the grouting support from the waves. The extreme value of wave load is affected by the depth of water, action time, and cycle time, the height of the wave and the height of grouting support. Considering the trend of each factor, the extreme value of wave load can be expressed as:

$$F_{\max} = K_M A_M \tag{5}$$

Choose the hydrodynamic parameters and take the parameters of the structural size of the grouting bag and the working conditions into Eq. (4) and Eq. (5), $C_M = 2.5$, $C_D = 2$, to find out the maximum value of F.

After finding out the load of the wave on the grouting support, it is necessary to find out the effect of the current load on the grouting support. The current load on the grouting support can be calculated still by the Morison equation.

$$F = \int_0^l \frac{1}{2} C_D \rho u_c^2 \left[b + \frac{(a-b)}{l} z \right] dz = \frac{C_D \rho u_c^2 (a+b) l}{4} \tag{6}$$

where A is the area of the surface perpendicular to the direction of ocean current; u_c is the horizontal velocity of the particle of the seawater. Based on Eq. (6), the load of the ocean current can be calculated, $C_D = 2$ $u_c = 0.6$ m/s. considering the calculation of the wave load, the relationship between the wave-current combined load and the height of grouting support is shown in figure 3.

As seen in figure 3, the wave-current load on the grouting support tends to be a constant with the increasing depth of water. To ensure the reliability of the support, the value of wave load and current load increases to the maximum appropriately in the calculation.

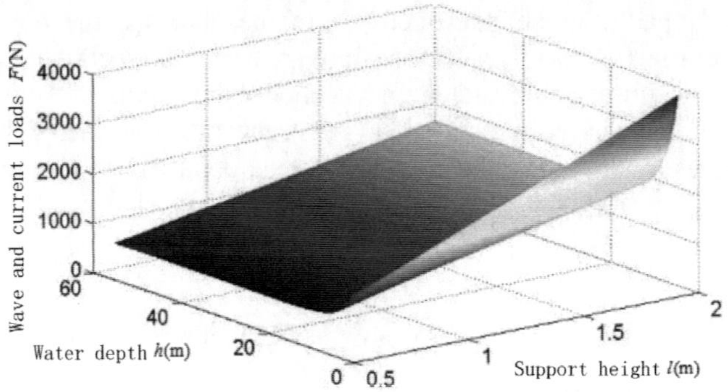

Figure 3. Variation of wave-current combined load with the depth of water and height of grouting support.

Assume that G_p is the gravity of the pipe; F_p is the force on the grouting support from the pipe; F_f is the friction on the support from the seabed; F is the wave-current on the support; F_1 is the component force in the direction along the inclined surface of the support. F_2 is the component force in the direction perpendicular to the inclined surface of the support; G_s is the gravity of the grouting support.
Then, G_p and F_p can be calculated by the microelement method as follows:

$$\begin{cases} G_P = \int_0^{\frac{\pi}{2}} p\frac{ad}{2}\cos\theta d\theta \\ F_P = \int_0^{\frac{\pi}{2}} p\frac{ad}{2}\sin\theta d\theta \end{cases} \tag{7}$$

where d is the diameter of the pipe.The maximum length of the spanning part of pipe is 10 m based on the investigation. The frictional force on the bottom of the grouting support from the seabed can be calculated via Eq. (8).

$$F_f = \mu(G_P + G_S) \tag{8}$$

where μ is the coefficient of friction of the sea bed.

The density of cement is 1.5, g/cm^3 and the coefficient of friction of the seabed is 0.1. The frictional force is calculated from Eq. (8). The surface of the seabed is rougher than the ideal situation, so the actual friction is larger than the calculation.

From the calculation, the frictional force on the grouting support is much larger than the horizontal combined load of wave-current. Therefore, the support will not slip on the seabed. The design pressure of the grouting bag is 0.8 MPa and is designed for working water depths of 60m. The grouting bag needs to be tough enough to bear the load of waves. The deformation of grouting bags without cement inside can be analyzed by the finite element method. The modulus of elasticity, the tensile strength, Poisson's ratio and density of the bag material (namely, composite nylon) are 6.2 GPa, 3.76 MPa, 0.35 and 1.15, respectively. The upper surface and the lower surface of the bag are fixed. The result of the simulation is shown in figure 4. It can be seen in figure 4 that the maximum deformation of the grouting bag is 163 mm and it is located in the middle of the side surface. The maximum strain is within the allowable range, and the grouting bag meets the engineering requirements.

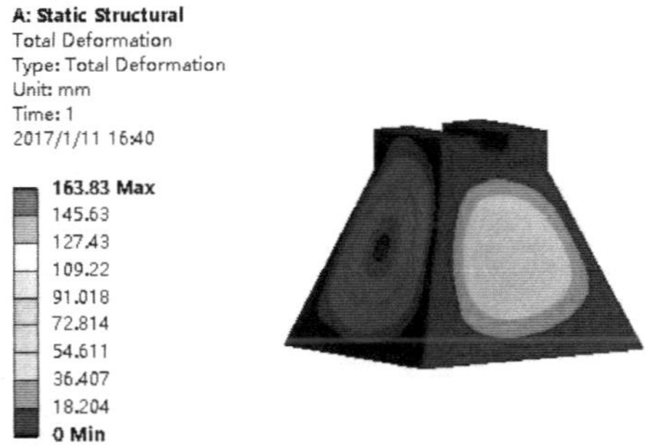

Figure 4. Total deformation of the outer surface of grouting bag.

3. The mechanical system design of the equipment

3.1. The structure and components of the device

The main aim of this study is the elaboration of three-dimensional design for the automatic bag-setting device. The latter consists of the frame, mechanism of propulsion, support; locking, and balance, controller, as well as other components. The container of the grouting bag is a drawer-shaped slider, which facilitates bag-setting and keeps their stability. The device can carry six sliders so that it can deal with six working positions in one process. The mechanism of propulsion is utilized to push the lowest slider out of the frame to the working position under the spanning part of the pipe. The whole structure is shown in figure 5.

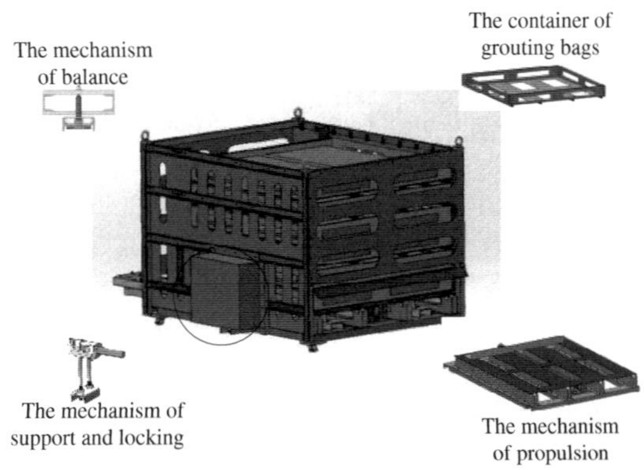

Figure 5. The structure of the automatic bag-setting device.

3.2. The working process of the device

The fabricated equipment is sent to the working position by the lifting equipment on the moving ship and the help of ROV (remote operated vehicle) and divers. In the beginning, the mechanism of support and locking was working, and the support part of the mechanism was holding up the sliders except the one at the bottom of the device. Then, the mechanism of propulsion was to push out the bottom slider and to set in the working position. After that, the mechanism of support and locking was working to let the rest of sliders fall down. The next slider was pushing out to be in position. Engineers on the working ship are supposed to inject cement to the grouting bag through the pump and pipelines. The grouting bag is expanding while the bag is filled with cement. When the support is formed, the diver will cut off the pipeline of transporting cement and the device will be taken to another working place. When the sliders are running out, the process is over, and the engineers will bring back the device. The whole work process can be expressed by the figure 6, where (a) shows that the device reaches the working position, (b) the device is setting grouting bag, (c) The final effect of the treatment.

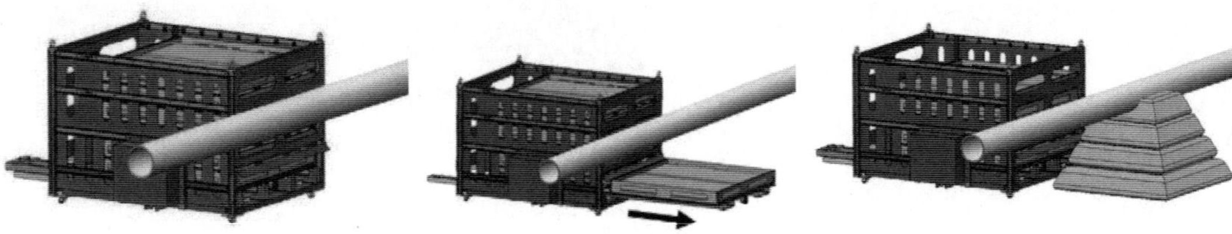

(a) (b) (c)

Figure 6. The working process of the device.

4. The prototype device test

To verify the motion of the device, a prototype was fabricated for the study. The experiment is intended to simulate the work condition of the device in the seawater to check its reliability. There are five steps in the experiment.

(1) The height of the pipeline was adjusted by the adjusting screw of the test application. It was used to simulate different heights of spanning pipes. The pipe is needed to be horizontal during the experiment.

(2) Each slider was set into the device through the crane, and the grouting bags were placed into the sliders.

(3) With the help of the controller to control the device for finishing the expected motion. All the mechanism work was undertaken in turn to push out the slider to the spanning position. The time of each motion of the device was recorded.

(4) The level of strength of the cement was 52.5, and the proportion of the slurry of the cement was 1.9. This kind of slurry of the cement had the shortest time of solidification. Then, the slurry was injected into grouting bag by the grouting pump.

(5) After recording the experimental data, the experiment was terminated.

The picture of the details of the experiment is illustrated in figure 7.

(a) (b)

Figure 7. Photos of the prototype test.

By the observation of the experiment, the return trip was faster than the work trip, and the motion of the mechanism of propulsion was steady. The grouting support could hold up the pipe deflection of about 2~3 cm, and the support was reliable to solve the problem.

5. Conclusions

This paper shows the improvement in the traditional grouting support mode of construction work by developing the automatic grout bag-laying device instead of the diver. This reduces the risk in the project and improves the efficiency. According to the wave theory, it can be concluded that the wave load on the grouting support is less than the frictional force of the seabed on the support, and the grouting support is reliable for the submarine-suspended pipeline treatment project. Through the finite element method, it can be concluded that the deformation of the grout bag is elastic under the wave load, and sufficient toughness can meet the strength requirement. It was experimentally verified that the motion of the device was stable and reliable, and it could efficiently support the suspended pipeline to ensure the stability of the pipeline without breaking. Therefore, the scheme is feasible for engineering applications.

Acknowledgments

This paper was funded by NSFC contracts "Research on ultimate bearing capacity and parametric design for the grouted clamps strengthening the partially damaged structure of jacket pipes, Contract number 51879063" and "Research of analysis and experiments of gripping and bearing mechanism for large-scale holding and lifting tools on ocean foundation piles, Contract number 51479043". The views of the authors do not necessarily coincide with the sponsors' policy.

Reference

[1] Fumes B 2003 *Ocean Eng.* **30** pp. 1553-1577
[2] Olso E, Nyhus B, Ostby E, *et al.* 2009 *ASME International Conf. on Ocean* pp. 841-850
[3] Wang Z T 2012 *Petroleum Engineering Construction* **6** pp. 21-23
[4] Chen Q Q 2008 *Flow and Sediment Transport Characteristics around Suspended Curtain of Harvesting Bedload* (Shanghai: Shanghai Jiao Tong University Press) pp. 25-42
[5] Feng H, Wang Y H, Wang J 2012 *Shipbuilding of China* **53** pp. 74-81
[6] Cao J 2011 *Shipbuilding of China* A**02** pp. 221-228
[7] Tang X X, Hu G H, 2011 *Oil-Gasfield Surface Engineering* **07** pp. 38-39
[8] Bertorelli S 1994 *13th Int. Conf. on Offshore Mechanics and Arctic Engineering* (New York) 12 pp. 3-10
[9] Liu J K, Zhang Z H 2009 *Petroleum Engineering Construction* **3** pp. 20-22
[10] Cheng L 2003 *Ocean Eng* **30** pp. 1595-1611
[11] Si F， Yu G L 2013 *Water Resources and Hydropower Engineering* 0**6** pp. 130-134
[12] Faltinse O M, Greco M and Landrini M 2002 *J Offshore Mech Arct* 1**24** pp. 97-103
[13] Lin Z, Teng B, Ning D Z, *et al.* 2011 *Port & Waterway Engineering* **2** pp. 11-15

Mechanical test and buckling analysis of X-type grouted clamps for the offshore platform structures

Z Wang[1], B Zhang[1,*] and T Wang[2]

[1]College of Mechanical and Electrical Engineering, Harbin Engineering University, Harbin 150001, China
[2]School of Mechanical Engineering, Hebei University of Technology, Tianjin 300401, China

*E-mail: wangzhuo_heu@hrbeu.edu.cn

Abstract. Grouting clamps, especially X-type ones, as typical jacket nodes, have been widely used in underwater engineering for repairing marine damaged jackets as a kind of reinforcement technology. This article is devoted to studying the X-type grouting clamp and conducted the relevant experiments to achieve the purpose of strengthening the strength of the jacket. At first, X-type grouting clamps were designed and processed. An experimental system was fabricated to carry out the experiment to test the bearing capacity which loads axial pressure on the X-type grouting clamp. Secondly, the stress and strain data are obtained experimentally. From the analysis of the experimental results, the X-shaped tube can be derived. The value of stress and displacement after reinforcement are reduced, as compared to earlier reinforcements, which indicates that the bearing capacity of the jacket is enhanced. Finally, the buckling analysis of the bearing capacity of the X-type grouting clamp is carried out by the finite element method, and the ultimate load of the grouting joint is obtained. It is proved that the X-type grouting clamp has the bearing capacity of a pipe joint.

1. Introduction

The whole world has caught the attention of development of marine oil and gas resources due to the depletion of land and oil resources [1-2]. The marine oil and gas resources depend on the structure of the offshore platform. The offshore platform is bulky, complex and very costly. The marine environment is distinct from the land structure such as sea currents, ice, breeze, and waves [3]. The long-term service of the jacket of offshore platform results in different types of damage forms in the harsh marine environment [4]. Due to the severe environment of the ocean and the damage of the jacket and other factors, the accidents of offshore platform frequently occur [5], which causes economic losses and adverse effects to the society. The maintenance and strengthening of an offshore platform play an essential role in ensuring the safe operation of the offshore platform. According to the different types of defects, different maintenance methods need to be used for maintenance and reinforcement. Samindi et al. summed up the marine maintenance and reinforcement technology [6]. In this paper, the grouted clamp is discussed, which is a mature high-tolerance simple and effective reinforcement technology [7].

Grouted clamp repair damaged jacket research began in1980s, straight pipe and K type joint grouted clamp has been applied [8], but the use of grouted clamp in marine engineering is still few in China. The grouted clamp involves only straight, *T*- and *K*-type pipes, the way of locking has the hydraulic telescopic type, the long and short bolt type, for other types (such as L/X/Y, etc.) grouted clamp structure is not found in the literature. Some scholars have used the grouted clamp to repair the offshore platform for some theoretical analysis, but most of the academic studies are focused on grouted sleeves casing , for example, the effect of shear bond shape on ultimate bond strength in grouted casing[9], the effect of temperature on the bearing capacity of the grouted sleeve [10], the force mechanism of grouted casing connection section, the relative displacement between the steel tube and the sleeve is measured by the axial load test [11-12]. The effect of cement compressive strength on the ultimate bearing capacity of the grouted sleeve is mainly obtained by experiment [13], and the theoretical results are also scarce.

In the light of the above findings, this study analyses the mechanical properties of X-type grouted clamp. Experimental results are compared to the results of the finite element buckling analysis of X-type to verify the reinforcement effect of grouted clamp on the offshore platform structure.

2. The experimental method of X-type grouted clamp

2.1. X-type grouted clamp device

To verify the reinforcement effect of different types of pipe joints, the X-type pipe was fabricated. Figure 1. is the state of the X-type pipe before and after the installation of the clamp. The diameter of X-type pipe joint is 50mm, the diameter of X-type grouted clamp is 76mm, and the wall thickness is 5mm. The X-type grouted clamp was filled with cement slurry of density, 1.85g/cm3, until the density of the outlet and inlet. Figure 1 (a), (b) and (c) shows X-type pipe before and after the installation of the clamp, and the grouting process of X-type grouted clamp, respectively. In the grouting process, there was no leakage of the cement slurry, which proved that the sealing effect of the structure was sound. The number of bolts and arranged location were reasonable. After grouting, the mechanical experiment was carried out.

(a) (b) (c)

Figure 1. X-type pipe clamp.

2.2. Mechanical experimental system

The experimental mechanical system of the grouted clamp is mainly composed of the press, X-type pipe, grouted clamp device, strain gauge and so on, as shown in figure 2.

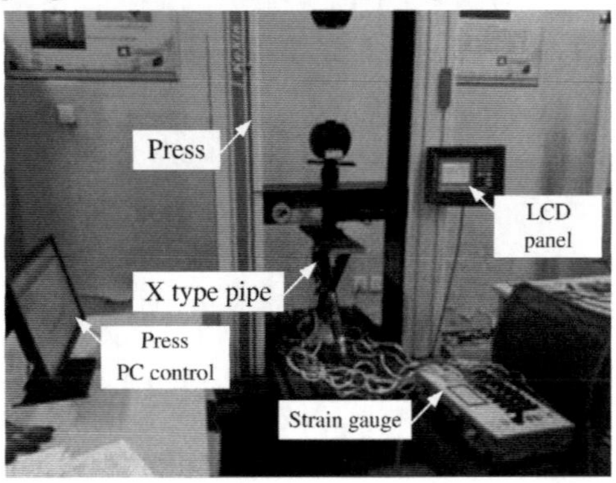

Figure 2. The composition of X-type pipe mechanics experimental system.

The yield limit is determined by the finite element simulation. The critical load corresponding to the yield limit of the X-type pipe is 2t, and that of X-type grouted clamp is 8t. The load should be controlled below the critical level to avoid the device damage. Given applied load range, the WDW3100 micro control electronic universal testing machine with the limit loading pressure of 10t was used. The computer software controlled the rise and fall of the moving beam, recorded and saved the experimental data in the real time scale, which where visualized on the LED display.

2.3. The mechanical experiment of the grouted clamp

Before the beginning of the mechanical experiment, the strain gauges were pasted. The X-shaped pipe joint has a diameter of 50 mm, the grouting clamp 76 mm, and the wall thickness is 5 mm. The height is 350mm, and the width is 400mm with an angle of 90°. The four branches of the root end of the X-type pipe were posted by the three pieces of biaxial strain gauge at 45° interval, two biaxial strain gauges were attached to the branch pipe with the distance of 80mm, and the biaxial strain gauge could measure the axial and circumferential strains. Position and number of strain gauges are shown in figure 3(a). Due to the lack of space for pasting strain gauges, their number was limited. To measure easily, the lead wire on the strain gauge was welded to the crocodile clip, and the latter was clamped in the position to be measured. Since no strain gauges could be pasted on the pipe outside surface, biaxial strain gauges were pasted on the surface of the X-type grouted clamp at the position shown in figure 3 (b).

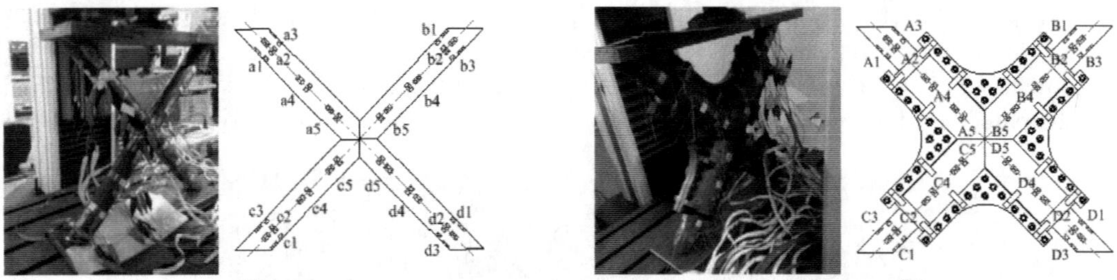

(a) (b)

Figure 3. Installation of position number of strain gauges.

To obtain a strain of the strain gauges accurately, it is necessary to compensate the strain gauge in parallel, in the strain gauge measuring circuit, as shown in figure 4.

Figure 4. Strain gauge compensation circuit wiring.

Because the size of X-type pipe end face was larger than that of the pressed chuck, the rectangular steel plates were placed above and under the X-type pipe, as shown in figure 3 on the left. In the software operation interface set, the initial applied load of 0.25t was raised with the interval of 0.25t until the load level of 1.5t was reached. The particular force and strain data were recorded and listed in table 1. The X-type pipe device was removed, and the X-type grouted clamp device was installed in the press using the same method, as shown in figure 5. An initial load of 0.25t was applied and raised stepwise until the load of 2T was attained. Specific force and strain data were recorded and listed in table 2.

3. Results and Discussion

3.1. Experimental results

For brevity sake, the data in tables 1 and 2 are reduced to the values of the X-type pipe and the X-type grouted clamp on the same side, respectively.

Table 1. Experimental data on loads and strains in the X-type pipe axial compression test.

Load, t	Measured strain ($\varepsilon=\Delta L/L$)															
	a_1 axis	a_1 cycle	a_2 axis	a_2 cycle	a_3 axis	a_3 cycle	a_4 axis	a_4 cycle	a_5 axis	c_1 axis	c_1 cycle	c_2 axis	c_2 cycle	c_3 cycle	c_4 axis	c_4 cycle	c_5 axis
0.25	-19	2	-41	11	-24	12	-26	-28	-1	-12	15	-7	30	21	-21	12	-18
0.50	-20	4	-54	38	-36	23	-42	-19	-3	-15	20	-31	41	48	-24	14	-19
0.75	-22	10	-69	41	-37	25	-58	-3	-5	-25	35	-56	60	72	-25	15	-20
1.00	-26	17	-77	45	-56	36	-67	17	2	-34	37	-87	82	81	-31	21	-25
1.25	-37	21	-94	54	-66	45	-88	14	12	-37	52	-96	100	84	-33	25	-27
1.5o	-42	22	-107	71	-73	67	-108	16	20	-42	62	-100	102	92	-36	30	-37

Table 2. Experimental data on loads and strains in the X-type grouted clamp axial compression test.

Load, t	Measured strain ($\varepsilon=\Delta L/L$)															
	A_1 axis	A_1 cycle	A_2 axis	A_2 cycle	A_3 axis	A_3 cycle	A_4 cycle	A_5 cycle	C_1 axis	C_1 cycle	C_2 axis	C_2 cycle	C_3 axis	C_3 cycle	C_4 axis	C_4 cycle
0.25	-7	1	-2	14	-1	6	2	3	-3	6	-5	7	-15	11	-1	0
0.50	-9	3	-4	29	-2	11	4	8	-5	10	-12	10	-19	13	-3	0
0.75	-15	7	-7	30	-2	12	10	10	-10	15	-13	12	-20	16	-4	0
1.00	-23	11	-11	31	-5	12	13	10	-14	17	-13	12	-21	17	-7	0
1.25	-25	11	-15	38	-6	13	14	17	-17	22	-18	13	-25	17	-8	1
1.50	-38	12	-15	39	-10	16	17	21	-19	25	-25	15	-26	18	-8	1
1.75	-43	14	-16	39	-11	21	19	23	-20	28	-29	23	-31	19	-9	2
2.00	-45	15	-19	42	-13	22	19	23	-21	36	-32	29	-35	23	-10	3

3.2. Analysis and discussion of experimental results

The analysis of data in tables 1 and 2 reveals that in the same position of comparative strain, the strain values decreases after reinforcement of X-type pipe, because the direction of the applied force is perpendicular to the cross-section of the pipe. Thus, each branch pipe is loaded by the axial force combined with the additional moment. But some data are significantly affected by tensile stress. When the tensile stress becomes compressive, the compressive strain is observed. The displacement after reinforcement is reduced considerably, indicating the improved bearing capacity of the einforced X-type grouted clamp.

4. Study on bearing capacity of X-type grouted clamp

4.1. X-type grouted clamping device

Grouted clamp structure is composed of two semi-circular clamps, as shown in figure 5. It is weld with the corresponding connecting plates on the edge of the grooved clamp structure, these are reinforced with ribs, and the two clamps are locked by bolts to achieve closure effectively. Due to the annular space formed between the grouted clamp and the steel, it is necessary to weld a thick annular shape at the end of the two ends of the grouted clamp. The annular diameter is the diameter of the pipe, the outer diameter of the ring is the diameter of the clamp, circular inner side is provided with a sealing groove, the upper and lower surfaces of the grouted clamp are designed with a slurry outlet and a slurry inlet, respectively. The grout is injected into the annular gap formed by the clamp and the damaged pipe through the grouting port until the density of the outlet and inlet is same, the filling is stopped. If the repair of the underwater damaged pipe is to be done. It is necessary to fill gas into the room formed by the clamp, draining the sea water, and then grout the work.

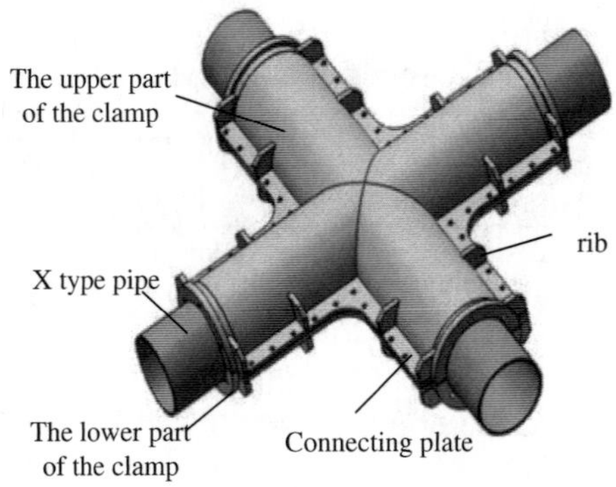

Figure 5. X-type grouted clamp.

4.2. Buckling analysis of X-type grouted clamp

In the stable equilibrium state, according to the principle of potential energy, the equilibrium equation of static structural calculation is expressed as [14-15]:

$$([K_g]) + ([K_G])\{U\} = \{P\} \tag{1}$$

where $[K_g]$ is elastic stiffness matrix of structures; $[K_G]$ is geometric stiffness matrix of structures; $\{U\}$ is nodal displacement vector; $\{P\}$ is nodal load vector.

When the structure reaches the equilibrium state, the two-order variation of the potential energy of the structural system is equal to zero. Thus, we get the following formula:

$$([K_g]) + ([K_G])\{\delta U\} = 0 \tag{2}$$

Eventually, this yields:

$$\left|[K_g] + [K_G]\right| = 0 \tag{3}$$

Assuming a set of external loads $\{P^0\}$, the geometric stiffness matrix of the structure is $[K_G^0]$. When the structure is unstable, considering load is $\lambda\{P^0\}$, so that $\left|[K_g] = \lambda[K_G]\right|$, and formula (3) can be reduced to the following form:

$$\left|[K_g] + \lambda[K_G^0]\right| = 0 \tag{4}$$

Formula (4) can be converted into the following eigenvalue equation:

$$([K_g] + \lambda_i[K_G])\{\phi_i\} = 0 \tag{5}$$

where λ_i is i-order eigenvalue; $\{\phi_i\}$ is the characteristic vector corresponding to i, that is the corresponding structural deformation shape, which is also an unstable mode.

To verify the correctness of the experimental results, that is the bearing capacity of the grouted clamp, the buckling analysis of inner tube finite element under axial compression before and after reinforcement is carried out. The critical load and stability of the X-type grouted clamp under a specific load is studied. Linear buckling analysis is based on the Eigenvalue as the object of study that is the buckling load factor, ten mode load factor is obtained before and after the reinforcement, to obtain the limit load of the grouted clamp. When the load reaches a critical value, the structure will jump to the equilibrium state of the event.

X-type pipe and installation of grouted clamp X-type pipe are subjected to the pressure of 1MPa, for which the ten-order models are shown in figures 6 and figure 7, respectively. These figures are deformation nephograms for X-type pipe before and after installing the grouted clamp. Further, we compare the deformation before and after reinforcement as shown in figure 6 (grouted clamp is hidden). The tenth-order buckling load factor is shown in table 3. The buckling load is the product of the external force 1MPa applied to the buckling analysis corresponding to the buckling load factor.

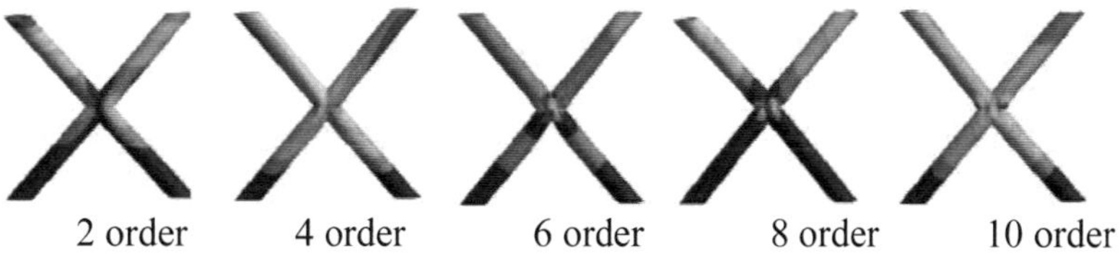

Figure 6. Deformation nephogram of X-type pipe before reinforcement.

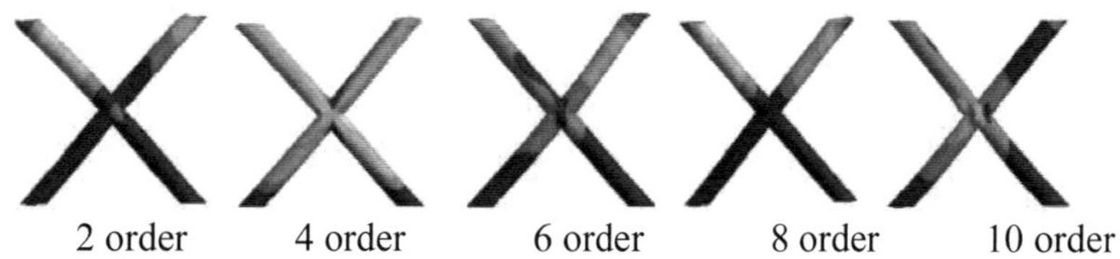

Figure 7. Deformation nephogram of X-type pipe after reinforcement.

Among these, the X-tube parameters are consistent with the previous ones. That is, the material is Q235-a, the constraint is applied to the bottom surface, the pressure load is applied to the top, and the node is the fusion treatment. Elastic modulus is E=2.05×10^{11} Pa.

Table 3. Tenth-order modal load factor of the X-type pipe before and after reinforcement.

	1 order	2 order	3 order	4 order	5 order	6 order	7 order	8 order	9 order	10 order
Pipe	453.46	570.94	639.82	668.57	853.35	1056.4	1116.3	1128.6	1211.7	1233.5
Clamp	674.84	1787.7	1873.7	2039.8	2238.8	3315.8	3403.4	3649.4	3734.4	3853.7

It can be seen in figures 5 and 6 that the grouted clamp can strengthen the reinforcement effect of the inner tube, especially near the node, and reduce the stress concentration. By order of the modal load factor, it can be found that the buckling pressure after reinforcement is higher than the reinforcement before with the increasing order. It is pointed out that the deformation of the damaged inner pipe would cause the contact between the cement ring and the inner pipe in the clamp, and the mechanical strength of the inner tube will be improved by this strengthening mechanism.

5. Conclusions

In this paper, the design of X-type grouted clamp and experimental research are carried out by taking the X-type grouted clamp in the form of a typical jacket node as an example. The theoretical and finite element simulation analyses of the X-tube joint and X-type grouted clamp were performed by the axial pressure test, buckling analysis theory, and buckling deformation assessment before and after the X-tube reinforcement. The finite element simulation results are verified by the axial pressure load tests, which proved that the X-type grouted clamp strengthen the jacket. This also solves the problem of insufficient strength of the damaged jacket and improves the bearing capacity of the pipe joint. Therefore, the analysed X-type grouted clamp has important engineering applications.

Acknowledgments
This paper is funded by NSFC "Research on ultimate bearing capacity and parametric design for the grouted clamps strengthening the partially damaged structure of jacket pipes," (Contract number 51879063) and "Research of analysis and experiments of gripping and bearing mechanism for large-scale holding and lifting tools on ocean foundation piles," (Contract number: 51479043). The views of the authors do not necessarily coincide with the sponsors' policy.

References

[1] Rathnayaka S, Khan F and Amayotte P 2013 *Journal of Risk and Reliability* **227** p 86-105
[2] Dingya W, Anda Z 2014 *Petroleum Machinery* **3** p 33-37
[3] Barltrop N and Adams A J 2013 *Dynamics of Fixed Marine Structures* (Oxford: Butterworth-Heinemann) 3 p 249-269
[4] Xu S, Yang S, Liu X Q 2011 *Petroleum Engineering Construction* **S1** p 65-68
[5] Ismail Z, Kong K, Othman S, *et al.* 2014 *Measurement* **51** p 18-33
[6] Samarakoon S and Ratnayake R 2015 *Reliability Engineering & System Safety* **135** p 15-26
[7] MSL Engineering Ltd 2014 *MSL Doc.,* Ref. C357R001, **Rev 1** p 83-112
[8] Gong S F, Jin W L, Wang G 2001 *China Offshore Platform* **2 p** 38-45
[9] Boswell L and Mello C 1985 The experimental behavior of grouted connections for construction and repair *Amsterdam: Elsevier Science Publishers* pp 835-844
[10] Zhao X, Ghojel J, Grundy P, *et al.* 2006 *Thin-Walled Structures* **44** p 751-758
[11] Lotsberg I, Serednicki A, Lervik A, *et al.* 2012 *Stahlbau* **81** p 695-804
[12] Dallyn P, El-Hamalawi A, Palmeri A, *et al.* 2015 Experimental testing of grouted connections for offshore substructures: A critical review *Structures* pp 89-96
[13] Jiang S, Wang Z and Zhao X 2011 *Procedia Engineering* **14** p 304-311
[14] Zou X J 2005 *Buckling theory analysis and experimental research of foundation piles* (Changsha: Hunan University Press) p 56-67
[15] Feng L 2003 *Analysis of typical structural strength based on buckling theory* (Xi'an: Chang'an University Press) p 80-96

IOP Conf. Series: Materials Science and Engineering **474** (2019) 012016 doi:10.1088/1757-899X/474/1/012016

The refinement of dynamic deformation equation for AISI1215 steel using the Zener-Hollomon parameter

C L Pu*, S N Fang, H Wang, Y Lu, J G Sun and F Gui

Huatian Engineering and Technology Corporation, Nanjing 210019, China

*E-mail: th19870922@126.com

Abstract. The flow stress of a metal being deformed depends on temperature, strain rate, and strain. The Johnson-Cook model describes these dependencies in a mathematical form with six material parameters, of which five have to be derived from forming experiments. However, the coupled interaction between parameters on flow stress has not been considered, and the solution of equation coefficients is difficult. This paper presents the development of a less complicated relationship between flow stress and forming parameters based on the Zener-Holloman parameter. In this paper, AISI1215 steel was chosen as experimental material and was deformed in the range of the temperatures of 750–850 °C and strain rate of 0.01–30 s^{-1} to resolve and verify the presented model. The investigation results show that the coefficients could be solved easily in the present equation through simple hot simulation experiments without complex and disputed mathematics. The experimental results were agreement with the modeling predictions. Here the dynamic deformation behavior was only considered before dynamic recrystallization.

1. Introduction

The flow stress of materials that undergo dynamic deformation due to high temperature and strain rate is affected by strain, strain rate, and deformation temperature [1]. Dynamic deformation behavior is an essential consideration in process simulations to determine the hot forming applications of materials; this parameter is usually described using constitutive models [2–4], such as Johnson–Cook (J–C).

Dynamic recrystallization (DRX) is promoted when the strain exceeds the critical strain (ε_c). Thus, the flow stress of materials should be divided into two stages of modeling: with or without DRX dependent on ε_c [5]. In the first stage, the flow stress of materials without DRX is an empirical and simple multiplication of strain, strain rate, and deformation temperature, such as in the J–C model [6]. The empirical equation has been improved from the J–C model [7-8]. However, the interaction between the strain rate and deformation temperature is not considered in these models. Therefore, calculating the coefficients in these equations is difficult and usually requires a disputed mathematical method [9]. Momeni [10] proposed a physically based model for the dynamic deformation behavior of materials, but this model is complicated and difficult to solve.

Equations with only a few parameters are easy to solve. For hot deformation processing, Sellar [11] combined strain rate and deformation temperature into the Zener–Hollomon (Z) parameter, i.e., strain rate–corrected temperature. In other words, the effect of strain rate and deformation temperature on flow stress can be described by the Z parameter. In the present work, a constitute model defined by the Z parameter was developed for the dynamic deformation behavior before DRX. The effect of DRX on the flow stress of materials during dynamic deformation should be considered in future works.

2. Hot compression experiments and results

Commercial AISI 1215 steel with a chemical composition of 0.07% C, 0.06% Si, 0.8% Mn, 0.05% P, and 0.28% S was utilized as the experimental material. The samples for thermal simulation examination were cut to 8 mm in diameter and 12 mm in length from the bar. Hot compression tests were conducted on Gleeble-1500 thermo-mechanical simulator at different temperatures of 700 °C, 750°C, 800°C, and 850°C at four strain rates of 0.1, 1, 10, and 30 s^{-1} because of the maximum rate limit of the facility. The specimens were first reheated to 1200 °C with 80 °C/s (15 s), held for 5 min, and then cooled to the deformation temperature at a cooling rate of 5 °C/s. The typical stress-strain curves obtained from hot compression tests are shown in figure 1.

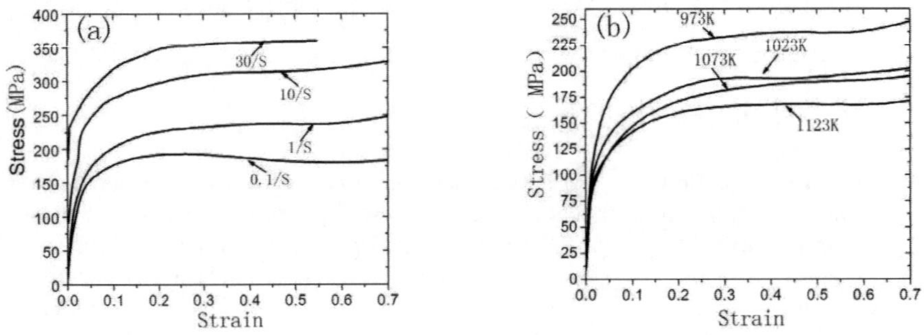

(a) 973 K (0.1, 1,10, and 30 s^{-1}) (b) 1 s^{-1} (973,1023,1073, and 1123 K)

Figure 1. Typical flow stress-strain curves in the thermal compression of AISI 1215.

As shown in figure 1, the flow stress increased as the deformation temperature was decreased from 850 °C to 700 °C for a given strain rate of 1 s^{-1} and decreased as the strain rate increased from 30 s^{-1} to 0.1 s^{-1} for a given deformation temperature of 700 °C. The hardening mechanism of strain rate was generally attributed to the tangled dislocation structures that hinder the dislocation movement at high strain rates. The softening mechanism of temperature was attributed to the decrease in P–N force of dislocation movement and dynamic recovery when no DRX occurred. When the strain exceeds the critical strain for DRX, recrystallization softening will overcome work hardening at the given strain rate and deformation temperature.

3. Dynamic deformation equation described by Z parameter for AISI 1215steel

3.1. Modeling of dynamic deformation behavior described by Z

The flow stress of materials during deformation at high temperature and strain rate can be expressed as the function of strain, strain rate, and deformation temperature:

$$\sigma = f(\varepsilon, \dot{\varepsilon}, T) \tag{1}$$

where ε, $\dot{\varepsilon}$ and T are the strain, strain rate, and deformation temperature, respectively.

The Z parameter can be expressed as:

$$Z = \dot{\varepsilon} \exp(Q / RT) \tag{2}$$

where $\dot{\varepsilon}$ is the strain rate, T is the deformation temperature, Q is the deformation activation energy, and R is the gas constant of 8.314 $Jmol^{-1}K^{-1}$.

Combining equations (1) and (2), we can express the flow stress of the materials during deformation at high temperature and strain rate as a function of strain and Z as:

$$\sigma = f(\varepsilon, Z) \tag{3}$$

The basic Z parameter, where the effect of strain rate and deformation temperature on flow stress can be ignored, is defined as Z0(293 K, 0.001 s^{-1}) in this study. The flow stress of materials during dynamic deformation can be described as:

$$\sigma = f(\varepsilon, Z/Z_0) \tag{4}$$

When the Z parameter is not high enough (i.e., Z = Z_0), the flow stress of the materials can only be the function of strain, called quasi-static deformation, which can be expressed as [12]:

$$\sigma = \sigma_0 + K\varepsilon^n \tag{5}$$

where σ_0 is the yield stress, *K* is the strength component, and n is the work-hardening exponent.

The existing model of the relationship between flow stress and Z in this work was not found in the references. It was established from the experimental database by section 2. The preliminary calculation showed that the strain range was 0.02–0.4 to avoid the effect of DRX on the flow stress. All of the experimental strain rate and deformation temperature databases were calculated as Z parameters with deformation-activated energy derived using the method described in the following chapter. The typical empirical relationship of the flow stress as a function of the Z parameter is shown in figure 2 at the given strains of 0.08 and 0.24.

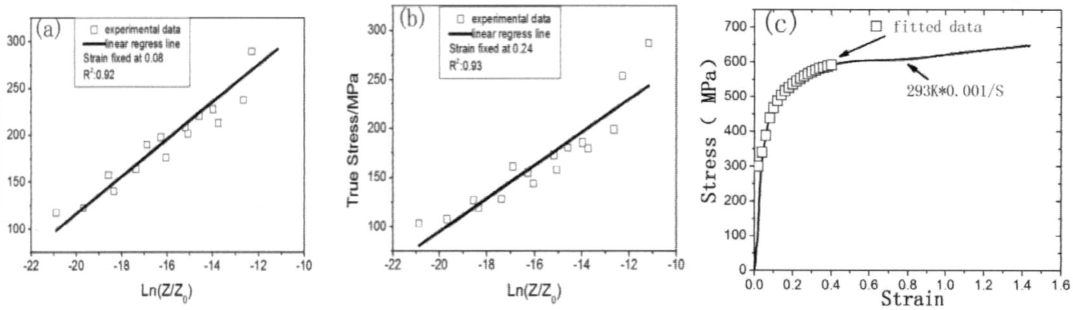

Figure 2. The relationship between flow stress and $Ln(Z/Z_0)$ at strains of (a) 0.08 and (b) 0.24, experiment and fitted flow stress-strain curves of quasi-static deformation conditions (c).

As shown in figure 2(a) and (b), the flow stress was well fitted linearly with $Ln(Z/Z_0)$. Therefore, the function of flow stress with (Z/Z_0) can be defined through a linear relationship as:

$$\sigma = A(\varepsilon) + B(\varepsilon)Ln(Z/Z_0) \tag{6}$$

where $A(\varepsilon)$ represents the basic flow stress at Z_0, i.e., without the influence of strain rate and deformation temperature. It was the only function of the strain, which should be the same as equation (5). $B(\varepsilon)Ln(Z/Z_0)$ consists of the combined contribution of strain and Z on flow stress. The flow stress of the material could not be zero even when the strain was equal to zero because of the existence of internal stress, such as P–N force of dislocation [13]. The function of $B(\varepsilon)$ can be expressed as:

$$B(\varepsilon) = C + D(\varepsilon) \tag{7}$$

where C is a material constant depending on the internal stress and $D(\varepsilon)$ is a function of flow stress and strain at a given Z parameter.

Combining equations (5–7), we can express the modeling of dynamic deformation behavior before

DRX as:

$$\sigma = \sigma_0 + K\varepsilon^n + CLn(Z/Z_0) + D(\varepsilon)\text{Ln}(Z/Z_0) \quad (8)$$

From the physical principle, σ_0, K, and n represent the contribution of work hardening of materials on flow stress, C represents the effect of Z (deformation temperature and strain rate) on flow stress, and $D(\varepsilon)$ describes the combined effect of strain and Z on the flow stress.

3.2. The solution of the proposed constitutive modeling parameters

A dynamic deformation modeling described by the Z parameter was promoted, as shown in equation (8). Then, the parameters of σ_0, K, n, C, and Z_0, and the function of $D(\varepsilon)$ in the equation were derived. σ_0 (197.6 MPa), K (539.8), and n (0.39) were fitted into equation (5) with the flow stress curve under quasi-static deformation conditions (Z_0). The fitted curve with open squares is shown in figure 2(c). The fitted curve was close to the flow stress curve of Z_0 (solid black line).

The parameter C and strain-related function ($D(\varepsilon)$) built up the slope item ($B(\varepsilon)$) together following equation (6) and (7). Therefore, these parameters can be obtained if the line fitted between $Ln(Z/Z_0)$ and flow stress is completed at a given strain. The deformation activation energy Q must be calculated first to get the Z parameter. Q is related to material composition and deformation parameters. The effects of deformation temperature and strain rate on the quasi-static deformation conditions (Z_0) can be ignored; thus, the value of Q under Z_0, named as $Q_{def}^{Z_0}$, can be calculated by the composition of AISI 1215 steel, and the output was 185 kJ/mol [14]. Under non-quasi-static deformation conditions, Q can be solved using multivariate linear regression with the peak flow stress (σ_p) of the compression test based on the three functions describing the relationship among flow stress, strain rate, and deformation temperature, divided by different stress levels as follows[15-16]:

$$\dot{\varepsilon} = A_1 \sigma_p{}^m \exp(\frac{-Q}{RT}) \quad (9)$$

$$\dot{\varepsilon} = A_2 \exp(\beta\sigma_p)\exp(\frac{-Q}{RT}) \quad (10)$$

$$\dot{\varepsilon}\exp(Q/RT) = A[\sinh(\alpha\sigma_p)]^n \quad (11)$$

where A_1, m, A_2, β, A, α, and n are the material constants.

In general, the power law of equation (9) is usually preferable for low stress at high deformation temperature or low strain rate, whereas the exponential function of equation (10) is suitable for high stress at low deformation temperature or high strain rate. However, the hyperbolic sine function of equation (11) can be used for a wide range of stress. The material constants m, β, and α are related to $\alpha = \beta / m$. Through the partial differentiation of equation (9) and (10) at constant deformation temperatures, we can use the following equations for low [figure 3(a)] and high [figure 3(b)] stress levels to obtain m and β. The effect of temperature on parameters m, β was small, and the average values of m, β, α were 3.36, 0.0238, and 0.0071, respectively.

Taking the natural logarithms of both sides of equation (11), we can obtain the following formula:

$$\ln[\sinh(\alpha\sigma_p)] = \frac{\ln\dot{\varepsilon}}{n} + \frac{Q}{nRT} - \frac{\ln A}{n} \quad (12)$$

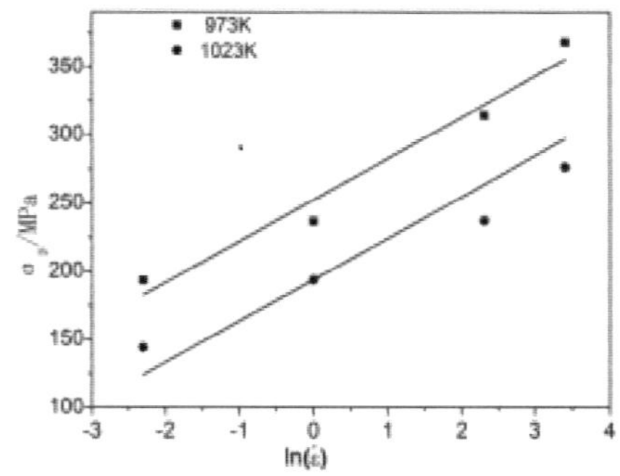

(a)

Figure 3 (a). The relationship between $\ln(\dot{\varepsilon})$ and σ_p at 973 and 1023 K.

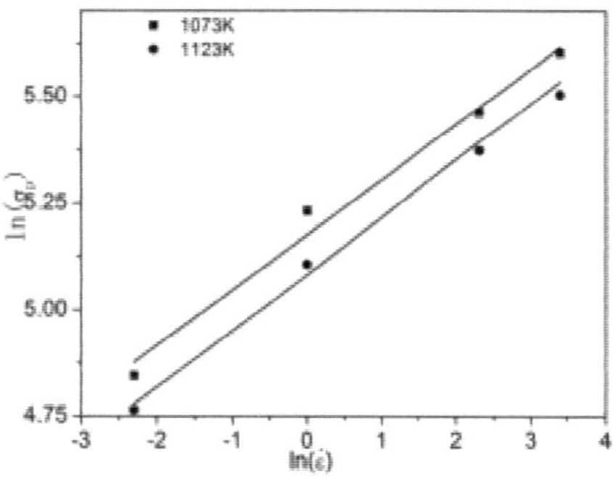

(b)

Figure 3 (b). The relationship between $\ln(\dot{\varepsilon})$ and $\ln(\sigma_p)$ at 1073 and 1123 K.

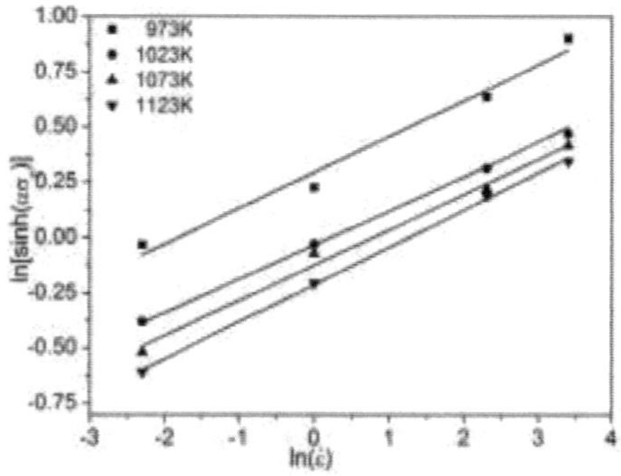

(c)

Figure 3 (c). The relationship between $\ln(\dot{\varepsilon})$ and $\ln[\sinh(\alpha\sigma_p)]$ at different deformation temperature.

When the deformation temperature in equation (12) is constant, the linear function can describe the relationship between $\ln[\sinh(\alpha\sigma_p)]$ and $\ln\dot{\varepsilon}$ well, and the slope yields $\frac{1}{n}$, as shown in figure 3(c). The average value of n=5.52 was determined by the calculation.

The functional relationship between $\ln[\sinh(\alpha\sigma)]$ and $1/T$ can be described by liner function well if strain rate was fixed in equation (12), differentiating the equation (12) tended to:

$$Q = Rn\left[\frac{\partial \ln[\sinh(\alpha\sigma)]}{\partial(1/T)}\right]_{\dot{\varepsilon}} \quad (13)$$

IOP Conf. Series: Materials Science and Engineering **474** (2019) 012016 doi:10.1088/1757-899X/474/1/012016

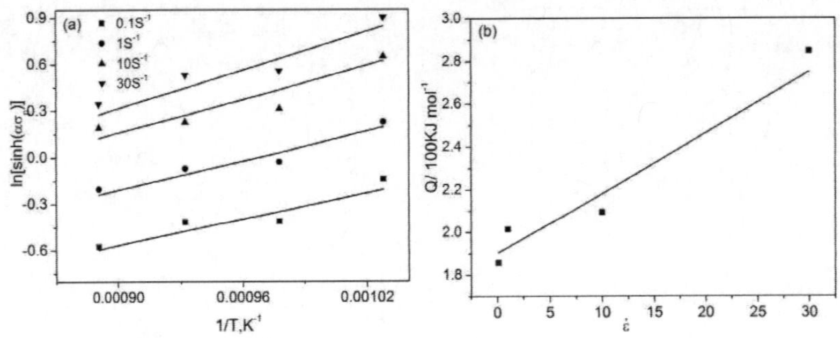

Figure 4. Evaluating the value of Q by plotting $\ln[\sinh(\alpha\sigma_p)]$ vs. $1/T$ with different strain rates (a); and at strain rate 10 s-1 with different strains (b).

Thereby, the value of Q could be decided with the slopes of $\ln[\sinh(\alpha\sigma)]$ vs. $1/T$ with different strain rate, multiplied by Rn according to equation (13). The output was shown in figure 4(a) where the linear slope increased with the increased in strain rate from 0.1 s-1 to 30 s-1. The trend is consistent with other studies [14-16]. As the value of n (5.52) have calculated, the fitting results of Q could be computed for different strain rate. The results of figure 4(b) indicated that the evolution of Q was closely linked to the strain rate from 0.1 to 30 s^{-1}, which could be described by a linear function as follows:

$$Q = 0.09\dot{\varepsilon} + 1.83 \tag{14}$$

After obtaining the value of Q, the linear effect of $Ln(Z/Z_0)$ on flow stress was investigated, and the mean value of R^2 was about 0.91. The scatter data of $A(\varepsilon)$ and $B(\varepsilon)$ could be plotted in figure 5. As seen in figure 5, the trends of scatter data of variable vs. function for $A(\varepsilon)$ and $B(\varepsilon)$ was similar. The modus of equation (5) was adopted to fit the functional relationship of $B(\varepsilon)$. According to the data in figure 5, the fitted equation of $B(\varepsilon)$ was obtained as flows:

$$B(\varepsilon) = C + D(\varepsilon) = 4.93 + 7.8 \times \varepsilon^{0.63} (R^2 : 0.96) \tag{15}$$

The fitted curves shown in figure 5 indicated the fitted curves closely agreed with the scatter data for $A(\varepsilon)$ and $B(\varepsilon)$.

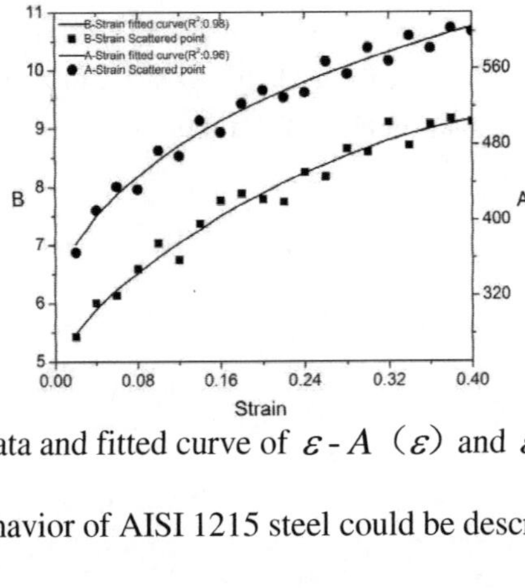

Figure 5. The scatter data and fitted curve of ε - A (ε) and ε - B (ε).

Finally, the dynamic deformation behavior of AISI 1215 steel could be described by the constitutive equation in the following:

$$\sigma = 197.6 + 539.8 \times \varepsilon^{0.39} + (4.93 + 7.8 \times \varepsilon^{0.63}) Ln(Z/Z_0) \tag{16}$$

$$Z = \dot{\varepsilon}_Z \exp(Q^Z_{def} / RT_Z) = \dot{\varepsilon}_Z \exp\left\{100000[0.09\dot{\varepsilon}_Z + 1.83] / RT_Z\right\} \tag{17}$$

$$Z_0 = \dot{\varepsilon}_{Z_0} \exp(Q^{Z_0}_{def} / RT_{Z_0}) \tag{18}$$

where $\dot{\varepsilon}_{Z_0}$ was 0.001 s^{-1}, T_{Z_0} was 298 K, $Q^{Z_0}_{def}$ was 185 kJ/Mol

3.3. Comparison of calculation and experiments

Though equations from (16) to (18), the flow stress could be computed at given deformation conditions for experimental steel. The comparison between the hot compression tests experimental data and the calculation value by the proposed model were shown in figure 6.

In figure 6, it can be seen that the calculation results are in good agreement with the most experimental results. However, the new constitutive model can not predict the whole flow stress curve after DRX. As can be seen in figure 6, the flow stress curves of 750 °C*0.1 s^{-1}, 800 °C*0.1 s^{-1}, and 850 °C*0.1 s^{-1} occurred DRX in evidence for the flow stress curves had a single peak. The value of ε_c calculated by the model of Jonas [3] was about 0.18, 0.17 and 0.15 for these three deformation conditions respectively. It could be seen that the fitted data marked as black solid deviated the experimental data shown as white open gradually from the point of ε_c in figure 6, and this could be accepted for the behavior of DRX had not been considered in the present model.

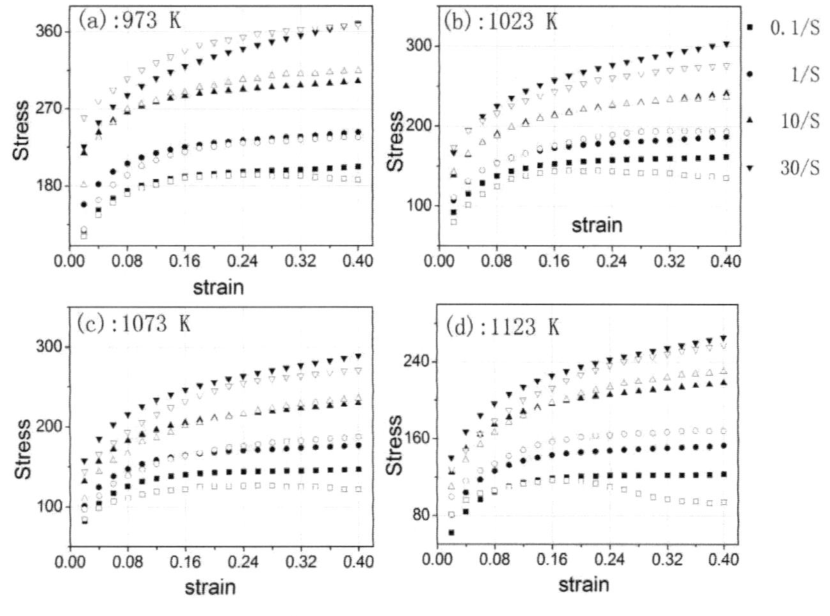

Figure 6. Comparison of the experimental and calculated flow stress.

4. Conclusions

A new constitutive model was proposed in the present work by applying the Zener-Hollomon parameter and considering the effect of strain rate on the activation energy Q_{def}. The new model coupled the interaction effect between three deformation parameters (strain, strain rate and temperature) on flow stress to characterize the hot deformation behavior of materials before DRX. The model could predict the behavior of flow stress well before the DRX, and the model parameters could be derived from limited experiments. The calculation using the proposed model shown a good agreement with the hot compression experimental data of AISI 1215 steel well before DRX.

Acknowledgment
This work was supported by a grant from the Key Research and Development Foundation of Anhui Province of China (1704a0902049).

References

[1] Ahamed H 2012 *Mater. Sci. Eng., A* **539** 349-59
[2] Joshi D 2011 *Mater. Des.* **32** 3655-70
[3] Han Y, Qiao G, Sun J and Zou D N 2013 *COMP MATER SCI* **67** 93-103
[4] Calamaz M, Coupard D and Girot F 2008 *INT J MACH TOOL MANU* **48** 275-88
[5] Jonas J J, Quelennec X, Jiang L, et al. 2009 *ACTA MATER* **57** 2748-56
[6] Ugarte A 2006 Fifth Int. Conf. on High-Speed Machining (HSM) Metz France 239-46
[7] Rhim O 2006. *J MATER PROCESS TECH* **171** 417-22
[8] Poliak E and Jonas J 1996 *ACTA MATER* **44** 127-36
[9] Kim S and Yoo Y 2002 *MATER SCI TECHNOL* **18** 160-4
[10] Momeni A, Dehghani K and Ebrahimi G 2011 *J ALLOY COMPD* **509** 9387-93
[11] Sellars C M 1985 *MATER SCI TECHNOL* **1** 325-32
[12] Maheshwari A 2013 *COMP MATER SCI* **69** 350-8
[13] Yu Y 2010 The principle of Metal Beijing: Metallurgical Industry Press 434-623
[14] Burachynsky V and Cahoon J R 1997 *Met and Mat Trans A* **28** 563-85
[15] Srivastava J and Uhlenwinkel J 2008 *Mater. Sci. Eng., A* **477** 86-95
[16] Pu C, Zhu G and Tao Y 2015 *INT J MATER RES* **106** 863-9

IOP Conf. Series: Materials Science and Engineering **474** (2019) 012017 doi:10.1088/1757-899X/474/1/012017

Analysis of corrosion and scaling in high-voltage direct-current valve cooling water system

X W Li[1] and D Ding[2,*], L Wang[2] and L W Sun[2]

[1] State Grid Shaanxi Electric Power Company, Xi'an, China.
[2] State Grid Shaanxi Electric Power Research Institute, Xi'an, China.

*E-mail: dingde5160@163.com

Abstract. In this paper, the problems of corrosion and scaling in the converter valve tower cooling water system based on the ± 500kV Baode HVDC transmission project with Siemens light-triggered thyristor HVDC transmission technology are analyzed in detail. The corrosion and scaling phenomenon can cause the closure or emergency shutdown of converter valves and bring potential dangers to the safe and stable operation of power grid. So far, there is no fundamental solution to this problem. Based on the investigation of scaling phenomenon of grading electrode and corrosion of thyristor aluminum alloy heat sinks, the scaling law of the grading electrode is proposed and the cause of scaling is put forward and analyzed. Some suggestions are given.

1. Introduction

Converter valve is the core equipment of a high-voltage, direct current (HVDC) electric power transmission system. Considering the valve loss, which inevitably exists, the heat generated by valve loss could largely harm the performance of converter valve. Thus, the internal water cooling system, which effectively helps dissipating the heat out of valve, is of great significance to ensure the normal operation of the converter valve [1]. Unfortunately, operating experiences show that the problem of corrosion and scaling is very common in inner cooling water system of converter valves [2-12]. Despite the use of mixed ion exchange resin for desalination and purification of the internal cooling water of the converter valve, corrosion and scaling still exist in the internal cooling water system of the converter valve, leading to an unsatisfied performance of the heat dissipation and insulation of the converter valve. Accordingly, the safe and stable operation of the thyristor, reactor, damper capacitor and water pipe of the converter valve could be negatively affected.

In this paper, the scaling problem of bipolar grading electrodes in the ±500 kV Baode HVDC power transmission project is taken as the research object and the corrosion and scaling issues of the internal cooling water system of the converter valve are analyzed and the relevant recommendations and solutions are proposed.

2. Valve cooling water path and grading electrode

2.1. HVDC transmission system

The HVDC transmission technology is one of the most advanced and energy-saving power transmission and transformation technologies in the world, and it is also the key developing technology in Chinese technical equipment field. It has the advantages of asynchronous operation between two AC systems, fast response, precise adjustment, convenient operation, stable voltage distribution along the line and low operation loss. Moreover, the investment and operation cost of HVDC transmission line are less than that of AC transmission system [13]. HVDC transmission system usually consists of rectifier station, inverter station and HVDC transmission line. Its wiring principle is shown in figure 1.

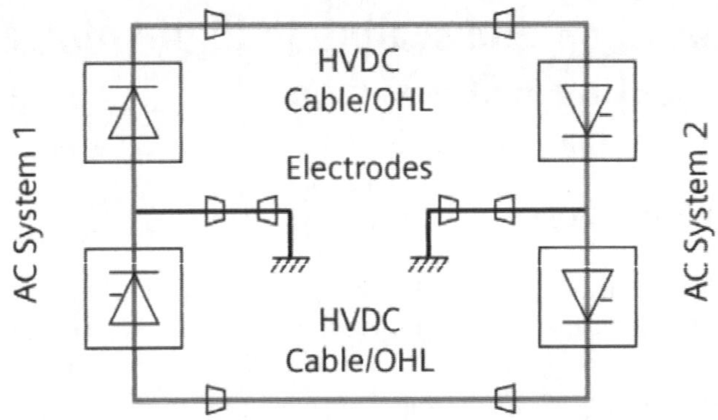

Figure 1. HVDC transmission system

2.2. Valve cooling water system

Converter valve is one of the most important equipment in HVDC transmission system. Under normal operation, the thyristor generates high heat with high current flowing, which causes the temperature of the thyristor to rise sharply. The thyristor must be cooled effectively, or it will be burnt out. The temperature of the thyristor of the converter valve is generally less than 90°C. The heat dissipation of thyristor is realized by the aluminum alloy heat sinks which is closely contacted with thyristor by internal cooling water cooling. The system flow chart is shown in figure 2.

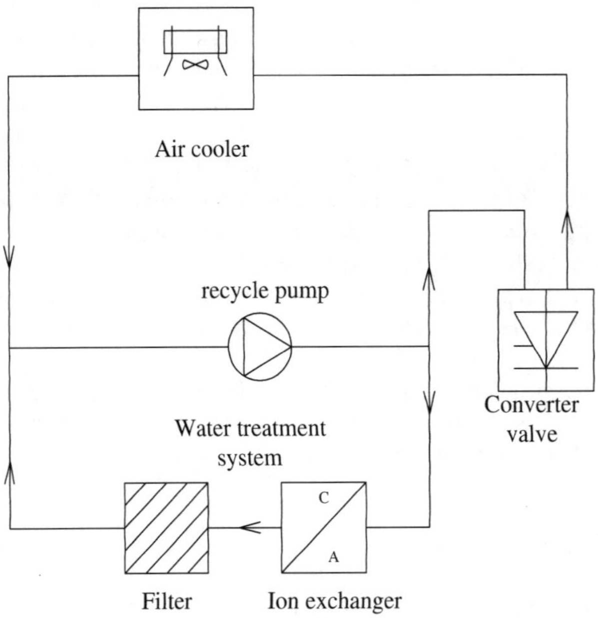

Figure 2. Air cooling water system

2.3. Valve assembly waterway

The valve hall of the converter valve is divided into three phases, namely A, B and C. Each phase is subdivided into left (L) and right (R) sides, each side being split into six layers. Each valve assembly is composed of series thyristors. The thyristors are separated by aluminum alloy heat sinks. The heat sinks are connected by waterways, and the heat of the heat sinks is derived by cooling water.

The connection modes of cooling water pipeline of valve assembly could be segmented to series type, parallel type and series-parallel type. The series type of connection mode is mainly based on ABB technology route. The flow rate in series type is large, but the temperature difference between the head and the tail is huge too. The parallel type is mainly based on SIEMENS technology route. The series-parallel type, mainly based on Zhonglian Puri technology route, combines the advantages of ABB and SIEMENS given water route arrangement of flow and temperature.

SIEMENS technology route was adopted in ±500 kV converter station. Its valve assembly waterway was parallel, A is thyristor, B is heat sinks, C is connecting water pipe. The schematic diagram is shown in figure 3, and the valve assembly is shown in figure 4.

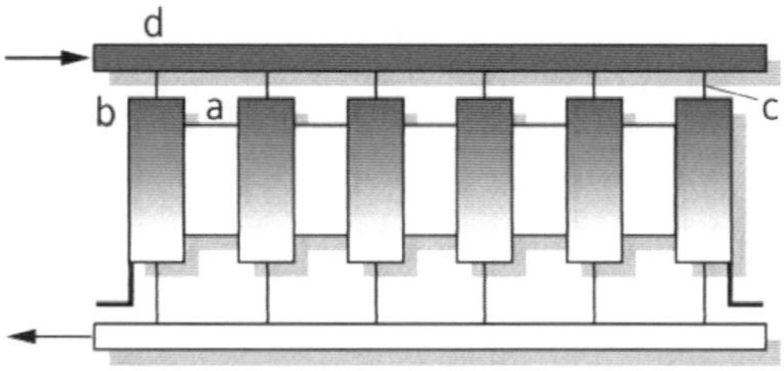

Figure 3. Parallel connection water line

Figure 4. SIEMENS Valve assembly

2.4. Grading electrode position

In order to avoid discharging unevenly on the inner and outer wall of cooling water pipeline in the converter valve due to uneven voltage distribution, and leakage current caused by the contact between metal equipment and water pipeline in the valve assembly, grading electrodes were installed on the main cooling water pipe of the valve tower and the water distribution and catchment pipes in the valve assembly. The appearance of the grading electrodes is shown in Figure 5. For the grading electrode, the effective part presented is platinum, of 23.8 mm in length and 2 mm in diameter; the remaining part is stainless steel. The installation of grading electrodes could help realizing even voltage distribution on the inner and outer wall of the main water cooling pipeline and help releasing leakage current, which now transfer from thyristor aluminum alloy heat sinks to inert platinum electrode. By doing this, the corrosion phenomenon of electrical equipment could be avoided [14].

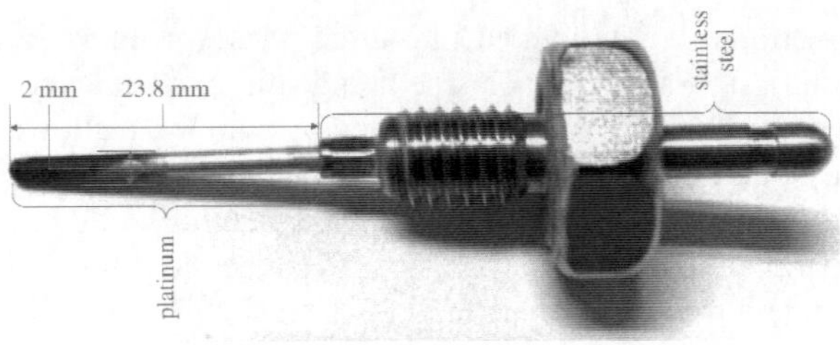

Figure 5. Grading electrodes

There were 432 grading electrodes installed in the single pole of Baoji Converter Station, including 24 zero potential electrodes. The number of grading electrodes installed on the T-type connection part of valve assemblies and main water pipe was 144, the number of grading electrodes installed on the anode side of thyristor valve assemblies was 288 and the number of grading electrodes installed on the cathode side was 288. The rest were electrode valve tower interlayer electrodes. The installation positions of grading electrodes could be divided into the following categories, as shown in table 1, L: left side, R: right side, L1: on the left and level 1, R1: on the right and level 1, R2\L3\L4\R4\R5\L6 and so on.

Table 1. Grading electrode positions in the valve tower

Valve tower electrode location (A phase, B phase, C phase)		Number of electrodes
Top part tube	L sides	2
	R sides	2
S-shape tube	L1 and R1 upper part	4
	R2 upper part	2
	L3 upper part	2
	L4 and R4 upper part	4
	R5 upper part	2
	L6 upper part	2
T-type tube	L and R sides	12
Water distribution pipe and confluence pipe	Cathode sides.	24
	Anode sides	24
Bottom part tube	L and R side	2

3. Scaling on Grading Electrodes

Based on the investigation of several HVDC transmission projects in China, it is found that the scaling phenomenon of grading electrodes exists in different degrees. Figure 6 (a) and (b) illustrate the difference in scaling for grading electrodes of ABB technical route and grading electrode of Siemens technical route, respectively. In addition, figure 6(c) shows a third type of grading electrode of Zhonglian Puri technical route with some scaling presented.

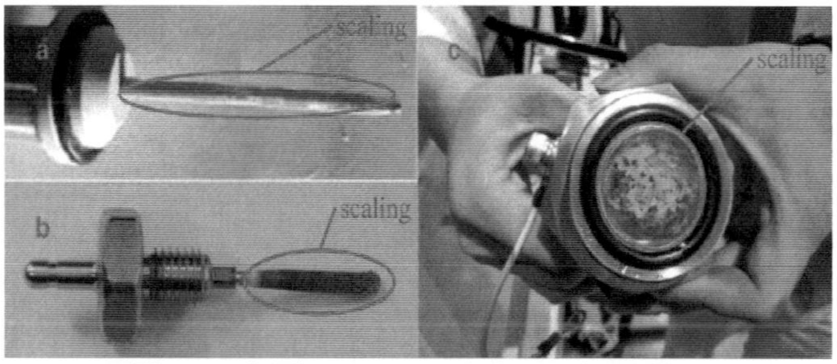

Figure 6. Three types of electrode scaling

3.1. Scaling phenomenon in Inverter station

The scaling phenomena in Baoji Converter Station were as following: the scaling of the grading electrodes on the anode side of thyristor on the valve assembly was slight and loose; the scaling of the grading electrodes on the cathode side was serious and firm; the zero potential point of the main water pipe of the valve tower was not scaling. The scaling of the grading electrode at the T-type connection between the water distributing pipe of the valve assembly and the main water pipe was moderate, but scaling was very firm.

The scaling on the anode side was slight. The thickness of the scaling of electrode could be calculated as the following: the diameter of the electrode with scaling minus the diameter of the electrode without scaling, and then divide the result by 2. The average thickness of scaling was 0.1-0.2 mm. The scaling distributed evenly, the scaling was crisp and easy to fall off. The scaling on the upstream side was slightly thicker than that on the back side. Only the root part was not covered by scaling, as shown in figure 7.

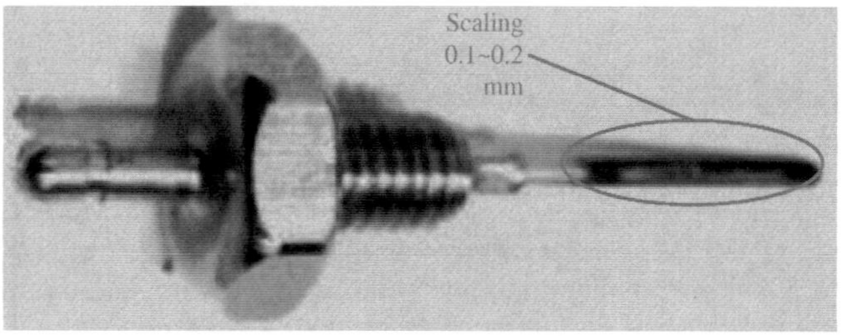

Figure 7. Deposit on grading electrodes at anode side of valve assembly of Invert station

The scaling of the grading electrodes on the cathode side is serious. Generally, the scaling thickness is 0.6-0.8 mm, the end is thick, the tail is thin, the scaling is firm, and it is not easy to fall off. There is no difference between the scaling on the front and back surfaces. Only the root is not covered by scaling, as illustrated in figure 8.

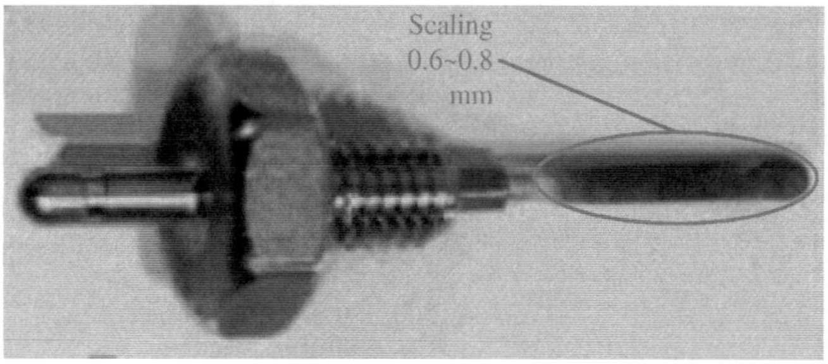

Figure 8. Deposition of grading electrodes at cathode side of valve assembly of Invert station

Observing the grading electrode at the T-type connection of valve assembly distributing pipeline and main water pipe, general scale thickness is 0.6-0.8 mm, scaling distributed evenly, scaling is very firm, firmly attached to the electrode surface. The scaling is no difference between water front side and back side. Only half of length of the platinum electrode part is covered by scaling, as seen in figure 9.

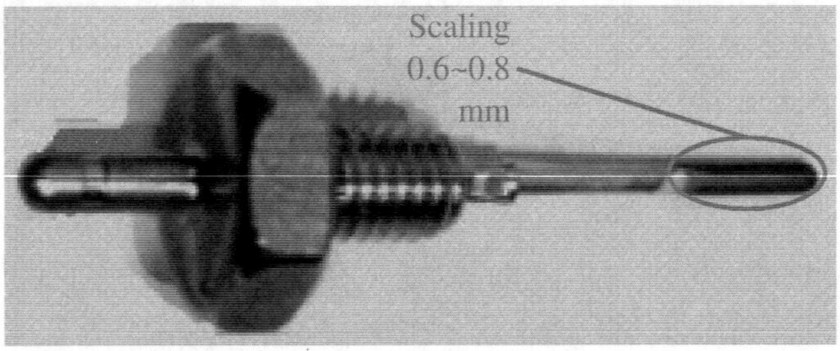

Figure 9. Deposition of grading electrodes at T-junction of the valve assembly and the main water pipe of Invert station

3.2. Scaling phenomenon in Rectifier station

Deyang converter station basically operates as a rectifier, and the scaling of anode side grading electrodes on valve assembly was serious, as shown in Figure 10, which was basically consistent with the scaling of cathode side of valve assembly in rectifier station. The scaling of grading electrodes on cathode side was slight, as shown in figure 11, which was basically consistent with the scaling of anode side of valve assembly in rectifier station; the zero potential point of main water pipe in valve tower was not scaling. The scaling degree of grading electrodes installed on T-type connection of water distributing pipe and main water pipe was at intermediate level, as seen in figure 12.

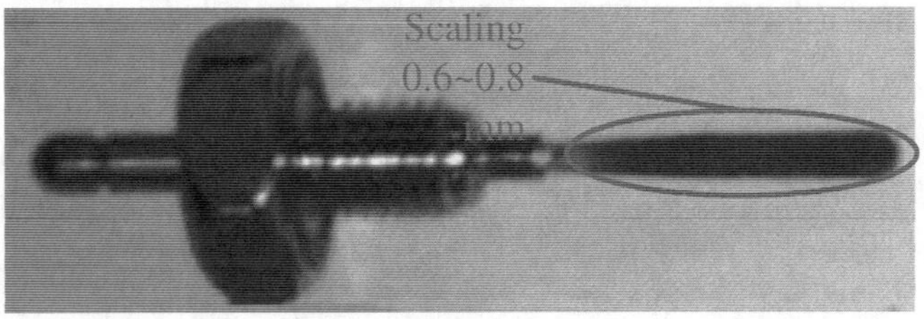

Figure 10. Deposit on grading electrodes at anode side of valve assembly of rectifier station

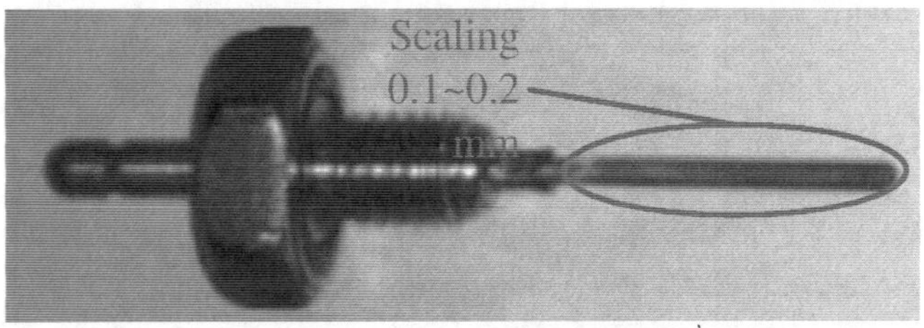

Figure 11. Deposit on grading electrodes at cathode side of valve assembly of rectifier station

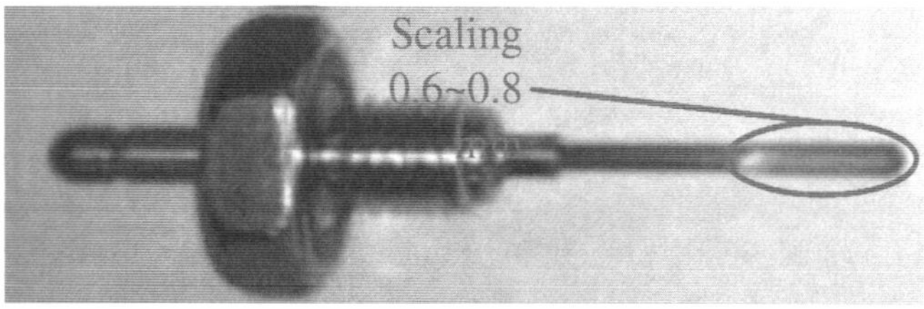

Figure 12. Deposit on grading electrodes at T-junction of the valve assembly and the main water pipe of rectifier station

4. Scale distribution principle

4.1. Distribution characteristics of scaling

According to the electrode scaling thickness, that is, the diameter of the electrode after scaling minus the diameter of the electrode without scaling, and then divided by 2, and position of grading electrodes scaling, the 864 branch electrodes of bipolar at the BAOJI converter station were all sampled for detection. The electrodes scaling were divided into 6 classes: valve assembly cathode side with serious scaling; anode side with slight scaling; valve assembly T-type tube electrodes with scaling in the middle; S-shape tube electrodes, top part electrodes, pole II bottom screen electrodes with scaling situation in agreement with those at anode side; and pole I bottom screen electrode with scaling situation in agreement with those at cathode side, as seen in tables 2 and 3, the diameter of the electrode after scaling is listed. note: The pole I converter valve has been operating for 44 months, the bottom voltage of the valve tower is -500 kV, the pole II has been operating for 39 months, and the bottom voltage of the valve tower is +500 kV. CL:CL represents the left side of the valve tower of the C phase converter valve, CR\BL\BR\AL\AR and so on.

Table 2. Pole I grading electrode with deposits' data analysis (unit: mm)

Average value	CL	CR	BL	BR	AL	AR
Top part	2.04	2.04	2.02	2.02	2.03	2.05
Mid-upper part	2.11	2.02	2.00	2.01	2.01	2.02
S-type tube	2.05	2.04	2.07	2.05	2.03	2.03
Cathode side	2.63	2.94	2.96	2.95	2.94	2.94
Anode side	2.13	2.23	2.23	2.23	2.23	2.23
T-shape tube	2.04	2.16	2.13	2.08	2.12	2.09
Bottom part	2.88	2.72	2.73	2.65	2.68	2.83

Table 3. Pole II grading electrode with deposits' data analysis (unit: mm)

Average value	CL	CR	BL	BR	AL	AR
Top part	2.04	2.05	2.04	2.04	2.04	2.04
Mid-upper part	2.45	2.52	2.49	2.72	2.62	2.02
S-shape tube	2.06	2.04	2.06	2.06	2.05	2.08
Cathode side	2.34	2.28	2.30	2.28	2.31	2.29
Anode side	2.99	2.93	3.09	2.89	2.95	3.05
T-type tube	2.24	2.10	2.23	2.14	2.19	2.07
Bottom part	2.01	2.01	2.02	2.01	2.02	2.03

4.2. Distribution and treatment of scaling

Baoji converter station functions as an inverter station, therefore, the scale distribution of 432 grading electrodes of pole I and pole II were basically identical. Thus, the scaling regularity of grading electrodes of Inverter station could be obtained: scaling on the cathode side of thyristor is serious and sturdy, the scaling on the anode side is slight and loose, while scaling degree at the T-type connection is moderate and very firmly attached.

As a rectifier station, the scale distribution of 432 grading electrodes of pole I and pole II in Deyang converter station were the same, but it was different from that in Baoji converter station. Therefore, the scaling regularity of grading electrodes could be concluded that: scaling on the anode side of thyristor is serious, scaling on the cathode side is slight and scaling on the T-type connection is moderate.

Therefore, according to the scaling distribution law of grading electrodes, scaling removal could be carried out purposefully. Inverter station could aim at removing scaling on the cathode side of thyristor of valve assembly, while rectifier station could only focus on scaling on the anode side of thyristor of valve assembly [15].

5. Related detection analysis and treatment recommendations

5.1. Scale analysis

The scales of all grading electrodes were broken and dissolved in hot dilute sulfuric acid with 35% mass concentration. By doing this, the content of metal elements was detected by ICP emission spectrometer. Aluminum elements accounted for 90-96%, iron elements 5-8%, calcium elements 0.1-0.5%, magnesium elements 0.1-0.5%, sodium elements 0.05-0.1%.Detecting the resistance values of the extreme and tail parts of the grading electrode after scaling ,the resistance values were found to be 280-350 kΩ.This indicates that the main harm of the scaling of the electrode lies in the gradual loss of the original function of grading voltage and transferring leakage current.

5.2. Water quality inspection

The main metal element found in scaling was aluminum. Analyzing the mass concentration of metal elements in inner cooling water and makeup water of pole I and pole II, seen in Table 4. It was found that the mass concentration of aluminum was very low, only 1.1-1.3 ug/L. This indicates that the medium aluminum in scaling of grading electrode may come from the makeup water of valve tower. Therefore, the concentration of Mg^{2+} and Ca^{2+} and the mass concentration of aluminum in the makeup water should be reduced.

Table 4. Water sample analysis of data on cation ion contents (μg/L)

Sample name	Na^+	NH^{4+}	K^+	Mg^{2+}	Ca^{2+}	Al
Pole I de-gas jar	2.6	0.2	1.5	0.3	5.9	1.3
Pole II de-gas jar	1.1	0.2	0.8	0.2	3.6	1.1
makeup water	1105.8	0.2	389.5	10.3	75.8	1.5

5.3. Structure of aluminum alloy heat sinks

The aluminum alloy material directly contacting the internal cooling water system is only the aluminum alloy heat sinks of the thyristor. Unfolding a piece of aluminum alloy heat sinks running for 3 years, it could be found that there were many black corrosion points in the water inlet and outlet. As shown in figure 13, the water inlet channel (the so-called "double mosquito perfume water path" section) was bright and clean, while the outlet channel had black corrosion spots, as shown in figure 14. The results show that the corrosion products of aluminum alloy may come from the scaling of the grading electrodes, and the corrosion effect exerts no obvious harm to the aluminum alloy heat sinks.

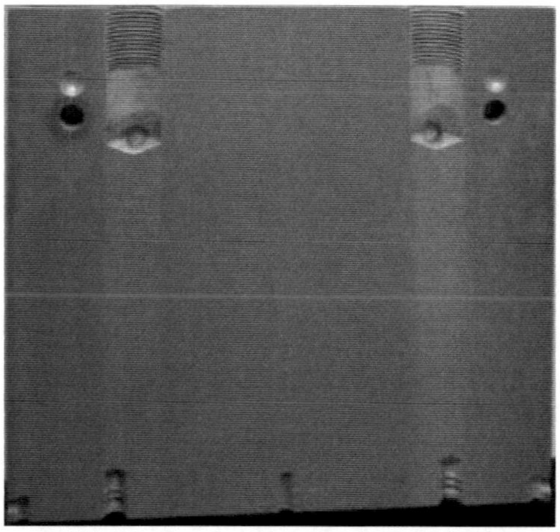

Figure 13. Inlet and outlet section of the heat sink

Figure 14. "Double mosquito perfume water path" section inside of the heat sink

5.4. Waterway voltage analysis

Since grading electrodes connect with heat sink, grading electrodes potentiality varies with the heat sink potentiality, the voltage difference among the grading electrodes is the sum of several heat sink voltage difference, the results listed in table 5.

Table 5. Voltage in the water circuit

Title	Voltage at two ends(V)	Valve condition	Current direction
Two electrodes	26	turn-on	Anode to Cathode
Neighboring heat sinks	2	turn-on	Anode to Cathode
Head-tail heat sinks	26	turn-on	Anode to Cathode
Two electrodes	41665	turn-off	Cathode to Anode
Neighboring heat sinks	3205	turn-off	Cathode to Anode
Head-tail heat sinks	41665	turn-off	Cathode to Anode

5.5. Electrolytic corrosion and scaling analysis

Based on the aluminum-hydroelectric potential-pH chart [16], the following patterns are possible. If the potential exceeds 1.35V above zone with oxygen reaction, the electrolytic water product is oxygen, see Eg.(1). If the potential is below 0.12V zone with the hydrogen reaction, see Equation (2). If the potential is between 0.12-1.35V, the absorption oxygen corrosion will take place, see Equation (3). If the potential is over -1.676V, aluminum loses electrons and is electrolyzed, see Equation (4).

$$4OH^- - 4e^- \rightarrow 2H_2O + O_2\uparrow \tag{1}$$

$$2H^+ + 2e \rightarrow H_2 \tag{2}$$

$$O_2 + 2H_2O + 4e^- \rightarrow 4OH^- \tag{3}$$

$$Al \rightarrow Al^{3+} + 3e \tag{4}$$

Through the calculation of $Al(OH)_3$ pks=32.89 and H_3AlO_3 pK1=11.2, at 18~25°C, the pH value is 5.6~5.8, the oxide film of alumina has the minimum solubility, that is, the aluminum alloy protective film is the best. When the pH value is less than 4.6 or more than 8.3, the protective film begins to dissolve, see Equations (5) and (6):

$$Al(OH)_3 \leftrightarrow Al(OH)^{2+} + 2OH^- \tag{5}$$

$$Al(OH)_3 \leftrightarrow H_3AlO_3 \leftrightarrow H_2AlO_3^- + H^+ \tag{6}$$

Therefore, the aluminum alloy heat sink is subject to electrolytic corrosion. The production of hydrogen or oxygen will generate $Al(OH)_4^-$. So, in the cooling water system, when the valve is turned on, the aluminum hydroxide is absorbed by the electrode surface of valve anode side. When the valve is turned off, the aluminum hydroxide is absorbed by the electrode surface of valve cathode side. In addition, the electrode at the positive potential can produce oxygen, with its volume being about 50% of that at the negative potential, and the electrolysis gas can remove the scaling. Therefore, both rectifier and invert stations have more severe scaling in the electrode at the positive potential than at the negative one.

6. Conclusions

Corrosion in HVDC valves cooling water system was analyzed. The chemical composition of the scaling on the electrodes revealed a large amount of aluminum. Further investigation showed that the aluminum detected was a corrosion product coming from the makeup water of internal cooling water and aluminum alloy heat sinks. It could be found that the surface conductive resistance of the grading electrode increased significantly and the function of grading voltage and transferring leakage current were lost after scaling. Observing the internal of the heat sinks of aluminum alloy, corrosion spots were found. But they are not enough to harm its safe operation. The experiments illustrated that the scaling at the positive potential was higher than that at the negative potential. The scaling principle was different in the rectifier and inverter stations. In the rectifier station, the scaling of grading electrodes at the anode side of valve is serious, while the scaling of grading electrodes at the cathode side is more severe in the inverter station. Based on these regularities of the distribution of the scaling on grading electrodes, targeted scale removal measurements could be carried out. On account of the corrosion of aluminum alloy heat sinks caused by leakage current electrolysis, the conductivity of the internal cooling water system should be lowered as far as possible, and the pH value should be adjusted to a reasonable range.

References

[1] Liu Z Y 2013 UHV AC/DC Power Grid (Beijing: China Power Press) pp 479-80
[2] Arinde S Dhaliwal and Fu Y 2008 Special Report for Group B4 (Cigre. Question 1.1-1.5) p 110
[3] Tian X W, Yang J M and Zhang W P 2010 *Modern Science* **8** pp 68-70
[4] Wang Y Y, Hao Z J and Lin R 2006 *Southern Power System Technology* **32** pp 80-3
[5] Li Y 2009 *Journal of Guizhou Electric Power Technology* **115** pp 5-6
[6] Zhang P D, Hao Z J and Liu H T 2008 *Journal of Electrical Engineering Technology* **12** pp 1-2
[7] Qian Y H, Zhou Y Y and Xu C W 2012 *Advanced Materials Research* **354-355** pp 1157-60
[8] Deng B F, Sun H M and Cao J F 2008 *Southern Power System Technology* **2** pp 73-6
[9] Wang C, Liu X, Wang X, Liu N and Jiao X Y 2015 *Electrical Insulation Conf. (EIC) (Seattle*

Washington USA) pp 7-10
[10] Neubert and Holweg 2007 *Siemens Investigation Report of Scaling Phenomenon on Grading Electrodes in Thyristor Valve Cooling Cirguit* (Erlangen, Simens Technical Note) p 7
[11] Lips H P 2002 *IEEE Transactions on Power Delivery* **9** (4) pp 1830-1837
[12] Jackson P O, Abrahamsson B, Gustavsson D and Igetoft L 1997 *IEEE Transactions on Power Delivery* **12** (2) pp1049-1052
[13] Zhao W J 2011 *HVDC Transmission Engineering* (Second edition) (Beijing: China Power Press) pp 26-27
[14] Zhang J W, Wang M H, Deng W B, Luo B and Zhu J 2012 *High Voltage Apparatus* **51**(32) pp 22-25
[15] Fu J H, Li Q Y, Gu Y G, Sun H T, Jiao Y F and Zuo K 2014 *Shaanxi Electric Power* **42**(10) pp 72-75
[16] Yang Z H, Xu N and Qiu Z X 2000 *Journal of Northeast University (Natural Science Edition)* **21**(4) pp 401-3

IOP Conf. Series: Materials Science and Engineering **474** (2019) 012018 doi:10.1088/1757-899X/474/1/012018

Effect of freezing and thawing cycle properties of cement-stabilized macadam materials with self-prepared early-strength agents

M Lin[1], Y C Guo[1], B S Tian[2] and Y L Song[1,*]

[1]School of Civil Engineering, Inner Mongolia University of Technology, Hohhot 010051, China
[2]Chang'an University, Xi'an 710064, China.

*E-mail: syllsj@sina.com

Abstract. Three types of cement-stabilized macadam materials and raw materials mixed with a self-made early-strength agent were prepared and subjected to freeze-thaw cycle tests. Based on the unconfined compression and freeze-thaw cyclic tests, the compressive strength and mass loss rate, and frost resistance coefficients of these samples were assessed and compared. The change of material properties caused by different content of early-strength agent was analyzed. The results show that three kinds of early-strength agents can improve the compressive strength and freezing resistance of cement-stabilized macadam. The compressive strength of cement-stabilized macadam with 2.0% anhydrous sodium sulfate and 1.0% triethanolamine is the best; the compressive strength is increased by 73.8%. Cement-stabilized macadam material with 1.5% anhydrous sodium sulfate and 1.0% triethanolamine has the best frost resistance, and its mass loss rate is reduced by 0.733%. The curves of mass loss rate and freeze-thaw cycle times of four groups of material are established, and the trend of positive correlation between the content of anhydrous sodium sulfate and the freezing resistance is obtained.

1. Introduction

At present, cement-stabilized macadam is used as subbase of most asphalt concrete pavement in China. This kind of material is between flexibility and rigidity, so it is called semi-rigid subbase. This subbase acts as main bearing structure, under the effect of reloading, fatigue performance of this material would significantly impact the pavement durability. Insofar as cement curing takes more than seven days to form the full strength, which seriously affects the construction progress, it is crucial to develop an early strength agent that can rapidly increase the strength of cement-stabilized macadam and reduce the curing time to speed up the construction progress. Zhang et al. [1] proved that the early-strength agents development which meet the low-temperature early strength requirements and get excellent operability is a vital direction of research now. [1]. Frigid zone (i.e., a zone between the Arctic Circle and the North Pole) in the North of China has as long as five months frost time in a year. This would make asphalt pavement cracking in this low-temperature environment. When the water permeates into the base from road surface course, the water freezes and expands at a temperature below 0°C, then internal structure of subbase course material would be damaged. Under repeated actions of vehicle load, the structure of semi-rigid subbase course material will be destroyed directly, mainly due to the temperature difference between day and night, and multiple freeze-thaw cycles during a day, the damage of semi-rigid subbase course material becomes aggravated [2-5]. Numerous research studies were devoted to the performance of semi-rigid subbase course materials [6-8]. By mixing an ordinary silicate cement and other types of cement in different proportions, several researchers, such as Péra and Ambroise [9], Janotka [10], Czernin [11], and Chatterjee [12], successfully prepared plenty of early strength cement, that could make the forming strength of cement faster in varying degrees. Through a series of experiments, Berger et al. [13, 14] revealed that gypsum could improve the hydration rate of sulphoaluminate cement to some extent, and thus enhance the early strength of sulphoaluminate cement. At the same time, the hydration rate of sulphoaluminate cement after 24 hours could be slowed down. Clark and Brown showed that lithium salt could also play a vital role in hydration of sulphoaluminate cement. They found a small amount of lithium carbonate could significantly improve the early strength of sulphoaluminate cement, while the setting rate of cement was accelerated to a certain extent [15,16]. By studying the structure and mechanical properties of triethanolamine (TEA), Ma et al. found that there is a double critical amount of TEA. A proper amount of TEA can effectively improve the compressive strength of cement and control the setting time of cement [17]. Sha Aimin got different materials fatigue life prediction models through fatigue indoor experiment of all kinds of semi-rigid base materials. Then he explains the influence of material on fatigue performance [18, 19]. In this paper, to improve the frost resistance of cement-stabilized macadam material, an early strength agent was given, the best dosage of early strength agent based on the freezing-thawing experiment was also provided, which has best anti-frost property. The curve relation between the dosage of early strength agent and the anti-frost coefficient is fitted.

2.Materials

2.1 Cement

We chose cement material according Technical Norm for Construction of Highway road bed (JTG-T-F20-2015) [20]. The result measured by testing all parameters of cement is recorded in Table 1.

Table1. Cement parameters

Project	Setting time, min		Compression strength, MPa		Rupture strength, MPa	
	Initial setting time	Final setting time	3d	28d	3d	28d
Technical requirements for cement raw materials	≥180	≥360 and≤600	≥16	≥32.5	≥3.5	≥5.5
experimental result	245	420	17.8	43	4.4	8.7

2.2 Aggregate、triethanolamine (TEA) and anhydrous sodium sulfat：Table 2

Table 2 Material origin and content

Material name	Source	Reagent content
aggregate	Mount Daqing stone plant in Hohhot	
TEA	Tianjin Chemical Co., Ltd. (China) was adopted	≥85%
anhydrous sodium sulfate	Tianjin Chemical Co., Ltd. (China) was adopted	≥99%

3.Empirical method

3.1 Unlimited compression test

We chose 3 effective implantation volume, which specific compounding ratio recorded in table 3, by doing complex hardening accelerator experiments. And compared them with control group. The testing period is 12h、24h、48h、72h、168h and 672h Every testing period needs one specimen.

Table 3 Each dosage group

Name	anhydrous sodium sulfate	TEA	Quantity
Volume I pressure	2.0%	1.0%	13
Volume II pressure	2.0%	0.75%	13
Volume III pressure	1.5%	1.0%	13

Through unconfined compression strength test, we find that the compressive strength of cement-stabilized macadam materials could be significantly improved by cement early-strength admixture of three trial groups, no matter the difference between three admixture [21]. During 24 hours, the strength results of unconfined compressive strength test respectively increased by 74% (dosage I), 62% (dosage II), 58% (dosage III), compared with the blank group compressive strength test. We used Friedman statistical analysis by SPSS software in order to study the difference of compressive strength between three admixture groups.

To analyze the difference between each group and eliminate the influence of the variation within the group, the data of each group was examined by the two-way grade variance analysis. Independent of the data of each group, the principal features in each group were ranked, the differences within the group were eliminated, and any differences between various dosage groups were checked. The statistical analysis of the compressive strength of 312 samples was carried out by using the SPSS software. The results showed no significant difference. If the probability (P) value was less than a given significance level of 0.05, the original hypothesis would be rejected. Therefore, it could be considered that there was a significant difference in the rank of each group of samples and that the distribution of the plurality of paired samples was significantly different; otherwise, the original hypothesis could not be rejected, and it could be considered that there was more or less same in the rank of each group of samples [22].

The statistical calculations are provided as in [22]

$$\aleph^2 = \frac{12}{bk(k+1)}\sum_{i=1}^{k}(R_i - \frac{b(k+1)}{2})^2 \quad (1)$$

where N^2 represents the chi-square statistic, b is the number of sample observations, k is the number of samples, and R_i is the sum of ranks of the ith sample.

Table 4 shows the statistics for the groups with different admixtures and their compressive strengths.

Table 4. Descriptive statistics and test statistics

	N	mean value	Std	minimum (min)	maximum (max)	Rank mean	x^2	df	Sig.
Volume 0 pressure	78	5.46	2.13	2.69	9.84	1.05	163.75	3	0.00
Volume I pressure	78	6.59	2.28	3.09	11.06	3.48			
Volume II pressure	78	6.40	2.07	3.23	10.51	3.13			
Volume III pressure	78	6.16	1.91	3.07	10.07	2.34			

In Table 4, N is the number of samples; the number of samples is the four mixed groups, with 78 samples in each group and a total of 312 samples; r is the rank of data; df is the degree of freedom; *Sig.* is significant level (P).

It can be seen from Table 4 that 0-admixture group has the minimum rank mean, so it has the minimum compressive strength. The samples mixed with early strength agent are all greater than the 0-admixture group. The rank's mean value of admixture I group is the largest, and its compressive strength is the highest. As shown in Table 4, the original hypothesis was rejected, because the progressive significance was $0.000 < 0.05$. So, there was a significant difference in the compressive strength of each mixed group.

It can be seen from figure 1 that the slopes of the four lines were almost identical, indicating that none of the three admixtures could change the time of intensity change. However, the admixture could improve the compressive strength of cement-stabilized macadam material. The admixtures with three kinds of dosages could all enhance the strength of cement-stabilized macadam material, and admixture I showed the best performance.

3.2. Freezing-thawing test

3.2.1 Preparation of the samples and preliminary procedure for the experiment.

The sample was prepared following the method of making the cylindrical sample of the stable material in the standard [23], and the concrete single side freeze-thaw test machine was used to conduct the freeze-thaw test.

There are four experimental groups: groups with the admixture of I, II, III and the control group. Each group has nine samples, numbered from 1 to 9. The mass of each sample was m0. The sample was placed in a freezer at a temperature of -18°C. After being frozen for 16h, the sample was taken out. The mass at each time was recorded. Then the sample was melted in a water tank for 8h at a constant temperature of 20°C. The water surface height in the same tank was 2.5 cm higher than the top of the sample. Afterward, the sample was removed, and the surface moisture was wiped with a wet cloth and the mass was weighed and recorded. This was the first freeze-thaw cycle. Then the sample was put into the freezer for the next freeze-thaw cycle. Finally, all the samples were subjected to five freeze-thaw cycles. A universal test machine was used to detect the maximum pressure (N) of samples submitted to compression failure after five freeze-thaw cycles, and to calculate the unconfined compressive strength.

3.3 Freeze-thaw mass loss rate

After five freeze-thaw cycles, the mass loss rates of the four experimental groups were recorded. To facilitate the analysis and summary of the experimental results, it would be convenient to compare the different experimental groups. The mean value of each experimental group was calculated. The mass loss rate of the first five freeze-thaw cycles of all samples was analyzed by Friedman statistics to further examine the difference in the mass loss rate of each group. Limited by the times of experiment, to further investigate the effect of this kind of early strength agent on the freeze-thaw resistance of cement-stabilized macadam material, the polynomial linear equation fitting was carried out from the existing five groups of data. According to the development trend of each set of fitting results, two freeze-thaw cycles were carried out. The mean mass loss rate of each group is shown in Table 5. The fitting equation is shown in Table 8. The relationship between the mean mass loss rate and the number of freeze-thaw cycles, and the two freeze-thaw cycles after the fitting is shown in Figure 1.

Table 5. The mass loss ratio of each group

Circulation	1	2	3	4	5
Group	Loss rate, %	Loss rate, %	Loss rate, %	Loss rate, %	Loss rate, %
Group 0	0.20	0.42	0.93	1.61	2.35
Group I	0.15	0.36	0.74	1.16	1.75
Group II	0.17	0.41	0.79	1.37	1.98
Group III	0.14	0.31	0.59	1.07	1.62

By analyzing Table 5, it could be concluded that zero-admixture group has the most considerable mass loss after five test cycles, with the loss rate of 2.353%. For the three groups with admixtures, the mass loss rate of the I-admixture group was 1.75%, and the loss rate of the II-admixture group was 1.98%. The mass loss rate of the III-admixture group was only 1.62%, And from Table 3, it was found that in the first two freeze-thaw cycles, the difference in mass loss rates among the four groups was very small. By the end of the second freeze-thaw cycle, the difference between the highest mass loss rate and the lowest mass loss rate in the four groups was 0.11%. After the end of the 5th cycle, the difference between the

IOP Conf. Series: Materials Science and Engineering **474** (2019) 012018 doi:10.1088/1757-899X/474/1/012018

highest mass loss rate and the lowest mass loss rate reached 0.73%. This indicated that the admixture could improve the freeze-thaw resistance of cement-stabilized macadam during the repeated freezing and thawing process.

Using the mass loss rate data of five freeze-thaw cycles in the four groups of samples (180), data enlisted in Table 6 were obtained by the Friedman statistics.

Table 6. Descriptive statistics and test statistics

	N	Mean value	Std	Minimum	Maximum	Rank mean	x^2	*df*	Sig.
Group 0	45	1.10	0.80	0.19	2.38	3.91	130.63	3	0.00
Group I	45	0.83	0.58	0.14	1.76	2.00			
Group II	45	0.94	0.66	0.16	1.99	3.09			
Group III	45	0.75	0.54	0.13	1.63	1.00			

In Table 6, *N is* the number of samples, which includes 45 samples in four mixed groups and equals to a total of 180 samples; x^2 is the chi-square statistic; *df* is the degree of freedom of the data; Sig. is significant level (P).

Because of the increasing significance of freeze-thaw cycle was 0.000<0.05, the original hypothesis was rejected, indicating that there were significant differences in the mass loss rate of the groups with different admixtures. It can be seen in Figure 1 that the mass loss rate increased with the increase of the number of freeze-thaw cycles, and the difference between the zero-admixture group became more and more apparent, showing that the anti-freezing performance was improved.

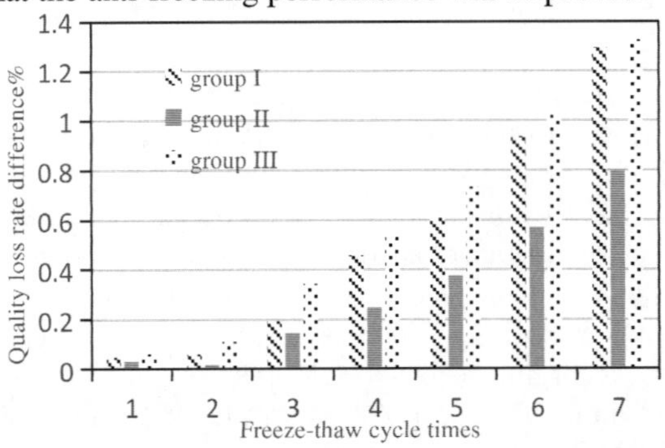

Figure 1. The difference of mass loss rate between the test and control groups

Table 7. Fitting equations and parameters

Group	Fitting relation y= ax2+bx+c	Correlation coefficient: R^2
Group 0	y=0.0868x2+0.0291x+0.0602	0.9982
Group I	y=0.0572x2+0.0548x+0.0371	0.9993
Group II	y=0.0676x2+0.0517x+0.0448	0.9992
Group III	y=0.0685x2-0.0392x+0.1097	0.9994

Table 7 tabulates the fitting formulas for the four admixture groups, which results are plotted in Figure 2. The obtained correlation coefficients R^2 for all fitting curves and the respective test results are above 0.99, suggesting their close correlation. The fitting curves also show the trend of the test results. After fitting two freeze-thaw cycles, the most significant mass loss rate was still in the zero-admixture group, and the mass loss rate was the lowest in group I. According to five tests and two fitting freeze-thaw cycles, it could be concluded that the admixture could not only improve the antifreeze property of materials at the initial stage of the cycle but also show a better trend of anti-freeze-thaw with the increase of freeze-thaw times. This occurs because TEA and ASS could enhance the antifreeze property of materials, and the antifreeze property could be further greatly improved on the original basis. Admixture III performed better in reducing the mass loss rate in the first five cycles. But the slope of the admixture I curve was the slowest, revealing that in a certain range in the case of equal dose, TEA and ASS dosage would directly affect the antifreeze property of the material.

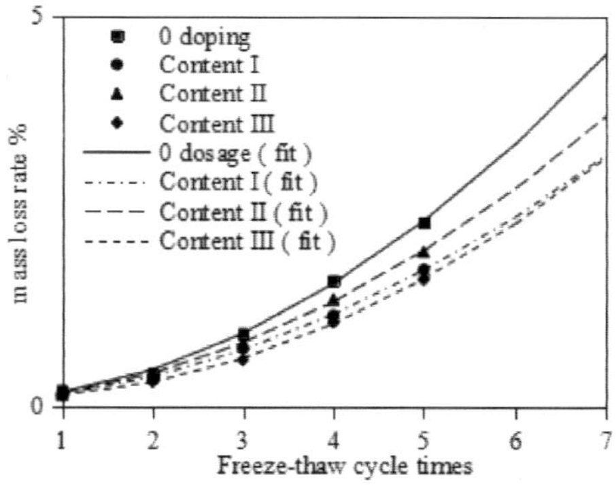

Figure 2. Mean mass loss rate of the freeze-thaw cycle

To verify the conclusions presented above, partial correlation analysis was performed on the data of the first five freeze-thaw cycles of 180 samples by the SPSS software. Time variables were controlled. The correlation between admixture group and mass loss rate was analyzed, and the correlation data were calculated by Eq. (2) [22]. The analysis results are listed in Table 8.

$$r = \frac{\sum XY - \frac{\sum X \sum Y}{N}}{\sqrt{\left(\sum X^2 - \frac{(\sum X)^2}{N}\right)\left(\sum Y^2 - \frac{(\sum Y)^2}{N}\right)}} \quad (2)$$

where r is the correlation coefficient between the admixture group and the mass loss rate; X is the number of admixture groups (four, in this study); Y is the mass loss rate of each sample in each admixture group; N is the number of samples.

Table 8. Correlation of each group and its mass loss rate

Controlled variable			Mass loss rate	Admixture group
Time	Mass loss rate	Relativity	1.00	-0.51
		Sig.	.	0.00
		df	0	177
	Admixture group	Relativity	-.51	1.00
		Sig.	0.00	.
		df	177	0

According to the analysis of Table 8, the partial correlation between the admixture group and the mass loss rate was -0.512. The relationship of each admixture group's mass loss rate was: group 0 > group I > group II > group III, in which the content of TEA in group I was the same as that in group III, but the amount of ASS in group I was higher than that in group III. It could be concluded that more ASS led to better antifreeze property of the material when the amount of TEA was between 0% and 1.25%.

3.4 Antifreeze coefficient

The antifreeze coefficient reflects the unconfined compressive strength of cement-stabilized macadam after many freezing-thawing cycles. The larger the antifreeze coefficient was, the smaller the strength loss caused by freezing-thawing of the cement-stabilized macadam material. Therefore, the antifreeze coefficient reflects the freezing-performance of materials on a certain level. According to the standards

[23], 72 samples were prepared and subdivided into four groups, with 18 samples in each test group. Nine samples of each test group were subjected to five freeze-thaw cycles, and then unconfined compression tests were conducted. Nine remaining samples of the same test group were subjected to unconfined compression tests without freezing and thawing.

According to the formula of antifreeze coefficient, the antifreeze coefficient of each group of cement-stabilized macadam was calculated and listed in Table 9. The calculation method is given by

$$BDR = \frac{R_{DC}}{R_C} \times 100\ \% \tag{3}$$

where *BDR is* the loss rate of compressive strength after *n* freeze-thawing cycles (%), R_{DC} and R_C are unconfined compressive strength values of samples of the same test group either subjected to *n* cycles of freezing and thawing or not, respectively (MPa).

Table9. Antifreeze coefficients of samples after five freeze-thaw cycles (%)

Group	Pressure after five freeze-thaw cycles, MPa	Non-freeze-thaw cycle pressure, MPa	The coefficient of freeze resistance
Group 0	7.92	9.34	84.80
Group I	9.76	10.82	90.20
Group II	8.97	10.14	88.46
Group III	9.16	9.56	95.82

Through the data analysis and calculation results in Table 9, four admixture groups could be obtained after five freeze-thaw cycles of the unconfined compression test. The mass loss rate of zero-admixture in control group 0 was the maximum, namely 15.20%, and the antifreeze coefficient was 84.80%. The mass loss rate of admixture in the test group I was 9.80%, with the antifreeze coefficient of 90.20%. The mass loss rate of admixture in group II was 11.54%, with the antifreeze coefficient of 88.46%. The mass loss rate of admixture in group III was only 4.18%, with the antifreeze coefficient of 95.82%. Thus, group III showed the largest antifreeze coefficient.

2. Conclusions

- ✓ All 3 new early-strength agents can make improvement on the compressive strength and frost resistance of cement-stabilized macadam material. The compressive strength also can be enhanced by TEA and ASS significantly within a short period.
- ✓ From the analysis of compressive strength and freeze resistance coefficient, it can be found that adding admixtures of compressive strength and frost resistance properties were superior to the control group. Thus, under the same conditions, with 2.0%-ASS and 1.0%-TEA, the cement-stabilized macadam material compressive strength values in all agents were optimal. Meanwhile, with 1.5%-ASS and 1.0%-TEA cement-stabilized macadam material antifreeze performance in five cycles were optimal.
- ✓ In the freeze-thaw cycle - mass loss curve, the slope growth rate of curve III is higher than that of curve I with the increasing times of freezing and thawing. It can be observed that 2.0% ASS and 1.0% TEA early strength agent has the best development prospects.
- ✓ Through the index of compressive strength and freezing resistance coefficient, it was found that when the content of TEA was 0.75, the effect of the three groups was worse than that of 1.0% TEA. The quality loss of rate of each dosage group was lower than that of the control group, and the freezing-performance of each dosage group was better than that of the control group.

References

[1] Zhang F, Bai Y, Cai Y, Du J, Ning F. 2017. Research status of low temperature early-strength agents for concrete. Materials Review (2017-21), http://en.cnki.com.cn/Journal_en/B-B020-CLDB-2017-21.htm.

[2] Kosmatka S H, Kerkhoff B, Panarese W C. 2002. *Design and Control of Concrete Mixtures* (Portland Cement Association, PCA R&D Serial Number SN2561, US).
[3] Yang J 2005. Study on performance of ultra-early strong cement-stabilized macadam. Chang'an University. 4-6.
[4] Gong L S, Liu C P 1990. *Durability of concrete and its protection and maintenance.* (Beijing: China Architecture Industry Press) 3.
[5] Wang Y Q, Tan Y Q, Wang X L 2017. *Highway 4. 239*-242.
[6] Yang Q, Zhu W X 2000. *Material and Structures.*33(4).229-234.
[7] Yan Y P 2012. Research on the technology of improving crack resistance of cement-stabilized macadam base in cold area. Chang'an University Master's Degree Thesis.
[8] Huang G X, Hui R Y 1990. *Shrinkage of Concrete.* (Beijing: China Railway Press) 1-79.
[9] Péra J, Ambroise J 2004. *Cement & Concrete Research.*34 (4) 671-676.
[10] Janotka I, Krajči L, Ray A, et al. 2003. *Cement & Concrete Research.*33 (4) . 489-497.
[11] Czernin W 1962. *Cement Chemistry and Physics for Civil Engineers* (Australia: Lockwood)
[12] Chatterjee A K 1991 Special and new cements. *India 9th International Congress on the Chemistry of Cement* (India) 26-9.
[13] Berger S, Coumes C C D, Bescop P L, *et al.* 2009. *Cement & Concrete Research.* 39(12). 1180-1187.
[14] Bernardo G, Telesca A, Valenti G L 2006. *Cement & Concrete Research*, 36(6). 1042-1047.
[15] Clark B A, Brown P W 1999. *Cement & Concrete Research,* 30(2).233-240.
[16] Clark B A, Brown P W 1999 *Cement & Concrete Research,* 30(2). 233-240.
[17] Ma B G, Xu Y H, Dong R Z 2006. *J. Building Materials,* 9(1). 6-9.
[18] Sha A M, Jia K, Li X G 2009. Fatigue performances of semi-rigid base course materials. *J. Transp. Engin.* 9(3). 29-33, www.researchgate.net/scientific-contributions/2090625835_A-M_Sha.
[19] Zhou H, Sha A M, Hu L Q 2012. *J. Chang'an Univ: Natural Science Edition* 32(3).6-10.
[20] Research Institute of Highway Ministry of Transport. *JTG-T-F20-2015. Road Base Construction Technical Specification*. Beijing: China Communications Press. 2015 (in Chinese)
[21] Research Institute of Highway Ministry of Transport. *JTG E42—2005 Test Methods of Aggregate for Highway Engineering*. Beijing: China Communications Press, 2005 (in Chinese)
[22] Xue W 2017. *Statistical Analysis and Application of SPSS (Fifth Edition).* (Beijing Renmin University of China Press) 150-6.
[23] Industry Standard of the People's Republic of China. *JTG E51-2009. Test Procedure for Inorganic Materials for Highway Engineering*. Beijing: China Communications Press. 2009 (in Chinese)

General expression of pressure angle and principal curvature of globoidal indexing cam mechanism

J Li [1], X Rao [1], M Tang[1*], Y J Li[2]

[1]School of Mechanical Engineering, Chengdu University, Chengdu, 610106, PR China
[2]School of Industrial and Systems Engineering, Georgia Institute of Technology, Atlanta, GA 30332, USA

*E-mail: 49660292@qq.com

Abstract. Given the lack of general mathematical model for the parameters of the globoidal indexing cam (GIC) mechanism, a method of modeling the contour for the GIC based on the unified mathematical expression is proposed by applying the envelope theory and the matrix transformation of the homogeneous coordinates. Taking the globoidal cams with cylindrical, conical, and hyperbolic meshing turret-rollers as the research object, a unified mathematical model of profile model for the GIC is derived. On this foundation, the general analytic expression of pressure angle and principal curvature is further obtained, which provides a theoretical basis for the precise design and manufacture of the GIC and provides a mathematical basis for the fatigue strength analysis of the GIC material.

1. Introduction

Globoidal indexing cam (GIC) is widely used in various intermittent motion mechanisms, which can meet the needs of substantial rigidity, massive torque, and accurate positioning. The pressure angle is one of the critical parameters affecting the performance, the force, and the structural size of this kind of mechanism. Therefore, the pressure angle of the camber cam mechanism must be analyzed. Besides, to prevent curvature interferences between the cam surface and the roller surface in the GIC mechanism and the root cutting phenomena when the cam is processed, it is necessary to analyze the principal curvature of the curved conjugate surface of the GIC. All along, scholars at home and abroad have done a lot of research on the profile equation, the pressure angle, and the principal curvature of GIC. A general contour equation of GIC is derived using the theory of generating method and the conjugate principle of the spatial surface [1-2]. A general equation for curved surfaces of GIC is established based on the meshing theory and differential geometry of spatial mechanisms [3]. A contour equation of globoidal cam of steel ball roller is determined using the approach of rotation transformation tensor, and the analytical expression of pressure angle is derived [4-5].

Furthermore, expressions of pressure angle and curvature of cylindrical roller follower globoidal cam mechanism are also derived from the principle of spatial meshing [6-7]. Besides, the pressure angle of globoidal cam mechanism with spherical cone roller is calculated and analyzed [8]. These studies are significant, but the derivation processes are complex and geometric intuition is not reliable. At the same time, different roller shapes and the angle between the camshaft and the rotating axis of the turret result in disunity of the final mathematical expressions in these references. This increases the difficulty and workload of formula deduction to some extent and further complicates the development of the globoidal cam software. At present, there has not been any study on the general expression of the pressure angle and the principal curvature of the GIC mechanism using the theory of envelopes for a one-parameter family of surfaces.

In this paper, based on the theory of envelopes for a one-parameter family of surfaces [9-12], a unified treatment of homogeneous coordinate transformation is performed for different roller shapes (cylindrical, conical, and hyperbolic ones) of the GIC and different angles between the camshaft and the turret rotating axis. The cam profile of the GIC is also derived via the same procedure. Moreover, a general expression of pressure angle and principal curvature of GIC mechanism is then established based on the unified mathematical model.

2. The Contour Surface Equation of the Globoidal Cam

The rotational angular displacement ϕ_c of the GIC is regarded as the parametric variable of the surface family. The outer surface equation of the roller and the curved cam profile is described by the roller angle θ and the roller height h. The unified mathematical model of the contour for the GIC is established using the theory of envelopes.

2.1 A unified mathematical expression of the surface equation of the outer surface of the rollers

The curve of the GIC mechanism is shown in figure 1. The mechanism consists of three parts, namely a globoidal cam, a turret, and a roller. In this paper, three kinds of rollers are meshed with the globoidal cam to generate the cam profile. As shown in figure 2, (a) is a cylindrical roller, (b) is a conical roller, and (c) is a hyperbolic roller. A coordinate system $(XYZ)_1$ is established. Specifically, r is the bottom radius of the roller (h=0), and L is the height of the roller. Here θ and h represent roller rotation angle and height respectively while γ represents the angle between Z_1 axis and the generating line of the ruled surface for the conical roller and the hyperbolic roller.

Let $\boldsymbol{R}_{cy}$, $\boldsymbol{R}_{co}$, and $\boldsymbol{R}_{hy}$ represent the surface equations of the cylindrical, the conical, and the hyperbolic rollers in the $(XYZ)_1$ coordinate system respectively. Assuming that the unit direction vectors of x, y, and z axes are $\boldsymbol{i}$, $\boldsymbol{j}$, and $\boldsymbol{k}$ respectively, the surface equations of these rollers are expressed in the coordinate system $(XYZ)_1$ as follows:

$$\boldsymbol{R}_{cy} = r\cos\theta\boldsymbol{i} + r\sin\theta\boldsymbol{j} + h\boldsymbol{k} \tag{1}$$

$$\boldsymbol{R}_{co} = (r + h\tan\gamma)\cos\theta\boldsymbol{i} + (r + h\tan\gamma)\sin\theta\boldsymbol{j} + h\boldsymbol{k} \tag{2}$$

$$\boldsymbol{R}_{hy} = (r\cos\theta - h\tan\gamma\sin\theta)\boldsymbol{i} + (r\sin\theta + h\tan\gamma\cos\theta)\boldsymbol{j} + h\boldsymbol{k} \tag{3}$$

We can unify Eqs. (1), (2), and (3) as follows:

$$\boldsymbol{R}_1 = [(r + W_1 h\tan\gamma)\cos\theta - W_2 h\tan\gamma\sin\theta]\boldsymbol{i} + [(r + W_1 h\tan\gamma)\sin\theta + W_2 h\tan\gamma\cos\theta]\boldsymbol{j} + h\boldsymbol{k} \tag{4}$$

where

$W_1=0,\ W_2=0$ for cylindrical roller

$W_1=1,\ W_2=0$ for conical roller

$W_1=0,\ W_2=1$ for hyperbolic roller

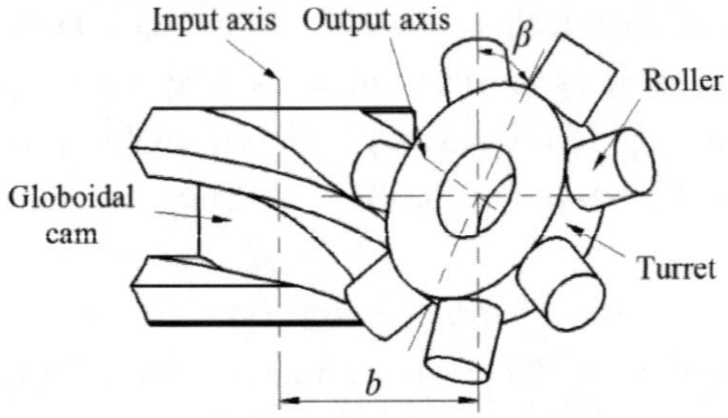

Figure 1. Diagram of a GIC mechanism

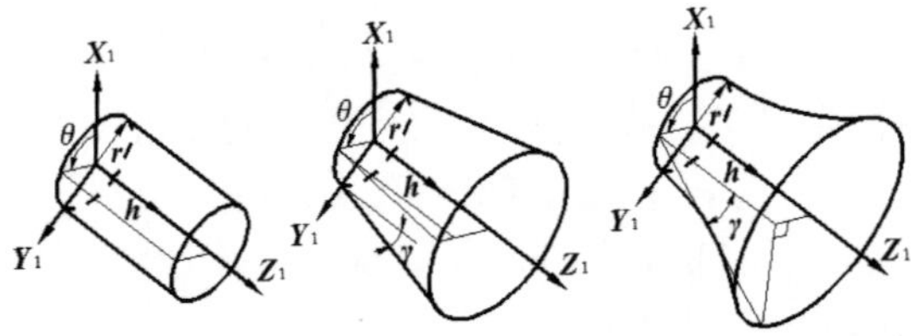

(a) Cylindrical roller (b) Conical roller (c) Hyperbolic roller

Figure 2. Diagrammatic sketch of the driven rollers

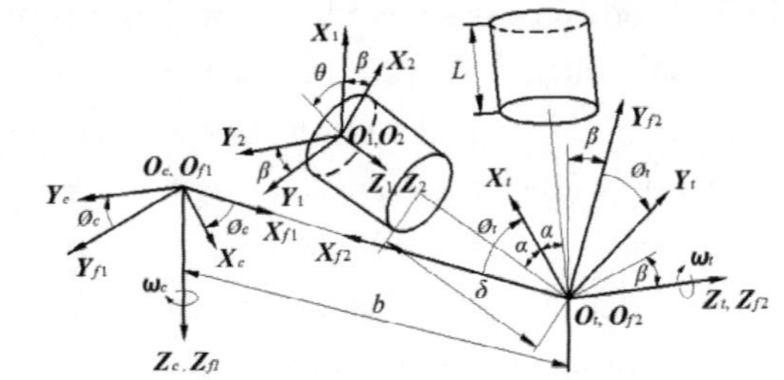

Figure 3. Coordinate system

2.2 Matrix transformation of homogeneous coordinates

The matrix transformation of homogeneous coordinates can be used to make the spatial positional relationship between the GIC and the follower more intuitionistic [13-14]. As shown in figures 1 and 3, the necessary design parameters and corresponding coordinate systems are given. Figure 3 shows the two fixed coordinate systems $(XYZ)_{f1}$ and $(XYZ)_{f2}$ with four relevant moving coordinate systems $(XYZ)_c$, $(XYZ)_t$, $(XYZ)_1$, and $(XYZ)_2$. Among them, the rotation axis of the globoidal cam is Z_{f1}, and the rotation axis of the turret is Z_{f2}. Angle β is defined between the positive axis Y_{f2} and the negative one Z_{f1}. The coordinate origins O_t and O_{f2} coincide at the turret center, while O_c and O_{f1} coincide at the globoidal cam center, respectively. Furthermore, the coordinate origins O_1 and O_2 are located at the center of the bottom plane of the roller (when h=0). Z_1 is collinear with Z_2 and X_{f1} is collinear with X_{f2}. The rotational angular displacements of the globoidal cam and turret are ϕ_c and ϕ_t, respectively. Positive ϕ_c and ϕ_t are measured counterclockwise for Z_{f1} and Z_{f2}, respectively. The distance between the globoidal cam and turret axes is b, while that between the turret center and the roller top surface is δ (when $h=L$). Here α is the half-angle between the axes of two adjacent rollers.

In the homogeneous coordinates, the coordinate transformation is described by 4×4 matrices, according to the coordinate system and set parameters. The detailed transformation process is as follows:

The transformation matrices providing the transformation between the following coordinate systems are as follows:

from $(XYZ)_1$ to $(XYZ)_2$:

$$\boldsymbol{T}_1=\begin{bmatrix}\cos\beta & -\sin\beta & 0 & 0\\ \sin\beta & \cos\beta & 0 & 0\\ 0 & 0 & 1 & 0\\ 0 & 0 & 0 & 1\end{bmatrix} \tag{5}$$

from $(XYZ)_2$ to $(XYZ)_t$:

$$\boldsymbol{T}_2=\begin{bmatrix}\sin\alpha & 0 & -\cos\alpha & (\delta+L)\cos\alpha\\ \cos\alpha & 0 & \sin\alpha & -(\delta+L)\sin\alpha\\ 0 & -1 & 0 & 0\\ 0 & 0 & 0 & 1\end{bmatrix} \tag{6}$$

from $(XYZ)_t$ to $(XYZ)_{f2}$ is:

$$\boldsymbol{T}_3=\begin{bmatrix}\cos\phi_t & -\sin\phi_t & 0 & 0\\ \sin\phi_t & \cos\phi_t & 0 & 0\\ 0 & 0 & 1 & 0\\ 0 & 0 & 0 & 1\end{bmatrix} \tag{7}$$

from $(XYZ)_{f2}$ to $(XYZ)_{f1}$:

$$\boldsymbol{T}_4=\begin{bmatrix}-1 & 0 & 0 & b\\ 0 & -\sin\beta & -\cos\beta & 0\\ 0 & -\cos\beta & \sin\beta & 0\\ 0 & 0 & 0 & 1\end{bmatrix} \tag{8}$$

from $(XYZ)_{f1}$ to $(XYZ)_c$:

$$\boldsymbol{T}_5=\begin{bmatrix}\cos\phi_c & \sin\phi_c & 0 & 0\\ -\sin\phi_c & \cos\phi_c & 0 & 0\\ 0 & 0 & 1 & 0\\ 0 & 0 & 0 & 1\end{bmatrix} \tag{9}$$

IOP Conf. Series: Materials Science and Engineering **474** (2019) 012019 doi:10.1088/1757-899X/474/1/012019

2.3 Establishment of the general expression for profile equation of GIC

Equation (4) can be further reduced as follows:

$$\boldsymbol{R}_1 = [(r+W_1 h\tan\gamma)\cos\theta - W_2 h\tan\gamma\sin\theta]\boldsymbol{i} + [(r+W_1 h\tan\gamma)\sin\theta + W_2 h\tan\gamma\cos\theta]\boldsymbol{j} + h\boldsymbol{k}$$
$$= [A_1\cos\theta - A_2\sin\theta]\boldsymbol{i} + [A_1\sin\theta + A_2\cos\theta]\boldsymbol{j} + h\boldsymbol{k} \tag{10}$$

where $A_1 = r + W_1 h\tan\gamma$ and $A_2 = W_2 h\tan\gamma$ for $0 \le \theta < 2\pi$ and $0 \le h < L$.

In homogeneous coordinates, the outer surface parameter equation $\boldsymbol{R}_1$ of the driven roller is expressed in a four-order matrix form as follows:

$$\boldsymbol{R}_1 = \begin{bmatrix} A_1\cos\theta - A_2\sin\theta \\ A_1\sin\theta + A_2\cos\theta \\ h \\ 1 \end{bmatrix} \tag{11}$$

The rotational angular displacement ϕ_c of GIC is the input motion parameter of the mechanism. The rotational angular displacement ϕ_t of the turret is the output motion parameter of the mechanism. It is assumed that the law of motion of ϕ_t varies with ϕ_c, where $\phi_t = \phi_t(\phi_c)$ has been given in advance. The parametric equation of the GIC contour surface is as follows:

$$\boldsymbol{r}_c = \boldsymbol{r}_c(\theta, h, \phi_c) = \boldsymbol{T}_5\boldsymbol{T}_4\boldsymbol{T}_3\boldsymbol{T}_2\boldsymbol{T}_1\boldsymbol{R}_1$$
$$= \{A_1[\cos(\theta+\beta)\sin(\phi_t-\alpha)\cos\phi_c - \cos(\theta+\beta)\sin\beta\cos(\phi_t-\alpha)\sin\phi_c + \sin(\theta+\beta)\cos\beta\sin\phi_c] - A_2[\sin(\theta+\beta)\sin(\phi_t-\alpha)$$
$$\cos\phi_c - \sin(\theta+\beta)\sin\beta\cos(\phi_t-\alpha)\sin\phi_c - \cos(\theta+\beta)\cos\beta\sin\phi_c] - (\delta+L-h)[\cos(\phi_t-\alpha)\cos\phi_c + \sin\beta\sin(\phi_t-\alpha)\sin\phi_c]$$
$$+ b\cos\phi_c\}\boldsymbol{i}$$
$$-\{A_1[\cos(\theta+\beta)\sin(\phi_t-\alpha)\sin\phi_c + \cos(\theta+\beta)\sin\beta\cos(\phi_t-\alpha)\cos\phi_c - \sin(\theta+\beta)\cos\beta\cos\phi_c] - A_2[\sin(\theta+\beta)\sin(\phi_t-\alpha)$$
$$\sin\phi_c + \sin(\theta+\beta)\sin\beta\cos(\phi_t-\alpha)\cos\phi_c + \cos(\theta+\beta)\cos\beta\cos\phi_c] - (\delta+L-h)[\cos(\phi_t-\alpha)\sin\phi_c - \sin\beta\sin(\phi_t-\alpha)\cos\phi_c]$$
$$+ b\cos\phi_c\}\boldsymbol{j}$$
$$-\{A_1[\cos(\theta+\beta)\cos\beta\cos(\phi_t-\alpha) + \sin(\theta+\beta)\sin\beta] - A_2[\sin(\theta+\beta)\cos\beta\cos(\phi_t-\alpha) - \cos(\theta+\beta)\sin\beta] + (\delta+L-h)$$
$$\cos\beta\sin(\phi_t-\alpha)\boldsymbol{k} \tag{12}$$

for $0 \le \theta < 2\pi$ and ϕ_c is an independent cam motion parameter.

$$\frac{\partial \boldsymbol{r}_c}{\partial \theta} \times \frac{\partial \boldsymbol{r}_c}{\partial h} =$$
$$\{[-A_1\sin\beta\sin\phi_c\cos(\theta+\beta) + A_2\sin\beta\sin\phi_c\sin(\theta+\beta) - (A_1A_3 + A_2A_4)\cos\phi_c]\cos(-\phi_t+\alpha) + [-A_1\cos\phi_c\cos(\theta+\beta) + A_2$$
$$\cos\phi_c\sin(\theta+\beta) + (A_1A_3 + A_2A_4)\sin\beta\sin\phi_c]\sin(-\phi_t+\alpha) + \cos\beta\sin\phi_c[A_2\cos(\theta+\beta) + A_1\sin(\theta+\beta)]\}\boldsymbol{i}$$
$$+\{[-A_1\sin\beta\cos\phi_c\cos(\theta+\beta) + A_2\sin\beta\cos\phi_c\sin(\theta+\beta) + (A_1A_3 + A_2A_4)\sin\phi_c]\cos(-\phi_t+\alpha) + [A_1\sin\phi_c\cos(\theta+\beta) - A_2$$
$$\sin\phi_c\sin(\theta+\beta) + (A_1A_3 + A_2A_4)\sin\beta\cos\phi_c]\sin(-\phi_t+\alpha) + \cos\beta\cos\phi_c[A_2\cos(\theta+\beta) + A_1\sin(\theta+\beta)]\}\boldsymbol{j}$$
$$+\{-\cos\beta[A_1\cos(\theta+\beta) - A_2\sin(\theta+\beta)]\cos(-\phi_t+\alpha) + (A_1A_3 + A_2A_4)\sin(-\phi_t+\alpha) - \sin\beta[A_2\cos(\theta+\beta) + A_1$$
$$\sin(\theta+\beta)]\}\boldsymbol{k} \tag{13}$$

where $A_3 = W_1\tan\gamma$ and $A_4 = W_2\tan\gamma$

The contour equation of GIC should satisfy both Eq. (12) and the following one:

$$\frac{\partial \boldsymbol{r}_c}{\partial \theta} \times \frac{\partial \boldsymbol{r}_c}{\partial h} \square \frac{\partial \boldsymbol{r}_c}{\partial \phi_c} =$$
$$\{[A_1 b\sin\beta + A_2\cos\beta(-A_1A_3 - A_2A_4 + \delta + L - h)]\cos(\theta+\beta) + [-A_2 b\sin\beta + A_1\cos\beta(-A_1A_3 - A_2A_4 + \delta + L - h)]\sin(\theta+$$
$$\beta)\}\cos(-\phi_t+\alpha) - \sin\beta b\left(A_1A_3 + A_2A_4\right)\sin(-\phi_t+\alpha) + [-A_1(-A_1A_3 - A_2A_4 + \delta + L - h)\sin\beta - A_2 b\cos\beta + \phi_t' A_1(-A_1A_3$$
$$-A_2A_4 + \delta + L - h)]\cos(\theta+\beta) + \sin(\theta+\beta)[A_2(-A_1A_3 - A_2A_4 + \delta + L - h)\sin\beta - A_1 b\cos\beta - \phi_t' A_2(-A_1A_3 - A_2A_4 + \delta$$
$$+L-h)] = 0 \tag{14}$$

where $\phi_t' = \dfrac{d\phi_t}{d\phi_c}$.

From Eq. (14), we can obtain:

$$\theta = 2\arctan\left[\frac{-H \pm \sqrt{H^2 + I^2 - J^2}}{J - I}\right] - \beta \tag{15}$$

where

$$H = [-A_1^2A_3 + A_1(\delta + L - h) - A_1A_2A_4]\cos\beta\cos(\phi_t+\alpha) - A_1 b\cos\beta - A_2 b\sin\beta\cos(\phi_t+\alpha) + [A_2^2A_4 - A_2(\delta+L-h)]$$
$$(\phi_t' - \sin\beta)$$
$$I = [-A_2^2A_4 + A_2(\delta+L-h)]\cos\beta\cos(\phi_t+\alpha) - A_2 b\cos\beta + A_1 b\sin\beta\cos(\phi_t+\alpha) + [-A_1^2A_3 + A_1(\delta+L-h) - A_1A_2A_4]$$
$$(\phi_t' - \sin\beta)$$
$$J = \left(A_1A_3 + A_2A_4\right)\sin\beta\sin(\phi_t+\alpha)$$

Equation (15) can be substituted into Eq. (12) to eliminate θ and yield the general expression of the GIC profile equation:

$$\boldsymbol{r}_c = \boldsymbol{r}_c(h, \phi_c) \tag{16}$$

Two values for θ can be obtained from Eq. (15). They represent the outline surface profile coordinates of two sides of the globoidal cam groove, respectively.

3. Pressure angle and principal curvature analysis of GIC mechanism

Based on the general equation of the GIC profile, general analytic expressions of the pressure angle and the principal curvature of the surface of the cam mechanism can be easily derived. This will lay a foundation for further studies of the machining error, the dynamic characteristics, and the design of the globoidal cam mechanisms.

3.1 The general expression of the pressure angle of GIC

The pressure angle of the globoidal cam mechanism refers to the angle between the moving direction of the driven part and the normal contour of the cam surface at the conjugate point of contact between the globoidal cam and the outer surface of the driven roller. As shown in figure 4, let the common unit normal vector of the cam and the driven roller at the contact point be $\boldsymbol{n}$, the pressure angle at this point be ψ, and the unit direction vector of the driven part be $\boldsymbol{p}$. Here $\boldsymbol{n}$ is defined by the following formula:

$$\boldsymbol{n} = \frac{\dfrac{\partial \boldsymbol{r}_c}{\partial \theta} \times \dfrac{\partial \boldsymbol{r}_c}{\partial h}}{\left|\dfrac{\partial \boldsymbol{r}_c}{\partial \theta} \times \dfrac{\partial \boldsymbol{r}_c}{\partial h}\right|} = \frac{B_1\boldsymbol{i} + B_2\boldsymbol{j} + B_3\boldsymbol{k}}{\left[\left(A_3^2+1\right)A_1^2 + 2A_1A_2A_3A_4 + A_2^2\left(A_4^2+1\right)\right]^{1/2}} \tag{17}$$

where

$B_1 = [-\sin\beta\sin\phi_c\cos(\theta+\beta)A_1 + \sin\beta\sin\phi_c\sin(\theta+\beta)A_2 - \cos\phi_c\left(A_1A_3 + A_2A_4\right)]\cos(-\phi_t+\alpha) + [-\cos\phi_c A_1\cos(\theta+\beta)$
$+\cos\phi_c A_2\sin(\theta+\beta) + \sin\beta\sin\phi_c\left(A_1A_3 + A_2A_4\right)]\sin(-\phi_t+\alpha) + \cos\beta\sin\phi_c[\cos(\theta+\beta)A_2 + \sin(\theta+\beta)A_1]$

$B_2 = [-\sin\beta\cos\phi_c\cos(\theta+\beta)A_1 + \sin\beta\cos\phi_c\sin(\theta+\beta)A_2 + \sin\phi_c\left(A_1A_3 + A_2A_4\right)]\cos(-\phi_t+\alpha) + [-\sin\phi_c A_1\cos(\theta+\beta)$
$-\sin\phi_c A_2\sin(\theta+\beta) + \sin\beta\cos\phi_c\left(A_1A_3 + A_2A_4\right)]\sin(-\phi_t+\alpha) + \cos\beta\cos\phi_c[\cos(\theta+\beta)A_2 + \sin(\theta+\beta)A_1]$

$B_3 = -\cos\beta[\cos(\theta+\beta)A_1 - \sin(\theta+\beta)A_2]\cos(-\phi_t+\alpha) + \cos\beta\left(A_1A_3 + A_2A_4\right)\sin(-\phi_t+\alpha) - \sin\beta[\cos(\theta+\beta)A_2 +$
$\sin(\theta+\beta)A_1]$

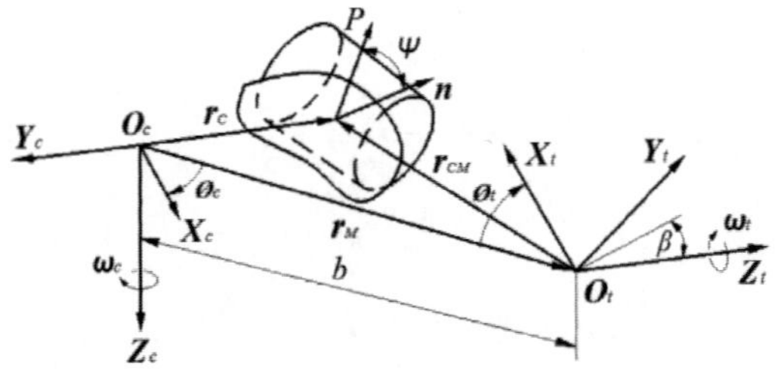

Figure 4. Diagrammatic sketch of the pressure angle

The unit direction vector $\boldsymbol{p}$ of follower motion can be determined by the following formula:

$$\boldsymbol{p} = \frac{\mathbf{W}_t \times \boldsymbol{r}_{MC}}{|\mathbf{W}_t \times \boldsymbol{r}_{MC}|} = \frac{\mathbf{W}_t \times (\boldsymbol{r}_C - \boldsymbol{r}_M)}{|\mathbf{W}_t \times (\boldsymbol{r}_C - \boldsymbol{r}_M)|} \tag{18}$$

where $\boldsymbol{W}_t$ is the velocity vector of the follower.

In the $(XYZ)_t$ coordinate system, $\boldsymbol{W}_t$ is represented by a matrix in homogeneous coordinates as follows:

$$\boldsymbol{W}_t = [0,\ 0,\ \omega_t,\ 1]^{\mathrm{T}} \tag{19}$$

After a series of coordinate transformations, the following equation can be derived:

$$\boldsymbol{W}_t = \begin{bmatrix} \cos\phi_c & \sin\phi_c & 0 & 0 \\ -\sin\phi_c & \cos\phi_c & 0 & 0 \\ 0 & 0 & 1 & 0 \\ 0 & 0 & 0 & 1 \end{bmatrix} \cdot \begin{bmatrix} -1 & 0 & 0 & 0 \\ 0 & -\sin\beta & -\cos\beta & 0 \\ 0 & -\cos\beta & \sin\beta & 0 \\ 0 & 0 & 0 & 1 \end{bmatrix} \cdot \begin{bmatrix} 0 \\ 0 \\ \omega_t \\ 1 \end{bmatrix}$$
$$= \omega_t(-\sin\phi_c\cos\beta\boldsymbol{i} - \cos\phi_c\cos\beta\boldsymbol{j} + \sin\beta\boldsymbol{k}) \tag{20}$$

where ω_t is the angular velocity of the driven part.

As shown in figure 4, $\boldsymbol{r}_M$ can be obtained as follows:

$$\boldsymbol{r}_M = \begin{bmatrix} \cos\phi_c & \sin\phi_c & 0 & 0 \\ -\sin\phi_c & \cos\phi_c & 0 & 0 \\ 0 & 0 & 1 & 0 \\ 0 & 0 & 0 & 1 \end{bmatrix} \cdot \begin{bmatrix} b \\ 0 \\ 0 \\ 1 \end{bmatrix} = b\cos\phi_c\boldsymbol{i} - b\sin\phi_c\boldsymbol{j} + 0\boldsymbol{k} \tag{21}$$

Thus, $\boldsymbol{p}$ can be further derived via Eq. (22).

$$\boldsymbol{p} = \frac{\mathbf{W}_t \times \boldsymbol{r}_{MC}}{|\mathbf{W}_t \times \boldsymbol{r}_{MC}|} = \frac{\mathbf{W}_t \times (\boldsymbol{r}_C - \boldsymbol{r}_M)}{|\mathbf{W}_t \times (\boldsymbol{r}_C - \boldsymbol{r}_M)|}$$

$$= \frac{C_1\boldsymbol{i} + C_2\boldsymbol{j} + C_3\boldsymbol{k}}{\left\{\omega_t^2\left[\left(A_1^2 - A_2^2\right)\cos(\theta+\beta)^2 - 2\cos(\theta+\beta)\sin(\theta+\beta)A_1A_2 + A_2^2 + \left(\delta+L-h\right)^2\right]\right\}^{1/2}} \tag{22}$$

where

$$C_1 = -\omega_t\{[-\cos\phi_c A_1\cos(\theta+\beta) + \cos\phi_c A_2\sin(\theta+\beta) + \sin\beta\sin\phi_c\left(\delta+L-h\right)]\cos(-\phi_t+\alpha) + [\sin\beta\sin\phi_c\cos(\theta+\beta)A_1$$
$$-\sin\beta\sin\phi_c\sin(\theta+\beta)A_2 + \cos\phi_c\left(\delta+L-h\right)]\sin(-\phi_t+\alpha)\}$$
$$C_2 = -\omega_t\{[-\sin\phi_c A_1\cos(\theta+\beta) - \sin\phi_c A_2\sin(\theta+\beta) + \sin\beta\cos\phi_c\left(\delta+L-h\right)]\cos(-\phi_t+\alpha) - [-\sin\beta\cos\phi_c\cos(\theta+\beta)$$
$$A_1 + \sin\beta\cos\phi_c\sin(\theta+\beta)A_2 + \sin\phi_c(\delta+L-h)]\sin(-\phi_t+\alpha)\}$$
$$C_3 = -\omega_t\cos\beta\{\cos(-\phi_t+\alpha)\left(\delta+L-h\right) + \sin(-\phi_t+\alpha)[\cos(\theta+\beta)A_1 - \sin(\theta+\beta)A_2]\}$$

Therefore, the general expression of pressure angle can be deduced via Eq. (23).

$$\cos\psi = \boldsymbol{n}\cdot\boldsymbol{p}$$
$$= \frac{\omega_t[\cos(\theta+\beta)A_1 - \sin(\theta+\beta)A_2](-A_1A_3 - A_2A_4 + \delta + L - h)}{[(A_3^2+1)A_1^2 + 2A_1A_2A_3A_4 + A_2^2(A_4^2+1)]^{1/2}\{\omega_t^2[(A_1^2 - A_2^2)\cos(\theta+\beta)^2 - 2\cos(\theta+\beta)\sin(\theta+\beta)A_1A_2 + A_2^2 + (\delta+L-h)^2]\}^{1/2}} \tag{23}$$

It can be seen from Eq. (23) that the pressure angle of the GIC is a complex function of the motion law of the follower and the geometric size of the globoidal cam mechanism, so the pressure angle can be adjusted by improving the motion law of the follower or the geometric size of the globoidal cam.

3.2 The general expression of principal curvature of GIC

The principal curvature of the profile surface of the globoidal cam must be calculated and analyzed to avoid the phenomena of over cutting and turning distortion when the cam is used. The principal curvature, *k*, can be evaluated by the following equation based on differential geometry [15] together with Eq. (16).

$$k^2 - 2Qk + S = 0 \tag{24}$$

where

$$Q = \frac{P_1P_6 - 2P_2P_5 + P_3P_4}{2(P_1P_3 - P_2^2)}, \quad S = \frac{P_4P_6 - P_5^2}{P_1P_3 - P_2^2},$$

$$P_1 = \boldsymbol{r}_{Ch}\cdot\boldsymbol{r}_{Ch}, \quad P_2 = \boldsymbol{r}_{Ch}\cdot\boldsymbol{r}_{C\phi c}, \quad P_3 = \boldsymbol{r}_{C\phi c}\cdot\boldsymbol{r}_{C\phi c}, \quad P_4 = \boldsymbol{r}_{Chh}\cdot\boldsymbol{r}_{Cn}, \quad P_5 = \boldsymbol{r}_{C\phi c}\cdot\boldsymbol{r}_{Cn}, \quad P_6 = \boldsymbol{r}_{C\phi c\phi c}\cdot\boldsymbol{r}_{Cn},$$

$$\boldsymbol{r}_{Ch} = \frac{\partial\boldsymbol{r}_C}{\partial h}, \quad \boldsymbol{r}_{C\phi c} = \frac{\partial\boldsymbol{r}_C}{\partial\phi_c}, \quad \boldsymbol{r}_{Chh} = \frac{\partial^2\boldsymbol{r}_C}{\partial h^2}, \boldsymbol{r}_{Ch\phi c} = \frac{\partial^2\boldsymbol{r}_C}{\partial h\partial\phi_c}, \quad \boldsymbol{r}_{C\phi c\phi c} = \frac{\partial^2\boldsymbol{r}_C}{\partial\phi_c^2}, \quad \boldsymbol{r}_{Cn} = \frac{\boldsymbol{r}_{Ch}\times\boldsymbol{r}_{C\phi c}}{\left|\boldsymbol{r}_{Ch}\times\boldsymbol{r}_{C\phi c}\right|}$$

Similar to the pressure angle general expression, the principal curvature of the globoidal cam profile is also a complex function of the motion law of the follower and the geometric size of the globoidal cam mechanism. Therefore, the principal curvature can be similarly adjusted by improving the motion law of the follower or the geometry of the globoidal cam.

4. Model verification

To verify the validity of the pressure angle mathematical model, the Mathematica numerical simulation tool is used. The parameters of the GIC mechanism are as follows: *r*=12.05 ㎜, *L*=15 ㎜, *b*=70 ㎜, *α*=30 deg, *δ*=31 ㎜, *β*=0 deg, and *γ*=0 deg. The number of rollers, z, is 6 and the dividing period, τ, is 270 deg, and the rest period, ϕ_d, is 90 deg. The motion law of the turret is an improved sine curve motion relationship, as shown in Eqs. (25), (26), and (27). The distribution of pressure angle along the contact line of the globoidal cam mechanism obtained by the above parameters is calculated as shown in figure 5 (only the stress surface is listed). Figure 5 shows that the pressure angle Ψ of the roller varies with the rotational angular displacement ϕ_c of cam and the contact point depth h of the roller surface at different contact points. Thus, the distribution of the pressure angle conforms to the motion law of the GIC mechanism. Therefore, the numerical results verify the validity of the pressure angle mathematical model.

$$\phi_t = h\ [0.43990\frac{\phi_c-\phi_d}{\tau} - 0.035014\ \sin(4\pi\frac{\phi_c-\phi_d}{\tau})], \qquad 0<\frac{\phi_c-\phi_d}{\tau}\le\frac{1}{8} \tag{25}$$

$$\phi_t = h\ [0.28005+0.43990\frac{\phi_c-\phi_d}{\tau} - 0.035014\ \sin(\frac{4}{3}\pi\frac{\phi_c-\phi_d}{\tau}+\frac{\pi}{3})], \qquad \frac{1}{8}<\frac{\phi_c-\phi_d}{\tau}\le\frac{7}{8} \tag{26}$$

$$\phi_t = h\ [0.56010+0.43990\frac{\phi_c-\phi_d}{\tau} - 0.035014\ \sin(4\pi\frac{\phi_c-\phi_d}{\tau})], \qquad \frac{7}{8}<\frac{\phi_c-\phi_d}{\tau}\le 1 \tag{27}$$

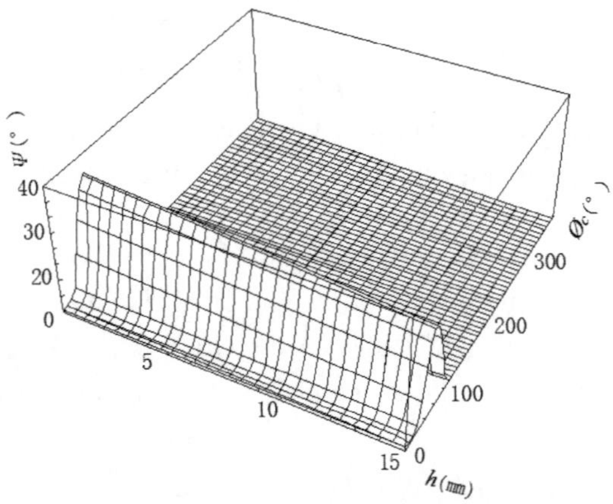

Figure 5. The distribution of pressure angle along the contact line of the bearing surface of the GIC

5. Conclusions

- ✓ Based on the theory of envelopes and the matrix transformation of homogeneous coordinates, the coordinate transformation process is described in detail. An equation of the curved surface of the GIC is established from a unified mathematical model, and general expressions of the pressure angle and the principal curvature of the mechanism are derived. The model can directly and accurately reflect the distribution of pressure angle along the contact line of GIC.
- ✓ The method used in the derivation process has a strong geometric intuition. The whole process is simple, straightforward, and requires no complex motion analyses. The final general expression applies to different roller shapes and different angles between the camshaft and the rotating axis of the turret.
- ✓ The unified expressions of the mathematical model of the pressure angle and the principal curvature can help verify the correctness of the design parameters and motion laws of the globoidal cam and distinguish whether the conjugate surfaces interfere or not. At the same time, they have specific practical values for the digital design and manufacture of the GIC mechanism

and provide a theoretical basis for the fatigue strength analysis of the globoidal cam material.

6. Acknowledgment

The financial support of the Sichuan Provincial Science and Technology Department of Chengdu University, China (2014JY0174) is gratefully acknowledged.

References

[1] Hai L 2015 *Analysis of Machining Profile Error and Design of Measuring Scheme for Globoidal Cam.* Hunan: Xiangtan University

[2] Shuting J 2016 *Study on the Manufacturing Error and Geometric Characteristic of Globoidal Cam Mechanisms.* Beijing: Beijing University of Technology

[3] Yan H S, Chen H H 1996 *J. Mech. Design* **118** 243-9.

[4] Fang D Z, Zhang C, Cao X J 2003. *Mech. Sci. Tech. Aero. Engin.* **22** 154-6.

[5] Li L, Feng X Y, Zhang Z P, *et al.* 2010 *Appl. Mech. Mater.* **979** 1608-14.

[6] Liu M, Ding J M. 2008 *J. Mech. Transm.* **32** 104-6+134.

[7] Yang W, Cao J J 2004 *Mech. Sci. Tech. Aero. Engin.* **23** 74-6.

[8] Zhao X N, Cao J J, Sun L X 2003 *J. Jilin University* (Engineering and Technology Edition) **33** 55-8.

[9] Liu Y J 2011 *The Machining Simulation and Solid Modeling of the Gear Based on the Gear Generation Principle.* Shanghai: East China University of Science and Technology

[10] Li J, Tang M, Zhang Y H, *et al.* 2009 *J. Machine Design* **26** 38-42.

[11] Zhang Y N 2015 *J. Shandong Agriculture and Engineering University* **32** 141+143.

[12] Li Y C 2016 *Space Analytic Geometry* [M]. Beijing: Science Press, 2016:20- 31.

[13] Liu D P 2012 *Computer Aided Design and Manufacturing* [M]. Beijing: China Railway Publishing House, 2012:78-91.

[14] Tang Z M, Huang M F, Sun Y H 2017 *Computer Integrated Manufacturing Systems* **23** 455-64.

[15] Chen W H 2017 *Differential Geometry.* Beijing: Peking University Press 132-58.

Effect of cooling process on mechanical properties of thin-gauge hot-rolled dual-phase steel

B Liu[1,2], R Wu[1*], J Hu[2]

[1]Key Laboratory for Ferrous Metallurgy and Resources Utilization of Ministry of Edu cation, Wuhan University of Science and Technology, Wuhan 430081, China
[2]Wuhan Branch of Baosteel Central Research Institute, Wuhan 430080, China

*E-mail: runwu@wust.edu.cn

Abstract. The high strength hot-rolled dual-phase steel suitable for a thin slab continuous casting and rolling (TSCR) process was developed by a low C and Nb micro-alloyed design with the controlled rolling and cooling technology. The effect of the isothermal temperature and holding time on the microstructure and mechanical properties of the dual-phase steel was studied using a uniaxial tensile test and optical microscope (OM). The results showed that with the decrease of the isothermal temperature in the ferrite phase transformation region, the thermodynamic conditions of the ferrite transformation were enhanced. The proportion of the ferrite increased while that of martensite decreased, resulting in the reduction of the yield strength and tensile strength and the increase of elongation. With the rise in the isothermal time, the kinetic conditions of the ferrite transformation were enhanced and the proportion of the ferrite increased. The ferrite transformation has completed after 8 seconds of the isothermal holding time, and it was not apparent for increased amount of the transformed ferrite to increase the isothermal time.

1. Introduction

With a rapid development of the lightweight work for automobile, the advanced high-strength steel (AHSS) has been widely used in body-in-white and structural parts in vehicles [1-4]. Dual-phase steel is one of the most commonly used advanced high strength steels in the automotive industry due to its excellent combination of strength and plasticity, low yield ratio, and high initial work-hardening rate [5-9]. In recent years, with the development of hot rolling equipment capacity and manufacturing technology, there is a significant improvement in a plate shape, size, surface quality and mechanical properties for the hot-rolled steel strip. Especially the thin gauge hot-rolled plate produced by thin slab continuous casting and rolling process has been able to meet the requirements of the manufacture of body-in-white and related structural parts. Through the "hot rolled steel on behalf of the cold rolled steel", the cold rolling and annealing processes in the strip steel production process can be reduced, which dramatically reduces the energy consumption of iron and steel materials manufacturing [10-12]. Therefore, the development of thin gauge hot-rolled dual-phase steels has become a new hot spot for the advanced high strength steels.

At present, the hot-rolled dual-phase steel was studied mainly for the traditional hot rolling process, and the carbon content was generally between 0.08% and 0.16% [13,14]. Due to the limitations of the peritectic region, there were only a few studies on the thin gauge dual-phase steel for the TSCR process, especially the tensile strength above 750 MPa. In this paper, the low-carbon and Nb micro-alloyed high-strength dual-phase steels were designed for the laboratory experiment with a controlled cooling process using salt bath cooling process, which provided a theoretical basis for the production of 750 MPa dual-phase steel in the TSCR process.

2. Experimental

The experimental steel was vacuum-melted and cast into a 50 kg ingot. The composition is shown in table 1.

Table 1. The chemical composition of tested steel (wt, %)

C	Si	Mn	P	S	Al	Nb	Cr
0.07	0.28	1.58	0.016	0.005	0.035	0.037	0.51

The experimental method was as follows. The ingot was heated to 1200 °C for 1 hour and then cast into a slab with a thickness of 30 mm. The slab was then hot-rolled to a thickness of 3.0 mm (30mm→18mm→12mm→8mm→5mm→3mm). To investigate the influence of the isothermal temperature and time of the ferrite transformation region on the microstructure and mechanical properties of the steel, the sample after hot-rolling was put into the salt bath furnace for ultra-rapid cooling, and the cooling process was controlled with different isothermal temperatures and isothermal time periods. After the salt bath, the sample was subjected to water quenching to obtain the martensite structure. The experimental process is shown in table 2.

Table 2. Processing parameters of controlled cooling of tested steel

No.	Final rolling temperature/°C	Isothermal temperature/°C	Isothermal time/s	Water quenching temperature/°C
1		700	8	
2		670	8	
3	810	640	8	$<$200
4		640	2	
5		640	15	

The mechanical properties were measured according to GB/T 228.1-2010, and the specimens were machined along the rolling direction with a gauge length of 50 mm after water quenching. After grinding, polishing and etching in 4% nital, the microstructure was observed by a Neophot-type optical microscope (OM). The relative amounts of each phase were analyzed quantitatively. The slices were first ground down to 80 μm thickness and then twin-jet polished by Tenupol-5 double-spray electrolyzer. Finally, the fine microstructure was observed by JEM-2100F transmission electron microscopy (TEM).

3. Experimental results

3.1 Microstructure

The microstructure features of the experimental steel at different isothermal temperatures and isothermal times are shown in table 3 and figure 1. The microstructure was ferrite and martensite dual-phase structure for all the experimental steels. With the decrease of isothermal temperature, the proportion of ferrite gradually increased, and the percentage of martensite gradually decreased. The proportion of martensite was 89.2% for sample 1#, 60.6% for sample 2# and 51.8% for sample 3#. With the increase of isothermal time, the proportion of ferrite increases gradually, and the percentage of martensite decreased gradually. The percentage of martensite was 87.1% for sample 4# and 50.3% for sample 5#.

Table 3. The quantitative metallographic results of the tested steels

No.	Microstructure	F/%	M/%
1	F+M	10.8	89.2
2	F+M	39.4	60.6
3	F+M	48.2	51.8
4	F+M	12.9	87.1
5	F+M	49.7	50.3

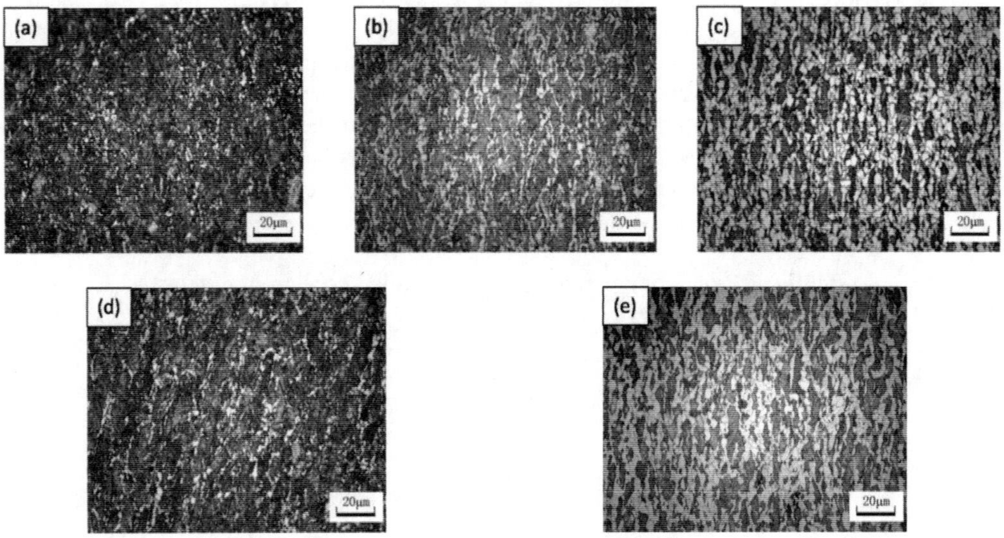

Figure 1. OM images of the steels microstructures (Black areas refer to martensite and gray/white ones to ferrite):
(a)Sample 1#; (b) Sample 2#; (c) Sample 3#; (d) Sample 4#; (e) Sample 5#

3.2 Mechanical properties

The mechanical properties of the experimental steels are listed in table 4. The tensile strength (TS) was over 750 MPa for all the samples, the yield strength (YS) was between 425 and 683 MPa, the elongation was between 10.5% and 18.0% and the yield ratio were below 0.70.

Table 4. Mechanical properties of the steels

No.	YS/MPa	TS/ MPa	A50mm/%	YS/ TS
1	693	1001	10.5	0.69
2	593	927	14.0	0.64
3	504	846	17.5	0.60
4	572	903	14.0	0.63
5	425	768	18.0	0.55

3.3 Discussion

The relationships between the isothermal temperature/time and the mechanical properties of experimental steels were shown in figure 2. The influence of the isothermal temperature on the microstructure and features of the experimental steel was mainly due to the decrease in the ferrite phase transformation temperature. The degree of the undercooling for ferrite transformation increased, and the thermodynamic conditions were enhanced. So, with the same holding time, the amount of the transformed ferrite increased. At the same time, the degree of undercooling (ΔT) for the retained austenite to martensite phase transformation decreased. According to the martensite phase transformation kinetic equation (1), the martensite volume fraction depended on the degree of undercooling (ΔT) but was irrelevant with the time [15], so that the proportion of martensite after water quenching decreased. For the mechanical properties, as the isothermal temperature dropped, the amount of the ferrite increased so that the yield strength and tensile strength decreased and the elongation increased.

$$f = 1 - 6.956 \times 10^{-5} \times (455 - \Delta T)^{5.32} \quad (1)$$

At the same isothermal temperature (640 °C), the kinetic conditions of the ferrite transformation were enhanced with the increase of the isothermal time. Thus the amount of transformed ferrite increased. Besides, since the isothermal temperature was set at 640 °C, the ferrite transformation degree of undercooling was high, and there were enough thermodynamic conditions to promote the ferrite transformation. Therefore, there was 12.9% transformed ferrite with a short isothermal time of 2 seconds. The short time ferrite transformation avoided the ferrite coarsening during the isothermal process, which made the elongation of the experimental steel reach 14.0%. The amount of ferrite was nearly the same for the isothermal time of 8 seconds and 15 seconds. Most ferrite transformation has been completed after 8 seconds of isothermal holding, and it was not apparent for the increased amount of the transformed ferrite to increase isothermal time. After 15 seconds isothermal holding, the amount of ferrite transformation increased a little, but the ferrite grain size increased (as showed in figure 1e), which resulted in the significant decrease of the yield strength.

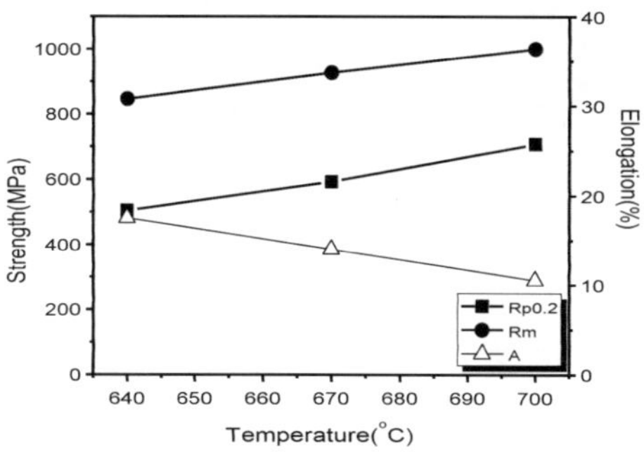

(a) temperature

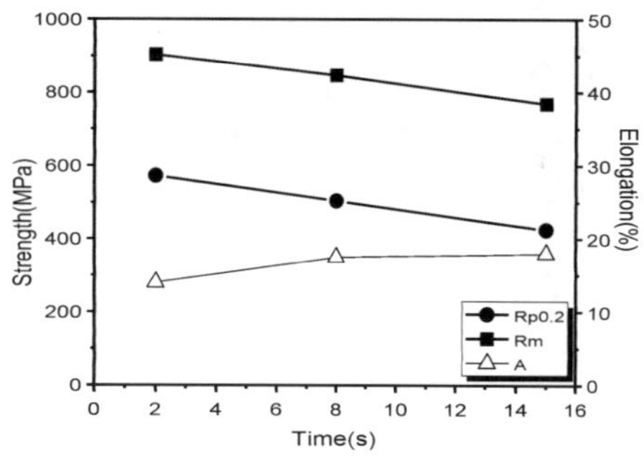

(b) time

Figure 2. The relationships between the isothermal temperature/time and the mechanical properties

The TEM microstructures of the test steels were shown in figure 3. It can be seen that the TEM microstructure changed obviously with the decrease of the isothermal temperature under the condition of isothermal time of 8 seconds (as shown in figure 3a-figure 3c). When the isothermal temperature was 700°C, the martensite proportion was very large in the TEM microstructure of sample 1#, and it revealed that the small ferrite was surrounded by massive block martensite. It can be seen that a large number of dislocation lines were accumulated between the ferrite and martensite phases due to the volume expansion of the martensite phase transformation. These dislocation lines can hinder the movement of grains and increase the yield strength of sample 1# to 693 MPa. There was a lot of block-like martensite in sample 2# at the isothermal temperature of 670°C. There was more ferrite with smaller size formed at the isothermal temperature of 640°C due to the relatively low isothermal temperature. Meanwhile, the martensite proportion was lower in sample 3#, and the size of the martensite was relatively small, as compared to samples 1 # and 2 #.

The effect of the isothermal time on the TEM microstructure was also investigated. It can be seen that the proportion of the ferrite increased and the percentage of the martensite decreased with the increase of the isothermal time under the condition of the isothermal temperature of 640°C (as shown in figure 3c-figure 3e). When the isothermal time was 2 seconds, due to the short ferrite phase transformation time, the amount of transformed ferrite was less, and the martensite proportion was as high as 87.1% in sample 4#, which was close to the martensite proportion in sample 1# experimental steel. However, the isothermal temperature was lower for sample 4# than sample 1# so that the grain size was finer in sample 4#. When the isothermal time was increased to 15 seconds, there was more ferrite formed by the ferrite phase transformation. The martensite proportion in sample 5# was the lowest, and the grain size was relatively coarse. Therefore, the yield strength and tensile strength of sample 5# were the lowest in all the test steels.

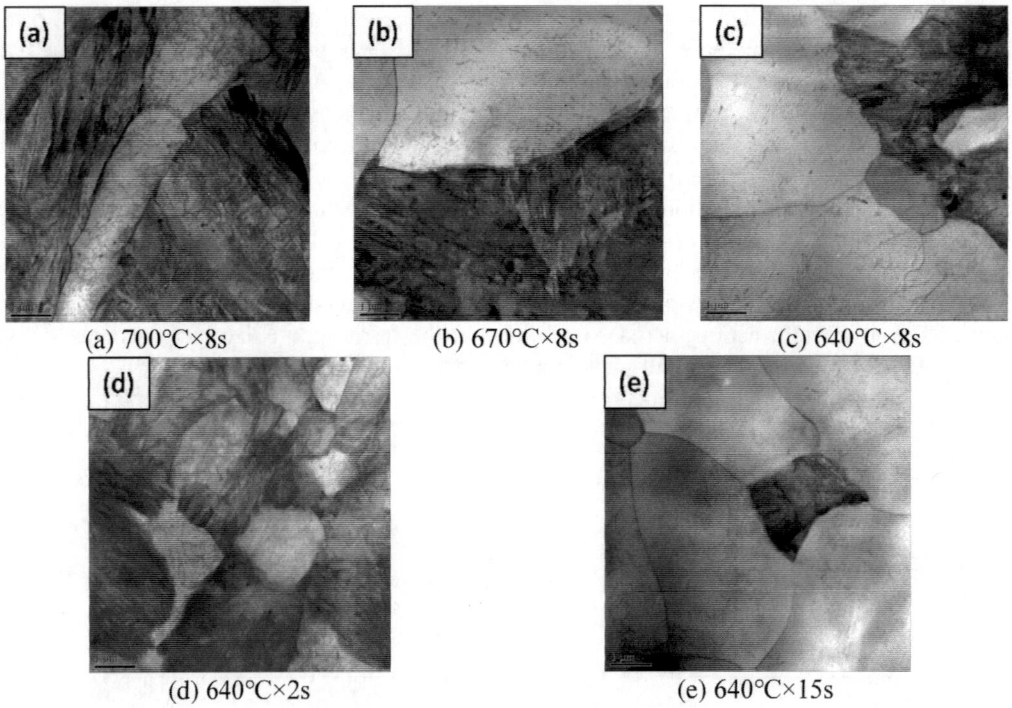

(a) 700°C×8s (b) 670°C×8s (c) 640°C×8s

(d) 640°C×2s (e) 640°C×15s

Figure 3. TEM images of the steels microstructures:
(a) Sample 1#; (b) Sample 2#; (c) Sample 3#; (d) Sample 4#; (e) Sample 5#

4. Conclusions

- ✓ The ferrite and martensite dual-phase structure of hot-rolled C-Mn-Nb-Cr steel with a tensile strength over 750 MPa was obtained through a controlled rolling and a cooling process with the salt bath isothermal holding-water cooling process
- ✓ With the decrease of the isothermal temperature in the ferrite transformation region, the thermodynamic conditions of the ferrite transformation were enhanced, the proportion of the ferrite increased, the proportion of the martensite decreased, the yield strength and the tensile strength decreased, and the elongation increased.
- ✓ With the increase of the isothermal time in the ferrite transformation region, the kinetic conditions of the ferrite transformation were enhanced, and the proportion of the ferrite increased. Most

ferrite transformation has been completed after 8 seconds of isothermal holding time, and it was not apparent for the ferrite transformation to increase the isothermal time.

Acknowledgments
This work was supported by the National Science Foundation of China (No. 51771139).

References

[1] Takahashi M 2003 *Nippon Steel Technol. Rep.*, **88** 8.
[2] Wang Z G, Zhao A M, Zhao Z Z, Ye J Y, Tang D, and Zhu G S 2012 *Int. J. Miner. Metall. Mater.***19** 915.
[3] Li Z, Wu D, Lu V, Yu H H, Shao Z Y and Luo L 2015 *J. Wuhan Univ. Technol.* **30** 156.
[4] Zhou L Y, Jiang B, Cui T H, Zhang D, He J Z and Liu Y Z 2014 *J. Iron Steel Res. Int.* **21** 1111.
[5] Li P, Li J, Meng Q G, Hu W B, and Kuang C F 2015 *Int. J. Miner. Metall. Mater.***22** 933.
[6] Molaei M J and Ekrami A 2009 *Mater. Sci. Eng. A* **527** 235.
[7] Onn I H, Ahmad N and Tamin M N 2015 *J. Mech. Sci. Technol.* **29** 51.
[8] Zhou L Y, Zhang D and Liu Y Z 2014 *J. Iron Steel Res. Int.* **21** 755.
[9] Davenport A T 1979 Proceedings of a Symposium. [in]TMS/AIME, New York 4.
[10] Mao X P, Huo X D, Sun X J and Chai Y Z 2010 *J. Mater. Process. Technol.* **210** 1660.
[11] Chen L, Zhang C, Zhu S and Kang Y L 2014 *Iron and Steel* **49** 57.
[12] Mao X P, Sun X J, Kang Y L and Lin Z Y 2006 *ACTA Metall. Sin.* **42** 1091
[13] Wang Z G, Zhao A M, Zhao Z Z, Ye J Y, Tang D and Zhu G S 2012 *Int. J. Miner. Metall. Mater.***19** 915-22.
[14] Zhou L Y, Zhang D and Liu Y Z 2014 *J. Iron Steel Res. Int.* **21** 755-65.
[15] Liu Z Z, Yuan Z X and Liu Y C 2010 *Solid Phase Transition*, Beijing, Machinery Industry Press.

Pressure-induced new spin-states in Ni-25at%Mn mechanical alloy under high pressure

N Q Sun[1], R Zach[2] and F Ono[3,*]

[1]Advanced Materials Developing Center, Research Laboratories, JFE Mineral Co. Ltd, 1 Niihamacho, Chiba 260-0826, Japan

[2]Institute of Physics, Cracow University of Technology, Cracow, Poland

[3]Okayama University of Science, 1-1 Ridaicho, Kitaku, Okayama 700-0005, Japan

*E-mail: fumihisaono@yahoo.co.jp

Abstract. AC-susceptibility measurements in mechanically alloyed Ni-25at%Mn were made under high hydrostatic pressures up to 7.5 GPa by using a cubic anvil press in the low-temperature range from 4.2 K to 300 K. At ambient pressure a reentrant spin glass (RSG) state was observed below 58 K, followed by a ferromagnetic (FM) state with the Curie temperature (T_C) at 120 K. T_C decreased with increasing pressure up to 4.3 GPa, while the transition temperature from RSG to FM increased to merge with T_C at 70 K, and then, with further increasing pressure the FM state disappeared, and a spin glass (SG) state appeared. The local moments of this high-pressure SG phase decreased with pressure. The results were compared with a bulk alloy made by the conventional method of melting. In the bulk alloy, a new high-pressure magnetic phase appeared having almost pressure-independent local magnetic moments.

1. Introduction

Ni-rich Ni-Mn alloys crystallize into a face-centered cubic structure. They have an ordered state of Cu_3Au-type structure around Ni_3Mn composition. It is well known that the magnetic structure of these alloys is sensitive to the short-range structure of atomic ordering [1-7]. We have demonstrated that a nanoscale atomic diffusion caused a significant change in their magnetic structure [8-11].

The technique of mechanical alloying has been used to introduce a large variety of nanoscale atomic concentration fluctuation in Ni-Mn alloys [8-11] and Fe-Ni alloys [12-14], and various magnetic structures have been introduced in these alloys. On the other hand, high hydrostatic pressure is a technique to change atomic distances without introducing any changes in atomic ordering. We combined these two techniques and looked for any new magnetic phases in Ni-rich Ni-Mn alloys.

2. Experimental

To synthesize mechanical alloy (MA) samples, 100-mesh size Ni and Mn powders were weighted and mixed, and then, put into two stainless containers. The purity of the elemental powders was 99.9%. They were mechanically alloyed by milling for 75 hours using a planetary ball mill in the same way as described in the references [12-14].

To compare the magnetic properties of the milled alloy with those shown in the literature [1-7], a standard bulk Ni-25at%Mn alloy was produced by using an arc-melting furnace. The bulk alloy was homogenized and powdered using the same procedure as mentioned in [12-14]. The homogenization temperature was 1273 K and the annealing time was a week. The bulk sample was again homogenized at 1273 K for 3 hours. After the annealing procedure the bulk alloy sample was quenched to room temperature to keep the face centered cubic structure and the disordered state stable.

To obtain the Cu_3Au-type ordered state, the both MA and bulk alloys were annealed at 693 K for 1000 h. The crystal structure and the degree of order for MA and bulk alloy were confirmed by X-ray diffraction at room temperature. To investigate the magnetic properties for MA and bulk alloy at ambient pressure, magnetic field dependence and temperature variation of magnetization were observed by using a SQUID magnetometer.

To generate a hydrostatic pressure, a cubic anvil press system was operated by a 250-ton press. The cubic anvil press system was placed in a thermos bottle with which the temperature of the sample was varied from 4.2 to 400 K. AC-susceptibility measurements were performed under high pressures up to 7.7 GPa [8-10] by using a pick-up coil system which was wound around the sample. The intensity of the external AC-magnetic field for the susceptibility measurements was 5×10^{-5} T.

3. Results

3.1. Mechanically alloyed Ni-25at%Mn

The magnetization values M for the bulk and the MA samples observed at 1 T and 4.2 K under ambient pressure were plotted in figure 1 as a function of annealing time at 693 K.

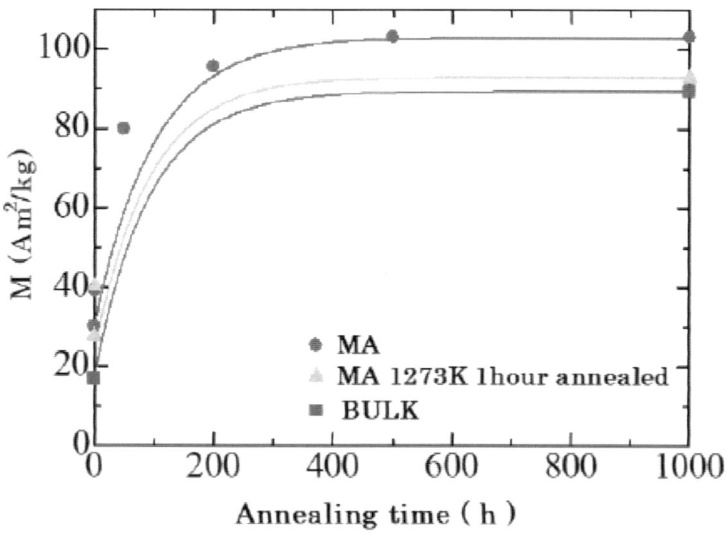

Figure 1. Annealing time variation of the magnetization during the ordering procedures observed at 4.2 K and 1 T for as milled MA, homogenized MA (annealed at 1273 K for 1 h) and the bulk Ni-25at%Mn alloy.

It was shown that the annealing time of 1000 h at this temperature was enough to get almost complete atomic ordering [11]. The M value in as milled MA was about 15% larger than that in the bulk alloy. However, after the homogenization (annealing at 1273 K for 1h) treatment it almost reached to the bulk value. This fact proved that two alloys produced in different ways (mechanical alloying and conventional melting) had nearly the same concentration of Ni-25at%Mn. In this figure, it is seen that the magnetization of the homogenized MA approaches to the value of the bulk alloy. The curves shown in this figure were calculated by solving the diffusion equation as explained in ref. [11].

The observed temperature dependences of the real part χ' and the imaginary part χ'' of AC-susceptibility in Ni-25at%Mn as milled MA under various pressures are shown in Figure 2 (a) and (b), respectively. In both figures ferromagnetic signals could be seen in a narrow temperature range about 65-120 K. A reentrant spin glass (RSG) state could be seen in the curves up to 3.5 GPa below about 65 K. With further increasing pressure the ferromagnetic signal disappeared, and a spin glass (SG) state came to appear.

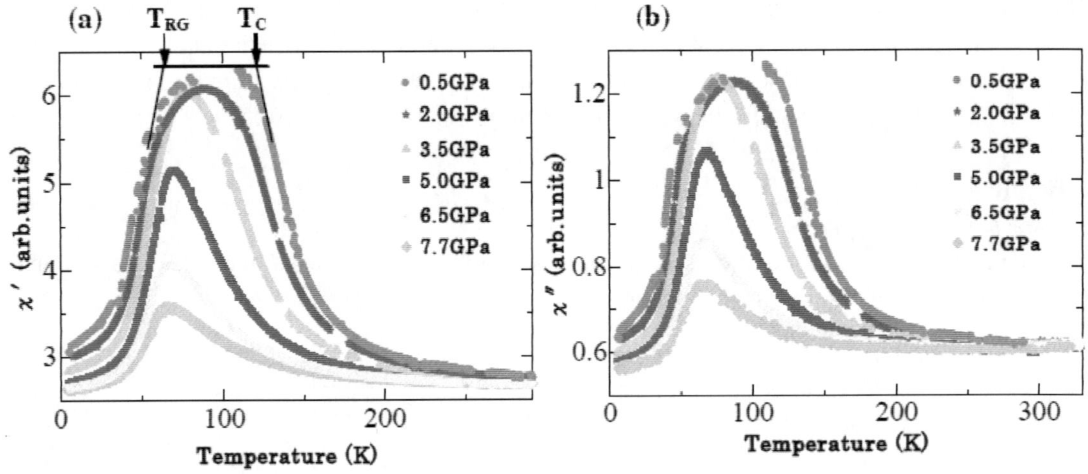

Figure 2. Observed temperature variation of magnetic susceptibility χ' (a) and χ'' (b) for Ni-25at%Mn as milled MA under various pressures up to 7.7 GPa. In both figures, a ferromagnetic signal can be seen in a narrow temperature range around 65-120 K below 3.5 GPa.

From the observed AC-susceptibility curves shown in figure 2 (a) and (b) a magnetic phase diagram was drawn for Ni-25at%Mn as milled MA and shown in figure 3 (a). As pointed out by Tange et al. [6], there is a difficulty in determining the Curie temperature from AC-susceptibility-temperature curves due to the complicated magnetic interactions in Ni-Mn alloys. In this paper, we need only the relative shift of T_C with pressure, and therefore, for convenience, T_C was determined by using the same method described in work [9].

In figure 3 (a), T_C is the Curie temperature, T_{RG} is the transition temperature from RSG to FM state and T_G from SG to paramagnetic state. The FM state disappeared at about 4.3 GPa, and with further increasing pressure a new SG phase came to appear. In figure 3 (b), the phase diagram determined for Ni-20at%Mn as milled MA [9] is also shown for comparison. The phase diagram (a) for Ni-25at%Mn MA was well fitted to the higher pressure region between 15 to 23 GPa (hatched area) in the phase diagram (b) for Ni-20at%Mn MA. From this result, we could estimate the chemical pressure acting on the disordered Ni-25at%Mn MA to be 15 GPa.

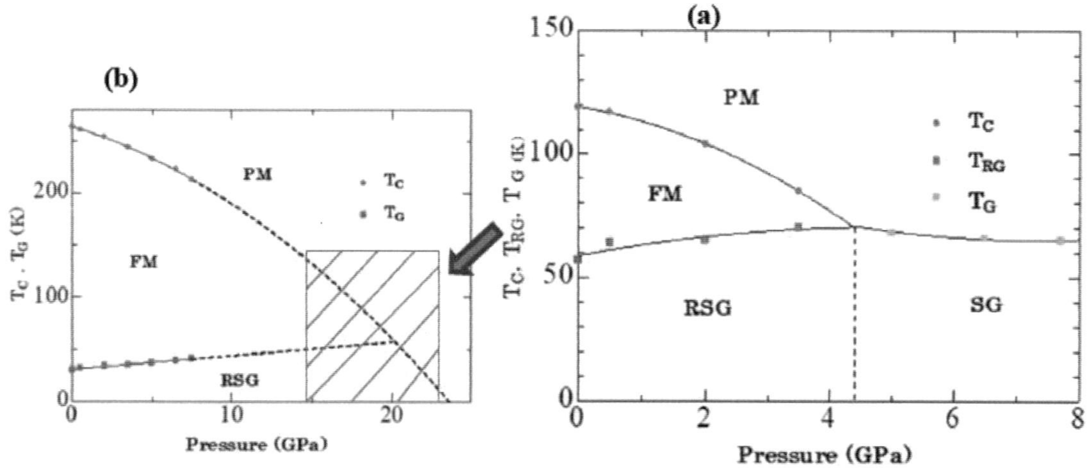

Figure 3. (a) The magnetic *P-T* phase diagram for Ni-25at%Mn as milled MA determined from the present experiment. This phase diagram corresponds to the hatched area in the phase diagram (b) obtained in [9] for Ni-20at%Mn as milled MA.

3.2. Ni-25at%Mn bulk alloy produced by conventional melting

Observed χ'-T and χ''-T curves in Ni-25at%Mn disordered bulk alloy under various pressures are shown in figure 4 (a) and (b), respectively.

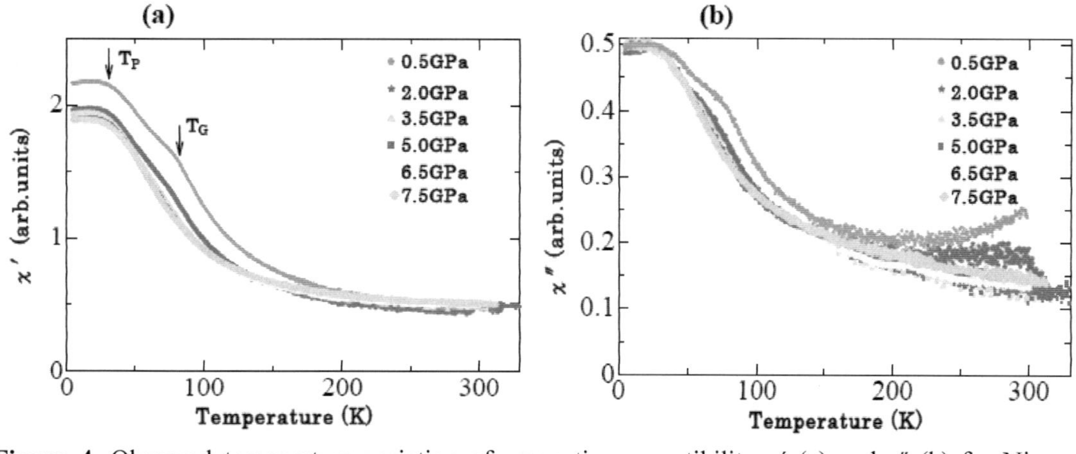

Figure 4. Observed temperature variation of magnetic susceptibility χ' (a) and χ'' (b) for Ni-25at%Mn disordered bulk alloy produced by conventional melting under various pressures up to 7.5 GPa.

As seen in figure 4 (a), there are two shoulders in the curve for P=0.5 GPa, one is about 30 K (shown as T_P), and the other is about 80 K (shown as T_G). Below the transition temperature T_P, the susceptibility did not change mainly with temperature. Hence the low-temperature phase below 30 K seems ferromagnetic. However, it seems difficult to define the magnetic structure of this low-temperature phase at this moment. This new magnetic phase persisted up to the maximum pressure of 7.5 GPa. Here, we refer to this phase as the high-pressure magnetic (HPM) phase.

To clarify the low-temperature magnetic structure, we performed squid-magnetization measurements at ambient pressure. The results are shown in figure 5 (a) and (b). In figure 5 (a) the magnetization at the very low field of 5×10^{-5} T is plotted as a function of temperature. We set the intensity of this weak magnetic field equal to that of AC-susceptibility measurements. In figure 5 (a) no such transition as T_P could be seen. Therefore we concluded that the appearance of the transition T_P was due to the application of the pressure of 0.5 GPa. In figure 5 (b) the magnetization at the field of 1×10^{-3} T is plotted against temperature. In this figure an existence of SG state is seen below 81 K. This SG state disappeared above 2 GPa.

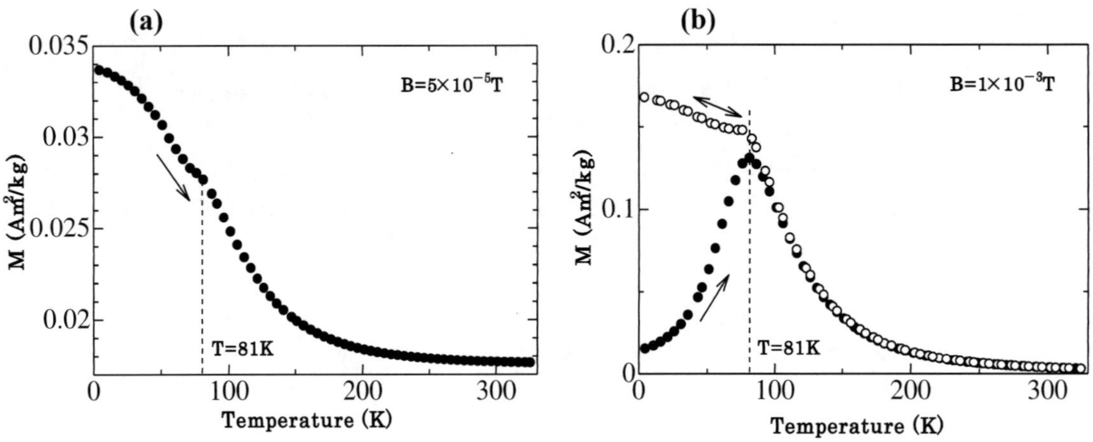

Figure 5. Temperature dependence of the magnetization for Ni-25at%Mn disordered bulk alloy; (a) at the field of 5×10^{-5} T and (b) at 1×10^{-3} T at ambient pressure. In (b), an SG state can be seen below 81 K, where the full circles show on heating from 4.2 K for the first time, and the open circles show on cooling from 320 K to 4.2 K and then again heating for the second time.

From the observed AC-susceptibility curves shown in figure 4, a magnetic phase diagram was drawn for the Ni-25at%Mn disordered bulk alloy and shown in figure 6 (a). The SG state observed at ambient pressure disappeared above 2 GPa. On the other hand, a new high-pressure magnetic phase (HPM) appeared above 0.5 GPa below 30 K. The transition temperature from HPM to a paramagnetic state remained constant with pressure up to 7.5 GPa. In this figure, T_P denotes the transition point from the new HPM phase to SG or paramagnetic state and T_G from SG to paramagnetic state.

In figure 6 (b) the phase diagram determined for Ni-20at%Mn disordered bulk alloy [9] is shown for comparison. The phase diagram (a) for Ni-25at%Mn bulk alloy corresponds well to the hatched area in the phase diagram (b) for Ni-20at%Mn bulk alloy. From this result we estimated the chemical pressure acting on the Ni-25at%Mn disordered bulk alloy to be 15 GPa. This value agrees well with the case of the Ni-25at%Mn disordered MA.

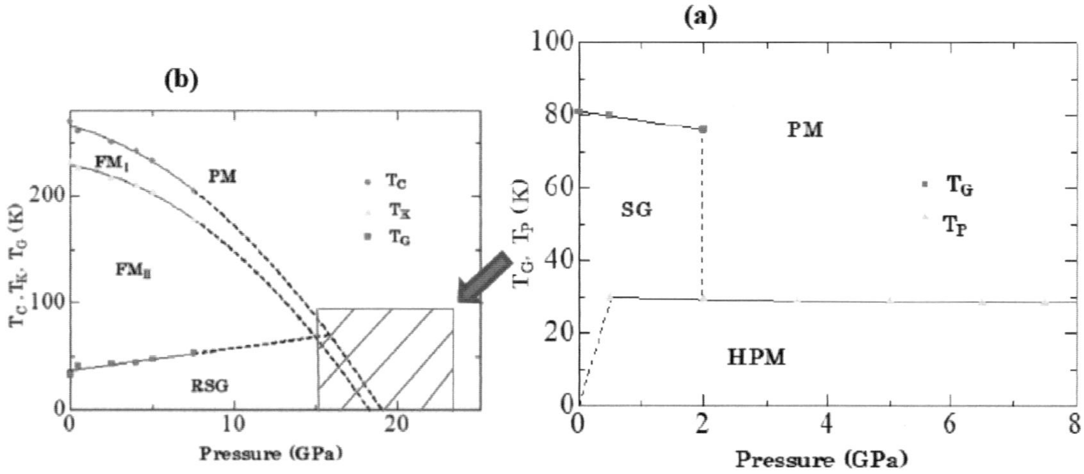

Figure 6. (a) Magnetic *P-T* phase diagram for Ni-25at%Mn disordered bulk alloy determined from the present experiment; (b) the phase diagram established in ref. [9] for Ni-20at%Mn disordered bulk alloy. The area to be compared with the present study is shown with the hatched square.

4. Discussion

It is well known that Ni-25at%Mn disordered bulk alloy made by conventional melting has no FM state, but has an SG state in the low-temperature range below 80 K [15]. On the other hand, an FM state was found in Ni-25at%Mn disordered MA in the low-temperature range between 58 K and 120 K. Furthermore, below 58 K, an RSG state was observed from the present study.

To understand the difference of the magnetic behavior between the two alloys, as milled MA and the bulk alloy, we may assume a broad distribution of concentration fluctuation of the constituent atoms in as milled MA [9]. The distribution function is expressed as,

$$f(x) = \frac{1}{\sigma\sqrt{2\pi}} e^{\frac{-(x-\bar{x})^2}{2\sigma^2}} \quad (1)$$

with the value of the standard deviation σ=10 for as milled MA [9]. The existence of the ferromagnetic state can be explained by considering the large concentration fluctuation introduced in the alloy by mechanical alloying [9]. By studying the well-defined phase diagram for Ni-Mn alloys [15], it is clear that Ni-rich nanoscale grains or particles in MA contribute to the appearance of ferromagnetism.

In Ni-25at%Mn disordered as milled MA the Curie temperature decreased with pressure at the rate of 12 K/GPa, while the RSG range increased, and both the RSG and FM phases disappeared at 4.3 GPa to merge into an SG state. From the present study, it was found that the *P-T* phase diagram determined for Ni-25at%Mn MA was well fitted to the phase diagram for Ni-20at%Mn MA in the pressure range between 15 and 23 GPa. From this fact we estimated the chemical pressure acting on the disordered Ni-25at%Mn MA to be 15 GPa. Here, we set the reference of the chemical pressure at the Ni-20at%Mn disordered MA. This chemical pressure, however, is too large and in the opposite sense as usually estimated from the difference of the lattice parameters between the two alloys [16]. This disagreement may come from the simple idea that magnetic properties depend only upon the lattice parameter. In Ni-Mn alloys, the magnetic properties depend more strongly upon the atomic configuration and the band structure [17]. The present result shows that change in the atomic configuration and the band structure with increasing the Mn-concentration by 5 % exceeds the effect of a simple change in the lattice parameter.

As to the Ni-20at%Mn disordered bulk alloy, we predicted in our previous study by merely extrapolating the phase diagram determined below 7.5 GPa that the FM phase disappears around 15 GPa and an SG state would come out [9]. In figure 6 (b) it is seen that the high-pressure region above 15 GPa fits almost exactly to that of Ni-25at%Mn disordered bulk alloy. Thus, the same value of the effective chemical pressure of 15 GPa was also confirmed in the Ni-25 at%Mn disordered bulk alloy.

5. Conclusions

P-T magnetic phase diagrams were determined for Ni-25at%Mn disordered as milled MA and bulk alloys. An FM state was found in MA in the low-temperature range between 58 and 120 K. Below 58 K, an RSG state was observed. It was found that the Curie temperature decreased with pressure, while the transition temperature from RSG to FM state increased gradually with pressure up to 4.3 GPa, where the FM state disappeared. Above 4.3 GPa a new spin glass state came to appear. The transition temperature from SG to paramagnetic state remained almost constant up to 7.7 GPa. The *P-T* phase diagram determined for Ni-25at%Mn MA was well fitted to that for Ni-20at%Mn MA in the pressure range between 15 and 23 GPa.

In Ni-25at%Mn disordered bulk alloy, the SG state observed at ambient pressure disappeared at 2 GPa. A new HPM phase appeared above 0.5 GPa below 30 K. The transition temperature from HPM to paramagnetic state remained constant with pressure up to 7.5 GPa. In Ni-25at%Mn disordered bulk alloy the P-T phase diagram could also be well fitted to that of Ni-20at%Mn disordered bulk alloy by shifting the P-axis to 15 GPa. The effective chemical pressures acting on both Ni-25at%Mn disordered MA and bulk alloys were estimated to be 15 GPa.

References

[1] Kaya S and Kussmann A 1931, *Z. Phys.,* 72, 293-309.
[2] Hahn R and Kneller E 1958 *Z. Metall.*, 49, 426-41.
[3] Abdul-Razzaq W and Kouvel J S 1987, *Phys. Rev. B* 35, 1764-7.
[4] Satoh T, Goldfarb R B and Patton C E 1978, *J. Appl. Phys.*, 49, 3439-43.
[5] Hennion B, Hennion M, Hippert F and Murani A P 1984, *J. Phys. F* 14, 489-504.
[6] Tange H, Tokunaga T and Goto M 1978, *J. Phys. Soc. Japan*, 45, 105-9.
[7] Okazaki T, Miyanaga T, Tsushima M, Kondo Y, Nakahata M, Yuza M and Nakamura N 1997, *J. Magn. Soc. Jpn.* 21, 1138-42.
[8] Sun N Q, Yao G Y and Ono F 2005, *J. Alloys Compd.* 398, 152-5.
[9] Sun N Q, Matsushima Y, Nishida D, Kobayashi T, Matsushita M, Endo S and Ono F 2006, *J. Magn. Magn. Mater.* 301, 37-43.
[10] Sun N Q, Chimi Y, Ishikawa N, Iwase A, Kambara T, Matsushima Y and Ono F 2007, *Nucl. Instrum. Methods B* 256, 442-5.
[11] Sun N Q, Zach R, Matsushima Y and Ono F 2009, *J. Alloys Compd.* 480, 87-90.
[12] Ono F, Hayashi K, Wei S and Maeta H 1998, *Proc. Powder Metallurgy World Congress* Vol.1, (Granada, 1998) 449-54.
[13] Wei S, Hayashi K and Ono F 1999, *J. Magn. Soc. Japan* 23, 391-3.
[14] Wei S, Duraj R, Zach R, Matsushita M, Takahashi A, Inoue H, Ono F, Maeta H, Iwase A and Endo S 2002, *J. Phys.; Condens. Matter* 14, 11081-4.

IOP Conf. Series: Materials Science and Engineering **474** (2019) 012021 doi:10.1088/1757-899X/474/1/012021

[15] Wasserman E F, *Ferromagnetic Materials*, edited by K. H. J. Buschow and E. P. Wohlfarth (North-Holland, Amsterdam, 1990) Vol. 5, 253.

[16] Ryazhkin A V, Babanov Y A, Miyanaga T, Ogasawara T, Nitta K, Ohwada S, Pilyugin V P, Patchelov A M and Chernyishov E G 2005, *Photon Factory Activity Report 2004*, 22 Part B 12C/2003G234.

[17] Kulkova S E, Valujsky D V, Kim J S, Lee G and Koo Y M 2002, *Physica B* 322, 236-47.

IOP Conf. Series: Materials Science and Engineering **474** (2019) 012022 doi:10.1088/1757-899X/474/1/012022

Comparison of SnO_2-carbon nanotubes composite and the SnO_2-carbon black mixture as an anode for Li-ion batteries

G Q Diao[1*], H Li[1], I Ivanenko[2], T Dontsova[2], I Astrelin[2], F Y Xie[1] and X Z Guo[1]

[1]School of Chemistry and Materials Engineering, Huizhou University, Huizhou 516007, China
[2]Department of Inorganic Substances Technology, Water Treatment, and General Chemical Engineering, National Technical University of Ukraine "Igor Sikorsky Kyiv Polytechnic Institute," Kyiv 03056, Ukraine

*E-mail: dguiqiang@hzu.edu.cn

Abstract. This article reports a comparative study on the electrochemical properties between SnO_2-carbon nanotubes (CNTs) composite and SnO_2-carbon black (CB) mixture. The SnO_2-CNTs composite was directly synthesized by a sol-gel method with purified multi-walled carbon nanotubes and tin chloride. The SnO_2-CB mixture was obtained by mixing carbon black with the previously prepared SnO_2. XRD profiles of SnO_2-CNTs and the obtained SnO_2 have shown the pure tin dioxide phase for the composite and the mixture. TEM images confirmed that SnO_2 nanoparticles with average particle size of 3-20 nm are deposited on the outer surface of carbon nanotubes, while the SnO_2 obtained for the mixture indicate larger average particle size than that of in the SnO_2-CNTs composite. These samples were used as anode materials for the lithium ion battery. The SnO_2-CNTs composite showed an initial charge capacity of 810 mAg^{-1} which slightly reduced to 500 mAg^{-1} after 100 cycles. The SnO_2-CB mixture had a smaller capacity, *i.e.,* an initial charge capacity of 350 mAg^{-1}, and decreased to 230 mAg^{-1} after 100 cycles. A reversible capacity at 0.2 Ag^{-1} was recovered for both samples after the cycling at different current densities, which implies excellent rate performance of the materials.

1. Introduction

The development of new anode materials for lithium-ion batteries has become one of the focus areas of materials science. In the past decade, tremendous progress has been made due to the significant advances in nanotechnology and materials science, and various nanostructured materials have been applied as the electrode materials for lithium-ion batteries. Among those materials, tin dioxide (SnO_2) is regarded as a promising anode candidate material for lithium-ion batteries because of its high theoretical capacity [1]. However, in the electrochemical processes, SnO_2 base anode materials exhibit an irreversible reduction process from SnO_2 to metallic Sn [2]. They also have some defects on high Li^+ voltage, low initial coulomb efficiency, and significant volume change during the charge and discharge process.

Several strategies have been proposed to overcome the above shortcomings, while two methods are widely studied recently [3-10]. Creating unique SnO_2 nanostructures, such as 1D and 2D nanostructures [11, 12] is one of the effective ways, while introducing carbonaceous materials to produce SnO_2-C composites or mixtures into electrodes is another beneficial approach. The formation of composites or mixtures blend with tin oxide and carbon is facile in execution. The composites or mixtures could also bring forwards many advantages [13-15], such as carbon acted as supports with elastic properties further provide a buffer effect against volume strain. Carbon-containing materials improved the electrical conductivity of the composites or mixtures as well [1].

The presented work aims to overcome the above problems of SnO_2 presented in the lithium battery application, further to explore high-performance SnO_2 base anode materials in energy storage filed. Therefore, the composite of tin dioxide with carbon nanotubes (SnO_2-CNTs) and a mechanical mixture of tin dioxide with a carbon black (SnO_2-CB) were prepared, their electrochemical properties were evaluated.

2. Experimental

2.1. Material fabrication

Synthesis of SnO_2: The SnO_2 was synthesized by the Sol-gel method according to [16].

Synthesis of multiwalled carbon nanotubes (CNTs): CNTs was obtained by the CVD method over the bimetallic catalyst ($NiO/Co_2O_3/MgO$) from acetylene at 600°C, according to our previous work [17]. The as-prepared CNTs were purified by refluxing in concentrated nitric acid for 1 hour. Then it was cooled down to room temperature naturally. After washing and centrifugation, CNTs were dried at 60°C for 12 h.

Synthesis of SnO_2-CNTs composite: In the presented procedure, the initial weight ratio of SnO_2 to CNTs was calculated in 50： 50. Typically, 0.8 g of $SnCl_2{\cdot}2H_2O$ was added into 50 mL of isopropyl alcohol under magnetic stirring for 2 h. Then 0.5 g purified CNTs obtained in above section was added and ultrasonicated for 1 h. The resulting suspension was evaporated at 120°C overnight and consequently calcinated in the air atmosphere at 400°C for 2 h [18].

Synthesis of SnO_2-CB mixture: The SnO_2-CB mixture was prepared by mechanical grounding method in a mortar with carbon black and SnO_2. In this way, the initial weight ratio of SnO_2 and CB is 50:50. The SnO_2 was prepared in the above procedure, the carbon black (Super-P) was purchased from Bonap Co., Ltd. (Shenzhen, China).

2.2. Structural characterization

Samples (SnO_2-CNTs composite and SnO_2) for transmission electron microscopy (TEM) were sonicated and dispersed in HPLC ethanol and processed in a FEI Tecnai 12 TEM with accelerating voltage of 120 KV. TG analysis was performed on a NETZSCH TG 209F1. Approximately 5 mg of material was heated from 25 °C to 900 °C at the ramping rate of 10 °C min^{-1} under air atmosphere. X-ray diffraction (XRD) patterns were collected by X-ray powder Diffractometer TTR3 Rigaku, Japan with a Cu *Ka* radiation (λ=1.5406 Å), 30 mA and 40 V.

2.3. Electrochemical measurements

The working electrode (anode) was prepared by mixing of 80 % (weight ratio) of active materials (SnO_2-CNTs composite or SnO_2-CB mixture), and 20 % of CMC binder. The electrode was coated on Cu foil. The active mass loading on the electrode was about 0.6 mg/cm^2. Lithium foil was used as the counter electrode. 1 M $LiPF_6$ in EC/DMC (1:1 in volume) was used as an electrolyte. Electrochemical experiments of the half-cell were carried out on CR 2025 coin cells, which were assembled in the Ar-filled glove box (Mikrouna, China). The charge and discharge tests were performed on a Land cell test system (Land CT2001A, China) in the potential window of 0.01 to 3.0 V at different current densities (mAg^{-1}).

3. Results and Discussion

3.1. Characterization of materials for the anode

The X-ray diffraction (XRD) was carried out to confirm the structures of SnO_2-CNTs composite and SnO_2 obtained by the sol-gel method. For SnO_2-CNTs composite, those peaks were located at 2Θ values of 26.5°, 33.8°, and 51.3° (figure 1a) are attributed to the SnO_2 (110), SnO_2 (101) and SnO_2 (211) face (ICDD 01-088-0287). The diffraction peak in the spectra located at 25.9◦ (figure 1a) are corresponded to C (002) face [19-21]. The average particle size of SnO_2 in the SnO_2-CNTs composite is about 3 nm which was calculated by Scherer Equation. The result is consistent with the following TEM results (figure 3).

On the other hand, the XRD profile of SnO_2 obtained by the sol-gel method has similar peaks (figure 1b). However, SnO_2 is more crystalline in the SnO_2-CB mixture compared with SnO_2-CNT composite. It indicated that the SnO_2 obtained in SnO_2-CNTs composite process weaker crystallinity and smaller size than that of SnO_2 obtained by the sol-gel method.

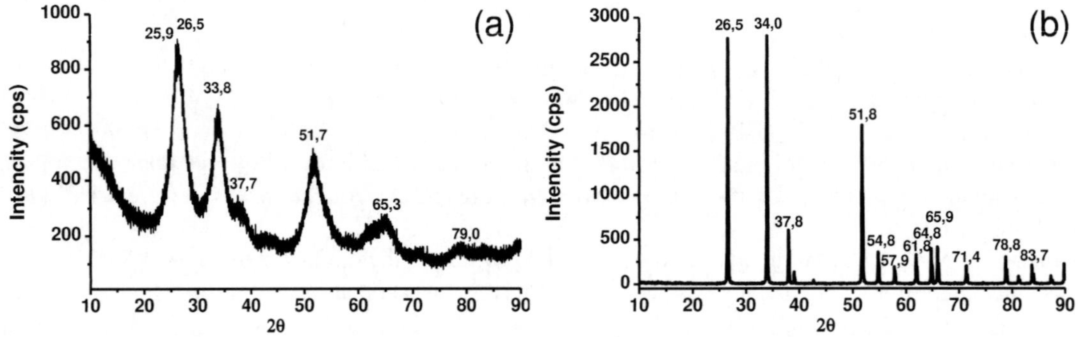

Figure 1. XRD profiles of (a) SnO_2-CNTs composite and (b) SnO_2.

Thermal analysis was performed to further reveal the water content and carbon content within SnO_2-CNTs composite and SnO_2-CB mixture. The measurements were carried out in an air atmosphere, and the temperature was raised at a rate of 10 °C/min to 900 °C. As shown in figure 2, the SnO_2-CNTs composite and the SnO_2-CB mixture showed a weight loss of 45.8% and 54.6% in the procedure, respectively. The weight loss of 1.6% for SnO_2-CNTs composite and 4.7% for SnO_2-carbon black mixture below 180 °C should be attributed to the removal of surface-adsorbed water. Above 200 °C, the CNTs and carbon black should have been removed entirely, which was also consistent with the starting weight ratios of 50:50 for the CNTs to SnO_2 and carbon black to SnO_2 in the synthesis procedurals detailed in section 2.1.

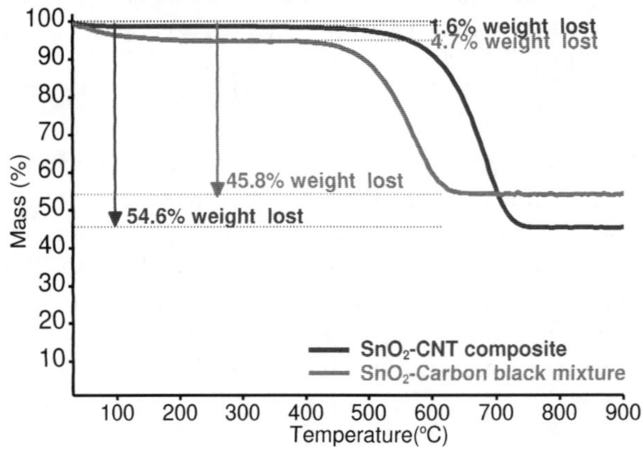

Figure 2. TG profiles of SnO_2-CNTs composite and SnO_2-CB mixture.

The morphology of the SnO_2-CNTs composite and SnO_2 obtained by sol-gel method are revealed by TEM. TEM-images (figure 3a) show that tin dioxide was deposited uniformly in the SnO_2-CNT composite. The SnO_2 particles are distributed arbitrarily on the surface of CNTs in the form of spherical particles about 3-20 nm in size (round red dash with the size of about 3 nm, and square yellow dash in size about 20 nm). It is consistent with the XRD results and our previously obtained results [21]. The specific surface area of SnO_2-CNTs is 310 m^2 g^{-1}. Such high specific surface area is derived from the initial CNTs. The SnO_2-CNTs with such structural adsorption characteristics could have a high potential in the preparation of nanocomposites as anode materials in lithium-ion batteries [16]. Figure 3b displays TEM images of the synthesized sample of SnO_2 in the sol-gel method, in which the individual crystal sizes are in the range of 50-100 nm. The particles showed a pronounced hexagonal shape. These findings are consistent with the XRD results.

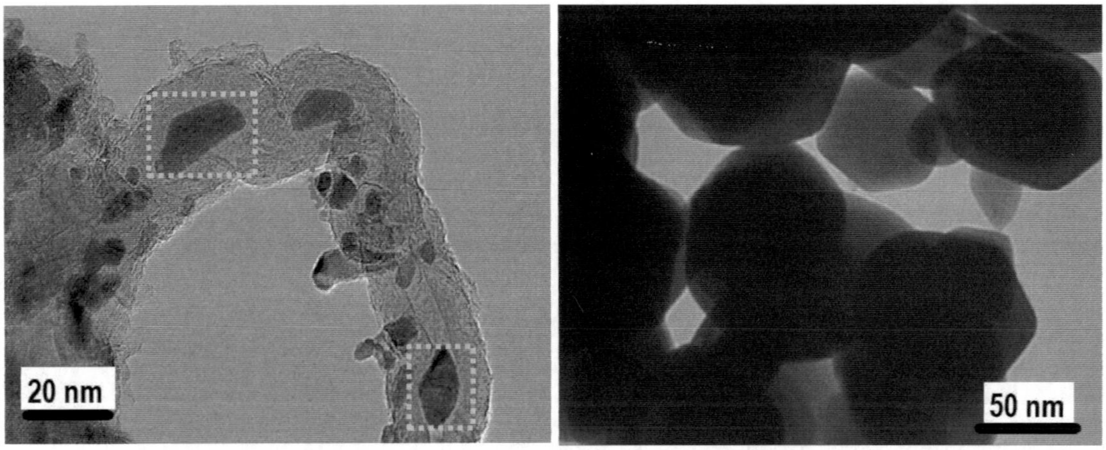

(a) (b)

Figure 3. TEM images of (a) SnO_2-CNTs electrode displayed continuous superiority in lithium performance SnO_2-CNTs composite and (b) SnO_2.

3.2. Electrochemical measurements

To determine the rate tolerance of the materials, galvanostatic charge/discharge performance of SnO_2-CNTs and SnO_2-CB was carried out at different current densities. As shown in figure 4, the discharge capacity of SnO_2-CNTs electrode is 810 mAh g^{-1} which is continuously higher than that of the SnO_2-CB electrode (200 mAh g^{-1}). Furthermore, SnO_2-CNTs electrode displayed superiority in lithium performance when the current densities were increased from 200, 400, 800, 1600 to 2400 mAg^{-1} respectively, indicating excellent rate capacity. Moreover, a reversible capacity about 700 mAh g^{-1} for SnO_2-CNTs can be recovered after cycled at high current densities. The capacity is higher than that of SnO_2-CB (350 mAh g^{-1}) which implied better rate performance for SnO_2-CNTs electrode than that of SnO_2-CB.

The 1st, 10th and 60th of the charge/discharge profiles of SnO_2-CNTs and SnO_2-CB anodes are presented on figure 5. The initial capacities for the SnO_2-CNTs (1380 mAh g^{-1}) are almost twofold than that of SnO_2-CB anodes (630 mAh g^{-1}). It should be noted that high initial Coulombic efficiency (ICE), about 65.1% for the SnO_2-CNTs composite and 52.4% for SnO_2-CB can be observed as well. Such ICE could be ascribed to the buffer effect against volume strain produced by the multi-walled carbon nanotubes or carbon black, at the same time, carbon-containing materials increase the electrical conductivity of the composite.

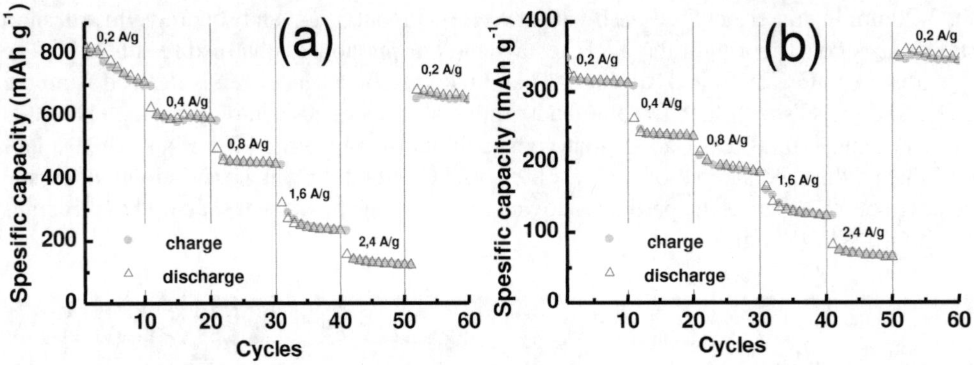

Figure 4. Rate capacity studies of SnO_2-CNTs (a) and SnO_2-CB (b).

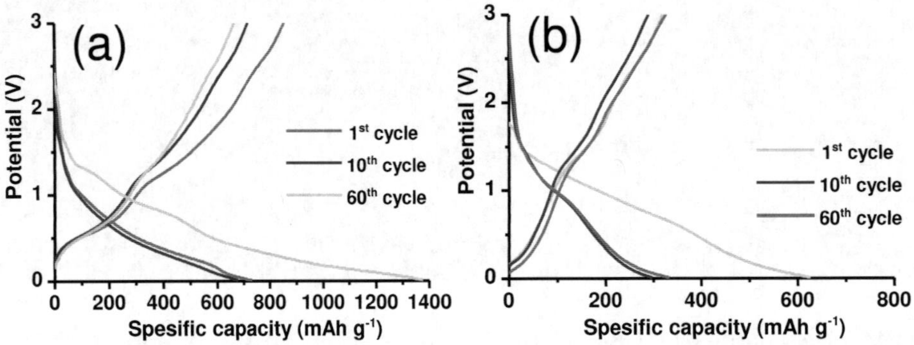

Figure 5. Charge/discharge voltage curves for SnO_2-CNTs (a) and SnO_2-CB (b).

Finally, the cycling performances of samples were carried out at a current density of 200 mAg^{-1} for 100 cycles (figure 6). The SnO_2-CNTs anode showed an initial charge capacity of 810 mAh g^{-1} which slightly reduced to 500 mAh g^{-1} after 100 cycles. The initial charge capacity for SnO_2-CB anode corresponds to 350 mAh g^{-1} and drops to 230 mAh g^{-1} after 100 cycles. These results indicated the excellent stability of those carbon materials' samples.

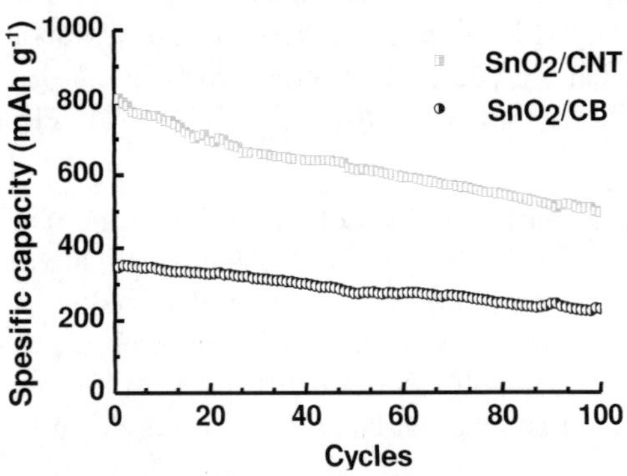

Figure 6. Charge/discharge cycling studies of SnO_2-CNTs and SnO_2-CB at a current density of 200 mAg^{-1} for 100 cycles.

4. Conclusion

In this article, the influences of the formation way of SnO_2, *i.e.*, in a composite with multi-walled carbon nanotubes and a mixture with carbon black, on their electrochemical properties are considered. The

SnO_2-CNTs composite was synthesized by the facile sol-gel method. The SnO_2-CB mixture was prepared by mechanical mixing of two phases. The formation of SnO_2 phase was confirmed by XRD and thermal analysis. The SnO_2 particles in the SnO_2-CNTs composite had a size of 3-20 nm, the size of SnO_2 in the SnO_2-CB mixture was 50-100 nm, the smaller size of the SnO_2 particles and the presented of CNTs have great contribution of their electrochemical performance for lithium ion batteries, *i.e.*, the SnO_2-CNTs composite demonstrates excellent initial capacity (810 mAh g^{-1}), it is higher than the theoretical capacity of carbon (370 mAh g^{-1}) and tin dioxide (790 mAh g^{-1}). Despite the differences in the initial capacity, both anodes exhibit the high initial Coulombic efficiency and excellent cycling stability.

The comparative study on the SnO_2-CNTs composite and SnO_2-CB mixture reveals that the SnO_2 exhibits smaller particle size, more uniform distribution and more abundant in chemical bond with the carbon surface on the SnO_2-CNTs composite could be the main contribution on its superior electrochemical properties than that of SnO_2-CB mixture. This work created an opportunity and a proper perspective for the development of the high performance of carbon-based electrode materials for lithium-ion batteries.

5. Acknowledgment

We would like to acknowledge the support provided by the Science and Technology Platform Project (China-Ukraine, 2014C050012001), National University Students' Innovation and Entrepreneurship Training Program(201810577006), Natural Science Foundation of Guangdong (2014A030310130, 2016A030313118 and 2014A030313642), Foundation for Distinguished Young Talents in Higher Education of Guangdong (2014KQNCX211), Natural Science Foundation of Huizhou University (No. 2015JB004 156020026), Science and Technology Project of Huizhou (2016X0421036, 2017X0202012).

References

[1] Chen J S and Lou X W 2013 *Small* **9** 1877
[2] Lou X W, Chen J S, Chen P and Archer L A 2009 *Chem Mater* **21** 2868
[3] Cui D, Zheng Z, Peng X, Li T, Sun T and Yuan L 2017 *J Power Sources* **362** 20
[4] Wang M S, Wang Z Q, Jia R, Yang Z L, Yang Y, Zhu F Y, Huang Y and Li X 2018 *J Electroan Chem* **815** 30
[5] Lei D, Zhang M, Hao Q, Chen L, Li Q, Zhang E and Wang T 2011 *Mater Let* **65**, 1154
[6] Zhang S, Yue L, Wang M, Feng Y, Li Z and Mi J 2018 *Solid State Ionics* **323** 105
[7] Kweon S C, Oh I H, Lee J K and Khodin A A 2012 *Mater Let* **83** 88
[8] Zhou H, Zhong Y, He Z, Zhang L, Wang J, Zhang J and Cao C N 2014 *Appl Surf Science* **314** 1
[9] Wua W, Zhao Y, Li J, Wu C and Guan L 2014 *J Energy Chem* **23** 376
[10] Huang Z, Gao H, Wang Q, Zhao Y and Li G 2017 *Mater Let* **186** 231
[11] Lee S H, Ryun Y, Noh J Y, Kim B J and Kim W B 2017 *J Power Sources* **367** 1
[12] Chen J S and Lou X W 2012 *Mater Today* **15** 246
[13] Yang J, Xi L, Tang J, Chen F, Wu L and Zhou X 2016 *Electrochim Acta* **217** 274
[14] Zhang X, Huang X, Geng X, Zhang X, Xia L, Zhong B, Zhang T and Wen G 2017 *J Electroan Chem* **794** 15
[15] Zhou X, Geng Z, Zhang C, Wang J and Wang D 2015 *J Energy Chem* **24** 529
[16] Nagirnyak S V, Lutz V A, Dontsova T A and Astrelin I M *Nanoscale Research Letters* (2016) **11** 343.
[17] Diao G, Li H, Liang H, Ivanenko I, Dontsova T and Astrelin I 2018 *Nano* **13** 1850036
[18] Ivanenko I M, Dontsova T A, Astrelin I M and Kuzmenko V V 2013 *Compos Theor Pract* **12** 114
[19] Elizabeth I, Naird A K, Singh B P and Gopukumar S 2017 *Electrochim Acta* **230** 98.
[20] Thirumal V, Pandurangan A, Jayavel R, Krishnamoorthi S R and Ilangovan R 2016 *Curr Appl Phys* **16** 816
[21] Dontsova T A, Ivanenko I M, Astrelin I M and Nagirnyak S V 2014 *J Electr Eng* **2** 34.

Characterization of the actuating response of the graphene oxide/MEO$_2$MA hydrogel layer

H-Y Hsiao, N A M Razali and W-C Lin*

Department of Mechanical and Electromechanical Engineering, National Sun Yat-sen University, Kaohsiung 80424, Taiwan.

* E-mail: wc.lin@mail.nsysu.edu.tw

Abstract. This study is aimed to characterize the actuating responses of the soft polymeric multilayer, which is composed of a coated graphene oxide (GO) layer and di (ethylene glycol) methyl ether methacrylate (MEO$_2$MA) hydrogel layer. The actuating mechanism was based on the swelling and deswelling behaviors of the MEO$_2$MA hydrogel that was triggered by the heat generated from the GO layer irradiated with near-infrared (NIR) light. As the swelling and deswelling properties of the MEO$_2$MA hydrogel were crucial factors in the performances and responses of the GO/MEO$_2$MA multiplayer, the lower critical solution temperature (LCST) and mechanical properties of three different cross-linked MEO$_2$MA hydrogels were measured. The measured LCST indicated that the trigger temperature of the GO/MEO$_2$MA multilayer was between 20 and 26°C. The swelling capability of the MEO$_2$MA layer could reach around 12 times, whereas the surrounding temperature was higher than LCST. Three different cross-link ratios of MEO$_2$MA samples could afford approximately 1.5 N of force loading. Moreover, the coated GO layer could generate heat to exceed the LCST temperature in a few minutes using 450 mW NIR irradiation. The bending responses of the GO/MEO$_2$MA layer were temperature responsive and reversible. Hence, the GO/MEO$_2$MA hydrogel might be a promising solution for several applications, such as the artificial muscles and soft gripper.

1. Introduction

Robotic arms have been widely employed in industries or even in medical applications for the past several decades. These arms are popular because of their ability to offer an assistance force during the working time. However, they are still limited with less biocompatibility and dimensions, leading to inconvenient movement. The weight issue and bulky body are still hardly conquered even after using alloy muscles. Therefore, soft artificial muscles are used to resolve this shortcoming [1]. They perform a similar movement and propose a flexible property compared with robot arms used nowadays. Polymer materials are rapidly being used due to their flexible actions. They are used not only as artificial muscles but as bio-scaffolds in tissue engineering.

Indeed, numerous characteristics, such as softness, lightweight, capability of drug loading, and biocompatibility, make the polymeric materials a good candidate for medical application. Indeed, soft actuators performing variant responses are fabricated using various kinds of polymeric materials [2-4]. Polymeric materials are employed for soft actuators due to their multiple dimensions for actuation. High mechanical properties of polymeric muscles could induce the primary action of a human body, including the bending behavior, lifting stretching, and twisting movements. For instance, Li *et al.* [5] revealed a poly(vinyl alcohol) film grafted with a photomechanical response. Kim *et al.* [6] employed an electric field–responsive film made with a commercialized Nafion film. Although many approaches applied the in the fabrication of actuators have been revealed in recent decades, most of them are limited with biocompatibility and idealess mechanical properties. Hence, hydrogels, which are a kind of polymeric materials, have been used in the last ten years to resolve this issue. Hydrogels could provide the variant machine property due to gelatinization with different cross-linkers, resulting in a unique condition needed [7-9].

Graphene oxide (GO) is commonly employed in several applications, such as bioengineering [10]. It is regarded as a heat source to increase the temperature attributed to the excitation of electrons by near-infrared (NIR) light within a specific wavelength (808 nm). Moreover, GO could strengthen the mechanical properties. Using these characteristics, numerous researchers fabricated actuators composed of GO. Cheng *et al.* [3] made a thin film containing GO, which had the ability of NIR light-stimulated control. The remote control stimulated by light also simplifies the action of actuators, and hence wireless equipment has become the trend in this century. Ultraviolet or NIR light helps to in making hydrogels with different patterns, including deswelling or structural change [7]. Without contact between the base material and the wire, the remote device would simplify the experiments. The di(ethylene glycol) methyl ether methacrylate (MEO_2MA) hydrogel is a temperature-responsive scaffold that has demonstrated reversible hydrophilic/hydrophobic properties at lower critical solution temperature (LCST). This temperature is close to the surface temperature of the human skin. Hence, the MEO_2MA hydrogels are also used as the bio-scaffold material owing to their adhesion onto the skin.

This study reported the fabrication and characterization of the soft GO/MEO_2MA hydrogel actuator. Under NIR irradiation, the atoms in GO were excited to generate heat, which was then transferred to the MEO_2MA hydrogel. Subsequently, temperature-responsive MEO_2MA hydrogel exhibited the deswelling behavior resulting in bending response as shown in figure 1. When the NIR light was removed, the MEO_2MA hydrogel turned to the original state. Repeating the irradiation steps, the GO/MEO_2MA hydrogel attained a reversible actuating behavior.

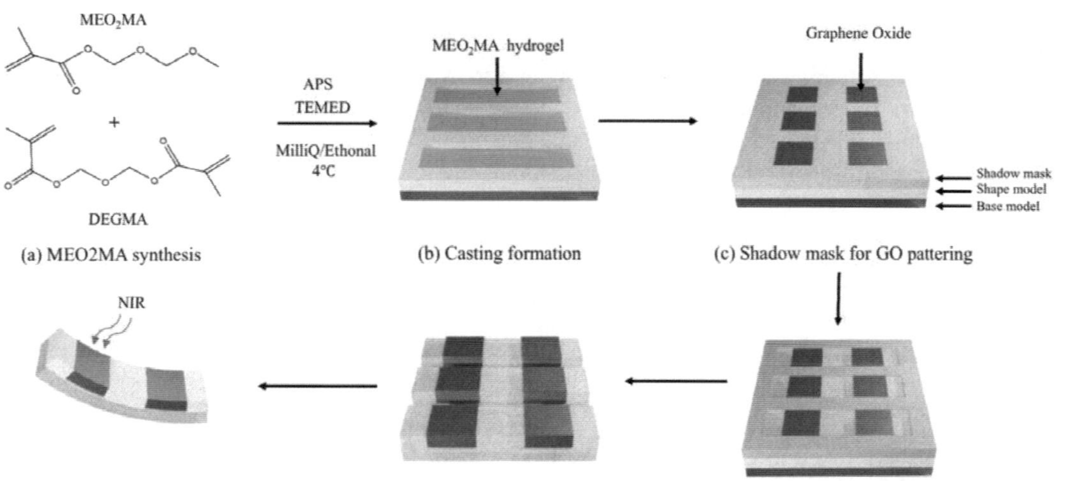

Figure 1. Scheme of preparation of the GO/MEO_2MA hydrogel samples and actuating operation via the irradiation with near-infrared light.

2. Materials and Methods

2.1 Materials
The monomers of di(ethylene glycol) methyl ether methacrylate (MEO_2MA, MW = 188.22, 95%), cross-linker, di(ethylene glycol) dimethacrylate (DEGMA, 95%), activator, N,N,N',N'-tetramethylethylenediamine (TEMED, 99%), and GO (powder, 15–20 sheets, 4%–10% edge oxidized) were purchased from Sigma–Aldrich(the USA). The activated alumina, which was used to purify the monomer of cross-linker and initiator, ammonium persulfate (APS) was provided by Wako (Japan). The last but not the least, the solvents for all materials were ethanol (99.5%) and distilled water, which were purified by Millipore System (Rephile Bioscience, Ltd).

2.2 Fabrication of GO/MEO₂MA hydrogel
The MEO_2MA composite hydrogels comprised the MEO_2MA monomer, and GO was fabricated by the gelatinization method. Specific amounts of DEGDMA cross-linker (i.e., 10, 20, and 30 mg) were added to each MEO_2MA solution (3.26 g). Around 100 mg APS was dissolved in 0.5 mL of water/ethanol mixture (V_{water}:$V_{ethanol}$, 1:1). The APS solution was then poured into the MEO_2MA solution as an initiator and placed in a refrigerator at 4°C. After freezing for 2 h, 0.02 mL of TEMED was added to the MEO_2MA solution to activate the reaction in all chemicals. During the reaction process, the mixture of MEO_2MA solution was filled into a customized die ($1 \times 5 \times 0.2$ cm^3) and kept at 4°C for 3 h. Finally, 5 mg GO was cast onto the MEO_2MA hydrogels.

2.3 A preliminary study of GO/MEO₂MA hydrogel properties
The temperature-responsive GO/MEO_2MA hydrogel exhibited various properties. According to the proportion of the cross-linker, each quantity would contribute different amounts of free radicals. Besides, these free radicals could control the properties of the hydrogel, such as transmittance, hydrophilicity and hydrophobicity, load behavior, and degree of bending movement. Therefore, numerous tests were carried out for analyzing the properties of the temperature-responsive MEO_2MA hydrogel.

2.4 Characteristics of swelling and deswelling on MEO₂MA hydrogels
The swelling/deswelling behavior is an essential property of MEO_2MA hydrogels associated with bending force and actuation. Q_t relative to the temperature was measured to observe the influence, which was caused by different amounts of the cross-linker in each MEO_2MA hydrogel. In brief, various MEO_2MA samples ($n = 3$) were cut into circular segments with a diameter of 0.5 cm. The swelling ratio and deswelling ratio were exhibited at a specific temperature and interval time. The Q_t of each sample was measured with an electronic balance (BSA224S-CW; Sartorius, Gottingen, Germany) calculated using Eq.(1).

$$Q_t = (m_t - m_0)/m_0 \quad (1)$$

where m_t is the mass of swollen hydrogel at time t and m_0 is the weight of the dried hydrogel.

2.5 Determination of LCST of GO/MEO₂MA hydrogel and tensile test
To determine the temperature interval for the LCST of MEO_2MA hydrogels, the ultraviolet-visible spectroscopy (V-770, JASCO, Tokyo, Japan) was used and operated at 500 nm with a scan rate of 2°C min^{-1}. However, the one-way tensile test was used to measure the mechanical properties of the MEO_2MA hydrogel. The samples were fabricated in a standard specification tensile scale based on ASTM D638. All samples were tested using a tensile test machine (EX-SX, Shimadzu, Japan) under a loading rate of 100 mm · min^{-1}.

2.6 Measurement of actuating responses of GO/MEO₂MA hydrogel
About 10 mg GO/MEO_2MA hydrogel was prepared as shown in figure 2. The NIR light machine (MDL-III-808 nm, Changchun New Industries Optoelectronics Technology Co., Ltd., Changchun, China) was used to stimulate GO within the specific wavelength (808 nm) and increase the temperature. The

distance between the output light source and the GO/MEO_2MA hydrogel was 5 cm. The power was set at 450 mW, and the current was 0.91 mA. The reversible test was similarly conducted by switching on/off the NIR light system at specific time intervals.

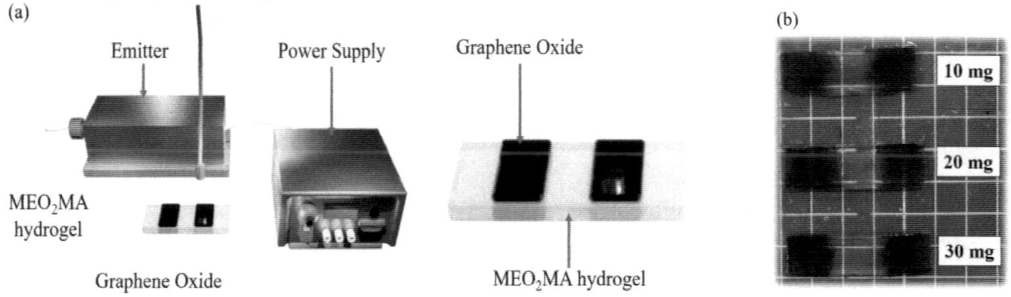

Figure 2. (a) Experimental setting for measuring actuating responses; and (b) photo images of the fabricated GO/MEO_2MA hydrogel film.

3. Results and Discussion

3.1 Transmittance and mechanical properties of MEO_2MA hydrogels

The results indicated that the LCST of each sample ranged from 24°C to 28°C (figure 3). Once the GO/MEO_2MA hydrogel was heated, it changed from a colorless state to a nontranslucent state. The temperature interval increased as the amount of added cross-linker was increased. As shown in figure 4, the GO/MEO_2MA hydrogels exhibited remarkable mechanical properties evaluated using the tensile test. Especially, a less amount of the cross-linker resulted in the higher displacement of the GO/MEO_2MA hydrogel but less force for the load. For instance, the GO/MEO_2MA hydrogel gelatinized using 10 mg of cross-linker showed the softest property, leading to the most flexible hydrogel compared with other samples. Three different cross-linked ratios of samples were characterized to study the machinal property by using the tensile machine. As shown in figure 4, the lower the added amount of cross-linker, the higher the displacement performed. The samples of 10, 20 and 30 mg demonstrated the displacements of approx. 75, 65 and 55 mm, respectively. This finding can be attributed to various amounts of the cross-linker reacting during the gelatinization process. The more cross-linker is used, the more free radicals are generated, resulting in a closer association with MEO_2MA molecules, and, hence, higher LCST temperature interval and better mechanical properties.

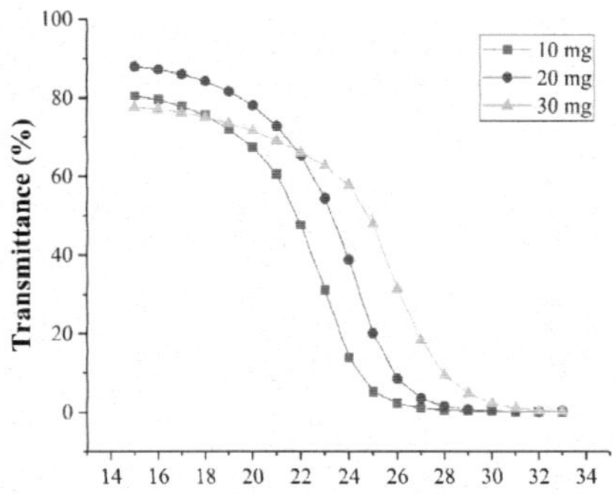

Figure 3. The transmittance of MEO_2MA with different cross-linker concentrations.

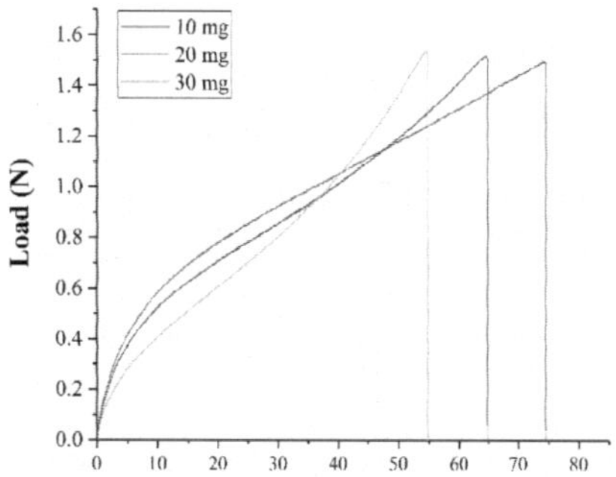

Figure 4. Tensile test of MEO_2MA with different cross-linker concentrations.

3.2 Swelling and deswelling behavior patterns of MEO_2MA hydrogels

The GO/MEO_2MA hydrogels exhibited different swelling and deswelling behaviors, as shown in figures 5 and 6, respectively. As the temperature increased beyond the LCST, the GO/MEO_2MA hydrogels began to shrink and desorb the water. With different ratios of the cross-linker used in the gelatinization process, the water was released with different slopes. For example, 10 mg of cross-linker used in the GO/MEO_2MA hydrogel demonstrated lower LCST, cliffy swelling/deswelling slope, and higher water content. The Q_t as the swelling ratio in the swelling test was about 11.5, which surpassed that of the other samples. Overall, the findings revealed that the GO/MEO_2MA hydrogel fabricated using 10 mg of cross-linker could rapidly and reversibly react within the changing temperature interval, and hence regarded as a good candidate for actuator application. Based on all conditions mentioned in Section 3.1, the GO/MEO_2MA hydrogel gelatinized using 10 mg of cross-linker was used to demonstrate the reversible bending behavior, as shown in figure 7. The results showed that the GO/MEO_2MA hydrogel shrank and then recovered to the original state if the NIR light system was switched off. The recorded weight in every 5 min was used to confirm that the swelling and deswelling behaviors did not alter during the repeated operation. Hence, the GO/MEO_2MA hydrogel extend and shrink, acting as an actuator.

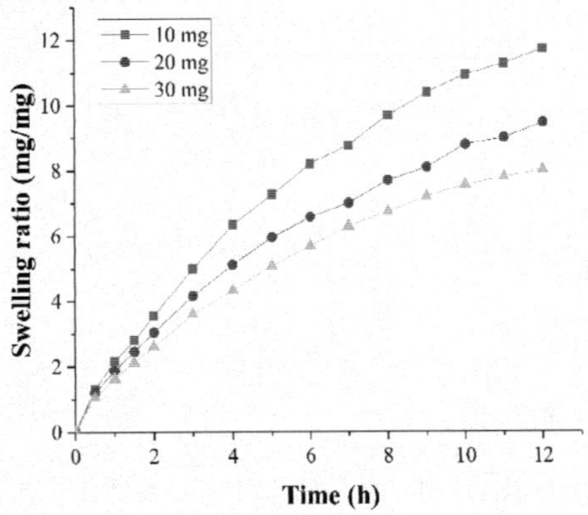

Figure 5. Swelling ratios of three different cross-linker concentrations of MEO_2MA hydrogels.

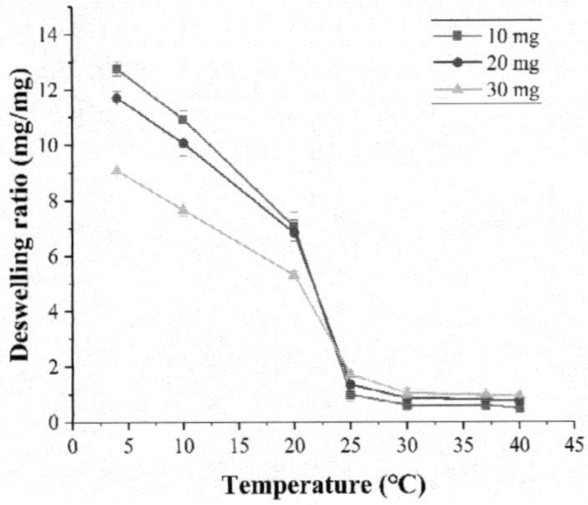

Figure 6. Deswelling ratios of three different cross-linker concentrations of MEO_2MA hydrogels.

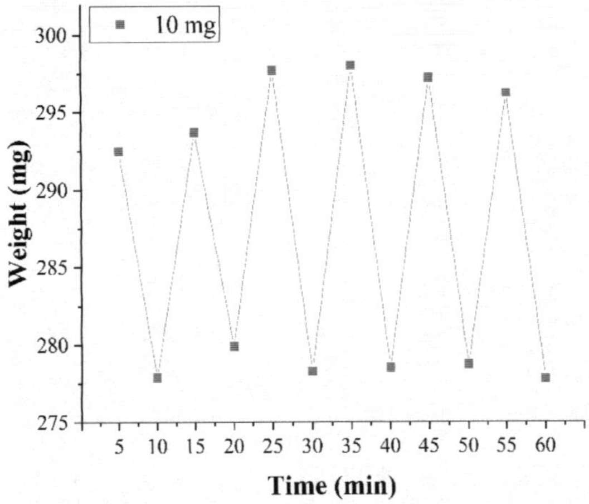

Figure 7. Reversible deformation of GO/MEO_2MA hydrogel actuated in six cycles.

3.3 Bending and reversibility test of MEO_2MA hydrogels

In the bending and reversibility tests, the MEO_2MA and GO/MEO_2MA hydrogels were compared and separately irradiated with NIR light. As shown in figure 8a and 8b, the shape of the hydrogel without GO was not altered despite irradiation for a long time. On the contrary, the GO/ MEO_2MA hydrogel exhibited an apparent bending movement on NIR irradiation after a few minutes, as shown in figure 8c and 8d. On NIR light irradiation, the left side of the MEO_2MA hydrogel coated with GO started to shrink and release the water, but no water release occurred on the right side. This resulted in the bending movement. It could be attributed to the increase in temperature caused by the irradiated GO. Once the temperature increased and reached the critical point of LCST, the GO/MEO_2MA hydrogel changed from the swelling state to a deswelling state, releasing the water. Therefore, the bending behavior of the GO/MEO_2MA hydrogel was believed to be controlled by NIR light. These findings might help in designing the shape of the hydrogel [10-12], which might be applied to the body for assisting in human behaviors. Hence, the GO/MEO_2MA hydrogel has excellent potential in soft artificial muscles.

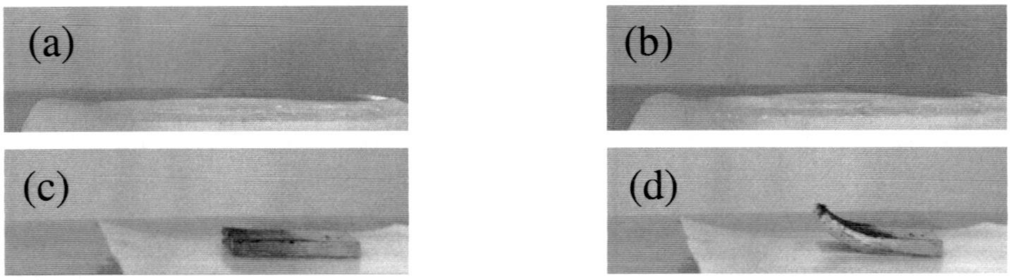

Figure 8. Bending behavior of the GO/MEO_2MA hydrogel stimulated with NIR light. (a) Sample without GO and (b) sample irradiated with NIR, which did not alter the shape. (c) Sample without NIR light irradiation and (d) sample irradiated with NIR light. The current was set as 0.91 mA, and the applied power was 450 mW.

4. Conclusions

In summary, the soft GO/MEO_2MA hydrogel actuator was fabricated and characterizedand, demonstrating the appropriate mechanical and temperature-responsive properties. The GO/MEO_2MA hydrogel preformed reversible actuating responses under a repeated switch-on/off NIR irradiation. The GO/MEO_2MA actuator was composited of the MEO_2MA hydrogel and the thin coated GO layer. The GO layer could generate the heat via the NIR irradiation to operate the actuating responses. While the generated temperature was higher than the LCST of the MEO_2MA hydrogels, the deswelling behavior occurred. The LCST of MEO_2MA hydrogel was characterized in the range between 24°C and 28°C. The DEGMA cross-linker can affect the swelling, deswelling and mechanical properties of fabricated hydrogel actuators. For example, the GO/MEO_2MA hydrogel cross-linked with 10 mg DEGMA demonstrated approximately 12-fold of Q_t in the swelling behavior. Overall, the GO/MEO_2MA hydrogels might be considered a high potential candidate as an actuator and could be applied for the application [13] of soft artificial muscles.

Acknowledgments

This study was supported by the Research Grant (MOST105-2622-E-110-004-CC2) from the Ministry of Science and Technology of Taiwan. All experiments in this study were conducted in the BEing Laboratory, National Sun Yat-sen University, Taiwan. Also, the authors would like to thank for the technical support, experimental assistance, and suggestions from Dr. Kotsuchibashi, Shizuoka Institute of Science and Technology, Japan, Dr. Lin, and Mr. Yeh from National Sun Yat-sen University, Taiwan.

References

[1] Miriyev A, Stack K and Lipson H 2017 *Nat Commun* **8** 596
[2] Cheng Z, Wang T, Li X, Zhang Y and Yu H 2015 *ACS Appl. Mat. Interfaces* **7** 27494-501K.
[3] Hou K, Wu W, Xia M and Zhu M 2017 *Macromol. Mater. Eng.* **302** 1700213
[4] Wang E, Desai M S and Lee S-W 2013 *Nano Lett.* **13** 2826-30
[5] Yu L, Cheng Z, Dong Z, Zhang Y and Yu H 2014 *J. Mater. Chem. C* **2** 8501-6
[6] Kim B, Kim B M, Ryu J, Oh I-H, Lee S-K, Cha S-E and Pak J 2003 *International Society for Optics and Photonics.* **5051** 486-96
[7] Haraguchi K and Li H-J 2006 *Macromolecules* **39** 1898-905
[8] Xiang H, Xia M, Cunningham A, Chen W, Sun B and Zhu M 2017 *J. Mech. Behav. Biomed. Mater.* **72** 74-81
[9] Zheng W J, An N, Yang J H, Zhou J and Chen Y M 2015 *ACS Appl. Mater. Interfaces* **7** 1758-64
[10] Zhu Y, Murali S, Cai W, Li X, Suk J W, Potts J R and Ruoff R S 2010 *Adv. Mater.* **22** 3906-24
[11] Techawanitchai P, Ebara M, Idota N, Asoh T-A, Kikuchi A and Aoyagi T 2012 *Soft Matter* **8** 2844-51
[12] Zhang X, Pint C L, Lee M H, Schubert B E, Jamshidi A, Takei K, Ko H, Gillies A, Bardhan R and Urban J J 2011 *Nano Lett.* **11** 3239-44
[13] Zhao Q, Liang Y, Ren L, Yu Z, Zhang Z and Ren L 2018 *Nano Energy* **51** 621-31

IOP Conf. Series: Materials Science and Engineering **474** (2019) 012024 doi:10.1088/1757-899X/474/1/012024

Recent research progress on CALPHAD-based ICME modeling of magnesium alloys

S S Song*

Shandong Normal University Library, Shandong Normal University, Jinan 250014, China.

*E-mail: shshsong@outlook.com

Abstract. Magnesium alloys combine several favorable attributes, including low density, high strength-to-weight ratio, good damping capability and so on. However, their application is limited by their relatively poor high-temperature performance and creep resistance. These deficiencies can be alleviated by the alloying-induced solid solution and precipitation strengthening. The integrated computational materials engineering (ICME) approach based on the calculation of phase diagram (CALPHAD) software is shown to accurately quantify the effect of alloying elements on the strengthening precipitates. Thus, it is quite instrumental forin enhancing the alloy development process, alloy design, and manufacturing optimization. This paper summarizes the theoretical background of CALPHAD-based ICME, outlines the current research progress in this direction and provides some insights into future development.

1. Introduction

Magnesium (Mg) is a lightweight structural metal, which density (1.74 g/cm^3) is lower by 20% than that of aluminum (2.70 g/cm^3) and more than four times lower than that of carbon steel (7.89 g/cm^3). It is been widely applied in industries where weight reduction is critical, including aerospace, automotive and portable electronic devices, given its favorable attributes such as high strength-to-weight ratio, good heat and electrical conductivity, excellent castability and good damping properties [1-5]. However, due to its lower high-temperature strength, as compared to aluminum and steel alloys, the application of Mg has been limited to room-temperature applications [6-8]. Alloying has become one of the most common methods for the strengthening of Mg alloys, introducing a solid solution strengthening and precipitation hardening effects. Modern Mg alloys usually are complex chemical systems including Al, Zn, Mn, Si, Sn, and rare earth elements like Gd, Nd, Y, and Sm [9-11]. To more efficiently develop multi-component Mg alloys, the equilibrium phase constructions, precipitation hardening mechanisms, and the process-structure-property relationship of the systems must be fully understood quantitatively.

Computational simulation such as CALPHAD (Calculation of phase diagrams) modeling holds promise to understand and predict structures and properties beyond known information [12]. Recently, the CALPHAD approach has been broadened to a holistic ICME (Integrated Computational Materials Engineering) framework in the design and development of new materials and products [13, 14]. The application of this approach in the development of novel alloys, including Al, steel and high-entropy alloys, simplifies the composition design procedure and reduces the development cost [15-18].

This article presents a summary of recent research progress on the application of CALPHAD-based ICME modeling on Mg alloy development and discusses the future trend of this promising alloy development approach. It aims to provide a brief introduction of the fundamentals of the above-mentioned modeling method and to explore the application of this integrated simulation tool in the development of novel Mg alloys.

2. Optimization of Mg alloys

Current development of novel Mg alloys has been focused on the development of creep-resistant Mg alloys, high-strength high-ductility Mg alloys, and Mg-based composite.

Due to its hexagonal crystal structure, Mg alloys have been widely known to have poor mechanical properties at elevated temperatures. The addition of rare earth elements, including Y [19, 20], Ce [21], Nd [22, 23], Gd [19, 24], and Sm [25, 26], introduces dense and uniformly distributed nano-scale precipitate phases, which act as practical obstacles for dislocations and grain boundary motions. It is thus one of the essential methods to develop creep-resistant Mg alloys for elevated-temperature applications. Mg-RE-based alloys have been an increasing research and engineering focus in recent Mg alloys' community. For example, Mg-RE-based Elektron 21 from Magnesium Elektron [27] exhibited high creep resistance at 200°C and was used in engine parts of the military Apache helicopter.

The strength and ductility of Mg alloys need further improvement for increasing industrial needs. Mg-Li dual-phase alloy is an alloy that is lighter than pure Mg. Further addition of Y, Al, Si, Sn improves thermal-stability of the precipitate phase and thus further increase strength levels of Mg-Li alloys [28].

Mg-based composites reinforced by ceramic particles (SiC, Al2O3, etc.) combine high characteristics of strength, strength-to-weight ratio, and wear resistivity. Chen et al. [29] reported a new kind of SiC-reinforced Mg-Zn alloy that made a record-yield strength exceding 700 MPa. On the other hand, in-situ Mg-based composite makes use of physical chemistry reactions of matrix and solute atoms to produce evenly distributed second phase, making it a great candidate for brake block applications [30].

It can be seen that nano-scale precipitate particle strengthening is the major approach for current Mg alloys strengthening. Combination of phase diagram calculation (CALPHAD) and ICME simulates the size, density, and distribution of nano-scale particles. Being supplemented by a quantitative understanding of the structure-property relationship of these materials, this information enables us to quantitatively calculate and predict the effects of different alloying elements on the final products' properties.

3. Fundamentals of CALPHAD-based ICME

CALPHAD-based ICME relies on the combination of CALPHAD and classical nucleation and growth theory. In classical nucleation theory,

$$J = \frac{dN}{dt} = Z\beta^* N_v exp\,(-\frac{\Delta G^*}{k_B T})\, exp\,(-\frac{\tau}{t}) \tag{1}$$

where Z is the Zeldovich factor, which represents the probability that the nucleus, once having reached a critical size, grows into a stable phase; β^* is the rate at which atoms attach to nuclei; N_v is the number of available nucleation sites; ΔG^* is the free energy barrier for nucleation, which depends on thermodynamic driving force ΔG_v and interfacial energy γ; τ is incubation time, describing the time for nuclei to grow to critical size; while k_B is the Boltzmann constant.

At the growth stage,

$$v = \frac{dR}{dt} = \frac{K}{R}(\frac{1}{R^*} - \frac{1}{R}) \tag{2}$$

$$\overline{R^3} - \overline{R_0^3} = \frac{4}{9}Kt \tag{3}$$
$$K = \frac{2\gamma V_{at}^{p}}{(C^{\alpha\beta}][M]^{-1}[C^{\alpha\beta})}.$$

where $(C^{\alpha\beta}]$and $[C^{\alpha\beta})$ where are the row and column vectors of the solute concentration differences between matrix α and precipitate β, while $[M]$ is the chemical mobility matrix.

From Equations (1-3), the nucleation and growth of nano-scale particles rely on the thermodynamic energy, interfacial energy, diffusion coefficient and so on. While CALPHAD provides thermodynamic energy and diffusion coefficient, the other parameters can be obtained through crucial experiments and other simulation methods. Simulation results include the precipitate kind, size and number density under different aging stages and conditions. Figure 1 is a flowchart for CALPHAD-based ICME.

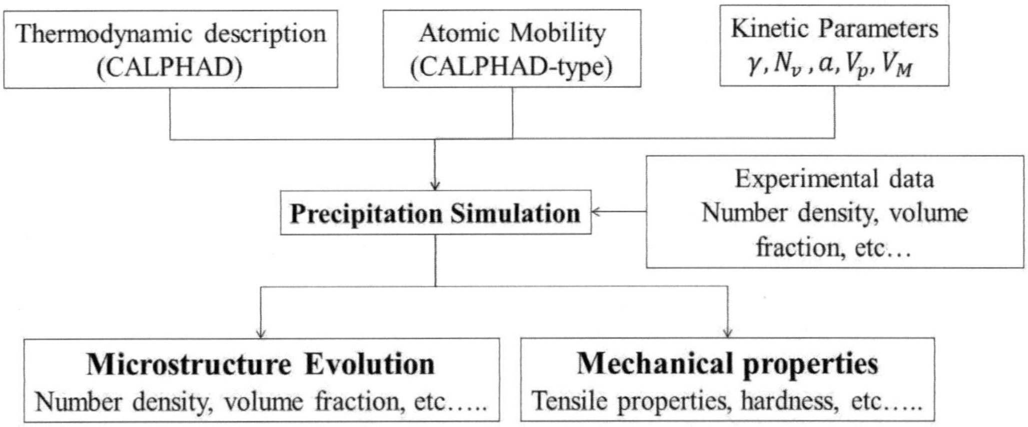

Figure 1. Schematic Procedure for CALPHAD-based ICME modeling [31]

The CALPHAD approach is a phenomenological thermodynamic methodology to obtain thermodynamic descriptions of multicomponent systems. The term thermodynamic description represents the parameters for Gibbs energies of all phases in the target system. The idea of classical CALPHAD approach is that: based on available information such as phase equilibria, heat capacities, formation enthalpies and structural information the Gibbs energy of each phase within the system is obtained and then used to predict phase constructions where data are lacking. The Gibbs energy model can be described as follows:

$$G_m^{\varphi} = G^{ref} + G^{id} + G^{ex} \tag{4}$$

Here G^{ref} defines a surface of reference. G^{id} is the molar configurational entropy. G^{ex} is the excess Gibbs energy, for which the models are determined according to the phase's crystal structure.

CALPHAD approach also gives atomic mobility of elemental species. It is obtained based on experimental information such as tracer diffusion, inter-diffusion and composition profiles of diffusion couples and used to extrapolate diffusion data beyond known conditions. In a system containing n elements, the diffusion coefficient of element B can be expressed as:

$$D_{Bj}^{n} = \sum_i (\delta_{iB} - x_B) x_i M_i \left(\frac{\delta\mu_i}{\delta x_j} - \frac{\delta\mu_i}{\delta x_n}\right) \tag{5}$$

where δ_{iB} is Kronecker delta, x_B is mole fraction of element B, M_i is mobility of element i, μ_i is chemical potential. $M_i = \exp(\frac{\Psi_i}{RT})\frac{1}{RT}$. Here Ψ_i can be described using a similar way as Equation (4) and, thus, can be calculated by the CALPHAD approach. Added with the chemical potential which is another form of thermodynamic energy, the diffusion coefficient can be modeled.

4. Current CALPHAD-based ICME database development for Mg alloys

4.1. *CALPHAD Thermodynamic Database*

The development of CALPHAD thermodynamic database is based on the key experimental data such as phase equilibria or phase transformation temperature. In phase equilibria identification experiments, Mg alloys are sealed in Ar/Hr protected quartz or glass tubes to avoid contact with air, then annealed at target temperatures for a sufficiently long time for equilibrium, followed by room-temperature quench. SEM, TEM or XRD then determine the phase constructions. Zhang et al. [32], Xia et al. [33] used such methods to evaluate phase equilibria of Mg-RE systems Mg-Nd-Zn and Mg-Sm-Zn. The determination of phase transformation temperatures relies on DSC or DTA. Schmid-Fetzer [34] was the first to seal Mg alloys in Ta tubes for DSC tests to avoid oxidation. Similar methods have been applied in the development of Mg-Al-Sn [34], Mg-Mn-Ni [35], Mg-Ca-Mn [35] systems' phase diagrams.

The development of Mg alloys' CALPHAD database has been an active research topic. At present, there have been several commercial Mg alloy CALPHAD databases, such as PanMg from Computherm (USA) [36]. It includes Al, Zn, Mn, Y, Gd and other in total 25 elements. In addition to phase diagram calculation, it is also capable of Scheil calculation to simulate solidification experiments and volume fractions [37, 38]. However, most of current Mg CALPHAD databases is focused on ternary or even binary systems. Quarternary systems are only reported for Mg-Mn-Fe-Ni [35], Mg-Ce-Gd-Y [39], Mg-Sn-In-Zn [40] and several other rare cases. Multi-component systems' descriptions are rarer, for which researchers usually extend from ternary systems, where critical data such as liquidus temperatures cannot be as accurate. Furthermore, the information of new Mg alloys is lacking. For example, Mg-RE-RE' alloys have drawn great interests in Mg community [41, 42]. However, to the best of the author knowledge, no CALPHAD databases for this group of alloys are available yet.

4.2. *CALPHAD-Type Mobility Database*

CALPHAD-type mobility database is developed according to diffusion coefficients that are experimentally measured. Das et al. [43] prepared Mg-Gd and Mg-Y diffusion couples, wrapped with Ta coil and annealed under Ar protection. They identified the compositional profiles near interfaces and simulated the mobility databases for the two systems. Zhong et al. [44] measured diffusion coefficients of Ca and Zn in Mg by a liquid-solid diffusion couple. Meanwhile, simulation is an increasingly effective way to obtain diffusion parameters. Zhou and Liu [45] applied high-throughput first principle calculations for the diffusion coefficients of 47 elements in the Mg matrix.

Mg alloys' mobility database is currently limited to binary systems such as Mg-Al [46], Mg-Zn [46], Mg-Ca [43] and several ternary systems Mg-Al-Sn [15]. MOBMG1 database from Thermocalc (Sweden) [47] simulates the diffusion of Al, Ca, Zn, Sn and other in total 23 elements in the Mg matrix. Other than diffusion coefficients, mobility databases can also be used to predict annealing time/temperature, diffusion couple compositional profile, solidification volume fraction, etc. However, compared to other mature alloys systems like Al or steel, there are shortages of Mg mobility database: not only there are limited systems reported, but also the interaction of solute atoms is not included. The accuracy and small inclusion have limited their wide applications.

4.3. *Mg Alloys' ICME Database*

Classical nucleation and growth theory have been used for simulation and calculation of precipitate hardening of Mg alloys for decades. Hutchinson [48] calculated the number density and volume fraction of β-$Mg_{17}Al_{12}$ precipitates in AZ91 (Mg-9Al-Zn) during a 200°C isothermal annealing. The resulting simulation data correspond very well to 200°C experimental measurements. However, due to the lack of general thermodynamic and kinetic parameters, classical nucleation and growth model is limited to the system composition and temperature range that the key experimental information was obtained. It has limited predication ability when alloy composition or aging temperature changes drastically. Such shortages are overcome when classical nucleation and growth model is combined with CALPHAD. As a comparison, Zhang C [49] integrated PanMg and Hutchinson's model within the ICME framework. The results are shown in figure 2. It is clear that Zhang's simulation yields a good agreement and prediction for the age-hardening response of A9, AZ31, AZ91 alloys at three different temperatures (150, 165, and 200°C).

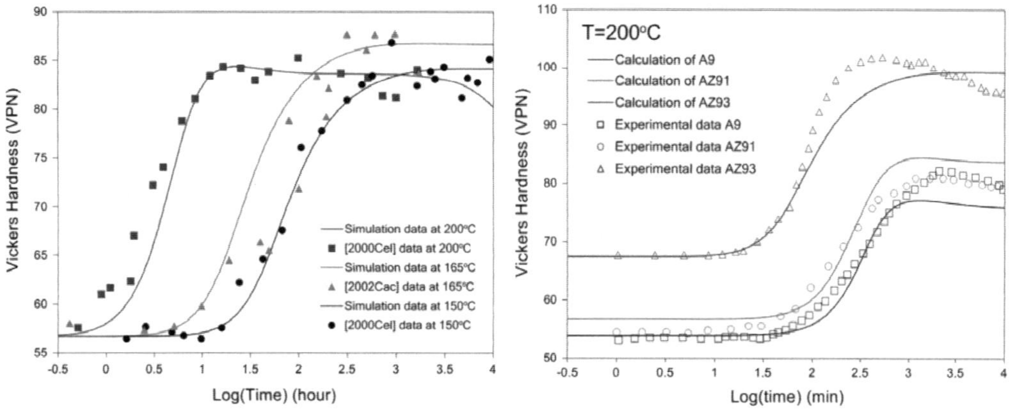

Figure 2. Simulated Age-Hardening Response of AZ91 Alloy at 150°C, 165°C, 200°C (a); Simulated Age-Hardening Response of A9, AZ31, AZ91 Alloy at 200°C (b)

Xia et al. [50] were the first to simulate precipitate particles in Mg-RE-based alloys (Mg-Nd-Zn and Mg-Sm-Zn). For Mg-Nd-Zn, they started with Mg-3Nd-0.2Zn as the target system. The critical experimental parameters (number density, volume fraction) of the significant strengthening particle β' were first obtained via TEM characterizations. The CALPHAD thermodynamic and mobility databases were built, accordingly. Eventually, the integration of CALPHAD databases and nucleation parameters were developed and optimized to experimental measurements. The obtained databases not only model and predict the evolution of β' particles, but also calculate the age-hardening response. The yield strength of various Mg-Nd-Zn alloys after different aging conditions was also calculated and compared to experimental data, from which good agreement was found.

Luo [51] applied ICME calculations for the Mg-Al-Sn system. Two kinds of precipitate phases were considered, $MgSn_2$ and $Mg_{17}Al_{12}$, either type of which dominates in different stages of aging. Luo developed Mg-Al-Sn ternary system first and included both strengthening phases for calculation. By comparing the Gibbs energies of $MgSn_2$ and $Mg_{17}Al_{12}$, the primary strengthening particles were determined under different aging conditions. The simulation results correspond well to experiments.

It is noteworthy that CALPHAD-based ICME can also be applied to the design and optimization of in-situ Mg-composites, where strengthening particles are produced by chemical reactions that are thermodynamically controlled. For example, Fan [52] calculated the effects of Mn, Nd, Ce, La, Ni on the thermodynamic reaction of AlN particles in Mg composite. The number density and volume fraction of strengthening particles can thus be predicted in these systems. Similar, yet more accurate, calculations can be performed for other Mg-based composites upon their integration with the CALPHAD database.

5. Conclusions

Mg alloys, because of their low density and high strength-to-weight ratio, are widely used in the industries where lightweight is crucial, including aerospace, automotive and consumer electronics. Strengthening of Mg alloys relies on careful control of nano-scale precipitate particles. CALPHAD-based ICME provides essential information for Mg alloys' composition and heat treatment condition design and optimization. Such databases can minimize experimental cost by quantitative modeling and predicting the evolution of materials' properties under different compositions and heat treatment conditions. From the above discussion, current Mg alloys' ICME databases can model and predict ternary Mg alloys' thermodynamic, mobility and precipitation quite well. However, more effort is in need for future development of Mg alloy ICME databases:

- ✓ Higher-order systems' databases are still lacking and require further research. For this purpose, more detailed experimental data are necessary, including unknown phases' crystal structure, chemical compositions, and thermodynamic stabilities;
- ✓ Sub-databases like thermodynamic and mobility need further optimization;
- ✓ A systematic understanding of structure-property relationship is required when combining CALPHAD databases and strengthening particle data. The final integration of CALPHAD, precipitate, and structure-property databases gives a reliable and highly efficient way to design novel Mg alloys to meet increasing industrial needs.

References

[1] Hao S M 2002 *Journal of Materials and Metallurgy* **1** 166.
[2] Yan Y Q, Zhang T J, Deng J, Zhou L 2004 *Rare Metal Mater. Eng.* **33** 561.
[3] Xu W, Birbilis N, Sha G, *et al.* 2015 *Nature Materials* **14** 1229.
[4] Hort N, Huang Y, Fechner D, *et al.* 2010 *Acta Biomaterialia* **6** 1714-25.
[5] Luo A A, Mishra R K, Powell B R, *et al.* 2012 *Materials Science Forum.* Trans Tech Publications **706** 69-82.
[6] Luo A A 2013 *Journal of Magnesium and Alloys* **1** 2-22.
[7] Tian S K, *et al.* 2011 *Casting Technologies* **32** 1174-7.
[8] Wu Y J, *et al.* 2011 *Materials China* **30** 1-4.
[9] Ding W J, *et al.* 2010 *Materials China* **8** 37-45.
[10] Yu K, Li W X, *et al.* 2003 *Chinese Journal of Nonferrous Metals* **13** 277-88.
[11] Rokhlin L L 2014 *Magnesium Alloys Containing Rare Earth Metals: Structure and Properties.* CRC Press.
[12] Chang Y A, Chen S, Zhang F, *et al.* 2004 *Progress in Materials Science* **49** 313-45.
[13] Horstemeyer M F 2012 *Integrated Computational Materials Engineering (ICME) for Metals: Using Multiscale Modeling to Invigorate Engineering Design with Science*. John Wiley & Sons.
[14] Luo A A 2015 *CALPHAD* **50** 6-22.
[15] Allison J, Backman D, Christodoulou L 2006 *JOM* **58** 25-7.
[16] Panchal J H, Kalidindi S R, McDowell D L 2013 *Computer-Aided Design* **45** 4-25.
[17] Zhang W, Liu Y, *et al.* 2007 *Rare Metal Letters* **26** 1-6.
[18] Zhao J C 2014 *Nature* **36** 89-104.
[19] Apps P J, Karimzadeh H, King J F, *et al.* 2003 *Scripta Materialia* **48** 1023-8.
[20] Li Y G, Duan J H, *et al.* 2003 *Modern Machinery* **5** 86-8.
[21] Mishra R K, Gupta A K, Rao P R, *et al.* 2008 *Scripta Materialia* **59** 562-5.
[22] Li Q, Wang Q, Wang Y, *et al.* 2007 *Journal of Alloys and Compounds* **427** 115-23.
[23] Sanaty-Zadeh A, Xia X, Luo A A, *et al.* 2014 *Journal of Alloys and Compounds* **583** 434-40.
[24] Gao L, Chen R S, Han E H 2009 *Journal of Alloys and Compounds* **481** 379-84.

[25] Zheng J, Wang Q, Jin Z, *et al.* 2010 *Materials Science and Engineering: A* **527** 1677-85.
[26] Zhang Q, Li Q A, Jing X T, *et al.* 2010 *Foundry* **3** 260-2.
[27] Lyon P, Syed I, Heaney S. *Advanced Engineering Materials* **9** 793-8.
[28] Zheng X D, Dong T S, *et al.* 2017 *Light Alloy Fabrication Technology* **45** 6-11.
[29] Chen L Y, Xu J Q, Choi H, *et al.* 2015 *Nature* **528** 539.
[30] Chen Xiao 2005 *In-Situ Particle Reinforced Magnesium-Based Composites,* Central South University
[31] Xia X Y, Sun W H, Luo A A, Stone D S 2016 *Acta Mater.* **111** 335
[32] Zhang C, Luo A A, Peng L M, Stone D S, *et al.* 2011 *Intermetallics* **19** 1720-6
[33] Xia X Y, Sanaty-Zadeh A, Zhang C, et al. 2014 *Journal of Alloys and Compounds* **593** 71-8.
[34] Doernberg E, Kozlov A, Schmid-Fetzer R 2007 *Journal of Phase Equilibria and Diffusion* **28** 523-35.
[35] Wang P S 2012 *Phase Diagram and Thermodynamic Description of Mg-Mn-Fe-Ni, Mg-Mn-Ca-Zr, Mg-Mn-Al-Li Systems*, Central South University
[36] PanMagnesium. 2013 *Thermodynamic Database for Magnesium Alloys CompuTherm LLC*, Madison, Wis., USA.
[37] Chen S L, Daniel S, Zhang F *et al.* 2002 *CALPHAD* **26** 175-88.
[38] Cao W S, Chen S L, Zhang F *et al.* 2009 *CALPHAD* **33** 328-42.
[39] Grobner J, Schmid-Fetzer R 2010 *Scripta Materialia* **63** 674-9.
[40] Wang J, Hudon P, Kevorkov D, *et al.* 2014 *Journal of Alloys and Compounds* **588** 75-95.
[41] Schmid-Fetzer R, Joachim G 2012 *Metals* **2** 377-98.
[42] Schmid-Fetzer R 2015 *Magnesium Technology* **2015** 283-7.
[43] Das S K, Kang Y B, Ha T K, *et al.* 2014 *Acta Materialia* **71** 164-75.
[44] Zhong W, Zhao J C 2017 *Scripta Materialia* **127** 92-6.
[45] Zhou B C, Shang S L, Wang Y, *et al.* 2016 *Acta Materialia* **103** 573-86.
[46] Das S K, Kang D H, Jung I H 2014 *Metallurgical and Materials Transactions A* **45** 5212-25.
[47] MOBMG1. 2016. *Mobility Database for Mg alloys, Thermo-Calc Software*, Stockholm, Sweden
[48] Hutchinson C R, Nie J F, Gorsse S 2005 *Metallurgical and Materials Transactions A* **36** 2093-105.
[49] Zhang C, Cao W, Chen S L, *et al.* 2014 *JOM* **66** 389-96.
[50] Xia X, Sun W, Luo A A, *et al.* 2016 *Acta Materialia* **111** 335-47.
[51] Sun W, Zhang C, Klarner A D, *et al.* 2015 *Simulation of Concurrent Precipitation of Two Strengthening Phases in Magnesium Alloys Magnesium Technology*, Springer, Cham **2015** 289-93.
[52] Fan T, Zhang C, Chen J, *et al.* 2009 *Metallurgical and Materials Transactions A* **40** 274.

The behavior of laced built-up columns made of aluminum alloy

I J Baláž[1,*] and Y P Koleková[2]

[1] Department of Metal and Timber Structures, Faculty of Civil Engineering, Slovak University of Technology, 810 05Bratislava, Slovakia
[2] Department of Structural Mechanics, Faculty of Civil Engineering, Slovak University of Technology, Bratislava 810 05, Slovakia

*E-mail: ivan.balaz@stuba.sk

Abstract. The second-order analysis is used to examine the laced built-up member with two extruded I-sections under combined compression and bending. The member bottom end is fixed, and the upper one is free in the case of in-plane buckling. The member is restrained against out-of-plane buckling at both ends. The global initial sway imperfection is taken into account instead of the local bow imperfection.

1. Introduction

Eurocodes EN 1993 [1] and EN 1999 [2] provide the guidance limited to the design of uniform built-up member under compression with hinged ends laterally supported only. This way of calculation follows the concept published in [3] based on equivalent bending and shear stiffness. Eurocodes [1, 2] use the local bow imperfection $e_0 = L / 500$. The numerical examples of built-up steel members were solved according to the former Czechoslovak standard ČSN 731401 in [4], according to the former German code in [5, 6] and according to Eurocode [1] in publications[7, 8]. However, the calculations in [6] are not complete, while those in [7] contain significant errors. This paper investigates a more general case described in the abstract. The numerical example of the built-up member made of aluminum alloy is only part of the large parametrical study of the battened and laced built-up members. In the parametrical study, the analysis was used based on the Rubin analytical solutions and formulas [4]. All numerical results were verified by the independent program IQ 100 [9], which revealed zero deviations.

2. Analyzed laced built-up column

The analysis of the second order with imperfection is used in calculations. The geometrical equivalent global initial sway imperfection is taken according to [1, 2]. The column is made of EN-AW 6061-T6 aluminum alloy, of buckling class A. The material possesses the yield strength $f_0 = 240$ MPa and elastic modulus modulus $E = 70$ GPa. The material safety margin $\gamma_{M1} = 1.1$ is used, while the design values of the external forces and moments applied to the column top (figure 1) are

$$F_{Ed} = 430\,kN, \quad H_{Ed} = 11\,kN, \quad M_{Ed,e} = 175\,kNm \tag{1}$$

The height of the column equals to $L = 7.2$ m. The boundary conditions are defined by the buckling lengths: $L_{cr,y} = L$ and $L_{cr,z} = 2L$ (table 6.8 in [2]). The distance between the posts of the lacing is $a = 1.2$ m. The initial sway imperfection is determined according to clause 5.3.2 [1, 2]:

$$\phi = \phi_0 \, \alpha_h \, \alpha_m = \frac{1}{200} 0.745 \; 1.0 = 3.727 \; 10^{-3}, \quad \phi = \frac{1}{268.328} = 3.727 \; 10^{-3} \tag{2}$$

The total design value of the horizontal force at the top of the column after the replacement of the initial imperfections by the equivalent horizontal forces:

$$H_{Ed,tot} = H_{Ed} + \phi N_{Ed} = 11kN + 1.603kN = 12.603kN \tag{3}$$

The properties of the cross-section class 3 extruded I section (Class 3 cross-section: flanges 120 x 12 mm, section height 240 mm, web thickness 9 mm, radius 16 mm), which creates a chord of the built member are as follows:

$$A_{ch} = 50.44 \; cm^2, \; I_{ch,y} = 4742 \; cm^4, \; I_{ch,z} = 348.5 \, cm^4, \; W_{ch,el,y} = 395.2 \; cm^3, \; W_{ch,el,z} = 58.09 \; cm^3 \tag{4}$$

The distance between the centroids of the chords is $h_0 = 40$ cm. The properties of the built-up member (two I sections):

$$A = 100.88 \; cm^2, \quad I_y = 9484 \, cm^4, \quad I_z = 162099 \; cm^4, \quad i_y = 9.7 \; cm, \quad i_z = 40.1 \; cm \tag{5}$$

$$\mu = 0, \quad I_{eff} = 2\left[\mu I_{ch,z} + A_{ch}(h_0/2)^2\right] = 1614 \; 10^6 mm^4, \quad I_{eff} = 161402 \, cm^4 \tag{6}$$

The two chords (I-sections) of the member are connected by the lacing consisting of diagonals L 40 x 4 mm and posts L 40 x 4 mm, Class 4 cross-section.

The discrete structure of the built-up column smeared to a continuum has the bending stiffness EI_{eff} = 112 981 kNm^2 and shear stiffness S_v = 45 721.5 kN.

$$n = 2, \; A_d = 3.08 cm^2, \quad A_v = 3.08 \; cm^2, \; d = \sqrt{a^2 + h_0^2} = 1.442 \, m, \; S_v = \frac{nEA_d a h_0}{d^3\left[1 + \frac{A_d h_0^3}{A_v d^3}\right]} = 9429.9 \, kN \tag{7}$$

The parameter of the member ε used in the analysis of the 2nd order takes into account the influence of the shear deformations through the parameter γ

$$\gamma = \frac{1}{1 - \frac{N_{Ed}}{S_v}} = \frac{1}{1 - \frac{430kN}{9429.9\,kN}} = 1.048, \quad \varepsilon = L\sqrt{\frac{\gamma N_{Ed}}{EI_{eff}}} = 7.2m\sqrt{\frac{1.048 \;\; 430\,kN}{112981\,kNm^2}} = 0.455 \tag{8}$$

3. Internal forces of the smeared continuum

The internal forces N_{Ed} (constant), V_{Ed} (non-linear) and M_{Ed} (non-linear) are calculated according to the theory of the 2nd order with the sway initial imperfection and the influence of the shear deformations (S_v = 9429.9 kN, γ = 1.048, ε = 0.455, ξ_H = 1.0). Their distributions are drawn in figure 1 with solid lines and their values are written in bold. The internal forces were calculated via Eqs. [5], which are valid for any point of horizontal force H_{Ed} application x_H from the interval (0 m $\leq x_H \leq L$) or ($0 \leq \xi_H = x_H / L \leq$ 1.0) and for any point of bending moment $M_{Ed,e}$ application x_M from the interval (0 m $\leq x_M \leq L$) or ($0 \leq \xi_M = x_M / L \leq 1.0$)

The bending moment due to horizontal force $H_{Ed,tot}$ at the top of the column:

$$M_{Ed}^H(\xi) = \left\{\frac{\gamma \sin[(1-\xi_H)\varepsilon]\sin(\xi\varepsilon)}{\varepsilon \sin(\varepsilon)} - \frac{\gamma \sin[(1-\xi)\varepsilon]}{\sin(\varepsilon)} \frac{\sin(\varepsilon) - \sin[(1-\xi_H)\varepsilon]}{\varepsilon \cos(\varepsilon)}\right\} LH_{Ed,tot}, \quad \text{for } \xi \leq \xi_H \tag{9}$$

$$M_{Ed}^H(\xi) = \left\{\frac{\gamma \sin(\xi_H \varepsilon)\sin[(1-\xi)\varepsilon]}{\varepsilon \sin(\varepsilon)} - \frac{\gamma \sin[(1-\xi)\varepsilon]}{\sin(\varepsilon)} \frac{\sin(\varepsilon) - \sin[(1-\xi_H)\varepsilon]}{\varepsilon \cos(\varepsilon)}\right\} LH_{Ed,tot}, \quad \text{for } \xi > \xi_H \tag{10}$$

The bending moment due to external bending moment $M_{Ed,e}$ at the top of the column:

$$M_{Ed}^M(\xi) = \left\{-\frac{\cos[(1-\xi_M)\varepsilon]\sin(\xi\varepsilon)}{\sin(\varepsilon)} - \frac{\sin[(1-\xi)\varepsilon]}{\sin(\varepsilon)} \frac{\cos[(1-\xi_M)\varepsilon]}{\cos(\varepsilon)}\right\} M_{Ed,e}, \quad \text{for } \xi \leq \xi_M \tag{11}$$

$$M_{Ed}^{M}(\xi)=\left\{\frac{\cos[(\xi_M\varepsilon)]\sin[(1-\xi)\varepsilon]}{\sin(\varepsilon)}-\frac{\sin[(1-\xi)\varepsilon]}{\sin(\varepsilon)}\frac{\cos[(1-\xi_M)\varepsilon]}{\cos(\varepsilon)}\right\}M_{Ed,e}\,, \qquad \text{for } \xi>\xi_M \quad (12)$$

The total bending moment and shear force are as follows:

$$M_{Ed}(\xi)=M_{Ed}^{H}(\xi)+M_{Ed}^{M}(\xi),\quad V_{Ed}(\xi)=[\mathrm{d}M_{Ed}(\xi)/\mathrm{d}\xi]/L \qquad (13)$$

The distributions of the internal forces N_{Ed} (constant), V_{Ed} (non-linear) and M_{Ed} (non-linear) calculated according to the theory of the 2nd order with the sway initial imperfection but without the influence of the shear deformations ($S_v = \infty$ kN, $\gamma = 1.0$, $\varepsilon = 0.444$, $\xi_H = 1.0$, $\xi_M = 1.0$) are written in figure 1 in italic.

The distributions of the internal forces N_{Ed}, V_{Ed} (constant) and M_{Ed} (non-linear) calculated according to the theory of the 1st order with the sway initial imperfection ($\varepsilon = 0$, $\xi_H = 1.0$, $\xi_M = 1.0$) are drawn in figure 1 with dotted lines. Their values are given in the brackets. If these values are multiplied by the ratio 11/12.603, one obtains the results valid for the case without imperfection.

Figure 1 also shows the values of the horizontal deflections at the top of the built-up column.

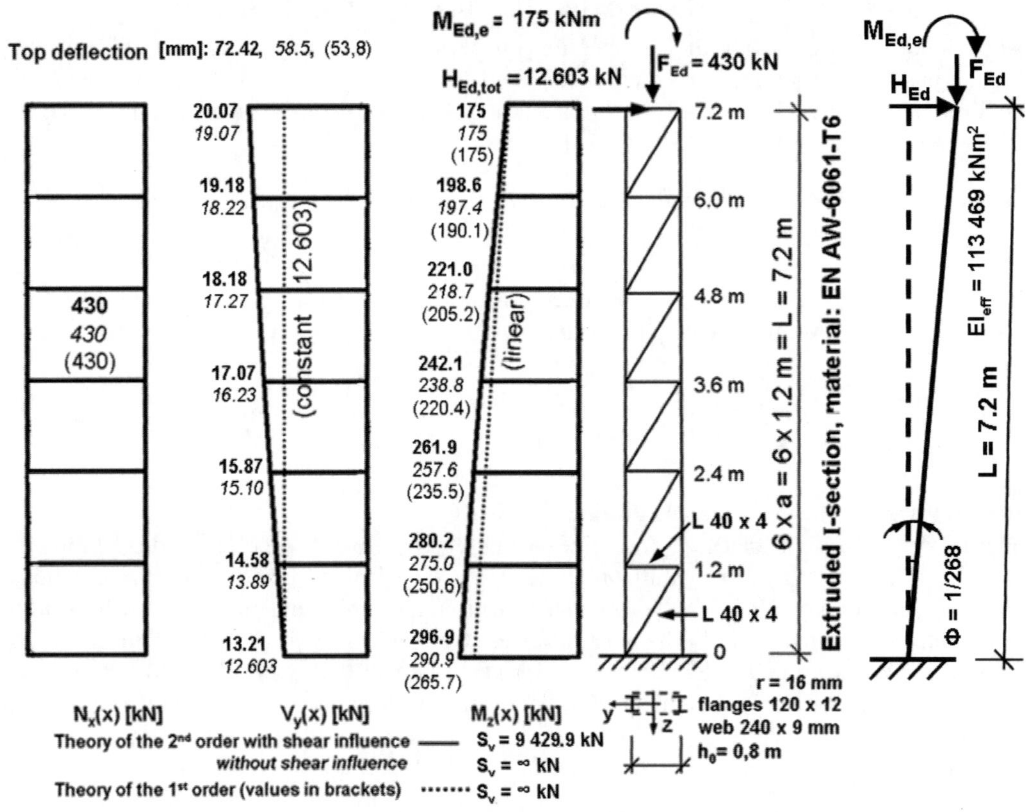

Figure 1. Column geometry, actions and distributions of the internal forces $N_x(x)$, $V_y(x)$, $M_z(x)$.

4. Internal forces of the components of the built-up column

Internal forces in the chords, diagonals, and posts may be calculated from the internal forces of the continuum with smeared stiffness (figure 2a). This approximate calculation according to [1, 2] was performed twice with bending stiffness: (i) $EI_z = 162099$ cm^4 ($\mu = 1$), (ii) $EI_{eff} = 161400$ cm^4 ($\mu = 0$). The differences are negligible. The values in figure 2a are compared with the internal forces in the chords, diagonals, and posts calculated more exactly on the calculation model represented by the frame structure (figure 2b). The frame structure was calculated by the second order theory twice: (i) all compression and tension forces were taken into account. These values are more exact and are given in

IOP Conf. Series: Materials Science and Engineering **474** (2019) 012025 doi:10.1088/1757-899X/474/1/012025

figure 2b in brackets, (ii) only compression forces were taken into account. The differences between (i) and (ii) values are from the practical point of view negligible. The comparison of the values in figure 2a and 2b confirmed that approximate Eurocode way of internal forces calculation on the model with smeared bending and shear stiffnesses is possible to use in the practical design of laced built-up columns.

4.1. Verification of the in-plane buckling of the column

The axial forces in the chords $N_{ch,Ed}$ used for the verification of the in-buckling resistance of the chord determined at the fixed bottom end

$$N_{ch,c,Ed} = \frac{N_{Ed}}{2} + \frac{M_{Ed}(x=0)}{W_{eff}} A_{ch} = 586.26\,kN \quad \text{(right chord in compression)} \tag{14}$$

$$N_{ch,c,Ed} = \frac{N_{Ed}}{2} - \frac{M_{Ed}(x=0)}{W_{eff}} A_{ch} = -156.26\,kN \quad \text{(left chord in tension)} \tag{15}$$

The in-plane buckling resistance of the chord in compression at the bottom fixed end (figure 2a)

$$N_{cr,ch,z} = \frac{\pi^2 EI_{ch,z}}{a^2} = 1672.2\,kN,\ \lambda_{ch,z} = 45.6,\ \bar{\lambda}_{ch,z} = 0.851,\ \alpha = 0.2,\ \bar{\lambda}_0 = 0.1 \tag{16}$$

$$\phi_{ch,z} = 0.937,\quad \chi_{ch,z} = 0.752,\quad N_{ch,b,z,Rd} = \chi_{ch,z} A_{ch} \frac{f_0}{\gamma_{M1}} = 827.65\,kN \tag{17}$$

Verification equation of the in-plane buckling resistance of the chord in compression:

$$U_{ch,b,z} = \frac{N_{ch,c,Ed}}{N_{ch,b,z,Rd}} = \frac{586.26}{827.65} = 0.708 < 1.0 \tag{18}$$

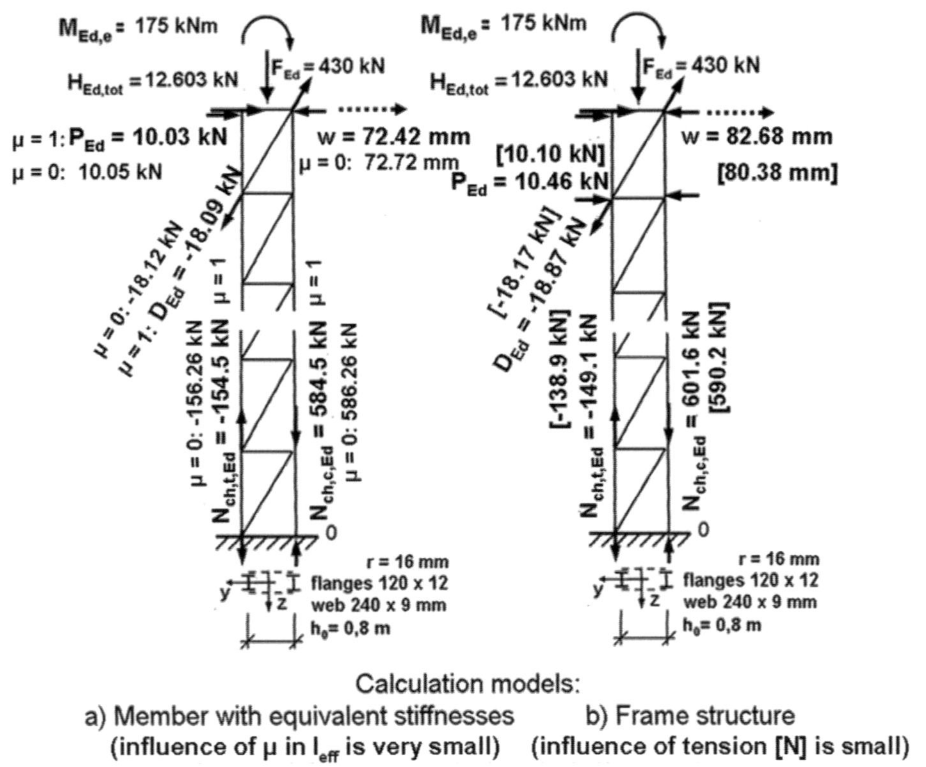

Figure 2. Comparisons of axial forces per 1 chord, 1 diagonal and 1 post calculated by the model

of: (i) continuum with smeared stiffnesses, according to [1, 2] and (ii) discrete frame structures

4.2. Verification of the out-of-plane buckling of the column

$$N_{cr,y} = \frac{\pi^2 EI_y}{L^2_{cr,y}} = 1264\,kN,\ \lambda_y = 74.3,\ \bar{\lambda}_y = 1.384,\ \alpha = 0.2,\ \bar{\lambda}_0 = 0.1 \tag{19}$$

$$\phi_y = 1.586,\quad \chi_y = 0.424,\quad N_{b,y,Rd} = \chi_y A \frac{f_0}{\gamma_{M1}} = 932.2\,kN \tag{20}$$

Utility grade due to the axial force only is

$$U_{b,y} = \frac{N_{Ed}}{N_{b,y,Rd}} = \frac{430}{932.2} = 0.461 \tag{21}$$

Utility grade due to bending moment only. The bending moment $M_{z,Ed.H,Me}$ = 235.565 kNm in the section x = 3.6 m due to the horizontal force H_{Ed} = 22 kN and $M_{Ed,e}$ = 175 kNm is taken into account:

$$M_{z,Ed,Htot,Me}(x = 3.6m) = 242.236\,kNm,\quad M_{z,Ed,H,Me}(x = 3.6m) = 235.565\,kNm \tag{22}$$

$$W_{eff} = \frac{I_{eff}}{0.5h_0} = 4035\,cm^3,\ N_{ch,M} = \frac{M_{z,Ed,H,Me}}{W_{eff}} A_{ch} = 294.46\,kN,\ U_M = \frac{\gamma_{M1} N_{ch,M}}{A_{ch} f_0} = 0.268 \tag{23}$$

Verification of the out-of-plane buckling resistance of the column is done in the middle of the column height. Utility grade due to the axial force and the bending moment is

$$U_y = U_{b,y} + U_M = 0.461 + 0.268 = 0.729 < 1.0 \tag{24}$$

4.3. Verification of the cross-section resistance of the diagonal in tension

The diagonals L 40 x 4 are in tension. The maximal axial force is at the top of the column (figure 2a)

$$D_{Ed} = \frac{V_y(x = 7.2m)}{n} \frac{d}{h_0} = \frac{20.10}{2} \frac{1.442}{0.8} = 18.12\,kN \tag{25}$$

Utility grade

$$U_{D,t} = \frac{\gamma_{M1} D_{Ed}}{A_d f_0} = 0.27 < 1.0 \tag{26}$$

If diagonals would be in the compression

$$\lambda_v = \frac{d}{i_d} = \frac{1442}{7.8} = 184.9 \approx \lambda_{\lim} = 180,\ \chi_d = 0.13,\ N_{b,d,Rd} = \chi_v A_{eff} \frac{f_0}{\gamma_{M1}} = 7.989\,kN \tag{27}$$

$$U_{d,c} = \frac{D_{Ed}}{N_{b,d,Rd}} = \frac{18.12}{7.989} = 2.27 > 1.0 \tag{28}$$

4.4. Verification of the buckling resistance of the post in compression

The maximal axial compression force in the post L 40 x 4 (figure 2a)

$$P_{Ed} = \frac{V_y(x = 7.2m)}{n} = \frac{20.10}{2} = 10.05\,kN \tag{29}$$

The buckling length and the slenderness about the weakest axis

$$L_{cr,v} = h_0 = 800\,mm,\quad \lambda_v = \frac{h_0}{i_v} = \frac{800}{78} = 102.56 \tag{30}$$

The post L 40 x 4 is the class 4 cross-section

$$\beta_v = \frac{b - t - r}{t} = \frac{40 - 4 - 6}{4} = 7.5 > 6\,\varepsilon_m = 6(250MPa/f_0)^{0.5} = 6.214 \rightarrow \text{Class 4} \tag{31}$$

Reduction factor for the local buckling

$$\rho_c = C_1 \frac{\varepsilon_m}{\beta_v} - C_2 \left(\frac{\varepsilon_m}{\beta_v} \right)^2 = 10 \frac{1.021}{7.5} - 24 \left(\frac{1.021}{7.5} \right)^2 = 0.916 \quad (32)$$

Effective area

$$A_{v,eff} = \rho_c A_v = 0.916\ 3.08 = 2.822 cm^2 \quad (33)$$

Critical force of the flexural buckling

$$N_{cr,v} = \frac{\pi^2 EI_v}{L_{cr,v}^2} = \frac{\pi^2\, 70GPa 1.86 cm^4}{(80\, cm)^2} = 20.078\, kN \quad (34)$$

Relative slenderness and effective relative slenderness (formula (BB.1) in [1])

$$\bar{\lambda}_v = \sqrt{\frac{A_{v,eff} f_0}{N_{cr,v}}} = 1.837\,, \qquad \bar{\lambda}_{v,eff} = 0.35 + 0.7 \bar{\lambda}_v = 1.636 \quad (35)$$

$$\Phi_v = 0.5 \left[1 + 0.2 \left(\bar{\lambda}_{v,eff} - 0.1 \right) + \bar{\lambda}_{v,eff}^2 \right] = 1.991\,, \quad \chi_v = \frac{1}{\Phi_v + \sqrt{\Phi_v^2 - \bar{\lambda}_{v,eff}^2}} = 0.32 \quad (36)$$

The design flexural buckling resistance of the post and utility grade are

$$N_{b,v,Rd} = \chi_v A_{eff} \frac{f_0}{\gamma_{M1}} = 19.692\, kN\,, \qquad \frac{P_{Ed}}{N_{b,v,Rd}} = \frac{10.05}{19.692} = 0.51 < 1.0 \quad (37)$$

5. Conclusion

It was confirmed that the general procedure proposed by the authors is possible to use for in-plane buckling verification of the laced built-up members with any boundary conditions under any combination of actions. Eurocodes [1, 2] give no guidance for such special cases.

The out-of-plane flexural buckling resistance verification of the built-up column took into account also influence of the bending moment due to in-plane actions. This was not solved in [3-9].

Acknowledgment

The project No. 1/0603/17 was supported by the Slovak Grant Agency VEGA.

References

[1] EN 1993-1-1 March 2005 and Corrigendum AC February 2006 Eurocode 3 - Design of Steel Structures Part 1-1 General rules and rules for buildings CEN Brussels

[2] EN 1999-1-1 May 2007 + A1: July 2009 + A2: December 2013 Design of Aluminium Structures Part 1-1 General Rules and Rules for Buildings CEN Brussels

[3] Ramm W and Uhlmann W 1981 *Der Stahlbau* **50**. Jahrgang Heft 6 pp161-172

[4] Baláž I J and Agócs Z 1985 1st edition *Metal Structures Part I* (in Slovak) ES STU Bratislava, p 1- 496

[5] *Stahlbau Handbuch Für Studium und Praxis Band I* 1982 Stahlbau-Verlags-GmbH Köln

[6] Lindner J, Scheer J and Schmidt H (Editors) 1998 *DIN Beuth-Kommentare Stahlbauten Erläuterungen zu DIN 18800 Teil 1 bis Teil 4* Beispiel 8.8.2 p 222-226 3. Auflage Beuth Verlag GmbH Ernst & Sohn Berlin Köln

[7] European project 2008 "Facilitating the market development for sections in industrial halls and low rise buildings (SECHALO) RFS2-CT-2008-0030" *Single-Storey Steel Building* Part 6: Detailed design of built-up columns p 6-22 – 6-33

[8] *Stahlbau Kalender* 2013 Eurocode 3 Anwendungsnormen Stahl im Industrie- und Anlagenbau Ernst & Sohn A Wiley Brand bauforumstahl p 242-244

[9] Rubin H, Aminbaghai M and Weier H 2010 IQ 100 Vollversion Feb. 2010 TU Wien

IOP Conf. Series: Materials Science and Engineering **474** (2019) 012026 doi:10.1088/1757-899X/474/1/012026

Damage investigation of pre-stressed cables in segmental box girder concrete bridge ev. No. 324-018 in Pardubice, Czech Republic

S Rehacek*, **D Citek, P Bouska and J Kolisko**

Klokner Institute, Czech Technical University in Prague, Solinova 7, Prague 6, Czech Republic

*E-mail: stanislav.rehacek@cvut.cz

Abstract. Complete reconstruction of the reinforced pre-stressed bridge Pavel Wonky in Pardubice was carried out in 2006. During this reconstruction specialists of the Klokner Institute made a complete revision of linear free external cables in bridge chambers, namely from the aspect of corrosion. During the ordinary visual inspection of the bearing structure of this bridge, in July 2017 [1], the SUS Pardubice staff members found out that in the places above both the pillars in the right-hand chamber of the Pavel Wonka bridge central beam in Pardubice, the blocks of deviators on the left-hand side of the chamber were deflected against the direction of stationing, towards Hradec Kralove. Some of the findings obtained during this revision are presented in this article.

1. Brief data concerning the bridge under study

Bridge ev. No. 324-018 was built between 1956 and 1959. The load-bearing structure of the bridge is composed of a continuous beam with a field length of 50, 70 and 50 m, shown in figure 1. In the cross-section, the bridge's load-bearing structure consists of three chamber beams interconnected with transverse pre-stressing at the level of the bridge board. Each beam has three chambers. The height of the cross-sectional area of the load-bearing structure is variable and ranges from 1.6 m in the field to 2.9 m above the support. The load-bearing structure is pre-stressed both longitudinally and transversally.

The longitudinal pre-stress is induced by cohesion cables placed in the chamber ribs as well as by free pre-stressed cables. The cohesion cables are made of 12 x P7 mm patent wire and are positioned and injected in 42 mm diameter armored tubes or 36 mm flexible tubes. The free pre-stressed cables originally consisted of 168 profiles of P7 mm patent wire. During the reconstruction in the early 1980s, drawbars were replaced by loose cables of 42 ropes with 15.5 mm in diameter. There are two free pre-stressed cables in each chamber, one on each side of the chamber. Each rope is covered by a guard made of PVC pipe segments with a diameter of 330 mm along the length of the bridge. The tube is filled with PUR foam inside. In the area of the bridge ends where the rope bundle is branched for anchoring, the protection is made of PVC boards forming a rectangular chamber filled with PUR foam again. All free pre-stressed cables are anchored to the cross member at the head of the bridge beams. Each cable is composed of 42 ropes with Lp 15.5 mm with a length of about 175 m. Due to the implementation and leading of the cables in deviators, the cables were divided into three-rope bundles which are lead separately and anchored into the anchor sleeve in the crosspieces at the end of the bridge.

IOP Conf. Series: Materials Science and Engineering **474** (2019) 012026 doi:10.1088/1757-899X/474/1/012026

Each three-chamber beam is pre-stressed across both the top and bottom plates. At the pre-stress in the top plate, some cables run through the boards of all three beams. Transverse pre-stressing is ensured by 16 x P4.5 mm cables lead in injected armored tubes.

2. History of the diagnostic survey

The Klokner Institute was surveying the bridge in 2006 [2], the aim of which was an assessment of the corrosion condition of the system of steel free pre-stress cables. The level of corrosion attack of the cables was assessed by using the scale:

1 - Reinforcement without any signs of corrosion,
2 - Reinforcement with starting surface corrosion of local nature to a small extent,
3 - Reinforcement with developed surface corrosion of local nature without peeling off of the corrosion products,
4 - Reinforcement with developed all-area surface corrosion and possibly with the peeling off of corrosion products.

The following conclusions were stated at that time by the visual inspection of free pre-stress cables carried out:

- ✓ The corrosion condition of the pre-stress reinforcement is at the level of degrees 1 and 2 in free length (i.e., out of the untwisting areas), which means without corrosion or with minimal symptoms. The existing protection system of the PVC conduits filled with PUR foam fulfills its anti-corrosion function sufficiently.
- ✓ A problematic area is formed virtually of all untwisting areas which are valued, from the corrosion point of view in 1/2 with the degree 3, and in 1/6 they are valued with the corrosion degree 4. It is evident that the existing implementation of the anticorrosion protection of untwisting areas with foam wrapping of PUR foam and protection of PVC panels is not systematic and fails to fulfill its protective function sufficiently. The most problematic aspect from the corrosion point of view seems to be the area of transition of cables from the cross-beam to the chamber and immediately downstream of this transition where the climbing-down holes are situated. For this reason, it is necessary to ensure the complete major repair of these untwisting areas (see the recommendations).
- ✓ On the whole, it is possible to state that besides the local issues in the untwisting areas the system of free cables is attacked in a relatively small extent and after implementation of additional anticorrosion measures it will continue to fulfill its function.

Another visual diagnostic inspection was carried out by the Klokner Institute in 2015 [3] intending to assessing a possible change in the corrosion condition of the steel free pre-stress cables against the previous visual inspection from July 2006 [2]. The following was stated from the survey carried out:

- ✓ There were carried out altogether four probes – 3 probes in the untwisting areas and one probe in the free-length cable, i.e., out of the untwisting area. The probes were carried out in the places where the corrosion of cables of the degree 3 or 4was stated during the visual inspection in 2006.
- ✓ The same degree of corrosion like the one discovered during the previous visual inspection in 2006 was diagnosed in all the implemented probes after the removal of the existing protection (PVC conduits or PVC panels and PUR foams).
- ✓ During the visual inspection of the pre-stress cables in the entire length of chambers 1 – 9 there were not discovered any spots where it would be possible to register any significant visible degradation of the existing protection of cables or leakage into the area of chambers, or onto the protective elements of pre-stressing cables.
- ✓ On the whole, it was stated that the attacking of the system of free cables from the viewpoint of corrosion in the place of the probes carried out for nine years has not changed since the initial visual inspection of 2006.

3. Inspection of the position of deviators and the state of rope corrosion

Within the framework of the survey, an investigation was carried out for deviators in the right-hand chamber of the central beam, and a change in their position and shifts of the plastic cover of cables were found out. In this chamber, there are seven deviators on each side, two deviators above the pillars, and two deviators in the first and third spans (each) of the bridge. The marking of their position and numerical identification of the cables are shown in figure 2.

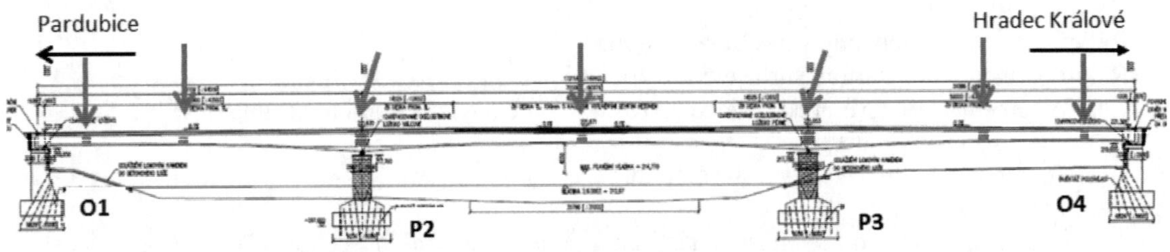

Figure 1. Chart of the placement of deviators

The deviators in the place above pillars P2 and P3 were not anchored to the crossbeams in any way, and therefore they may have deviated from their original condition. According to the data provided, this deviation was not discovered during the previous regular inspections.

Above the P2 pillar, plastic protection of the cable was cut out, and PUR foam was partly removed. No symptom of rope damage by corrosion was found out on the ropes uncovered.

In the next step, PUR foam was gradually removed above the O1 support on both the right-hand and left-hand sides of the chamber in the place of the untwisting areas. According to the expert report [3], there was found out a corrosion level at the cable No.7 in a small extent, at the cable no. 8 with developed corrosion of local nature. After the uncovering, there has not yet been found out any corrosion damage to an increased extent.

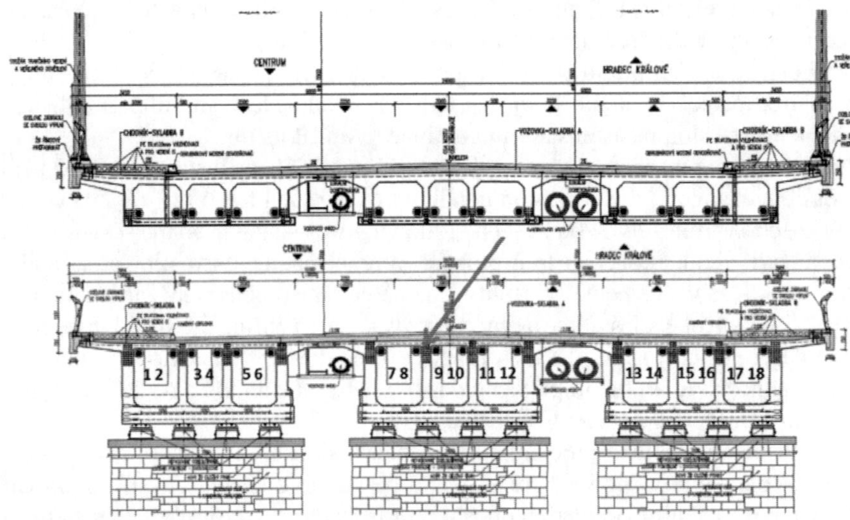

Figure 2. Cross-sections in the bearing structure with numerical marking of the cables: a view from the stationing direction. Stress deviators inclined to the right (Hradec Kralove) side of the bridge are marked by the red arrow.

Some rope strands of cable No. 8 are partly or entirely released. This finding was confirmed by the tapping with a hammer on the strand. At damaged and undamaged strands, different acoustic responses are apparent. A logical explanation of this fact is that a partial or complete failure of this cable has occurred.

The edge of the cross-beam above the P2 pillar was damaged, it was broken off in a width of approx. 80 mm under the shifted block of deviators, figure 3.

From the above-stated findings, it is possible to conclude that the cable failure probably occurred in the first bridge span.

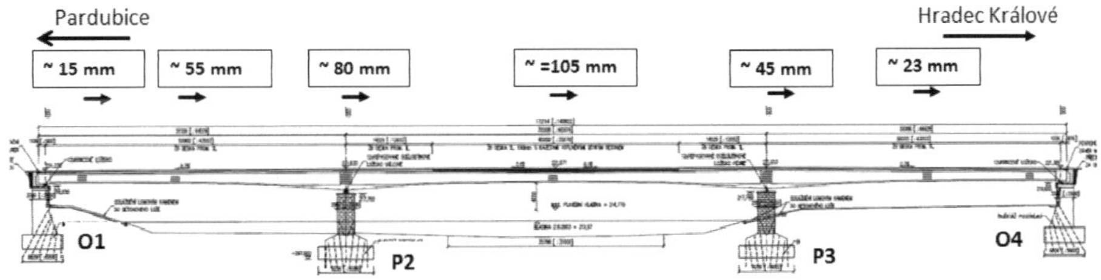

Figure 3. The approximate shift of the cable found out in the direction against the stationing

At the request of the customer, only the visual inspection of the cables was performed at the first phase. The extension of the diagnostic range was considered at the second phase, based on the results of the first phase. At the second phase, a stress analysis on the cables as a function of the separation between the pillars to determine the susceptibility to fail by corrosion-fatigue was planned.

4. Additional Inspection of the rope corrosion

Two inspections were carried out in February-March 2018 after the strands of the ropes were made accessible, and these inspections were focused on their corrosion state.

The following conclusions were found out from the survey carried out:

- ✓ At the support no. O1: 35 ropes out of the total number of 42 ropes were damaged by fracture.
- ✓ At the support no. O4: 1 rope was damaged, and one rope was heavily corroded, out of the total number of 42 ropes.
- ✓ In the places of failure, massive corrosion of pre-stressing ropes was identified, with peeling-off layers of corrosion products.
- ✓ The massive corrosion of pre-stressing ropes in the places of failure is in the areas of the original assembly holes in the bridge deck of the chamber at the O1, and O4 supports. At a certain distance (several meters) from this place, the reinforcement was without any corrosion, or with significantly lower surface corrosion. This evaluation must also be confirmed for other cables.
- ✓ The damage of the pre-stressed ropes was called out without any doubt by the localized form of the corrosion damage caused by moisture and stimulated by chloride anions. The damage was further supported by the mechanism of the diameter weakening due to ordinary electrochemical corrosion, and partly also the corrosion under stress stimulated again by the presence of chloride anions.
- ✓ According to the preliminary evaluation by the staff members who ensure the removal of the PUR foam, similar rope failures are also registered in other chambers, in connection with assembly holes.
- ✓ No visible signs of corrosion were found on the ropes in the area in front of the deviator at the O1 support, and no symptoms of corrosion in the ropes are registered behind the deviator either.

5. Passport of the corrosion state of the ropes of all the cables

At the next phase, it was recommended to verify the corrosion state of the pre-stressing ropes in all the places where the so-called historical handling holes are situated above the untwisting areas of the cables and where they could be affected by long-term water leakage [4, 5]. Another visual inspection in all the chambers on both sides of the bridge was held in March 2018. Within the framework of this visual inspection, the corrosion state of the ropes was photographically documented and visually assessed in the places where technological holes were made in the upper slab of the chambers, providing access for the assembly of the free pre-stressing cables. These holes are at a distance of approx. 7-9 m from the internal front of the cross-beam. It was clear that long-term leakage must have occurred in this area before the repair. The testing place was described through localization of the place assessed, by means of the position of the exposed cable from the front part of the cross-beam, through existence of the concrete-covered breakdown above the test place, number of damaged ropes and subjective evaluation of the corrosion state according to the scale selected (corrosion degrees).

The following can be stated from the results of visual inspections and corrosion state assessment:

- ✓ In the untwisting area at the O1 support the ropes were characterized:
  - o Sixty-four ropes were damaged, i.e., approx. 8.5 %,
  - o 14 % rope corrosion state 2,
  - o 26 % rope corrosion state 3,
  - o 59 % rope corrosion state 4.
- ✓ In the untwisting area at the O4 support the ropes were characterized:
  - o Forty-nine ropes were damaged, i.e., approx. 6.5 %,
  - o 9 % rope corrosion state 2,
  - o 24 % rope corrosion state 3,
  - o 67 % rope corrosion state 4.
- ✓ It is clear that in the place of the original assembly holes, through which water leakage into the structure and onto the ropes was very likely to occur in the past, there is a local area in a length of approx. 2 m, in which the pre-stressing cables and their ropes are affected by massive corrosion (corrosion degree 4) which is already manifested through the peeling-off of the corrosion products and diminishing of the area of the pre-stressing reinforcement. Already 2/3 of all pre-stressing ropes of the free pre-stressing system are affected by corrosion this way.
- ✓ The very unfavorable corrosion state which was detected, even though in a local scale only, must be considered, from the viewpoint of the further identified influence on mechanical characteristics of the pre-stressing reinforcement as highly risky and hardly acceptable on a long-term basis.

6. Non-destructive determination of the reinforcement position

The HILTI PS 1000 radar was (figure 4) used for determination of the thickness of the pre-stressing reinforcement in the walls of the chambers (external surface). The instrument works on the principle of transmission of electromagnetic impulses into the structure. The output of the registration of reflections of the electrical impulses from non-homogeneities of the material is the area scanning with a cross section of the structure in question. The orientation determination of the position and thickness of coverage on external surfaces of individual structures was made with the use of the Hilti PROFIS PS 1000 software interface. The HILTI PS 1000 instrument works with a maximum detection depth of 300 mm. The precision of indication of a depth smaller than 100 mm is ±10 mm. At a depth exceeding 100 mm, the accuracy is ± 15%. The localization precision is ±10 mm.

It was selected for execution of individual area scanning operations that always one area image would be scanned with a spacing of approximately 3 m. Scanning was carried out in the direction from the O4 support to ½ of the middle span. This was justified by the presumption of the bridge symmetry. The non-destructive diagnostics presumed that it would make it possible to determine the courses of the pre-stressing reinforcement for individual structural assessments.

The illustration output from the HILTI PROFIS PS 1000 graphics interface is provided for in figure 5. For the reason of thick reinforcement using soft reinforcement material as well as because the cables of the pre-stressing reinforcement are conducted in as many as four layers it was always possible to identify the first reinforcement layer (considered from the external face of the structure).

The aim of the additional diagnostics was that a structural analysis of the load-bearing capacity of the bridge would be subsequently drawn upon the basis of the information found out about the course of the cohesion reinforcement, corrosion state, and corrosion risks at free ropes.

Figure 4. Non-destructive determination of the reinforcement position by radar HILTI PS 1000

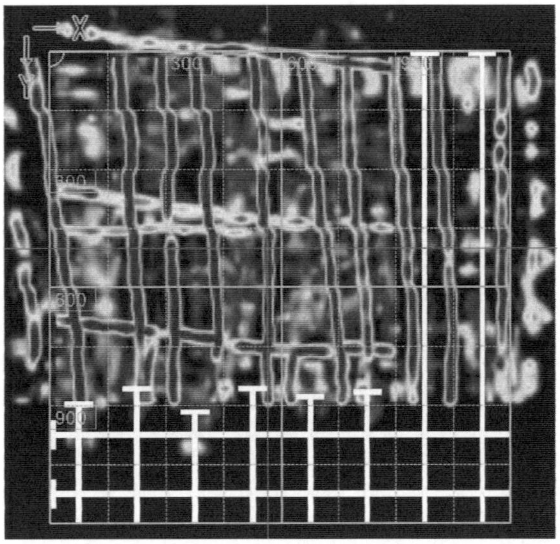

Figure 5. Illustration output from the HILTI PROFIS PS 1000 graphics interface

7. Conclusions

The visual inspections and laboratory analyses carried out in this study made it possible to state:

- ✓ It is evident that in the place of original assembly holes, through which water leakage into the structure and onto the ropes was very likely to occur in the past, there is a local area in a length of approx. 2 m, in which the pre-stressing cables and their ropes are affected by massive corrosion (corrosion degree 3-4) which is already manifested through the peeling-off of the corrosion products and diminishing of the area of the pre-stressing reinforcement. Already 2/3 of all pre-stressing ropes of the free pre-stressing system are affected by corrosion this way.
- ✓ From the above-mentioned findings, it is, therefore, possible to derive and basically it is not possible to exclude that the mechanical strength of the free pre-stressing ropes is approximately (for 2/3 of ropes) lower than at the time of installation by as much as 1/3, and at the same time they do not have ductile capacity in the case of damage.
- ✓ The damage of the pre-stressed ropes was called out, without any doubt, by the localized form of the corrosion damage caused by moisture and stimulated by chloride anions. The damage was

further supported by the mechanism of the diameter weakening due to ordinary electrochemical corrosion and partly also the corrosion under stress, stimulated again by the presence of chloride anions.

- ✓ PUR foam is not a source of contamination of the rope surface with corrosion stimulants (chloride and sulfate anions). With high probability, these are chloride anions originating from road salts used during winter maintenance of roads.
- ✓ The very unfavorable corrosion state which was detected, even though in a local scale only, especially under the assembly holes, must be considered, from the viewpoint of the further identified influence on mechanical characteristics of the pre-stressing reinforcement as highly risky and unacceptable on a long-term basis.
- ✓ If the condition identified in 2018, as well as the condition of the corrosion of the ropes, detected already at the time of renovation (2006), but in entirely different places, are taken into consideration, then from the viewpoint of corrosion of the pre-stressing ropes the entire untwisting operation is risky. The ropes are protected from corrosion with PUR foam, it is true, but as it implies from visual inspections, the actual packing of the foamed area is not tight. It is clear that where effects of moisture can be observed and where moisture gets to the PUR foam, then moisture further penetrates closer to the ropes and causes local corrosion. This means that the anticorrosion system used (protection with PUR foam) has its certain limits. In the existing structural arrangement, it is not possible to verify the corrosion state of the ropes along the entire length. A failure and therefore also the loss of the functionality of a rope or even of a whole cable can appear anywhere, even though (of course) the riskiest place is the untwisting area at the O1 and O4 supports.
- ✓ In this situation, when already relatively small surface corrosion leads to "fragile" behavior, it is virtually possible to exclude any coupling of the ropes with surface corrosion at the level of degrees 3 and 4.
- ✓ Concerning the possibility of pit corrosion, which forms a bordered “groove” on the surface also the degree 2 can be risky from the viewpoint of coupling.

Acknowledgments
This article was written with the contribution of the project P105/12/G059.

References

[1] Mares V 2017 *Current visual inspection on 18 July 2017 – Pavel Wonka Bridge*, SUS Pardubice.
[2] Kolisko J and Stepanek D 2006 Expert Report 06 062-1, Klokner Institute of the Czech Technical University, Prague, 7/2006.
[3] Rehacek S and Kolisko J 2015 Expert Report 1500 J 077-4, Klokner Institute of the Czech Technical University, Prague, 25 April 2015.
[4] Documents from the archives of Doc. Ing. V. Hrdoušek, CSC.
[5] Rehacek S, Hunka P and Kolisko J 2014 *Adv. Engin. Mater. Technol.* 619-24, Volumes 919-921.

Pavement performance of environmentally friendly asphalt modifier in hot and humid environment

J Zhang[1], K Zhu[2,3,*], G Qiu[1], Z Yang[1], Y Zhang[3]

[1] Zhejiang University, Urban-Rural Planning & Design Institute Co, Ltd, Hangzhou, China

[2] College of Quality and Safety Engineering, China Jiliang University, Hangzhou, China

[3] Ningbo Communication Engineering Construction Group Co, Ltd, Ningbo, China

*E-mail: zhukai@cjlu.edu.cn

Abstract. The modified asphalt binders and mixtures were prepared by SBP directed vat set modifier with recycled polyethylene as the main component. The effect of SBP modifier on the high-temperature characteristics of asphalt was characterized and analyzed by dynamic shear rheometer and differential scanning calorimeter. Thus, the pavement performance of SBP modified asphalt mixture was evaluated, and the test road was paved. The results show that the high-temperature rheological characteristics of asphalt were significantly improved with the SBP modifier. The high-temperature performance grade (PG) can be increased by 2-3 classes, but the improvement in high-temperature sensitivity was not noticeable. The SBP modifier can improve the low-temperature performance of asphalt to a certain degree, but this is not enough to change the low-temperature performance grade of asphalt. The high-temperature and water stability of SBP modified asphalt mixture are better than that of SBS modified asphalt mixture, which is more suitable for high-temperature and rainy areas.

1. Introduction

The rise in the social economy and the continuous adverse climatic conditions like high-temperature and precipitation have resulted in massive traffic flow and vehicle loads on the pavements. These conditions have put forward higher requirements on the performance of asphalt pavement in Southern China [1].

As a result, polymer modified asphalt has been widely used to improve the pavement performances like high-temperature and fatigue resistance, water stability performance, etc. [2-4]. Polyethylene (PE) is a kind of commonly used polymer in the modifier, it can increase the viscosity of asphalt, and improves the high-temperature stability of asphalt pavement [2]. In the view of the economy and environment, the recycled and low-price polyethylene can be used as a modifier for asphalt in road engineering. This can reduce not only the cost of construction but also achieve the recycling of resources. Further, it reduces the pollution to the environment, which is the worldwide concern for the usage of PE.

Panda et al.[5] have studied the effect of recycled PE on the performance of asphalt mixture. The results show that the recycled PE can improve the rebound modulus of asphalt mixture, enhance the water stability and improve the fatigue life. Garc´a-Morales et al.[6] have investigated the effect of the four kinds of polymer waste including PE on the rheological properties of the modified asphalt, in which the modified polymer composite with PE and ethylene-vinyl acetate copolymer(EVA) can form a mesh structure, so high-temperature rheological properties of asphalt can be improved. The study of Punith et al. [7] has shown that the fatigue life of asphalt mixture can be increased by 2.5 times after adding the PE. Fang et al. [8] have studied the microstructure and pavement performance of modified asphalt with PE and organic montmorillonite. The study of Yu et al. [9] has shown that the basic performances (like the penetration) of the modified asphalt composite can be improved with the addition of waste powder and waste PE. At present, PE waste is mainly used to prepare modified asphalt, and the direct vat set (DVS) modifier, but the use of recycled PE is found less in research. The DVS modifier can directly be added to the aggregate together with base asphalt mixture, and the modified asphalt mixture can rapidly be produced. It can do away with the tedious procedure of processing, storage, and transportation of modified asphalt, which can save money and energy costs. By using PE waste as the DVS modifier, it will produce more environmentally friendly production and economic benefits. SBP is a novel DVS modifier of asphalt with the primary material of recycled PE.

In this paper, the effect of SBP modifier on the high-temperature and low-temperature rheological characteristics, temperature sensitivity, differential scanning calorimetry (DSC) analysis and the pavement performances of asphalt are characterized, and a test road was paved. It can provide a basis for the application of this novel environmental modifier in engineering.

2. Experimental

2.1. Materials

2.1.1. SBP modifier. DSV SBP modifier is a kind of flat black particles, mainly compounded of recycled PE, organic sealant and nano-polymer, and produced by the process of plasticizing production, water ring granulation, etc. It is provided by Zhejiang New Shuaibang Traffic Facilities Co, Ltd, with 14.0% ash ratio. The particle size, softening point, melting point, melt flow rate, and density are 4 mm, 130°C, 175 °C, 2.33 $g \cdot 10min^{-1}$, 1.067 $g \cdot cm^{-3}$, respectively. The microstructure of recycled PE is shown in figure 1(a). PE shows linear long chain molecular structure with more methyl branched chain and alkyl side chain, which formed a multi-branched dendritic structure [10]. The microstructure of SBP modifier is shown in figure 1(b). From figure 1(b), during the formation of SBP, the chain structure of PE changed and formed a more stable network structure.

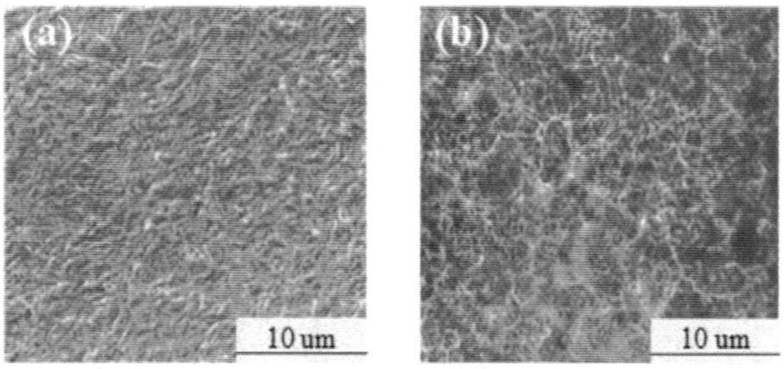

Figure1. The microstructure of modifiers (a) PE (b)SBP

2.1.2. Asphalt. The base asphalt is the grade used in the South Korean SK-70 road asphalt, while SBS(I-D) modified asphalt produced by Zhenhai Petrochemical Engineering Co, Ltd was used in the comparative test. The main technical indices, which meet all requirements [11], are shown in table 1.

Table 1. Technical indices of SK-70 and SBS modified asphalts

Test items		SK-70 asphalt		SBS modified asphalt	
		Technical requirements	Test results	Technical requirements	Test results
25°C penetration /0.1 mm		60~80	68.2	40~60	53.0
Ductility /cm		≥20(10°C)	27(10°C)	≥20(5°C)	31(5°C)
Softening point /°C		≥46	50	≥60	76
Dynamic viscosity / Pa.s		≥180(60°C)	219(60°C)	≤3.0(135°C)	1.46(135°C)
After TFOT	Change of mass /%	-0.8~0.8	0.036	-1.0~1.0	-0.020
	25°C penetration ratio /%	≥61	65.0	≥65	71.7
	Ductility /cm	≥6(10°C)	7(10°C)	≥15(5°C)	22(5°C)

2.1.3. Preparation of SBP modified asphalt. SBP modified asphalt was prepared by a high-speed shearing method with a small indoor high-speed shear emulsifier, as shown in figure 2. The base asphalt was heated at a constant temperature of 150 °C in a sand bath. The dried SBP modifier was added slowly to the base asphalt at the rate of 4%, 7%, 10% (mass ratio of the base asphalt). And the shearing was carried out at a rate of 500 r·min^{-1} for 15 min to make the SBP modifier dispersed in base asphalt uniformly. Then the temperature is raised to 180 °C, and shearing was carried at high speed of 4 000 r·min^{-1} for 60 mins; then developed at a low speed of 500 r·min^{-1} for 30 mins and the bubbles are removed which are generated during the high-speed shearing.

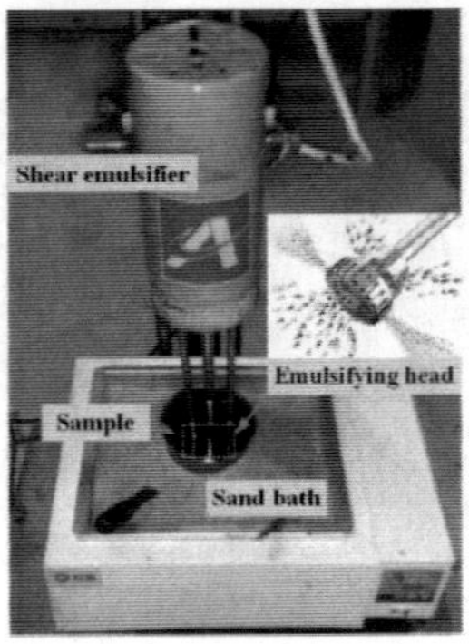

Figure 2. Preparation of SBP modified asphalt

2.1.4. Aggregate and Mineral powder. The coarse aggregate used the basalt gravel procured from Shengzhou in Zhejiang P.R. China, with the crushing value of 12.5%, the polished value of 47PSV, Los Angeles wear loss of 7.3% and apparent relative density of 2.908. The fine aggregate has used artificial sand. The mineral powder used limestone fillers procured from Lanxi in Zhejiang P.R. China, and the apparent relative density is 2.768.

2.2. Methods

2.2.1. Dynamic shear rheological test (DSR). The ADS fully automatic dynamic shear rheometer (Malvern Co, Ltd, UK) was used to carry out the DSR tests. Both the base asphalt and the asphalt after RTFOT short-term aging were used including SBP modified asphalt with three modifier dosages, SK-70 base asphalt, and SBS modified asphalt. The samples were poured into round cakes with 25 mm in diameter and 1 mm in thickness. The test was using strain control mode, and the loading frequency was 10 rad/s, with the initial temperature of 64 °C, and increased at 6 °C level until the rutting factor was not satisfied. The test was carried out under ASTM D7175[12].

2.2.2. Bending beam rheological test (BBR). The low-temperature BBR tests were performed using a bending beam rheometer (RHE-102, ATS Co, Ltd, US). Both the asphalt after RTFOT short-term aging and PAV long-term aging were used including SBP modified asphalt, SK-70 base asphalt, and SBS modified asphalt. The test temperatures were -12 and -18 °C. The tests were carried out under ASTM D6648[13].

2.2.3. Differential scanning calorimetry test (DSC). The DSC tests were performed using a Q100DSC analyzer (TA Co, Ltd, US). SK-70 base asphalt and 7% SBP modified asphalt were used. The samples were heated from -40 to 100 °C at a heating rate of 10 $°C \cdot min^{-1}$, in the nitrogen atmosphere.

2.2.4. Pavement performance test of the asphalt mixture. Asphalt mixtures (AC-13 median gradation) with SBP, SK-70 and SBS asphalt were tested. SBP asphalt mixture was prepared by directed vat set method with SBP modifier. The best dosage of SBP modifier is 0.35% of the mass of asphalt mixture. The high-temperature performance was evaluated by rutting test under two additional harsh conditions (60 °C, wheel pressure of 0.7 MPa, immersion and 70°C, 0.8 MPa), along with the standard test condition (60°C, 0.7 MPa). The low-temperature performance was evaluated by beam bending test with the standard test condition: -10 °C, a loading rate of 50 mm·min^{-1}. The water stability was tested by freeze-thaw split test under the standard test condition with a freeze-thaw cycle. The pavement performance test was carried out according to JTG E20 [14].

3. Results and Discussion

3.1 Analysis of SBP modified asphalt properties

3.1.1. High-temperature rheological property. The rutting or permanent deformation of the asphalt pavement is the result of the irreversible, partial accumulation of the deformation of the shear flow under the repeated load of the asphalt mixture at high-temperature, which is closely related to the viscoelastic characteristics of the asphalt at high-temperature. Therefore, it has been an effective method to evaluate the high-temperature performance of asphalt by studying the rheological property of asphalt under high-temperature by DSR test [15-17]. The complex shear modulus G^* and phase angle δ can be obtained from DSR test. G^* indicates the ability of shear deformation resistance, and this is enhanced with the increase of G^*. δ reflects the ratio of viscosity and elasticity of asphalt. The greater the value of δ, the greater the viscosity ingredient of asphalt is. The proportion of irreversible asphalt is increased with the value of δ. SHRP is planned to take the rutting factor $G^*/\sin\delta$ as the evaluation indicator of high-temperature stability. At the condition of high-temperature, if G^* increases, δ reduces. And hence the resistance of shear rheological deformation improves along with the high-temperature stability. It's specified in SHRP specification that the $G^*/\sin\delta$ of base asphalt should not be less than 1.0 kPa, the $G^*/\sin\delta$ of residual asphalt after RTFOT should be no less than 2.2 kPa on the designated high-temperature.

(1) Complex modulus G^* and phase angle δ

The results of the DSR test of SBP modified asphalt are shown in table 2. It can be seen from table 2 that the complex modulus G^* of asphalt at different temperatures is significantly higher than that of base asphalt after the modification of SBP modifier, and resistance of the shear deformation of asphalt is enhanced. The effect of G^* is related to the amount of SBP modifier. At the same temperature, G^* is improved with the amount of modifier. But the addition of SBP modifier decreases the phase angle of asphalt, and the decrease is attributed to the amount of modifier. δ is the measure of the ratio of the loss modulus G'' to the storage modulus G'. SBP modifier increases the storage modulus G' above the loss modulus G''. Elasticity component of asphalt increases, loss of viscosity component decreases, that is, the part of recoverable deformation increases and the permanent deformation resistance is improved.

Table 2. The results of DSR for SBP modified asphalt

Asphalt	Complex modulus G^* /Pa				Phase angle δ /°			
	64°C	70°C	76°C	82°C	64°C	70°C	76°C	82°C
SK-70	1923.4	774.7	500.0	-	87.9	89.0	89.8	-
4%SBP	5661.7	2689.8	1295.6	698.3	82.1	83.8	85.0	85.9
7%SBP	10795.6	4977.9	2398.5	1216.4	77.7	79.7	80.8	81.2
10%SBP	20009.9	9366.2	4679.8	2393.0	71.2	74.6	76.0	76.9

(2) Rutting factor $G^*/\sin\delta$

The results of the rutting factor of the base asphalt and RTFOT short-term aged asphalt are shown in table 3. At the same temperature, after adding the SBP modifier, the rutting factor of asphalt increases, the high-temperature stability is improved, and the high-temperature stability is increased with the amount of modifier. As shown at 64°C, after the modification of 7% SBP modifier, the rutting factor is improved to 11 050 Pa from 1 925 Pa, the high-temperature stability is improved by 4.74 times, which is 2.64 times higher than that of SBS modified asphalt. Five kinds of asphalt were classified by the high-temperature performance according to SHRP specification. Under three dosage types (4%, 7%, 10%), the high-temperature classification temperature of SBP modified asphalt is improved to 70, 76, 82°C, respectively. The high-temperature performance classification of 7% SBP modified asphalt can achieve that of SBS modified asphalt. The high-temperature performance classification of 10% SBP asphalt is higher than that of SBS modified asphalt. That is because the cross-linked network structure acts as a constraint on the flow of asphalt at high-temperatures after the addition of SBP modifier. Then, the viscosity and the resistance of shear rheological deformation at high-temperature of asphalt are increased, and the cross-linked network structure is improved. Hence, the modification is enhanced by the amount of modifier. Considering the high-temperature stability, the sufficient amount of SBP modifier should be above 7%.

Table 3. The rutting factors of SBP modified asphalt

Asphalt	Base asphalt ($G^*/\sin\delta$) /Pa				Short-term aged binder($G^*/\sin\delta$) /Pa					PG high-temperature range /°C
	64°C	70°C	76°C	82°C	64°C	70°C	76°C	82°C	88°C	
SK-70	1925	775	500	---	2483	1185	602	---	---	64
SBS	4179	2581	1725	1287	6446	3778	2336	1526	---	76
4%SBP	5716	2705	1301	700	7981	3814	1794	---	---	70
7%SBP	11050	5060	2430	1231	15371	7384	3689	1916	---	76
10%SBP	21135	9717	4823	2457	31800	14980	7493	3994	2134	82

3.1.2. Temperature sensitivity in the high-temperature range. The temperature sensitivity of high-temperature range directly affects the pavement performance of high-temperature of the asphalt mixture. If the sensitivity towards the temperature is strong, with the increase of temperature, the viscosity of asphalt decreases significantly, and the resistance of the shear rheological deformation is reduced. And it is easy to prepare high-temperature permanent deformation such as rutting. At the high-temperature range of 64 ~82 °C, asphalt rutting factor and temperature have a linear relationship in the semi-logarithmic coordinates, so we can use the (k) of this linear relationship to evaluate the temperature sensitivity of asphalt. As the absolute value of the slope increases, the change of the rutting factor with temperature is increased, and hence the temperature sensitivity of asphalt is enhanced. The corresponding fitting results are shown in figure 3. It can be seen from figure 3 that SBS modified asphalt has the least sensitivity towards the temperature, SK base asphalt has the strongest, and SBP modified asphalt is better than the base asphalt, but it is still sensitivity towards the temperature. The reason may be due to the degradation of recycled PE performance. With the rise in temperature, the SBP network structure is gradually deformed and lost the constraints to asphalt. Therefore, it reflects that the rutting factor declines rapidly with the increasing temperature, but in the range of high-temperature, the high-temperature stability of the amount of SBP asphalt above 7% is still better than that of SBS modified asphalt.

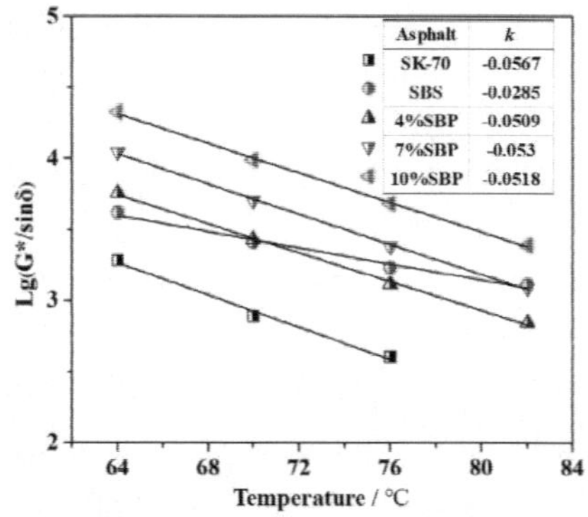

Figure 3. The rut factor-temperature relationships of asphalt

3.1.3. DSC analysis. DSC test results of asphalt are shown in figure 4 and 5. In the temperature range of 30 ~60 °C, there is an apparent endothermic peak in the DSC curves of the two kinds of asphalt. In the high-temperature range, the asphalt enters from the viscoelastic state into the viscous state. Some asphalt components show a change in solid-liquid phase. The more significant the solid-liquid phase shift, the more is the significant effect on the macroscopic properties of asphalt. And the temperature range is just the range of high-temperature permanent deformation of asphalt pavement like rutting [1]. Therefore, the study of heat absorption property of asphalt in this temperature range is instrumental in evaluating the macroscopic high-temperature stability performance of asphalt. The amount of heat absorption corresponding to the endothermic peak can be calculated by integrating the endothermic peak curve over time in the temperature range.

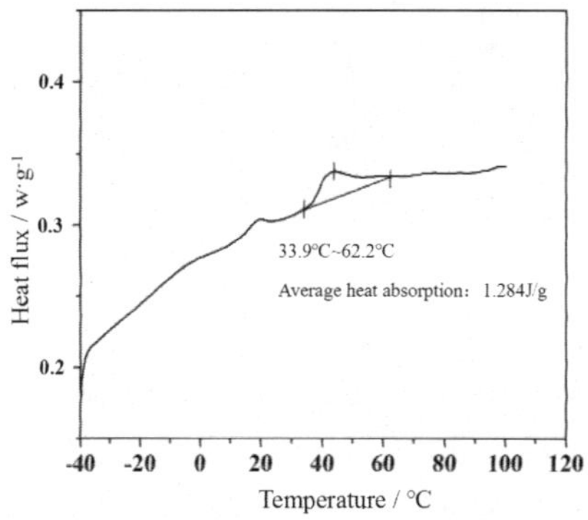

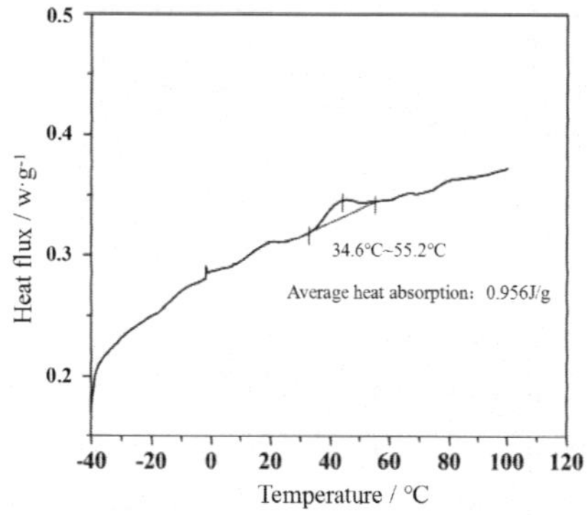

Figure 4. The DSC curve of SK-70 asphalt **Figure 5.** The DSC curve of SBP modified asphalt

It can be seen from figures 4 and 5 that within the high-temperature range of the asphalt from the viscoelastic state into the viscous state, the average heat absorption of SK base asphalt is more than that of SBP asphalt. The more heat absorption indicates more change of the aggregation state of the asphalt component and the more change in solid-liquid phase observed. After the addition of a modifier, in this temperature range, the aggregation state of the asphalt component prone to change is decreased, the multiphase system of asphalt is more stable than that of base asphalt, the thermal stability is improved, and high-temperature stability is enhanced. But the temperature range reflects the change of the aggregation state of the asphalt component. The temperature range of the endothermic peak of SBP modified asphalt is narrower than base asphalt. It indicates that the temperature range of the aggregated state is small, and the range of effect of temperature is reduced. The temperature range of the stable asphalt multiphase system is decreased, and the temperature sensitivity of the asphalt is diminished. The result is consistent with the DSR test.

3.1.4. Low-temperature rheological property. The effect of the BBR test is shown in table 4. In the BBR test, the bending creep stiffness modulus (S) and the rate of creep stiffness change (m) is taken as the evaluation indexes of the low-temperature rheological properties of asphalt. The higher the creep stiffness modulus S, the more is brittleness, and the easier is cracking. However, larger m values result in lower tensile stresses, and the less the possibility of low-temperature cracking. It can be seen from table 4 that the stiffness modulus (S) of asphalt becomes smaller, the m value increases, the stress relaxation ability is improved, and the low-temperature performance is improved when the SBP modifier is added. The asphalt shows the best performance when the dosage is 7%. The requirement of PG low-temperature grade in SHRP specification is $S<300$ MPa, $m>0.3$. According to this standard for low-temperature classification, the temperatures of SBP modified asphalt and SK-70 base asphalt are both -22°C. That is because of low-temperature PG classification at intervals of -6°C; the temperature interval is too large to describe the differences of the low-temperature property of asphalt effectively. Simultaneously, it indicates that SBP modifier can improve the low-temperature property of asphalt, but the effect is limited, it cannot improve the low-temperature property grade of base asphalt.

Table 4. The results of BBR for SBP modified asphalt

asphalt	-12°C		-18°C	
	S /MPa	m	S /MPa	m
SK-70	246	0.305	444	0.241
SBS	186	0.314	358	0.240
4%SBP	193	0.307	370	0.259
7%SBP	173	0.310	323	0.271
10%SBP	191	0.305	326	0.236

3.2 SBP modified asphalt mixture property analysis

3.2.1. High temperature stability. The result of rutting test is shown in table 5. It can be seen from table 5 that the dynamic stability of SBP asphalt mixture is 8.1 times higher than that of base asphalt mixture, 1.5 times higher than that of SBS modified asphalt under the standard conditions. The SBP modifier significantly improves the anti-rutting performance of asphalt mixture. The dynamic stability is decreased by 2.6 and 23.6% under the terms of immersion and 70 °C with the wheel pressure of 0.8 MPa. The anti-rutting performance of SBP modified asphalt and under the two harsh conditions is still better than that of SBS modified asphalt mixture under the standard requirements. The result of the high-temperature test of asphalt mixture is consistent with the result of the DSR test of asphalt, SBP modified asphalt shows excellent high-temperature stability.

Table 5. The results of the rutting test

Asphalt mixture	Dynamic stability /cycles·mm^{-1}	Rut depth /mm
SK-70	1136	3.608
SBS	6169	1.584
SBP standard	9232	1.175
SBP immersion	8996	1.245
SBP (70°C, 0.8MPa)	7050	1.179

3.2.2. Low temperature crack resistance. The results of the low-temperature bending test are shown in table 6. The flexural strength and the maximum flexural strain of SBP mixtures are both increased, but the amplitude is limited. The low-temperature performance of SBP mixture is better than the base asphalt but not as good as SBS modified asphalt mixture. The result is consistent with the result of the BBR test. The improvement of low-temperature performance of asphalt mixture by SBP modifier is not as significant as that of high-temperature performance.

Table 6. The results of the low-temperature bending test

Asphalt mixture	Flexural tensile strength/MPa	Maximum bending strain/με
SK-70	7.783	2091
SBS	9.231	2969
SBP	8.804	2561

3.2.3. Water stability. The results of freezing and thawing split test are shown in table 7. On comparing with the base asphalt, the tensile strength of SBP asphalt mixture at 25 °C is improved by 49.1%, which is higher than that of SBS modified asphalt. The mechanical properties of the mix could be improved by SBP modifier. The freeze-thaw splitting strength is 10.5% higher than that of the base asphalt, which indicates that the SBP modifier can adjust the water stability of the asphalt mixture. That is because of SBP modifier modified the interface of the aggregate. With the addition of SBP modifier in the heated aggregate, the high-temperature heat transfer of the aggregate and the mutual friction and shearing action of the mixture during the mixing process are carried out. The modifier is rapidly melted, dispersed and adhered to the aggregate surface, forming a layer of modifier film that can diffuse into the micro-void of the aggregate surface. After adding the asphalt, the modified asphalt is well wrapped around the aggregate surface to form a good aggregate asphalt interface. The cohesive force between the aggregate and the asphalt is greatly enhanced, and the water resistance of the mixture is improved.

Table 7. The results of freezing and thawing split test

Asphalt mixture	SK-70	SBS	SBP
Split tensile strength before freeze-thaw /MPa	0.865	1.168	1.290
Split tensile strength after freeze-thaw/MPa	0.734	1.025	1.182
Residual strength ratio /%	82.9	87.9	91.6

3.3. Engineering application. The experimental results show that the SBP modifier can improve the high-temperature and water stability of asphalt and mixture. To further verify the practical application of SBP modifier in the high-temperature and rainy areas, a test road was paved to verify it. The test road is in Jiaojiang Bridge, Taizhou City, Zhejiang Province. The AC-13C SBP asphalt mixture was used in the upper layer of K65+400~K65+600 (right), K65+600~K65+700 (left). The usage of SBP modifier is 0.35%, directed vat set, dry mixing time 15s, wet mixing time 40s. In addition to 15 s dry mixing time, there are no special requirements during the construction process of SBP mixture like transport, paving, rolling, etc. It shows good construction characteristics. From the field sampling, the pavement performance tests under the standard conditions were performed, the results are shown in table 8. All the performance can meet the requirements. It should be pointed out that it is necessary to observe the test road for a long time for the long-term performance of SBP asphalt mixture.

Table 8. The asphalt mixture test results of experimental road

Test items	Technical requirement	Test results
Immersed residual stability /%	≥85	91.1
Freeze-thaw splitting strength ratio /%	≥80	92.9
Dynamic stability /cycles·mm^{-1}	≥2 800	8 012

4. Conclusions

- ✓ With the addition of SBP modifier, the complex modulus of asphalt increases, and the phase angle decreases. The high-temperature performance is improved with the amount of SBP. However, the improvement of temperature-sensitive property in the high-temperature range is not as significant as the high-temperature stability.
- ✓ From the view of thermal analysis, the average heat absorption of the asphalt in the high-temperature range is reduced, and the temperature range of the endothermic peak is narrow down after the modification of SBP modifier. The high-temperature stability is improved, which is in good agreement with the DSR test.
- ✓ Concerning the low-temperature performance of asphalt, the best usage is 7%, it can improve the low-temperature performance of asphalt to a certain degree, but not enough to improve the low-temperature property grade of base asphalt.
- ✓ SBP modifier can significantly improve the high-temperature and water stability of asphalt mixture and shows a better effect than SBS, and the improvement of the low-temperature performance of asphalt mixture is limited. It can be concluded that the performance characteristics of SBP can meet the requirements of high-temperature rutting and water damage resistance of asphalt mixture in south China, which is more suitable for the high-temperature and rainy areas.

Acknowledgments

The research described in this paper was supported by the National Natural Science Foundation of China (Nos. 51078331, 51808521), Zhejiang Provincial Natural Science Foundation of China (Nos. LY13E080006, LQ17E080011), Science and Technology Project of Zhejiang Province Communication Department (No. 2016-2-8), Ningbo Transportation Science and Technology Program (No.201309) and Postdoctoral Foundation by Zhejiang Province of China (No.2017-44) which are gratefully acknowledged.

References

[1] Shen J 2001 *Road performance of asphalt and asphalt mixture* (Beijing: China Communication Press)
[2] Yang J 2007 *Polymer modified asphalt* (Beijing: Chemical Industry Press)
[3] Yildirim Y 2007 *Constr. Build. Mater.* **21** 66
[4] Sengoz B and Isikyakar G 2008 *Constr. Build. Mater.* **22** 1897
[5] Panda M and Mazumdar M 2002 *J. Mater. Civil. Eng.* **14** 527
[6] Garcia-Morales M, Partal P and Navarro F J 2006 *Fuel* **85** 936
[7] Punith V S, Veeraragavan A and Amirkhanian S N 2011 *Int. J. Pave. Res. Technology.* **4** 1
[8] Fang C and Yu R 2012 *Polym. Tes.* **31** 276
[9] Yu K Liu L and Yu Q 2010 *Chinese Journal of Environmental Engineering* **3** 689
[10] Zhang J 2009 *Bituminous material* (Beijing: Chemical Industry Press)
[11] JTG F40 2004 *Technical specifications for construction of highway asphalt pavements* (Beijing: China Communication Press)
[12] ASTM D7175 2008 *Standard test method for determining the rheological properties of asphalt binder using a dynamic shear rheometer* (West Conshohocken: American Society for Testing and Materials)
[13] ASTM D6648 2008 *Standard test method for determining the flexural creep stiffness of asphalt binder using the bending beam rheometer (BBR)* (West Conshohocken: American Society for Testing and Materials)
[14] JTG E20 2011 *Standard test methods of bitumen and bituminous mixtures for highway engineering* (Beijing: China Communication Press)
[15] Huang Z Hu X and Wang J 2015 *Journal of Zhejiang University (Engineering Science)* **49** 1448
[16] Tan Y Guo M and Cao L 2013 *China Journal of Highway and Transport* **26** 7
[17] Xiao F Amirkhanian S and Wang H 2014 *Constr. Build. Mater.* **64** 316

Phase structure and fracture toughness of the polyethersulfone-modified bismaleimide resin

X G Li[1], W Q Mu[1], Y T Li[1,2], H H Qian[1,2,*], L J Cheng[1,2] and J L Liu[1,2]

[1] AECC Beijing Institute of Aeronautical Materials, Beijing, 100095, China
[2] Aviation Key Laboratory of Science and Technology on Materials and Application Research for Vibration and Noise Reduction, AECC Beijing Institute of Aeronautical Materials, Beijing, 100095, China

*E-mail: huanghai.qian@biam.ac.cn

Abstract. Although bismaleimide (BMI) resins have been widely used for aerospace applications, owing to their good thermal stability, humidity resistance and excellent mechanical property, they are usually brittle due to high crosslinking density and vulnerable to cracking. Hence enhancement in fracture toughness is desired. Polyethersulfone (PES) is a kind of high-performance thermoplastic materials. The blends of PES and BMI were initially homogeneous solutions but become separated in phases during the polymerization of a thermoset. The morphologies of the BMI/PES resin system were studied. The effect of phase structure on the final mechanical properties was investigated

1. Introduction

Bismaleimide (BMI) resins are considered as a kind of high-performance thermoset materials for the matrix in advance composites, due to their good thermal stability, excellent corrosion resistance and humidity resistance, and good retention of mechanical properties at high temperatures [1]. However, bismaleimide resins suffer from brittleness owing to a highly crosslinked structure. To improve the fracture toughness of BMI, several approaches have been developed [2]. One approach is to use various thermoplastic to modify the fracture toughness properties of BMI. These thermoplastics included PESF [3], PEI [4], poly (phthalazinone ether ketone) [5], PEEK [6], polyimide [7] and poly (arylene ether) [8].

In most cases, the thermoplastic is initially dissolved in the thermoset prepolymer, developing a homogeneous solution. And then a phase separation occurs in the blend of thermoplastic and thermoset resin during the polymerization of a thermoset. This process is known as polymerization-induced phase separation [9,10]. The final properties of the thermoplastic/thermoset blend are primarily determined by the developed phase structures [11-14].

This paper will present the BMI/PES blends by investigating the morphologies and furthermore the relationship between morphologies and the mechanical properties was discussed in detail.

2. Experimental

2.1. Materials

The BMI used a combination of N, N'-4, 4'-bismaleimdodiphenylmethane (BDM), 0, 0'-diallyl-bisphenol A (DABPA), and phenol. It is an RTM-able BMIs developed by the National Key Laboratory of Advanced Composites, with the trademark of BMI 6421. Polyethersulfone (PES), employed as toughener in this study was hydroxyl-terminated polyether sulfone Virantage® VW-10200 RFP, supplied by Solvay Specialty Polymers [15]. All materials were commercial products and were used as received without further purification. Figure 1 shows the molecular structure of the BDM, DABPA, and PES, respectively.

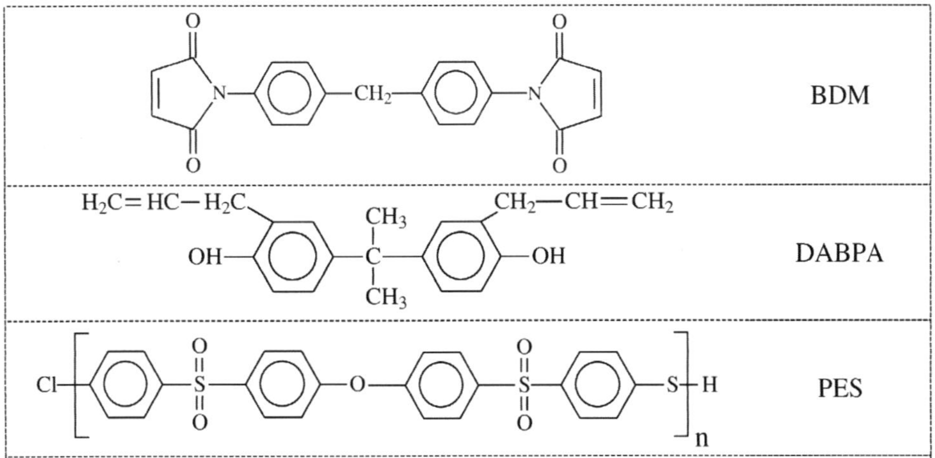

Figure 1. Chemical structures of BDM, DABPA, and PES.

2.2. Sample preparation

BMI/PES blends were prepared as follows. PES was dissolved in the mixture of DABPA and phenol at 120 °C, then BDM was added to the blend at 130 °C. The mixture was mechanically stirred continuously at 130 °C for 0.5 h until an optically clear and transparent blend was obtained, then cured under full vacuum for 30 min to remove any bubbles. After curing, the sample was allowed to cool slowly to room temperature. Table 1 demonstrates the composition of uncrosslinked blends of BMI/PES. Figure 2 gives the curing profile of BMI/PES blends.

Table 1. Designation of the blends of BMI and PES

Sample name	BMI (phr)	PES (phr)
BMI/PES-0	100	0
BMI/PES-5	100	5
BMI/PES-10	100	10
BMI/PES-15	100	15
BMI/PES-20	100	20
BMI/PES-25	100	25

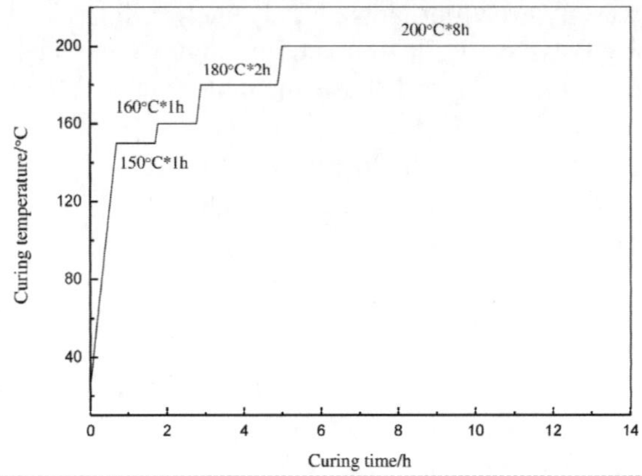

Figure 2. The curing profile of BMI/PES blends.

2.3. Measurements

2.3.1. Scanning electron microscope (SEM). The morphology of the cured resins was examined by SEM, using a Hitachi S-4800. The samples were fractured in liquid nitrogen, coated with a layer of gold and then observed by SEM.

2.3.2. Fracture toughness tests. Fracture toughness test was made using Single-Edge Notched Bending (SENB) geometry in three-point bending, following the ASTM D5045 Standard [16]. Figure 3 shows the adopted sample geometry.

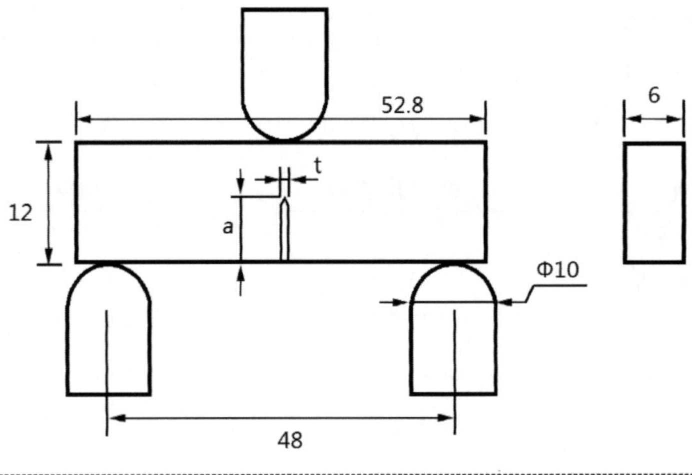

Figure 3. Dimensions of the SENB specimen.

Samples were cut from the cast resin plate with nominal dimensions of 52.8×12×6 mm. A 4.5 mm deep notch was machined by a diamond saw, and a fresh razor blade was lightly tapped into the tip of the notch by a hammer to create a natural pre-crack. The pre-cracked SENB specimens were loaded under three-point bending using a universal testing machine (Instron 3345) at a rate of 0.5 mm/min. Figure 4 shows an example of BMI/PES sample with the produced notch geometry.

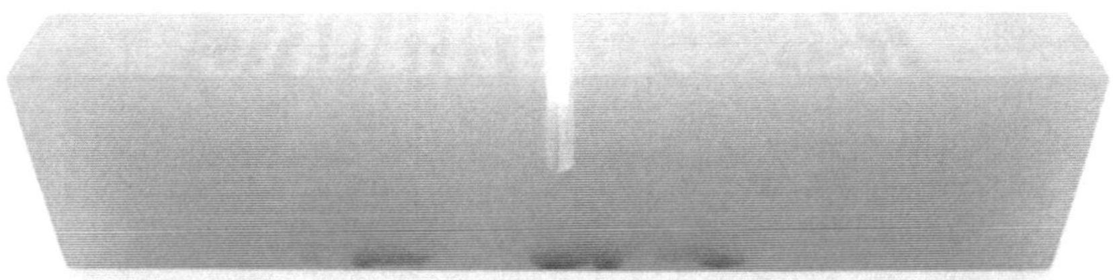

Figure 4. Dimensions of the single edge notched resin specimen.

Five samples were tested in each category. Failure load from each test was recorded. The values of K_{IC} and G_{IC}, respectively, the critical stress intensity factor also known as the critical plane strain fracture toughness and the critical strain energy release rate were calculated from the experimental curves.

The tentative value of fracture toughness, K_Q, is determined by Eq. (1):

$$K_Q = f(x)\left(\frac{P_Q}{BW^{1/2}}\right), \mathrm{W} = 2\mathrm{B}, x = \frac{a}{W} \tag{1}$$

where B is specimen thickness, W is specimen width, a is crack length.
with

$$f(x) = 6x^{1/2}\frac{[1.99 - x(1-x)(2.15 - 3.93x + 2.7x^2)]}{(1+2x)(1-x)^{3/2}} \tag{2}$$

The resulting K_Q is accepted as a valid critical stress intensity factor value K_{IC} if all characteristic lengths of the samples (X=B, a or the ligament (W-a) comply with the condition:

$$B, a, W - a \geq 2.5\left(\frac{K_Q}{\sigma_{YS}}\right)^2 \tag{3}$$

Then, GIC can be determined by equation 4:

$$G_{IC} = \frac{1-v^2}{E}K_{IC}^2 \tag{4}$$

where v represents Poisson's ratio and E is Young moduli.

The value of the yield stress (σ_{YS}), v and E required for the toughness calculations can be measured using cast dog-bone specimens tested in tension using ASTM standards D638.

3. Results and discussion

3.1. Morphology

Figure 5 shows the morphologies of BMI/PES resin system by SEM micrographs. The SEMs are shown at different magnifications for better observation.

The BMI resin systems without modifier exhibited a single phase morphology with a brittle fracture surface (figure 5a). Contrarily, all resin systems modified with PES showed heterophasic morphologies, indicating that a phase separated in these materials.

There were three different types of phase morphologies showed in the PES modified BMI resins: sea-island structure, co-continuous structure, and phase-inverted structure. The type of morphology varied with the loading of thermoplastic PES modifier in the blend.

The BMI/PES-5 resin system generated the sea-island morphology with PES-rich particles dispersed in a BMI-rich matrix (figure 5b). The size of spherical particles was in the range of 5-8 μm.

BMI/PES-15, BMI/PES-20, and BMI/PES-25 showed the phase-inverted morphology, which also called the phase-inverted nodular structure (figures 5f, 5g and figure 5h). This nodular morphology consisted of particles BMI-rich dispersed in a continuous PES-rich matrix. An increase in the loading of PES led to an increase in the loading of dispersed particles and a decrease in their average size(5–20μm). This trend can be ascribed to the increase in viscosity of the medium of BMI/PES blend with the amount of PES increasing. This hinders the diffusion of matter and favors nucleation, resulting in more dispersed particles but smaller size. The micrographs clearly showed the excellent interfacial bonding between BMI and PES with few bared BMI-rich particles.

The co-continuous morphology (figure 5d and figure 5f) was developed by BMI/PES-10. BMI/PES-10 also contained a complex morphology with sea-island morphology and phase-inverted structure irregularly generated in the resin.

A feature observed in some of the morphologies was a second phase separation structure (figure 5c).The second phase separation can be described as follow: the PES-rich phase after phase separation contains a vast number of low molecular weight monomers and oligomers of DABPA-BDM. As for curing of DABPA-BDM component proceeds, the PES-rich phase comes to be immiscible with the dissolved BMI component and eventually separating the BMI-rich domains.

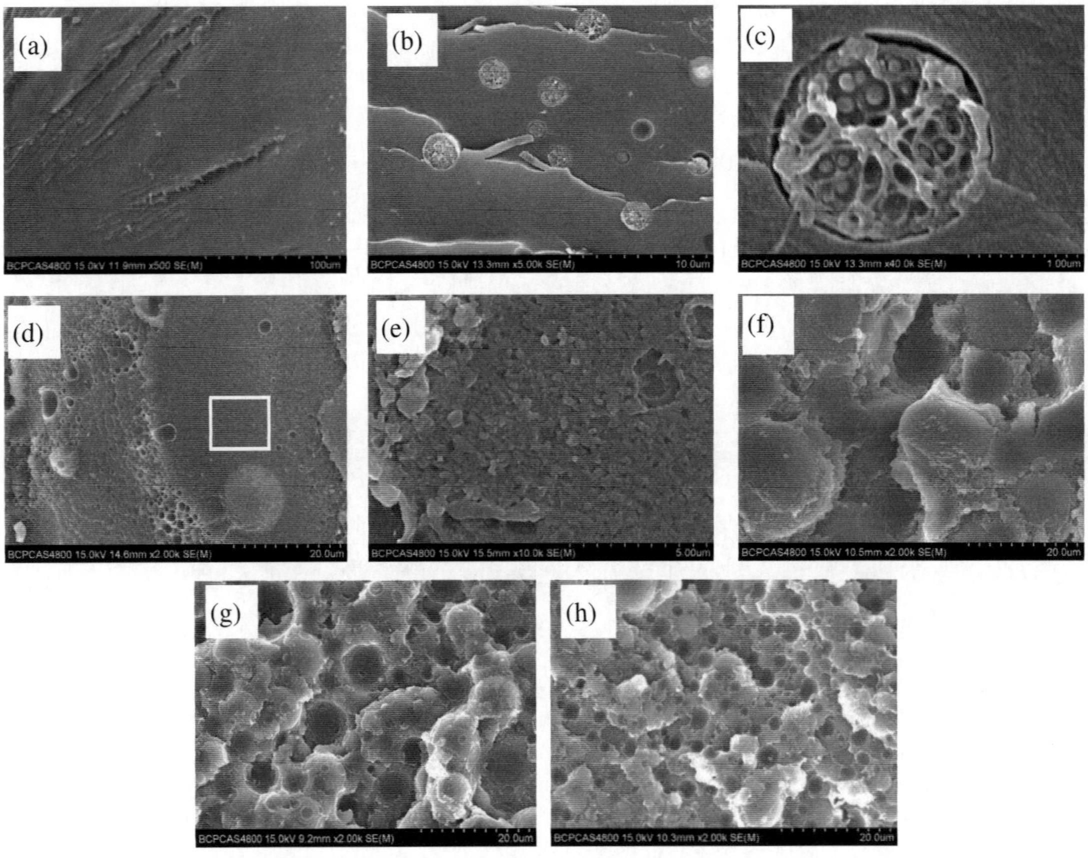

Figure 5. SEM of the morphologies of the BMI/PES resin system.
(a) BMI/PES-0; (b) BMI/PES-5 ; (c) high magnifications of BMI/PES-5; (d)BMI/PES-10; (e) high magnifications of BMI/PES-10; (f)BMI/PES-15; (g) BMI/PES-20; (h)BMI/PES-25.

3.2. Mechanical properties

Figure 6 shows the typical SENB curves of BMI/PES resin. The BMI resin systems without modifier showed unstable crack propagation. The load dropped suddenly, and the structural failure occurred quickly. It is worth to note that the BMI resin toughened with PES shows more non-linear load decrease after the maximum load than the neat BMI resin, indicating a gentle crack propagation. BMI/PES-15 exhibited the highest peak load.

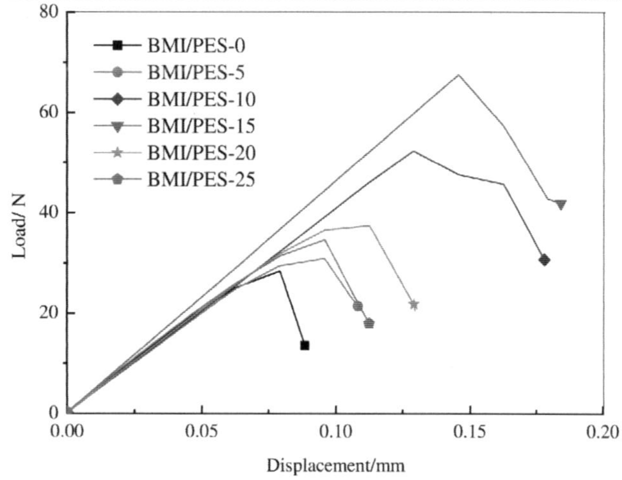

Figure 6. Typical load-displacement of BMI/PES resin under SENB tests.

The fracture toughness of BMI/PES resin characterized by K_{IC} and G_{IC} are given in figure 7. The modifier PES improved the fracture toughness of BMI matrix substantially. The fracture toughness approximately linear increased with the incorporation of PES, attained its maximum, and then was gradually decreased. The BMI/PES resin showed maximum fracture toughness at 15 phr. K_{IC} increased up to 1.22 MPa•m$^{1/2}$ from 0.53 MPa•m$^{1/2}$ of neat resin while G_{IC} increased from 58 up to 257 J•m^{-2}. This significant improvements of fracture toughness due to the incorporation of PES can be ascribed to the phase separated and inverted structure. Tougher resin increases the plastic yielding zone size ahead of the crack tip, giving greater load redistribution away from the crack tip as well as more crack-tip blunting through interfacial debonding of the BMI-rich particles phases along with a plastic drawing of the PES-rich phases and hence more energy absorption.

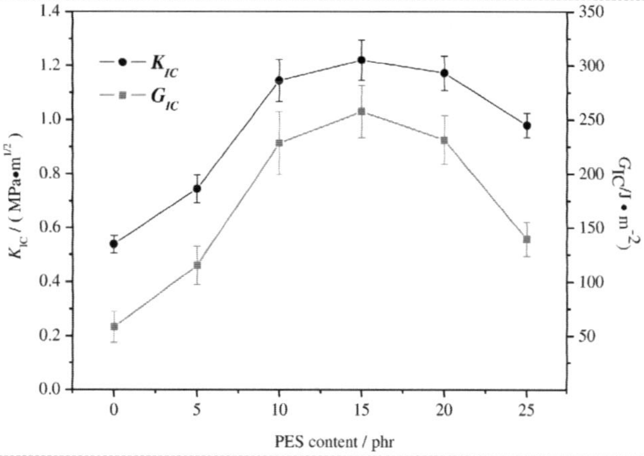

Figure 7. Fracture toughness of BMI/PES resin.

4. Conclusions

The morphology generated in the blend resin based on a polyethersulfone (PES) modifier and BDM/DABPA systems, were observed.

Three different types of morphologies were gained determined by the initial concentration of PES: sea-island morphology of PES-rich particles surrounded with the continuous BMI-rich phase for low PES content (PES 5phr), nodular morphology of BMI-rich particles surrounded with the continuous PES-rich phase for high PES content (PES 15, 20, and 25phr), and co-continuous morphology for intermediate PES content (PES 10phr). An increase in the loading of PES resulted in a decrease in the average size of BMI-rich particles (5–20μm) and an increase in their loading in the nodular morphology.

The fracture toughness increased with the incorporation of PES attaining the maximum then gradually decreases. Compared to neat BMI resin, K_{IC} of BMI/PES-15phr increased by about 130% while G_{IC} of BMI/PES-15phr improved by about 343%. This can be attributed to the phase separated and inverted structure of BMI/PES system. During crack growth, the particle phase acted as stress concentrator, which resulted in plastic yield and drawing of the continuous PES phase, deflecting the fracture path, and hence increasing energy absorption.

Acknowledgments

Authors wish to acknowledge the assistance of Dr. Dong H.M. at AECC Beijing Institute of Aeronautical Materials for the technical support.

References

[1] Dong H M, An X F, Yi X S, Yan L, Su Z T and Bao J W 2015 *J. Mater. Eng.* **43** 89-100.
[2] Dong H M, Yi X S, An X F, Zhang C Q, Yan L and Deng H 2014 *Acta Material Compositae Sinica* **31** 273-285
[3] Liu X, Yu Y and Li S 2006 *Eur. Polym. J.* **42** 835-842
[4] Zhao L, Li L, Tian J X and Zhuang J H 2004 *Compos Part A-Appl S* **35** 1217-1224
[5] Han Y, Liao G, Xu Y, Yu G and Jian X 2009 *Polym. Eng. Sci.* **49** 2301–2308
[6] Chu P P, Wu C S, Liu P C, Wang T H and Pan J P 2010 *Polymer* **51** 1386-1394
[7] Pascal T, Mercier R, Sillion B 1989 *Polymer* **30** 739-744
[8] Stenzenberger H D, König P 1993 *High Perform. Polym.* **5** 123-137
[9] Kim J Y, Cho C H, Palffy-Muhoray P and Kyu T 1993 *Phys. Rev. Lett.* **71** 2232-2235
[10] Luo K 2006 *Eur. Polym. J.* **42** 1499-1505
[11] López J, Ramírez C, Abad M J, Barral L, Cano J and Díez F J 2010 *J. Appl. Polym. Sci.* **85** 1277-1286
[12] Zucchi I A, Galante M J, Williams R J J 2005 Polymer **46** 2603-2609
[13] Prolongo M G, Arribas C, Salom C and Masegosa R M 2007 *J. Therm. Anal. Calorim.* **87** 33-39
[14] Rico M, López J, Montero B and Bellas R 2012 *Eur. Polym. J.* **48** 1660-1673
[15] Zhang C Q, An X F, Dong H M, Yan L and Gao J P 2012 *China Thermosetting Resin* 33-36
[16] Sham Prasad M S, Venkatesha C S, Jayaraju T 2011 *J. Miner. Mater.Charact.Eng* **10** 1263-1275

IOP Conf. Series: Materials Science and Engineering **474** (2019) 012029 doi:10.1088/1757-899X/474/1/012029

Electric and elastic singularity eigenanalysis of the V-notch in a piezoelectric Reissner plate

C Z Cheng*, H L Dou, Z Y Gao, SL Yao, D P Wang and Z R Niu

Department of Engineering Mechanics, Hefei University of Technology, Hefei 230009, China

*E-mail: changzheng.cheng@hfut.edu.cn

Abstract. The singular eigenstate with respect to the stress and electric displacement at the vertex of a V-notch in a piezoelectric plate is investigated. The displacement and electric potential near the vertex of a V-notch are expressed as series asymptotic expansions, and their typical terms are substituted into the governing equations to establish the singularity eigenequations, from which the singularity orders and angular eigenfunctions can be yielded by the interpolating matrix method. The coupling effect of elastic and electric singularities can be found by distinguishing the singularity orders from their angular eigenfunctions. There are most four singularity orders for the V-notch in a piezoelectric Reissner plate, which is one more than the one in a homogeneous plate. Two orders are corresponding to the singularity of bending moment and electric displacement in the thickness direction, and another two orders are with respect to the singularity of shear force and electric displacement in the mid-plane. The free and electrically opened, and clamped and electrically closed boundary conditions are respectively introduced to investigate the influence of boundary conditions on the singularity. It is found that the singularity orders with respect to bending moment and electric displacement in the thickness direction change with boundary conditions, while the singularity orders with respect to shear force and electric displacement in the mid-plane are identical under these two kinds of boundary conditions.

1. Introduction

Due to the direct-converse piezoelectric effect, the piezoelectric material has been extensively used in electromechanical transducers, sensors, etc. [1]. However, these piezoelectric structures or components always include V-notches, which are generated by the discontinuity of material property and/or geometry. There will present the singularities near the vertex of a V-notch, in which the stress singularity may lead to the crack initiation and extension, and the singularity of the electric displacement may lead to the electrical breakdown.

As the specific case of a V-notch, the singularity eigenanalysis for the crack in a piezoelectric structure had been thoroughly investigated [2-4]. A piezoelectric structure in the energized state will produce the heat, which will make the thermal stress singularity near the crack tip and weaken the structures. The fracture problem for cracks in piezoelectric material subjected to thermal loading attracted much attention [5-7]. In comparison with the crack problem, the work on the singularity analysis for a piezoelectric notch is much less [8-10]. The piezoelectric structures are often used in a plate or shell model [11,12]. According to the best knowledge of authors, the present research is mainly focused on the piezoelectric V-notch in plane, anti-plane or three-dimensional state [13]. There are scarce data on the characteristic singularity analysis for the V-notch in a piezoelectric plate V-notch. The difficulty is from the coupling of the stress singularity and electric displacement singularity.

The motivation of the present paper is to propose the singularity eigenanalysis of stress and electric displacement for a piezoelectric plate V-notch. The elastic and electrical equilibrium equations for a piezoelectric Reissner plate are firstly given out. Then, the typical terms in the asymptotic displacement and potential electric fields near the singular vertex are substituted to the governing equations together with radial boundary conditions to establish singularity eigen equations, which will be solved by the interpolating matrix method [14] to yield singularity orders and angular eigenfunctions. The coupled effect of the elastic singularity and electric singularity is investigated.

2. Equilibrium equations for a piezoelectric plate notch

Let's consider a piezoelectric plate weakened by a single edge V-notch. The electric potential is assumed to vary as a parabolic function along the thickness direction [15], according to the following equation

$$\psi(r,\theta,z)=\frac{h^2}{4-z^2}\psi_0(r,\theta) \tag{1}$$

where ψ and ψ_0 are the electric potentials in the plate and mid-plane, respectively, while h is the plate thickness. The equilibrium differential equations regarding generalized displacement components for a piezoelectric Reissner plate without body force can be written as follows

$$D(\frac{\partial^2\phi_r}{\partial r^2}+\frac{\partial\phi_r}{r\partial r}-\frac{\phi_r}{r^2}+\frac{1-\nu}{2r^2}\frac{\partial^2\phi_r}{\partial\theta^2}+\frac{1+\nu}{2r}\frac{\partial^2\phi_\theta}{\partial r\partial\theta}-\frac{3-\nu}{2r^2}\frac{\partial\phi_\theta}{\partial\theta^2})+C(\frac{\partial w}{\partial r}+\frac{d_{15}h^2}{5}\frac{\partial\psi_0}{\partial r}-\phi_r)$$
$$+\frac{d_{31}h^3}{6s_{11}(1-\nu)}\frac{\partial\psi_0}{\partial r}=0 \tag{2a}$$

$$D(\frac{1+\nu}{2r}\frac{\partial^2\phi_r}{\partial r\partial\theta}+\frac{3-\nu}{2r^2}\frac{\partial\phi_r}{\partial\theta}+\frac{1-\nu}{2}\frac{\partial^2\phi_\theta}{\partial r^2}+\frac{1-\nu}{2r}\frac{\partial\phi_\theta}{\partial r}+\frac{1}{r^2}\frac{\partial^2\phi_\theta}{\partial\theta^2}-\frac{1-\nu}{2}\frac{\phi_\theta}{r^2})$$
$$+C(\frac{1}{r}\frac{\partial w}{\partial\theta}+\frac{d_{15}h^2}{5r}\frac{\partial\psi_0}{\partial\theta}-\phi_\theta)+\frac{d_{31}h^3}{6rs_{11}(1-\nu)}\frac{\partial\psi_0}{\partial\theta}=0 \tag{2b}$$

$$\frac{\partial^2 w}{\partial r^2}+\frac{\partial w}{r\partial r}+\frac{\partial^2 w}{r^2\partial\theta^2}+\frac{d_{15}h^2}{5}(\frac{\partial^2\psi_0}{\partial r^2}+\frac{\partial\psi_0}{r\partial r}+\frac{\partial^2\psi_0}{r^2\partial\theta^2})-(\frac{\partial\phi_r}{\partial r}+\frac{\phi_r}{r}+\frac{\partial\phi_\theta}{r\partial\theta})=0 \tag{2c}$$

$$-5(\frac{d_{31}}{s_{11}+s_{12}}+\frac{d_{15}}{s_{44}})(\frac{\partial\phi_r}{\partial r}+\frac{\phi_r}{r}+\frac{\partial\phi_\theta}{r\partial\theta})+\frac{5d_{15}}{s_{44}}\nabla^2 w+h^2(\frac{d_{15}^2}{s_{44}}-\beta_{11})\nabla^2\psi_0$$
$$+10[\frac{2d_{31}^2}{s_{11}(\nu-1)}+\beta_{33}]\psi_0=0 \tag{2d}$$

where $\nu=-s_{12}/s_{11}$, $D=\frac{h^3}{12(1-\nu^2)s_{11}}$, $C=\frac{5h}{6s_{44}}$, $\nabla^2=\frac{\partial^2}{\partial r^2}+\frac{\partial}{r\partial r}+\frac{1}{r^2}\frac{\partial^2}{\partial\theta^2}$, ϕ_r and ϕ_θ are the bending rotations of the mid-plane normal in radial and circumferential directions, respectively, w is the displacement component along the z-direction. s_{11}, s_{12}, s_{44} are flexibility coefficients, d_{15} and d_{31} are piezoelectric coefficients, β_{11} and β_{33} are permittivity values. The piezoelectric material PZT-4 with a z-axial polarization is taken into consideration, with the following material properties [8]:

$s_{11}=10.90\times10^{-12}\,\mathrm{m^2N^{-1}}$, $s_{12}=-5.42\times10^{-12}\,\mathrm{m^2N^{-1}}$, $s_{44}=19.30\times10^{-12}\,\mathrm{m^2N^{-1}}$, $d_{15}=39.40\times10^{-3}\,\mathrm{VmN^{-1}}$,

$d_{31}=-11.10\times10^{-3}\,\mathrm{VmN^{-1}}$, $\beta_{11}=7.66\times10^{7}\,\mathrm{V^2N^{-1}}$, $\beta_{33}=8.69\times10^{7}\,\mathrm{V^2N^{-1}}$.

If two radial edges of the V-notch are supposed to be set traction-free and electrically opened, this boundary condition can be reduced to the following form

$$M_\theta=0,\ M_{r\theta}=0,\ Q_\theta=0,\ \int_{-\frac{h}{2}}^{\frac{h}{2}}D_\theta(h^2/4-z^2)\mathrm{d}z=0\quad(\theta=-\alpha,\alpha) \tag{3}$$

where α is half of the notch opening angle. These can be represented by the generalized displacement components $(\phi_r,\phi_\theta,w,\psi_0)$,

$$\frac{1}{6(s_{11}-s_{12})}\left[s_{12}\frac{\partial\phi_r}{\partial r}-s_{11}(\frac{1}{r}\frac{\partial\phi_\theta}{\partial\theta}+\frac{1}{r}\phi_r)\right]-d_{31}\psi_0=0\quad(\theta=-\alpha,\alpha) \tag{4a}$$

$$\frac{1}{r}\frac{\partial\phi_r}{\partial\theta}+\frac{\partial\phi_\theta}{\partial r}-\frac{\phi_\theta}{r}=0\quad(\theta=-\alpha,\alpha) \tag{4b}$$

$$5(-\phi_\theta+\frac{1}{r}\frac{\partial w}{\partial\theta})+h^2 d_{15}\frac{\partial\psi_0}{r\partial\theta}=0\quad(\theta=-\alpha,\alpha) \tag{4c}$$

$$\beta_{11}\frac{1}{r}\frac{\partial\psi_0}{\partial\theta}-\frac{5d_{15}}{s_{44}h^2}(-\phi_\theta+\frac{\partial w}{r\partial\theta})-\frac{d_{15}^2}{s_{44}}\frac{1}{r}\frac{\partial\psi_0}{\partial\theta}=0\quad(\theta=-\alpha,\alpha) \tag{4d}$$

If two radial edges of the V-notch are supposed to be set clamped and electrically closed, this kind of boundary conditions can be written as

$$\phi_r=0,\ \phi_\theta=0,\ w=0,\ \psi_0=0\quad(\theta=-\alpha,\alpha) \tag{5}$$

3. Singularity eigen equations for a piezoelectric plate notch

The elastic and electric displacement field near the vertex of a V-notch can be expressed by a series of asymptotic expansions as follows [16]

$$\phi_r=\sum_{k=1}^{N}A_k r^{\lambda_k}\tilde{\phi}_{rk}(\lambda_k,\theta),\quad \phi_\theta=\sum_{k=1}^{N}A_k r^{\lambda_k}\tilde{\phi}_{\theta k}(\lambda_k,\theta),\quad w=\sum_{k=1}^{N}A_k r^{\lambda_k}\tilde{w}_k(\lambda_k,\theta),\quad \psi_0=\sum_{k=1}^{N}A_k r^{\lambda_k}\tilde{\psi}_{0k}(\lambda_k,\theta) \tag{6}$$

where r is the radial distance to the notch tip, A_k is the amplitude coefficient, λ_k is the singularity order, N is the number of truncated series terms, $\tilde{\phi}_{rk}(\lambda_k,\theta)$, $\tilde{\phi}_{\theta k}(\lambda_k,\theta)$, $\tilde{w}_k(\lambda_k,\theta)$ and $\tilde{\psi}_{0k}(\lambda_k,\theta)$ are angular eigenfunctions of the generalized displacements.

Substituting the typical terms in Equation (6), i.e., $A_k r^{\lambda_k}\tilde{\phi}_{rk}(\lambda_k,\theta)$, $A_k r^{\lambda_k}\tilde{\phi}_{\theta k}(\lambda_k,\theta)$, $A_k r^{\lambda_k}\tilde{w}_k(\lambda_k,\theta)$ and $A_k r^{\lambda_k}\tilde{\psi}_{0k}(\lambda_k,\theta)$ into Equation (2) and letting the sum of the coefficients of the lowest order of r set as zero, yields the ordinary differential eigenequations for the V-notch in a piezoelectric plate, which can be written as

$$\frac{1-\nu}{2}\tilde{\phi}_r'' - \frac{3-\nu}{2}\tilde{\phi}_\theta' - \tilde{\phi}_r + \lambda\frac{1+\nu}{2}\tilde{\phi}_\theta' + \lambda^2\tilde{\phi}_r = 0 \tag{7a}$$

$$\tilde{\phi}_\theta'' + \frac{3-\nu}{2}\tilde{\phi}_r' - \frac{1-\nu}{2}\tilde{\phi}_\theta + \lambda\frac{1+\nu}{2}\tilde{\phi}_r' + \lambda^2\frac{1-\nu}{2}\tilde{\phi}_\theta = 0 \tag{7b}$$

$$\tilde{w}'' + \frac{h^2 d_{15}}{5}\tilde{\psi}_0'' + \lambda^2 w + \lambda^2\frac{h^2 d_{15}}{5}\tilde{\psi}_0 = 0 \tag{7c}$$

$$\tilde{w}'' + \frac{h^2}{5}\left(d_{15} - \frac{s_{44}\beta_{11}}{d_{15}}\right)\tilde{\psi}_0'' + \lambda^2 w + \lambda^2\frac{h^2}{5}\left(d_{15} - \frac{s_{44}\beta_{11}}{d_{15}}\right)\tilde{\psi}_0 = 0 \tag{7d}$$

where $(\cdots)' = \mathrm{d}(\cdots)/\mathrm{d}\theta$, $(\cdots)'' = \mathrm{d}^2(\cdots)/\mathrm{d}\theta^2$, λ, $\tilde{\phi}_r$, $\tilde{\phi}_\theta$, $\tilde{w}$ and $\tilde{\psi}_0$ are the simplified forms of λ_k, $\tilde{\phi}_{rk}(\lambda_k,\theta)$, $\tilde{\phi}_{\theta k}(\lambda_k,\theta)$, $\tilde{w}_k(\lambda_k,\theta)$ and $\tilde{\psi}_{0k}(\lambda_k,\theta)$, respectively.

Similarly introducing the typical terms from Equation (6) into Equation (4), the traction-free and electrically opened boundary conditions can be expressed by angular eigenfunctions and their derivatives like

$$\tilde{\phi}_\theta' + \tilde{\phi}_r - \frac{s_{12}}{s_{11}}\lambda\tilde{\phi}_r = 0,\quad \tilde{\phi}_r' - \tilde{\phi}_\theta + \lambda\tilde{\phi}_\theta = 0,\quad \tilde{w}' + \frac{h^2 d_{15}}{5}\tilde{\psi}_0' = 0,\quad \tilde{w}' + \frac{h^2}{5}\left(d_{15} - \frac{s_{44}\beta_{11}}{d_{15}}\right)\tilde{\psi}_0' = 0 \tag{8}$$

The clamped and electrically closed boundary condition can be rewritten as

$$\tilde{\phi}_r = 0,\quad \tilde{\phi}_\theta = 0,\quad \tilde{w} = 0,\quad \tilde{\psi}_0 = 0 \quad (\theta = -\alpha, \alpha) \tag{9}$$

Until now, the singularity analysis for the V-notch in a piezoelectric Reissner plate was reduced to solve the singularity eigenequation (7) under radial boundary conditions described by Equations (8) or (9). The interpolating matrix method, which was earlier proposed in [14] and used by some authors, is employed here to solve the singularity eigenequations (7, 8) or Equations (7, 9), from which the singular orders together with angular eigenfunctions for the V-notch in a piezoelectric plate can be derived.

4. Numerical examples and discussions

The influence of plate thickness on the singularity of the V-notch in a piezoelectric plate is firstly investigated. There are three different plate thicknesses (i.e., 0.001, 0.01, and 0.1m), taken into consideration. The variation of singular orders with plate thickness for the V-notch with $2\alpha = 300^\circ$ is listed in table 1, where traction-free and electrically opened boundary conditions are considered. It can be seen that the singularity orders do not vary dramatically with the plate thickness. Thus, the plate thickness is set at $h = 0.1\text{m}$ in the further evaluation.

Table 1. Variation of singularity orders with plate thickness for $2\alpha = 300^\circ$.

h(m)	λ_1	λ_2	λ_3	λ_4
0.001	0.512225	0.600000	0.600000	0.730924
0.010	0.512223	0.600000	0.600000	0.730912
0.100	0.512223	0.600000	0.600000	0.730912

The calculated singularity orders for V-notches with different opening angles are listed in table 2, where traction-free and electrically opened boundary conditions is adopted. It can be seen that there are three singularity orders for $180^\circ < 2\alpha \leq 260^\circ$, and four singularity orders for $260^\circ < 2\alpha \leq 360^\circ$. The singularity orders corresponding to bending moment, shear force and electric displacement can be distinguished by the angular eigenfunctions.

Table 2. Singularity orders for V-notches under traction-free and electrically opened boundary conditions.

$2\alpha(^\circ)$	Method	λ_1	λ_2	λ_3	λ_4
180	Present	/	/	/	/
	Ref.[17]	/	/	/	/
220	Present	0.697167	0.818182	0.818182	/
	Ref.[17]	0.697165	0.818182	-	
260	Present	0.562841	0.692308	0.692308	0.980489
	Ref.[17]	0.562839	0.692308	-	0.980474
300	Present	0.512223	0.600000	0.600000	0.730912
	Ref.[17]	0.512221	0.600000	-	0.730901
340	Present	0.500429	0.529412	0.529412	0.562017
	Ref.[17]	0.500426	0.529412	-	0.562007
360	Present	0.500000	0.500000	0.500003	0.500009
	Ref.[17]	0.500000	0.500000	-	0.500000

Note: the slash means no singularity and the dash means no corresponding item in the reference.

The singularity orders for the V-notch in a homogeneous plate by the eigen function expansion method [17] are listed in table 2 to provide the reference ones. It can be found out that the present results are in good agreement with the reference ones for λ_1, λ_2 and λ_4, except for λ_3, which was not provided in 17]. Thus, one singularity order will be increased for the piezoelectric plate notch against the homogeneous plate notch.

The calculated singularity orders for the V-notches under clamped and electrically closed boundary condition are listed in table 3, where the results from [17] are also listed for reference. It should be pointed out that, the singularity orders do not depend on Poisson's ratio for the V-notch under free boundary condition, while they depend on Poisson's ratio under clamped boundary condition for the V-notch in a homogeneous Reissner plate. Thus, the reference results are calculated from the V-notch with Poisson's ratio $\nu = -s_{12} / s_{11}$. From table 3, it can be seen that the present results agree well with the reference ones, and one more singularity order is derived for the V-notch in a piezoelectric plate, as compared to the V-notch in a homogeneous plate.

Table 3. Singularity orders for V-notches under clamped and electrically closed boundary condition.

$2\alpha(^\circ)$	Method	λ_1	λ_2	λ_3	λ_4
180	Present	/	/	/	/
	[17]	/	/	/	/
220	Present	0.742749	0.818182	0.818182	0.911443
	[17]	0.742747	0.818182	-	0.911441
260	Present	0.611143	0.692308	0.692308	0.800546
	[17]	0.611140	0.692308	-	0.800542
300	Present	0.545301	0.600000	0.600000	0.667441
	[17]	0.545299	0.600000	-	0.667434
340	Present	0.511738	0.529412	0.529412	0.548367
	[17]	0.511735	0.529412	-	0.548359
360	Present	0.500000	0.500000	0.500004	0.500008
	[17]	0.500000	0.500000	-	0.500000

Note: the slash means no singularity and the dash means no corresponding item in the reference.

It had been indicated in [17] that the singularity order with respect to shear force keeps the same for the V-notch under free boundary condition and clamped boundary condition. By comparing tables 2 and 3, it can be found that the singularity orders λ_2 and λ_3 are identical in these two tables, which means that the singularity orders concerning shear force and electric displacement in the mid-plane keep identical under free and electrically opened boundary condition, and clamped and electrically closed boundary condition. However, λ_1 and λ_4 are different in these two kinds of boundary conditions.

5. Conclusions

The singular orders for the V-notch in a piezoelectric Reissner plate were investigated. It was found that the number of singularity orders in a piezoelectric plate notch exceeds by one that in a homogeneous plate notch. The coupling effect of elastic singularity and electric singularity in a piezoelectric plate V-notch was found. To investigate the influence of the boundary conditions on the singularity of the V-notch, the free and electrically opened, and clamped and electrically closed boundary conditions were taken into consideration. Under these two different boundary conditions, the singularity orders with respect to bending moment and electric displacement in the thickness direction are different, while the singularity orders with respect to shear force and electric displacement in the mid-plane remain identical.

Acknowledgment

This work was supported by the National Natural Science Foundation of China, China (No. 11772114).

References

[1] Artel J and Becker W 2005 *Compo. Struct.* **69** 329-35
[2] Wang H T, Sze K Y and Yang X M 2006 *Comput. Mech.* **38** 551-64
[3] Li C, Man H, Song C M and Gao W 2013 *Compos. Struct.* **101** 191-203
[4] Ou Z C and Wu X J 2003 *Int. J. Solids Struct.* **40** 7499-511
[5] Shang F L and Kuna M 2003 *Comp. Mater. Sci.* **26** 197-201
[6] Rogowski B 2009 *Int. J. Pres. Ves. Pip.* **86** 384-94
[7] Zhang A B and Wang B L 2013 *Int. J. Solids Struct.* **50** 2962-69
[8] Chen C D 2006 *Int. J. Solids and Struct.* **43** 957-81
[9] Huang C S and Hu C N 2013 *Appl. Math. Model.* **37** 4517-37
[10] Cheng C Z, Yao S L, Han Z L, Recho N and Niu Z R 2015 *Theor. Appl. Fract. Mec.* **76** 50-59
[11] Viola E, Boldrini C and Tornabene F 2008 *Eng. Fract. Mech.* **75** 4542-67
[12] Sheng G G and Wang X 2010 *Appl. Math. Model.* **34** 2630-43
[13] Guo J H, Liu P, Lu Z X and Qin T Y 2011 *Appl. Math. Mech.-Engl.* **32(1)** 75-82
[14] Niu Z R, Ge D L, Cheng C Z, Ye J Q and Recho N 2009 *Appl. Math. Model.* **33** 1776-92
[15] Wang Q, Quek S T, Sun C T and Liu X 2001 *Smart Mater. Struct.* **10** 229-39
[16] Williams M L 1952 *J. Appl. Mech.* **19(4)** 526-28
[17] Huang C S 2003 *Int. J. Mech. Sci.* **45** 1-20

An experimental study on thermal contact conductance across the SS316L/CuCrZr interface

X B Liang[1], J Y Yao[2], W F Zhang[2,*], N Li[1], M Y Ma[2] and Y R Wang[1]

[1] School of Energy and Power Engineering, Beihang University, Haidian Dist., Beijing 100191, China

[2] School of Reliability and Systems Engineering, Beihang University, Haidian Dist., Beijing 100191, China

[*]E-mail: 08590@buaa.edu.cn

Abstract. A vacuum apparatus to measure thermal contact conductance (TCC) was introduced. A TCC test method and procedure was presented. By monitoring the temperature distribution on the test specimen, the heat flux and temperature drop of the contact interface were calculated, and then the TCC was obtained. The experimental research was conducted with pressed pairs SS316L/CuCrZr contacts in a range of 100~300°C and 1~10MP. The experimental data analysis results show that the TCC increases with the increase of contact temperature and the TCC also grows with the growth of contact pressure. The TCC appears as a linear relationship with the change trend of pressure at a given temperature. By uncertainty analysis, the measurement error of TCC decreases with the increase of contact pressure at a given temperature. And the variation rate of measurement error increases with the rise in contact temperature.

1. Introduction

Thermal contact conductance (TCC) is the ratio between the average heat flux and the temperature drop across the interface. The reciprocal of TCC is thermal contact resistance (TCR). Thermal contact resistance (TCR) is the ratio between the temperature drop and the average heat flux across the interface. TCC is a vital heat transfer property in many engineering applications, such as aerospace technologies [1], electronics [2] and nuclear reactors [3]. Thermal contact conductance depends on a lot of factors such as pressure, temperature, material properties, contact interface topography [4]. The factors affecting the thermal conductivity can be divided into macro-influence factors and micro-influence factors. Macroscopic factors include thermal properties, elastoplasticity, surface hardness and surface shape of materials. Microscopic factors include asperity, micro-morphology and elastoplastic deformation of contact peak. Therefore, the research on TCC involves three underlying problems: geometry, mechanics and thermal. And there are three kinds of contact mechanics problems: pure elastic contact, pure plastic contact and more complex elastoplastic contact [4]. To investigate TCC quantitatively, researches have carried out a lot of experimental and theoretical research work [5-7].

Contact thermal conductivity is a multidisciplinary field of geometry, heat, and machinery. Many scholars have conducted extensive and in-depth research on it. In recent decades, many researchers have proposed contact interface model based on experimental data and mathematical analysis to calculate TCC [8-10]. Yovanovich [11] introduced a plastic contact thermal resistance model with an average height Gaussian distribution of contact interface. Leung [12] compared the asperity of contact interface with the particles in statistical thermodynamics to investigate the thermal contact resistance. Many pieces of research applied the fractal theory to examine the contact surface morphology, and then obtained the relationship model between surface morphology and interface thermal resistance [13,14]. Based on the fractal theory TCR analysis method, the researches also established a multi-scale TCR model considering the scale effect and the material thermal properties [15]. Unfortunately, so far, there is no satisfactory theoretical model and no reliable empirical formula for experimental research.

In this paper, based on the TCC model proposed by Yovanovich, the experimental method was designed to calculate the TCC of pressed pairs SS316L/CuCrZr. Furthermore, the influence of temperature and pressure on the TCC of pressed couples SS316L/CuCrZr was analyzed to provide experimental data and method basis for the investigation of TCC.

2. Method and procedure

2.1. Test principle

A temperature drop is observed when two solid interfaces come into physical contact. Thermal contact conductance (TCC) is the ratio between the average heat flux and the temperature drop across the interface. The TCC of the contact interface can be expressed as

$$h_s = \frac{q}{\Delta T} = \frac{q}{T_{C1} - T_{C2}} \tag{1}$$

where h_s is the thermal contact conductance, ΔT is the interface temperature drop, q is the heat flux, T_{C1} and T_{C2} denote the temperature of different sides of the contact surface, respectively[11].

In this paper, the steady heat transfer method is used to test the TCC between the pressed pairs SS316L/CuCrZr contacts. As is shown in figure 1, the contact surface temperature of specimen 1 T_{C1} is calculated by the linear fitting of the temperature of four points on the specimen 1. In the same way, T_{C2} is obtained, and ΔT is derived as $\Delta T = T_{C1} - T_{C2}$. According to the heat conduction theory, the heat flux q is derived via the heat flux meter. As is shown in Fig. 1, it can be expressed as

$$q = \lambda \frac{T_i - T_{i+1}}{\delta} \tag{2}$$

where λ is the thermal conductivity coefficient of the heat flux meter specimen, T_i, T_{i+1} is the temperature measured by thermocouples of the heat flux meter ($i = 1, 2, 3, 5, 6$), δ is the distance of the adjacent thermocouples hole on the heat flux meter.

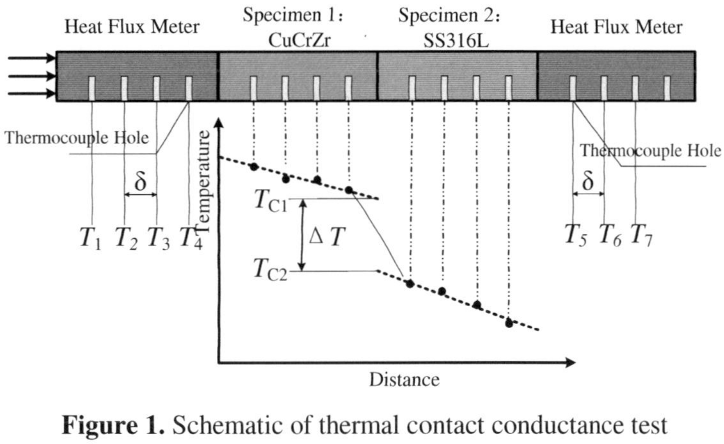

Figure 1. Schematic of thermal contact conductance test

2.2. Test setup

The schematic of TCC experimental apparatus is shown in figure 2, which mainly consists of a vacuum chamber, loading system, heating system, temperature, and pressure data acquisition system and molecular pump.

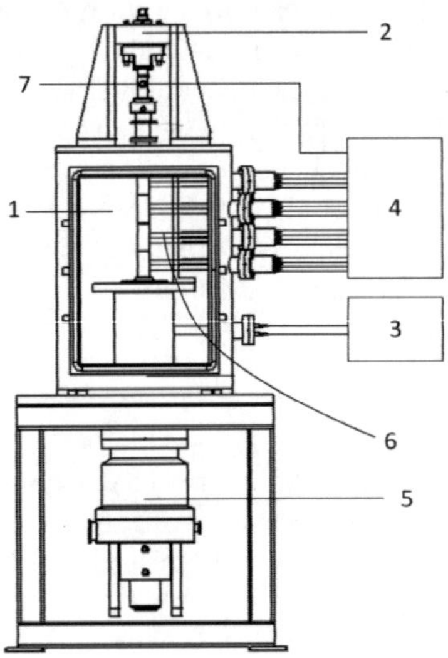

1—vacuum chamber; 2—loading system; 3—heating system; 4—temperature and pressure data acquisition system; 5—molecular pump; 6—thermocouple; 7—pressure sensor

Figure 2. Schematic of TCC experimental apparatus

The pressed pairs SS316L/CuCrZr specimens are cylindrical with 30mm diam and 65mm height. Four holes have been drilled into the specimens along the radial direction with a depth of 15mm for placing thermocouples. The TCC measurement device consists of two brass heat flux meters and pressed pairs SS316L/CuCrZr specimens (as is shown in figure 1). Sixteen thermocouples are mounted in the holes of specimens and heat flux meter to measure and collect the temperature data of the specimens. According to formula (1), the TCC of pressed pairs SS316L/CuCrZr is calculated based on the temperature data collected. By controlling thyristor voltage regulator of heating system, the TCC at different temperatures of the specimens can be obtained. Meanwhile, the TCC at different pressure values in specimens can be obtained by controlling the loading system.

3. Results and Discussion

Based on the TCC test principle and TCC experimental apparatus setup, the pressed pairs SS316L/CuCrZr specimens were used to verify the feasibility of the TCC test method. The TCC between pressed pairs SS316L/CuCrZr specimens were tested with the interface pressure from 1MP to 10MP at 100°C, 200°C and 300°C respectively.

3.1. Test results

Table 1 shows the TCC of pressed pairs SS316L/CuCrZr at different temperatures and pressures. For three different interface temperatures, the TCC results under pressures of 1, 2.5, 5, 7.5, and 10MP were calculated. As the temperature and pressure of the contact surface affect each other, the temperature cannot be accurately adjusted to 100, 200, and 300 °C, and the pressure also cannot be carefully adjusted to 1, 2.5, 5, 7.5, and 10MP at the thermal equilibrium state of the test process.

Table 1. TCC of specimens at different temperatures and pressures

102.87°C	201.98°C	300.9°C

Pressure, MPa	TCC, W·°C^{-1}·m^{-2}	Pressure, MPa	TCC, W·°C^{-1}·m^{-2}	Pressure, MPa	TCC, W·°C^{-1}·m^{-2}
1.06	2865	1.01	6934	1.04	14804
2.43	3424	2.66	10044	2.65	19954
5.01	5185	5.02	14055	4.93	30300
7.55	6527	7.43	17818	7.40	37793
9.96	7689	9.99	22532	9.97	45018

The TCC results listed in table 1 were fitted to reveal the tendency of TCC versus pressures at different temperatures. The fitting results shown in figure 3 reveal that the TCC of pressed pairs SS316L/CuCrZr gradually increases with pressure and exhibits a good linearity. All data fall in the range $10^{-4} \le P/H_e \le 10^{-2}$ (where P is the contact pressure, H_e is the effective contact micro-hardness)[11], whereas the dimensionless conductance and dimensionless contact pressure appear as a straight line according to CMY model. Contact pressure at 1~10MP and temperature at 100~300°C, the effective contact micro-hardness can be considered constant. Therefore, the TCC has a linear relationship with the change trend of contact pressure.

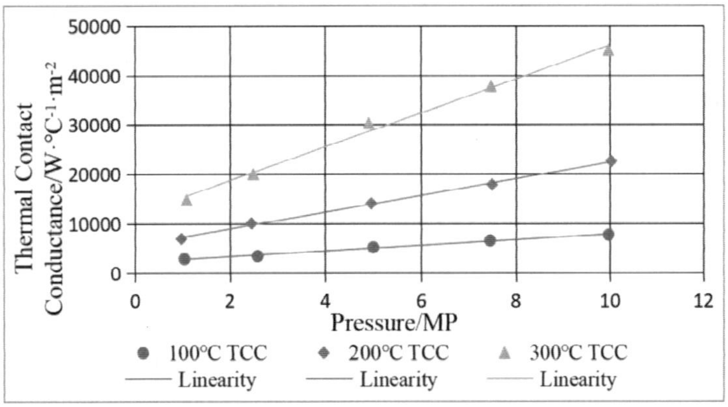

Figure 3. TCC curve of SS316L/CuCrZr with pressure at different temperatures

3.2. Uncertainty analysis

The errors in the test can be divided into systematic error and accidental error. The system error is mainly caused by the test equipment. According to the equipment parameters provided by the TCC experimental apparatus manufacturer, the uncertainty of temperature and pressure control system, temperature and pressure measurement system are 1.8 and 0.6%, respectively. The system error can be written as

$$error(h_s) = \sqrt{(1.8\%)^2 + (0.6\%)^2} = 1.9\% \tag{3}$$

and in the experiment, the random error comes from heat loss. According to error propagation law [16] and equation (1), the accidental error can be expressed as

$$error(h_a) = \frac{\Delta h}{h} = \frac{\sqrt{\left(\frac{\partial h}{\partial q}\right)^2 \Delta q^2 + \left(\frac{\partial h}{\partial t}\right)^2 \Delta t^2}}{h} = \frac{\sqrt{\left(\frac{1}{t}\right)^2 \Delta q^2 + \left(\frac{q}{t^2}\right)^2 \Delta t^2}}{h} \tag{4}$$

where h is the TCC of the contact interface, t is the temperature drop, q is the heat flux, Δt is the accident error of temperature drop and Δq is the random error of heat flux. According to figure 1, the accident error of temperature drop can be expressed as

$$\Delta t = \sqrt{\frac{T_{C1}^2 + T_{C2}^2}{2}}. \tag{5}$$

And according to Equation (2), the accident error of heat flux can be expressed as

$$\Delta q = \sqrt{\frac{q_1^2 + q_2^2 + q_3^2 + q_4^2 + q_5^2}{5}}. \tag{6}$$

The total error of TCC is written as

$$error(h) = \sqrt{error(h_s)^2 + error(h_a)^2} \tag{7}$$

According to Equations (3-7), the measurement errors of TCC at different temperature and pressure are shown in figure 4.

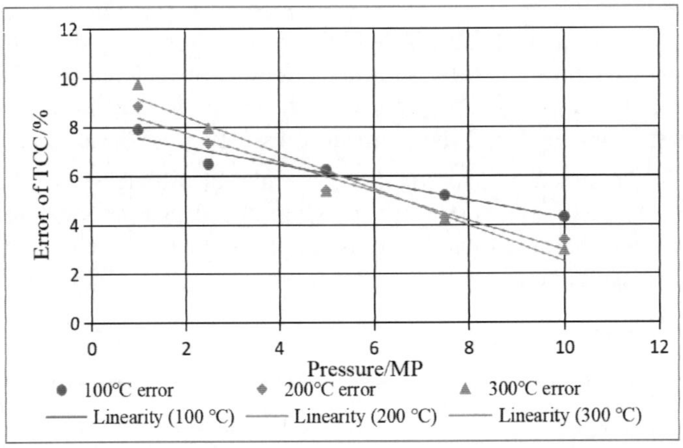

Figure 4. Measurement errors of SS316L/CuCrZr TCC versus pressure at different temperatures

As is shown in figure 4, all measurement errors are less than 10%. Meanwhile, as the contact pressure increases, the measurement error decreases. And the variation rate of measurement error increases with the increases of contact interface temperature.

4. Conclusions

✓ An accurate method was proposed to calculate the TCC between two contact interface. An experimental vacuum apparatus to test the TCC was introduced, and the TCC between pressed pairs SS316L/CuCrZr materials at different pressure and temperature was tested.

✓ It can be seen in figure 3 that the TCC is increasing as the temperature and pressure increase. And at a given temperature, the TCC appears as a linear relationship with the change trend of pressure. As shown in figure 4, the measurement error of TCC is decreasing as the pressure increases at a given temperature. And the variation rate of measurement error increases with the rise in contact interface temperature.

✓ In this paper, the effect of pressure and temperature on TCC was investigated quantitatively. The obtained results are consistent with the theoretical analysis of TCC and verify the feasibility of the designed TCC test method.

References

[1] Angirasa D 2015 *J. Thermophysics and Heat Transfer* **29** 412-5
[2] Maamir F, Guiatni M, Morsly Y and Kheddar A 2014 *Int. 22nd Mediterranean Conf. on Control and Automation (MED)* (Italy)
[3] Wu H, Gui N, Yang X, Tu J and Jiang S 2017 *ASME. J. Heat Transfer* **140**
[4] Madhusudana C V and Fletcher L S 1985 *J. AIAA* **24** 510-23
[5] Azuma K, Hatakeyama T and Nakagawa S 2015 *ICEP-IAAC Proc.*
[6] Kshirsagar B, Misra P and Jampana N 2005 *J. Heat Transfer* **127** 657-9

[7] Misra P 2010 *J. Heat Transfer* **132** 1-4
[8] Yuan C, Duan B and Li L 2015 *J. Heat Mass Transfer* **80** 398-406.
[9] Zheng J, Li Y Z, Chen P W, Yin G Y and Luo H H 2016 *Cryogenics* **80** 33-43
[10] Zhao Y S, Fang C, Cai L G and Liu Z F 2017 *J Process Mechanical Engineering* 1-12
[11] Yovanovich M M 2005 *IEEE Trans. Compon. Packag. Technol* **28** 182-206
[12] Leung M, Hsieh C K and Goseami D Y 1998 *Journal of Heat Transfer* **120** 51-7
[13] Xu R, Feng H, Zhao L and Xu L 2006 *J. Heat Transfer* **33** 811-8
[14] Zou M Q, Yu B M, Cai J C and Xu P 2008 *J. Heat Transfer* **130** 1-9
[15] Jackson R L, Sushil H B and Timothy P F 2008 *J. Heat Transfer* **130**
[16] Cacuci D 2003 *Sensitivity and Uncertainty Analysis: Theory*. Chapman & Hall/CRC Boca Raton FL

IOP Conf. Series: Materials Science and Engineering **474** (2019) 012031 doi:10.1088/1757-899X/474/1/012031

Structural design and simulation analysis of the carbon fiber-reinforced polymer main frame of a mapping camera system

Q L Wang*, W Z Wang, Q X Wang, H J Jiang, X H Zhou, and L Zhang

Beijing Institute of Space Mechanics & Electricity, China.

*E-mail: 13488723514@163.com

Abstract. The main frame is the primary load-bearing component of a mapping camera system. As the critical structure of the system, the main frame determines the mechanical properties of a camera system. The main frame design is hindered by its large size, complicated interface, as well as high-stability, high-frequency requirement, and lightweight requirements. During the launching process, the main frame should be able to withstand the mechanical action of the active section without affecting the performance of the camera system. This paper describes the design scheme of the main frame. The modal analysis, static analysis and frequency response analysis of the main frame are performed. The simulation analysis results prove that the natural frequency of the mapping camera system satisfies the design requirements and the main frame satisfies the required safety margin in the linear overload in X-, Y-, and Z-directions. Moreover, under condition of the frequency response in X-, Y-, and Z-directions, the main frame has excellent structural stiffness, which provides a proper mechanical environment for the camera system. The design scheme of the main frame is reasonable and feasible.

1. Introduction

The mapping camera is an integral part of the remote sensor for ground imaging observation. From the original film recoverable camera to the present transmission camera and from the frame type camera to the single-linear array, two-linear array, and three-linear CCD camera, the spatial resolution is reduced from tens of meters to 0.3m [1-2]. The two-linear array stereo mapping camera is the main load of the surveying satellite. It is composed of a front camera, a rear camera, and a main frame. The combination of the laser range finder and stereo mapping camera can obtain high accuracy of 3D positioning of ground targets [3]. The mapping camera system is installed on the satellite platform through the main frame [4].

During the launching process, the main frame should be able to withstand the mechanical action of the active section without affecting the performance of the camera system. In orbit operation, the main frame should maintain the stability of the angle of the mapping camera under the condition of a given temperature environment. The main frame should also avoid the influence of the structural deformation caused by the thermal environment factors on the geometric relationship between the mapping cameras [5].

To sum up, the main frame is the primary load-bearing component of a mapping camera system. As the key element of the system, the main frame controls the mechanical properties of the camera system. [6-7]. Combining the advantages of high specific stiffness, low density and low expansion coefficient of carbon fiber composites [8-10], the design scheme of a carbon fiber-reinforced polymer main frame is presented. To obtain a high stability composite material system main frame, the modal

IOP Conf. Series: Materials Science and Engineering **474** (2019) 012031 doi:10.1088/1757-899X/474/1/012031

analysis, static analysis, and frequency response analysis of the main frame is carried out.

2. Structural topology optimization design

2.1. Structural topology optimization theory

The purpose of structural topology optimization is to optimize certain properties of the structure or to reduce its weight by seeking the structural stiffness in the optimal distribution of the design space or to find the best way of structural force transmission path [11].

The material equivalent elastic modulus based on the variable density method can be expressed as in [12]:

$$E_p(x_i) = E_{\min} + f(x_i, p)(E_0 - E_{\min}) \tag{1}$$

Assume that

$$\Delta E = E_0 - E_{\min}$$

Then, Equation (1) can be expressed as

$$E_p(x_i) = E_{\min} + f(x_i, p)\Delta E \tag{2}$$

In Equation (2), ΔE is the equivalent elastic modulus after interpolation, E_0 is the elastic modulus of the solid material, $E_{\min}$ is the elastic modulus of the material of the hole portion, $f(x_i, p)$ is the density interpolation function. If the relative density of the *i*-th unit is 1, it means that the unit is a solid material, and if the relative density is 0, it is represented as space.

2.2. Structural topology optimization design

The topology optimization of the structure uses the Optistruct module of the Hypermesh software package. The constraint boundaries are as follows:

1) The main frame has an analog load, and the bottom plate mounting point is fixedly installed;
2) Optimization method: variable density method;
3) Optimization constraints: camera fundamental frequency exceeds 22Hz;
4) Optimization goal: The quality of the main frame is the smallest.

Through the topology optimization of the main frame, the main force path and topology configuration are obtained, as shown in figure 1.

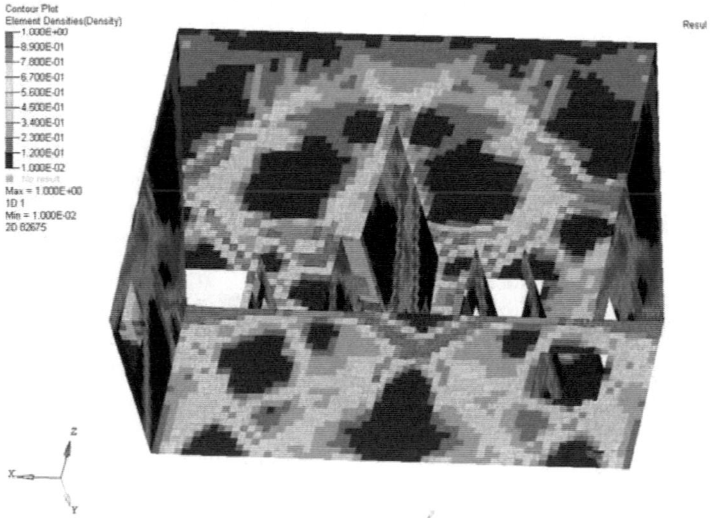

Figure 1. Optimization result for the main frame

According to the results of topology optimization, the main frame is lightweight, and the position of the ribs is reasonably arranged, which significantly reduces the structure weight. The main frame is shown in figure 2. The main frame adopts the box structure design idea, which is composed of carbon fiber composite honeycomb panels, and the inside of the honeycomb panel is provided with reinforcing ribs according to the force transmission path.

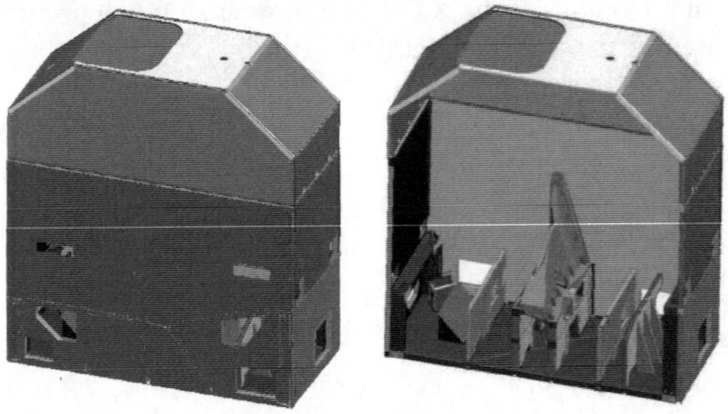

Figure 2. Schematic diagram of the main frame

3. Modal analysis

Modal analysis is used to derive the natural frequency and vibration mode of the mapping camera system. It is an integral part of the mapping camera system research process [13]. Through modal analysis, the inherent vibration mode of the surveying camera system can be obtained. The weak link of the stiffness of the main frame can also be determined, which can verify the rationality of its structural design and improve the design effectively.

3.1. Finite element modeling

According to the design requirements, it is necessary to calculate the fundamental frequency of the two stereo mapping cameras and the laser range finder under the condition of the mapping camera system. The modeling of the mapping camera system is completed in hyper mesh, in which the front and Rear cameras are modeled in detail according to the design results, and the laser range finder uses the quality point to simplify the modeling. The main frame of the system is modeled by shell element, solid element, and beam element. The finite element model of the surveying camera system is shown in figure 3.

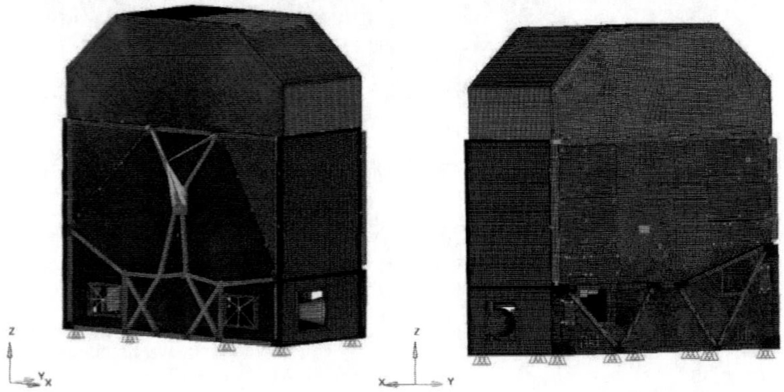

Figure 3. Finite element model of the mapping camera system

3.2. Modal analysis

According to the installation constraints of the surveying camera system, a total of 32 installation points in the eight groups of the main frame are fixed, and the modal analysis is completed. The third- and fourth-order modal analysis results are shown in figure 4.

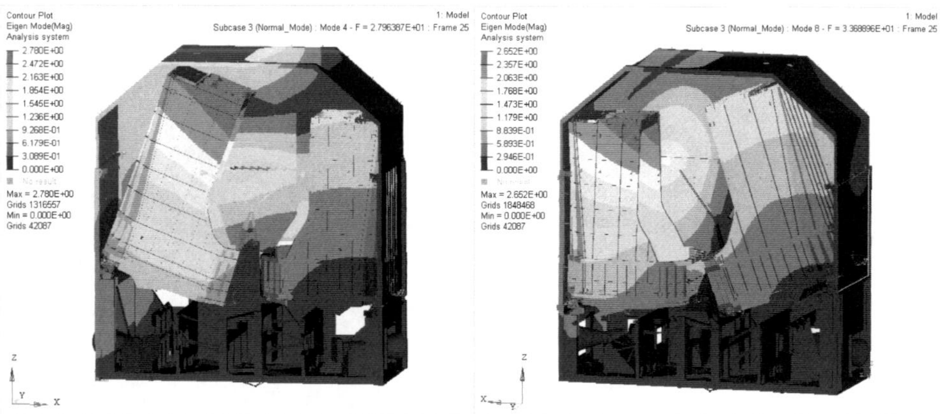

a) third-order mode front camera at 28Hz b) fourth-order mode rear camera at 33.5Hz

Figure 4. Model analysis of the mapping camera system

The result of the modal analysis shows that under the constraint conditions of the mapping camera system, the base frequencies of the front and rear cameras are 28 and 30.4Hz, respectively, which meets the requirements of no less than 25Hz in the assignment document.

4. Static Analysis

4.1. Finite element modeling

To accurately analyze the mechanical properties of all parts of the main frame, the front and rear cameras, and the laser range finder are calculated by the simplified model of the mass point. The finite element model of the mapping camera system is shown in figure 5.

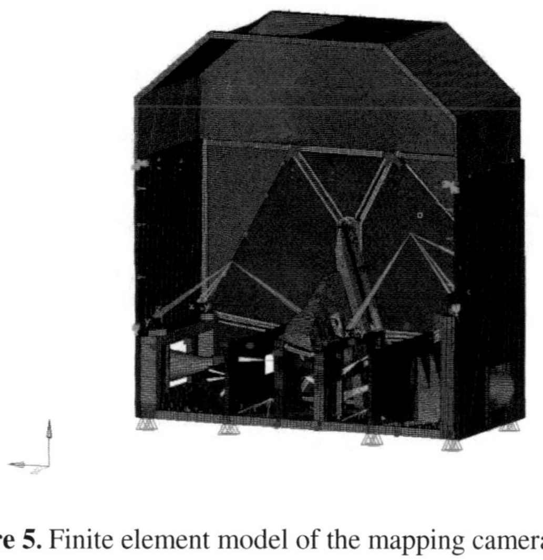

Figure 5. Finite element model of the mapping camera system.

4.2. Static analysis

4.2.1. X-direction 3g linear overload.

The calculation results of X-direction 3g linear overload are shown in figure 6. The maximum stress of the titanium alloy joint is 87.9MPa, which is located at the central support point of the front and rear cameras. The maximum value of the failure factor of the composite material is 0.07, which is located at the installation point of the main frame and satellite platform.

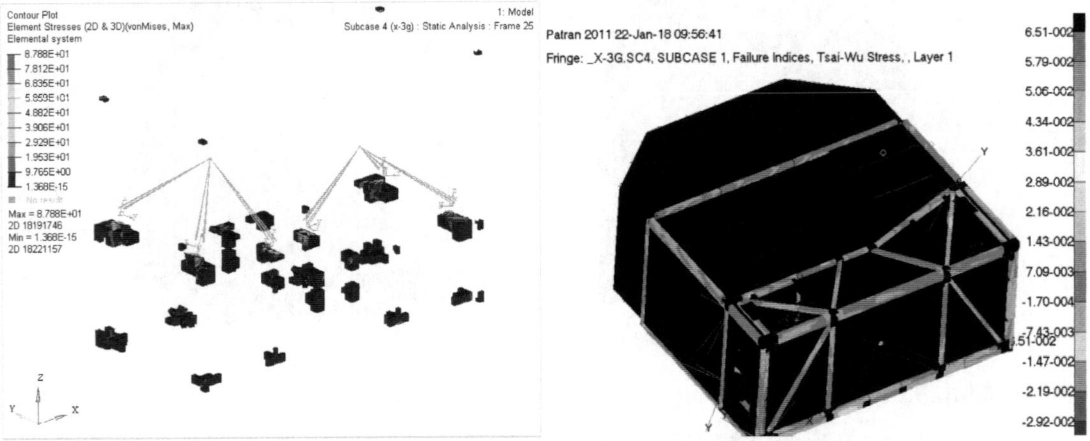

Figure 6. Cloud chart of stress distribution in titanium alloy joints and composite material failure indices by the Tsai-Wu criterion

4.2.2. 3g linear overload in Y-direction.

The calculation results of 3g linear overload in Y-direction are shown in figure 7. The maximum stress of the titanium alloy joint is 78.7MPa, which is located at the central joint of the +Y side honeycomb plate. The maximum value of the failure index of the composite material is 0.11, which is located at the installation point of the main frame and satellite platform.

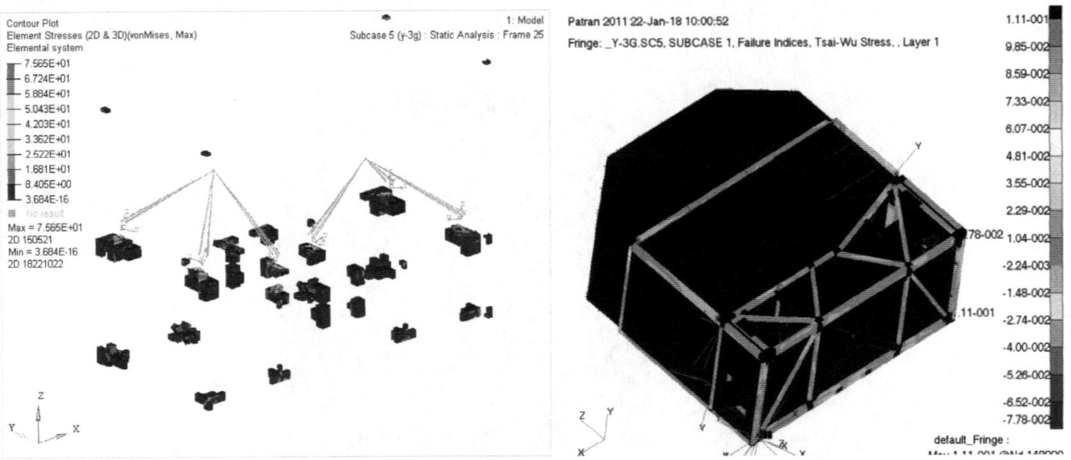

Figure 7. Cloud chart of stress distribution in titanium alloy joints and composite material indices by the Tsai-Wu criterion

4.2.3. 8g linear overload in Z-direction.

The calculation results of 8g linear overload in Z-direction are shown in figure 8. The maximum stress of the titanium alloy joint is 98MPa, which is located at the central joint of the +Y side honeycomb plate. The maximum value of the failure index of the composite material is 0.11, which is located at the installation point of the main frame and satellite platform.

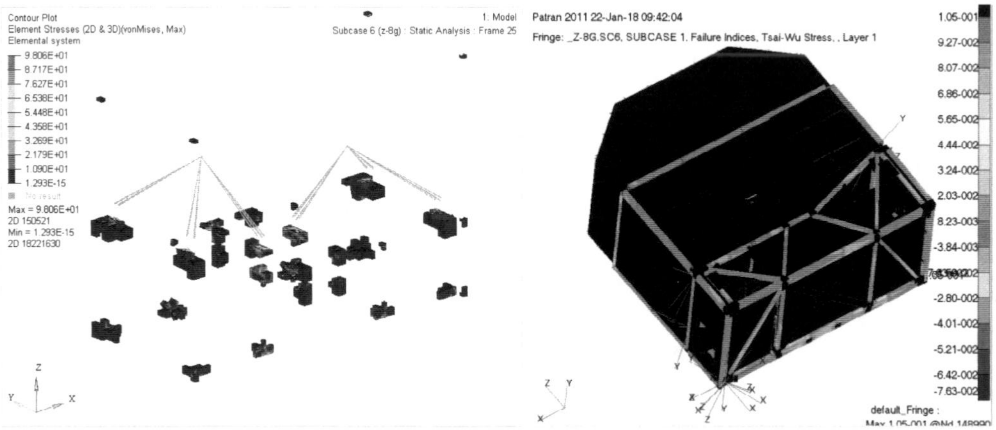

Figure 8. Cloud chart of stress distribution in titanium alloy joints and composite material failure indices by the Tsai-Wu criterion

5. Frequency response analysis

5.1. Analysis of X-direction sinusoidal response

The analysis result of X-direction sinusoidal response is shown in figure 9, where the following patterns can be observed.

1) The monitoring points with response exceeding 10g are mainly the centroid of the laser range finder and rear cameras, which are 16g@30Hz and 23g@40Hz, respectively.

2) The response of the centroid of the front and Rear cameras is amplified three times and eight times respectively. It shows that the structure of the main frame is very rigid, and it can provide an excellent mechanical environment for the mapping camera system.

3) The response of the centroid of the laser range finder is 17 times higher. Considering that the input point of laser range finder is smaller, it can meet the requirements of the mechanical environment.

4) The response of the centroid of the two stereo mapping cameras and the laser range finder is not much different from that of their input surface.

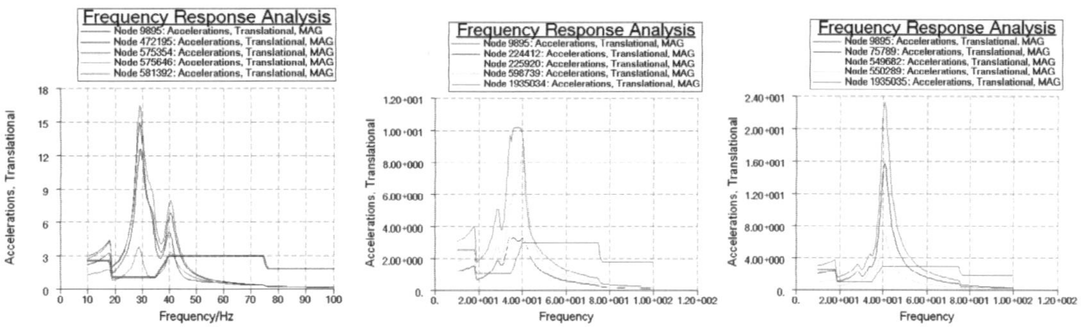

(a)Laser range finder (b) Front camera (c) Rear camera

Figure 9. X-direction sinusoidal response cures

5.2. Analysis of Y-direction sinusoidal response

The analysis result of Y-direction sinusoidal response is shown in figure 10, where the following patterns can be observed.

1) The monitoring points with larger response are mainly the centroid of the laser range finder and the front and Rear cameras.

2) The responses of the centroid of the front and rear cameras are 13g@28Hz and 24g@38Hz, respectively. Because the response of the camera's centroid is less than 30g, the stereo mapping cameras can meet the requirements of the resistance environment.

3) The response of the centroid of the laser range finder is 26g@20Hz. The peak response of the laser range finder is about 16 times higher than its input surface. Considering that the input point of laser range finder is smaller, the laser range finder has better mechanical conditions.

4) The response of the centroid of the two stereo mapping cameras and the laser range finder is not much different from that of their input surface.

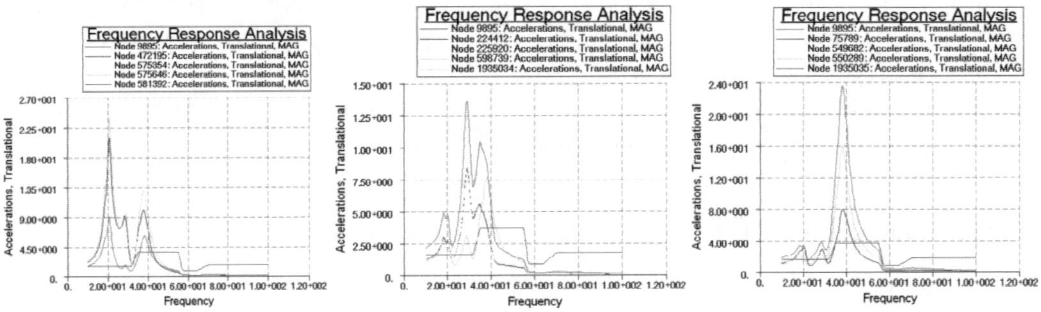

(a)Laser range finder (b) Front camera (c) Rear camera

Figure 10. Y-direction sinusoidal response cures

5.3. Analysis of Z-direction sinusoidal response

The analysis result of Z-direction sinusoidal response is shown in figure 11, where the following patterns can be observed.

1) The monitoring points with a larger response is mainly the centroid of the rear camera, which is 20g@40Hz.

2) The Z-direction mechanical environment test is easy to pass for the front and rear cameras and the laser range finder. Although the input excitation is large, the relevant measuring points have weak response.

3) The response of the centroid of the two stereo mapping cameras and the laser range finder is not much different from that of their input surface.

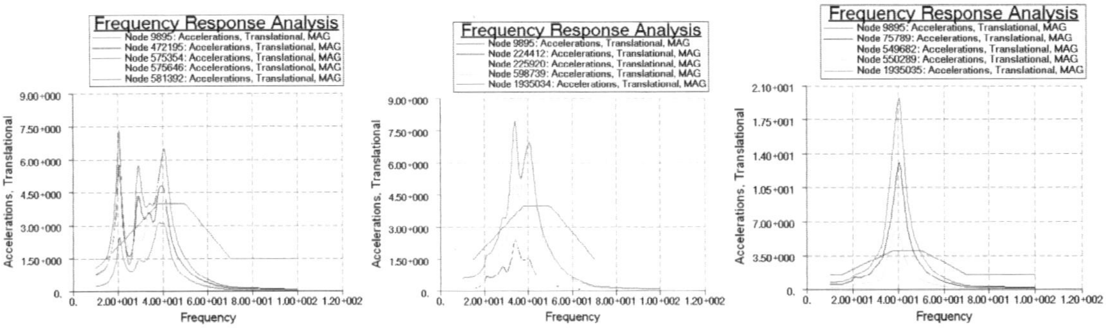

(a)Laser range finder (b) Front camera (c) Rear camera

Figure 11. Z-direction sinusoidal response cures

Thus, under the sinusoidal test conditions of X-,Y-, and Z-directions, the maximum peak response of the main optical instruments does not exceed 26g. This implies that the main frame of the system has an excellent structural stiffness. During the launching process, the main frame can provide an excellent mechanical environment for camera system without affecting its performance.

6. Conclusions

In this paper, the carbon fiber composite main frame is designed, which combines the advantages of high specific strength, low density, low expansion coefficient, and modified modulus of carbon fiber composites. The carbon fiber composite honeycomb panels, which constitute the main frame, consist of carbon fiber composite panels, aluminum honeycomb cores, and composite frames. The composite frame is made of composite rods and stiffeners connected by bonding titanium alloy joints. The simulation results show that the strength and stiffness of the main frame meet the requirements, and the design scheme is reasonable and feasible. The carbon fiber composite main frame has been applied to a particular type of remote sensor, which can provide new ideas for carbon fiber composite structure design of high-precision space optical instruments in the future.

References

[1] Sun G 2007 *J. Geomatic Science and Engineering*. **27** 46-51
[2] Miao J Y, Zhang L P and Tian T Y 2011 *J. Chinese Journal of Scientific Instrument*. **32** 2183-9
[3] Deng Y T, Li X and Yan F J 2015 *J. Spacecraft Recovery &Remote Sensing*. **36** 86-91
[4] Liu Y, Xiao R Q and Han D D 2015 *J.Opt. Precision Eng*. **23** 3083-9
[5] Chen SH P 2003 *Design and Test of Space Camera (Munich)* (Beijing: Yu-hang Publishing Company) p 125
[6] Zhang J Q, Yan CH X 2009 *J. Chinese Journal of Optics and Applied Optics*. **2** 79-84
[7] Li W, Liu H W 2008 *J. Opt. Precision Eng*. **16** 2173-9
[8] Lin Z W, Liu Y Q, Liang Y 2007 *J. Opt.Precision Eng*. **18** 1181-5
[9] Li W, Guo Q F 2011 *J. Chinese Optics*. **4** 201-12
[10] Zhao Y, Li L 2011 *J. Spacecraft Recovery &Remote Sensing*. **32** 68-74
[11] Jia ZH X, Jin G and Jia J Q 2011 *J. Chin. J. Space Sci*. **31** 395-400
[12] Guan Y J, Xin H W, Zhao G J 2008 *J. Chinese Journal of Optics and Applied Optics*. **15** 1157-63
[13] Wu J and Mao J J 2012 2*J. Spacecraft Recovery & Remote Sensing*. **33** 34-41

Experimental study on repairing of damaged cast iron cylinder heads by 3D printing arc welding system

H Y Wang[1*], C Y Li[1], Y M Ding[1], J Li[1], S C Chen[1]
[1]Guangdong Polytechnic Normal University, Guangzhou, Guangdong, 510665, China

*E-mail: why_gd@163.com

Abstract. To realize repairing of damaged cast iron cylinder heads by 3D printing arc welding system mainly composed of an industrial robot and an arc welding machine, the remanufacturing process flow of 3D printing arc welding system was introduced, and a single straight wall stacking experiment was designed. The effects of process parameters on the appearance quality and mechanical properties of welding pieces were studied. The forming laws of welding piece were obtained. The main conclusions include: The higher the preheating temperature, and interlayer temperature, the more obvious the effects of process parameters on the hardness of welding pieces, and the effects of preheating temperature are more evident than that of interlayer temperature. With the increase of layer number, the hardness of welding pieces decreases gradually. When welding piece is lower than eight layers, the rise of width and height of the welding piece is approximately proportional to the increase of layers. Those conclusions of this paper provide general guidance and technical support for the repairing of cast iron cylinder heads and welding of other cast iron parts.

1. Introduction

Due to advantages of low price, durability, and high strength, cast iron cylinder heads are still widely used in high-powered engines and turbocharged engines [1]. During handling, cast iron cylinder heads will be damaged for cracks, porosities, wear, and other problems. At present, the disposal method of rusted cast iron cylinder heads has been gradually changed from recycling to remanufacturing to reduce energy and resource waste and environmental pollution [2]. 3D printing arc welding system mainly composed of industrial robot and arc welding machine has the characteristics of high efficiency, good flexibility, energy saving and so on [3]. The application of the system to repair damaged cast iron cylinder heads can realize the high efficiency and high quality remanufacture of cylinder heads.

For cast iron having poor weldability and producing cracks efficiently in welding [4, 5], it is necessary to reasonably select welding specifications and process parameters to repair damaged cast iron cylinder heads by 3D printing arc welding system. Besides, when slicing treatment and path planning are carried out, it is necessary to know height and width of welding pieces under selected process parameters to determine rising height of welding gun and welded channels per layer [6]. Those process parameters need to be obtained by experiment.

Therefore, a single straight wall stacking experiment is designed, the influence of welding process parameters on the appearance quality and mechanical properties of welding pieces is studied, and the welding forming laws are obtained, which provide reference process parameters for 3D printing arc welding system to repair damaged cast iron cylinder heads. The research results in this paper can guide actual repairing of damaged cast iron cylinder heads by arc welding, and also provide technical guidance and technical support for welding of other cast iron parts.

2. Remanufacturing process flow of 3D printing arc welding system

The remanufacturing of 3D printing arc welding system is a process of repairing damaged metal parts by melting and depositing with welding robots. The remanufacturing process flow is shown in figure 1. Firstly, 3D digital model of welding piece is got through 3D digital models of the intact metal part and damaged metal part obtained by 3D scanning and modeling softs. Then, the 3D digital model of welding piece is sliced, and the geometric shape of each slice is formed along a particular path (path planning). Finally, the welding piece is welded, and the damaged metal part is repaired.

In the process of slicing, path planning and welding of the 3D model of welding piece, it is necessary to set process parameters including forming parameters (such as layer numbers, welding direction and so on) and the welding parameters (such as wire feeding speed, welding speed, damaged piece's height, etc.). Those forming parameters and welding parameters affect appearance quality, mechanical properties of repairing part and bonding strength between repaired part and substrate. Therefore, the forming welding experiment should be carried out before repairing to determine the forming laws and obtain the optimal repairing process parameters.

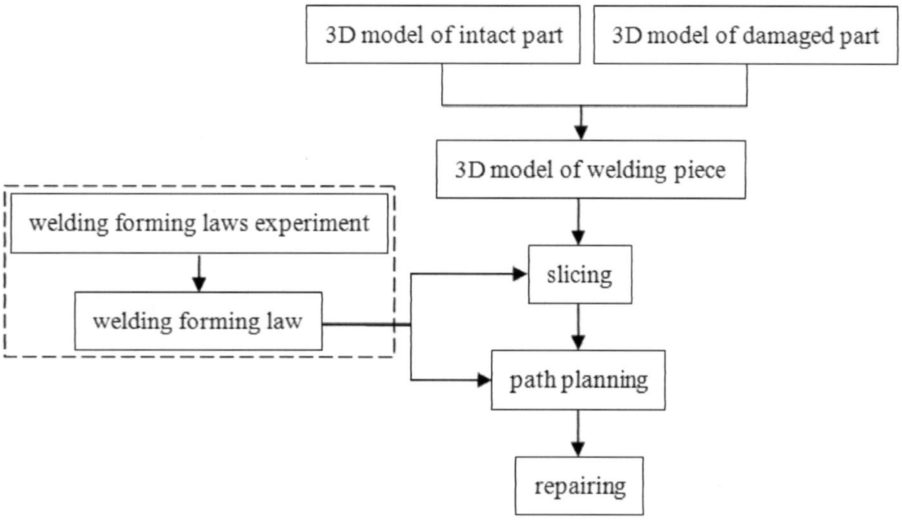

Figure 1. Remanufacturing process flow of 3D printing arc welding system

3. The experiment on welding forming laws

3.1. Experimental equipment and materials

Experimental equipment: 3D printing arc welding system selected by the experiment is mainly composed of MotoMan-Up20 industrial arc welding robot produced by Yaskawa Company of Japan and MotoWeld-S350 automatic arc welding machine. The wire feeding speed of the welding machine is automatically matched according to the current, and they are proportional to each other.

Table 1. The chemical composition of welding wire Z308 (mass percentage, %)

Chemical composition	C	Si	Mn	S	P
Z308	≤2.00	≤2.50	≤1.00	≤0.03	≤1.00

Experimental materials: welding wire Z308 with a diameter of 1.2 mm (see table 1 for chemical composition) is selected as the experiment material. The welding wire Z308 is a pure nickel-based cast iron welding wire which has strong crack resistance and good machinability and can be used for surface repairing of thin-walled workpiece [4].HT300 is selected as substrate.

Hardness testing of welding pieces is carried out by HVS-1000 digital microhardness tester.

3.2. Experimental scheme

To repair damaged cylinder heads, it is necessary to obtain the influence of process parameters on the geometric size and mechanical properties of welding piece. Therefore, a single straight wall stacking experiment is designed. In the experiment, welding voltage, welding current, welding speed, preheating temperature of substrate and interlayer temperature are selected as process parameters.

In the experiment, a single straight wall welding piece with a length of about 100 mm and eight layers are welded with each group of process parameters. The welding adopts one-way welding path, and the arc is extinguished after each layer is finished. The interlayer temperature is controlled within the set range. After a layer is completed, welding gun is raised, and the raising height kept consistent with that of the previous layer, width and height of welding piece are measured at the front, middle and rear positions of the middle part of the welding piece. The average value of the three values is recorded as welding width and height in this layer. After the 8th layer is finished, the welding piece is divided into three layers with the same interval along the vertical direction. The hardness of each layer is measured at the front, middle and rear positions. The average value is used as the hardness of this layer. The average value of the hardness of the three layers is taken as the hardness of the welding piece.

In order to prevent welding pieces from whitening, inclusion porosity, refractory oxides, cracks, or deformation, the following measures are adopted in the experiment: polishing substrate to eliminating oxide skin, removing impurities such as surface oil and dirt, preheating substrate before welding, increasing welding speed when welding current increase in order to control heat input and tapping welding pieces with hammer to reduce internal stress immediately after welding. Considering that repairing methods should be convenient and fast, preheating temperature of the substrate should be as low as possible and should be conveniently realized. In the experiment, the preheating temperature is controlled within the range of (60-120) °C.

Before the straight wall stacking experiment, the range determination experiment of welding parameters is carried out by observing the appearance quality of welding pieces. The experiment results show that when welding voltage is 50%, the welding current is within (230-270) A and the corresponding welding speed is within (40-60) (cm/min), the appearance quality of the welding pieces is the best, and the welding pieces are in good contact with the substrate.

4. Experimental Results and Analysis

4.1. Hardness

Table 2. The hardness of welding pieces

Group	Preheating temperature (°C)	Interlayer temperature (°C)	Welding current (A)	Welding speed (cm/min)	Hardness (HV)
1	100-120	60-70	230	40	237.7
			250	50	374.6
			270	60	310.5
2	60-70	60-70	230	40	334.9
			250	50	287.7
			270	60	359.2
3	60-70	30-40	230	40	324.1
			250	50	294.8
			270	60	334.7

Table 2 shows the hardness of some welding pieces which are higher than that of the substrate of 220HV and also higher than or within the range of (179-235) HB required by the design of cast iron cylinder heads [7].

The following conclusions are drawn from table 2 that when interlayer temperature, welding voltage, welding current and welding speed are the same and preheating temperature differences are about (30-60) °C, the minimum hardness difference is 48.7 HV, which indicates that the variation of preheating temperature has effects on hardness at the interval (30-60) °C.

From table 2, it is also concluded that when preheating temperature, welding voltage, welding current, welding speed are the same and interlayer temperature differences are about (20-40) °C, the maximum hardness difference is 24.5 HV, which indicates that the variation of interlayer temperature has effects on hardness at the interval (30-60) °C. However, the results are not apparent.

Table 3. Hardness variation amplitude of welding pieces

Groups	Preheating temperature (°C)	Interlayer temperature (°C)	Maximum hardness (HV)	Minimum hardness (HV)	Variation amplitude of hardness (HV)
1	100-120	60-70	374.6	237.7	136.9
2	60-70	60-70	359.2	287.7	71.5
3	60-70	30-40	334.7	324.1	39.9

The maximum and minimum hardness of each group in table 2 are subtracted to obtain the variation amplitude of hardness under this group of process parameters (where the welding current varies within the range of (230-270) A, and the welding speed varies within the range of (40-60)(cm/min), as shown in table 3. From table 3, it is concluded that the higher the preheating temperature and interlayer, the greater the hardness variation amplitude is. The results indicate that the effects of process parameters on hardness are more apparent when preheating temperature and interlayer temperature are higher.

Besides, it is also concluded that hardness tends to decrease with increasing of layers, which is the same as that in references [2, 8].

4.2. Welding Forming Laws

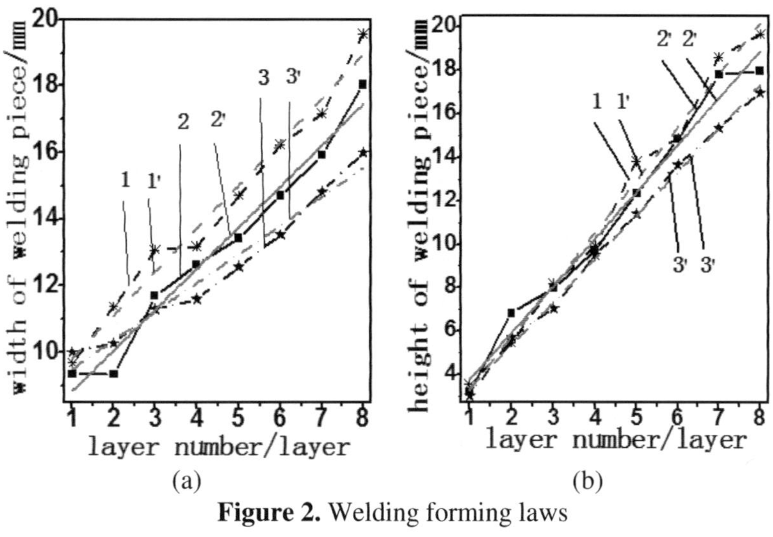

Figure 2. Welding forming laws

Figure2 shows the forming laws of the welding pieces when the preheating temperature is (100-120)°C, and the interlayer temperature is (60-70) °C. In figure 2, lines 1,2 and 3 are the curves of relationship between weldingpiece width, welding piece height and layer number when welding currents are 230 A,

250 A, and 270 A, and corresponding welding speeds are 40 (cm/min), 50 (cm/min), and 60 (cm/min), respectively. According to figure 2, the welding piece width and welding height increase approximately proportionally with increasing of layer number

Lines 1', 2', and 3' in figure 2 are linearly fitted curves corresponding to lines 1, 2, and 3, respectively. The equations for lines 1', 2', and 3' are as follows.

$$w_1 = 1.30n + 8.51,\ w_2 = 1.23n + 7.58,\ w_3 = 0.86n + 8.63\ . \quad (1)$$

$$h_1 = 2.42n + 0.87,\ h_2 = 2.16n + 1.60,\ h_3 = 2.00n + 1.35\ . \quad (2)$$

Here, w_i and h_i are the width and height of the welding piece i, respectively, while n is the layer number.

It was also found that after the 6th layer is finished, the new coating did not affect the width of the first layer of the welding piece. When the welding piece exceeds ten layers, the change in the width of the welding piece becomes smaller and tends to be fixed.

5. Conclusions

To successfully apply 3D printing arc welding system to repair damaged cast iron cylinder heads, a single straight wall stacking experiment is designed. The effects of process parameters on appearance quality and mechanical properties of welding pieces are studied. The welding forming laws of welding pieces are established and analyzed. The following conclusions are obtained.

- ✓ Preheating temperature of substrate and interlayer temperature have effects on the hardness of welding pieces, and the effects of the preheating temperature are more visible. The higher the preheating temperature and the interlayer temperature, the more obvious the effects on the hardness of welding pieces.
- ✓ Even though the hardness of welding pieces at different welding process parametersis different, the hardness variation law along the normal direction of the bonding interface is consistent, and the hardness decreases gradually with the increase of the number of layers.
- ✓ When the welding piece has less than eight layers, the welding piece width and height increase nearly proportionally with the number of layers. After that, the welding piece width tends to be fixed.

Acknowledgments

Sponsored by the Science and Technology Planning Project of Guangdong Province (2015B090920006), Science and Technology Planning Project of Guangdong Province (201300000104) and Natural Science Foundation of Guangdong Province (2015A030313662)

References

[1] Wang C G*et al.* 2012*Foundry***61**136-42
[2] Xu Q Z2012*Experimental Study on Application of Cold Welding Technology in Engine Remanufacturing*,Shandong University.
[3] Yang X Z *et al.* 2016*Journal of Hubei Polytechnic University***32**1-4, 9
[4] Wang Y H2016*Lanzhou University of Technology*
[5] Dong S Y*et al.* 2013*Journal of Academy of Armored Force Engineering***27**90-3
[6] Xiong J2014*Harbin Institute of Technology*
[7] Zhao S C2005*China Foundry Machinery & Technology***5**7-10
[8] Peng S D*et al.* 2006*Electric Welding Machine***36**33-5

Calculation and analysis of nonlinear dynamic model based on the characteristic line method

J Li[1], B A Jiang[1,*], Q L Ning[1] and S Xue[1]

[1]Department of Arms Engineering, Army Academy of Artillery and Air Defence Force, Hefei, Anhui, 230031, China

*E-mail: 2249861737@qq.com

Abstract. To solve the problem that the impact dynamics model is nonlinear and cannot be resolved the analytical solution, the differential equations of the characteristic line compatible relation are constructed, the initial value conditions and the boundary value conditions are determined. The dynamic state of the whole domain is gradually recurred by selecting the appropriate time step and the space step. The theoretical results are consistent with the experimental results, indicating that the method of solving the model is feasible. This method provides an effective theoretical analysis and calculation method for the study of dynamic characteristics of structures under impact.

1. Introduction

The solution of the nonlinear dynamic model is the basis of studying the dynamic characteristics of the structure under the impact environment. Because the impact dynamics model is nonlinear, the analytical solution cannot be found for the nonlinear problems. To solve this problem, the theory and application of nonlinear viscoelastic wave propagation are discussed in the reference [1]. The numerical analysis method is used to simulate the constitutive model constructed in [2-3]. Two-stage finite element modeling and analysis techniques are used in the reference [4], and this method is used to analyze the overall stability and local residual strength of structural frame systems and components. The impact analytical model based on Hertzian elastic contact law is proposed in reference [5], and the analytical method is given. The one-dimensional dynamic model is solved by a precise integration method in reference [6], and the structural dynamic response of missile-borne devices is obtained under a high overload environment. In this paper, the differential equations of the characteristic line compatible relation are constructed, the initial value conditions and the boundary value conditions are determined. The dynamic state of the whole domain is gradually recurred by selecting the appropriate time step and the space step. This method provides an effective theoretical analysis and calculation method for the study of dynamic characteristics of structures under impact.

2. The establishment of the compatible equation based on the characteristic method

Because of the coupling of stress wave and strain rate, the mechanical behavior of viscoelastic material is nonlinear. The propagation of force wave in one-dimensional nonlinear viscoelastic material can be determined by the following three equations, and the dynamic process of the whole viscoelastic material model can be obtained by the following equations [7-10].

(1) Continuous equation

$$\frac{\partial v}{\partial x}=\frac{\partial \varepsilon}{\partial t} \quad (1)$$

(2) Motion equation

$$\rho_0\frac{\partial v}{\partial t}=\frac{\partial \sigma}{\partial x} \quad (2)$$

(3) Constitutive equation

$$\frac{\partial \sigma}{\partial t}+\frac{\sigma}{\theta_2}=(\frac{d\sigma_e}{d\varepsilon}+E_1+E_2)\frac{\partial \varepsilon}{\partial t}+\frac{\sigma_e+E_1\varepsilon}{\theta_2} \quad (3)$$

In equations (1), (2) and (3), $\theta_2=\eta_2/E_2$, $\sigma_e=E_0\varepsilon+\alpha\varepsilon^2+\beta\varepsilon^3$, E_1, E_2, E_0, α, β and θ_2 are the material's property parameter, E_1, E_2 and E_0 are the elastic modulus, α and β are the nonlinear correlation, θ_2 is the relaxation time, η_2 is viscous parameter, ρ_0 is the material density, v is the particle velocity, σ is the stress , ε is and the strain.

The dynamic process of the whole viscoelastic material model can be described by the above three equations (1), (2) and (3), and the above three equations can be solved by using the characteristic line method.

The concrete practice is as follows:

The above three equations are multiplied by A_0, B_0, C_0, and then added together:

$$[A_0+C_0(\frac{d\sigma_e}{d\varepsilon}+E_1+E_2)]\frac{\partial \varepsilon}{\partial t}+(B_0\rho_0\frac{\partial}{\partial t}-A_0\frac{\partial}{\partial x})v+(C_0\frac{\partial}{\partial t}-B_0\frac{\partial}{\partial x})\sigma+f(\sigma,\varepsilon)=0 \quad (4)$$

In formula (4), $f(\sigma,\varepsilon)$ is a one-item function of σ, ε.

According to the compatibility of the characteristic lines, it can be got as follows:

$$\frac{dx}{dt}=\frac{0}{A_0+C_0(\frac{d\sigma_e}{d\varepsilon}+E_1+E_2)}=\frac{-A_0}{B_0\rho_0}=\frac{-B_0}{C_0} \quad (5)$$

The original partial differential equations can be transformed into ordinary differential equations consisting of the compatible relations of three families of characteristic lines by formula (5).

$$dx = \pm C_V dt \tag{6}$$

$$dv = \pm \frac{1}{\rho_0 C_V} d\sigma \pm \frac{\sigma - \sigma_e - E_1\varepsilon}{\rho_0 C_V \theta_2} dt \tag{7}$$

When $A_0 = B_0 = 0$, and $C_0 \neq 0$, then $dx = 0$, then the corresponding third family characteristic lines are:

$$d\varepsilon - \frac{d\sigma}{\frac{d\sigma_e}{d\varepsilon} + E_1 + E_2} - \frac{\sigma - (\sigma_e + E_1\varepsilon)}{\frac{d\sigma_e}{d\varepsilon} + E_1 + E_2} \frac{dt}{\theta_2} = 0 \tag{8}$$

In formula (8), $C_V = \sqrt{\dfrac{\frac{d\sigma_e}{d\varepsilon} + E_1 + E_2}{\rho_0}} = \sqrt{\dfrac{E_0 + E_1 + E_2 + 2\alpha\varepsilon + 3\beta\varepsilon^2}{\rho_0}}$.

From the above, as long as knowing the corresponding material parameters, σ, ε, and v can be solved by using equations(6),(7), and(8) together. Among them, the strength of the material is characterized by σ, the deformation degree of the material is characterized by ε, the particle motion of the material is characterized by v, the velocity of the particle and the compatibility condition of the strain are the embodiment of the conservation of mass. The relation of the stress and velocity is the embodiment of the conservation of momentum. The relationship between the strain and the stress is the embodiment of the constitutive model of the material, and the three parameters are known. The whole impact problem can be solved.

3. Numerical solution of nonlinear dynamic equations

3.1. Characteristic line method for solving nonlinear dynamic equations

Compared with the compatibility relation of linear elastomers, the propagation term of the nonlinear viscoelastic body contains terms related to strain rate and cannot be decoupled. Therefore, equations (6), (7) and (8) do not have analytic solutions at the same time. The only way to solve equations is by numerical means. In this paper, the above equations are solved by the characteristic line method, and the specific solution is shown in Figure 1.

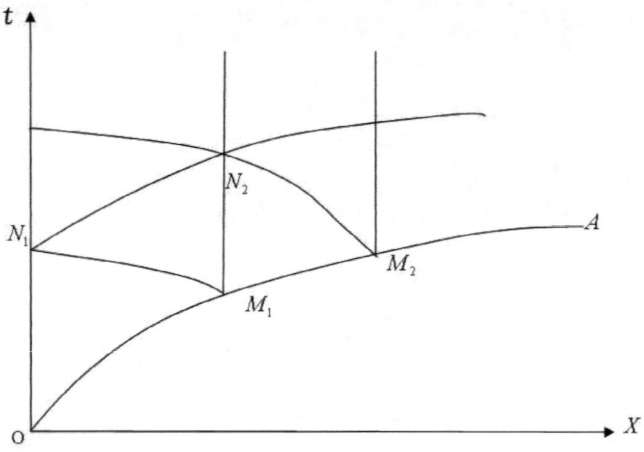

Figure 1. Three characteristic lines of nonlinear dynamic equations.

Figure 1 shows the propagation of stress waves on viscoelastic materials, which can be divided into three regions: the first region is the AOX region, which represents the undisturbed region and the data are all zero; the second region is on the t-axis, which represents the external disturbance of the viscoelastic material model and belongs to the boundary condition; the third one is in the AOX region. It is the AOt region, with the OA line as the firm discontinuous boundary, indicating the disturbance situation in front of and behind the wavefront when the first stress wave propagates, the disturbance area above OA, and the undisturbed area below OA. Thus, in the AOt area, it satisfies:

$$\begin{cases} v(N_2)-v(N_1)=\dfrac{1}{\rho_0 C_V}[\sigma(N_2)-\sigma(N_1)]-[\dfrac{\sigma(N_1)-(\sigma_e+E_1\varepsilon)}{E_0+E_1+E_2+2\alpha\varepsilon+3\beta\varepsilon^2}]\dfrac{[X(N_2)-X(N_1)]}{\theta_2} \\ v(N_2)-v(M_2)=-\dfrac{1}{\rho_0 C_V}[\sigma(N_2)-\sigma(M_2)]+[\dfrac{\sigma(M_2)-(\sigma_e+E_1\varepsilon)}{E_0+E_1+E_2+2\alpha\varepsilon+3\beta\varepsilon^2}]\dfrac{[X(N_2)-X(M_2)]}{\theta_2} \\ \varepsilon(N_2)-\varepsilon(M_1)=\dfrac{\sigma(N_2)-\sigma(M_1)}{\dfrac{d\sigma_e}{d\varepsilon}+E_1+E_2}-\dfrac{\sigma(M_1)-(\sigma_e+E_1\varepsilon)}{\dfrac{d\sigma_e}{d\varepsilon}+E_1+E_2}\dfrac{[t(N_2)-t(M_1)]}{\theta_2} \end{cases} \tag{9}$$

Thus, for any point N_2 in the AOt plane, the above three compatible relations can be obtained simultaneously. However, it is noticed that all the three compatible equations contain functions of ε. ε Changes with time and displacement, and cannot be solved directly. A difference equation which satisfies the CFL condition, i.e., the necessary condition for the convergence of the difference scheme, is constructed. The key idea is to characterize the curve. The solution is simplified to a linear elastic model by linearizing it into several small linear segments, and then the solution of the whole perturbation region is obtained by recursion along three characteristic lines in the entire perturbation region.

The specific measures are as follows:

To obtain the numerical solution of each point in the AOt domain, the points in the computational domain can be divided into: (1) discontinuous points (points on OA); boundary points (points on Ot);

and (3) interior points. The above analysis shows that the breakpoints and boundary points are unique and can be determined.

For any interior point, it is possible to set the internal pointer to A, as shown in figure 2. There must be three classes of characteristic lines passing through this point and all satisfy the compatible equations (6), (7) and (8). As long as the appropriate time and space steps are intercepted, the corresponding difference equations can be constructed along the three characteristic lines. In the area adjacent to the point A, three points B, C, and D having a time step of Δt and a space step of Δx are obtained. Then, the original characteristic line of a curve can be approximated as a straight line, and the complex nonlinear dynamic problem can be simplified as a recursive calculation of several linear problems.

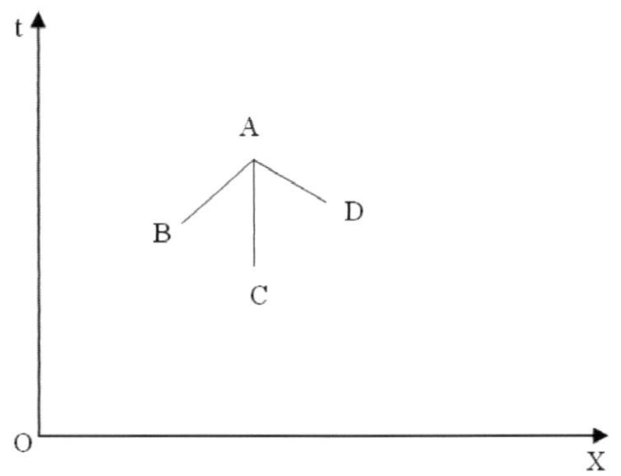

Figure 2. Schematic diagram of internal point calculation.

As long as the initial shock disturbance and loading conditions are known, the numerical solutions of each point in the whole disturbance region can be derived successively by the corresponding differential equations.

Thus, point A satisfies the compatibility equation along three characteristic lines:

Along AB

$$v_A - v_B = \frac{1}{\rho_0 C_B}[\sigma_A - \sigma_B] + F(B)\Delta x \tag{10}$$

Along AD

$$v_A - v_D = -\frac{1}{\rho_0 C_D}[\sigma_A - \sigma_D] + G(D)\Delta x \tag{11}$$

Along AC

$$\varepsilon_A - \varepsilon_C = \frac{\sigma_A - \sigma_C}{E_C} - H(C)\Delta t \tag{12}$$

Because the data of point B, C, and D are all obtained in the previous step, the solution of point A can be calculated by the solution of point B, C and D. By analogy, the solution of point A can be recursively extended to the whole disturbance region until the arrival of the boundary stops, and then the numerical solution of the entire AOt region can be obtained.

3.2. The establishment of difference scheme

When choosing the difference scheme, because the characteristic line is a curve, then select the appropriate space step Δx, and then let $\Delta t = \Delta x / C_V$, the specific feature line family as shown in figure 3. In this way, the AOt in the whole perturbation region is discretized into a lattice, and the solution corresponding to each point is the numerical solution of the entire plane. In particular, it should be pointed out that when solving along the $dx = 0$ characteristic line, the characteristic line $dx = 0$ is a straight line. Then, the nonlinear dynamic equation can be addressed directly by the OA characteristic line, which is consistent with the linear elastic model.

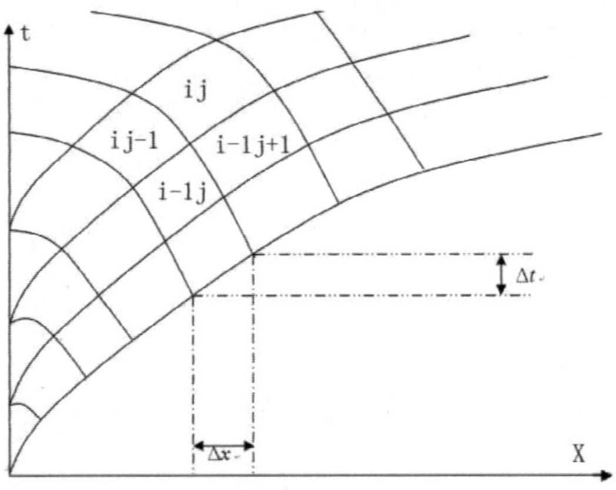

Figure 3. Characteristic lines in calculation area.

According to the physical meaning of the characteristic line, a certain space step and time step can be selected, and the computational domain can be divided into a small area with the characteristic line, as shown in Figure 1. In these regions, all stress states are equal, and the stress states in the adjacent areas satisfy the compatible relationship of stress waves. In this way, with known loading and boundary conditions, as long as the material parameters are known, the stress state of each small area can be calculated recursively from the differential format of the compatible relation. For ij the area, its stress state is ultimately obtained by the logical relation of $ij-1$, $i-1j+1$ and $i-1j$ regions, and the specific equations are as follows:

$$v_{ij} - v_{ij-1} = \frac{1}{\rho C_V}(\sigma_{ij} - \sigma_{ij-1}) - \left(\frac{\sigma_{ij-1} - \sigma_e}{E_0 + E_1 + 2\alpha\varepsilon_{ij-1} + 3\beta\varepsilon_{ij-1}{}^2}\right)\frac{(X_{ij} - X_{ij-1})}{\theta_1} \quad (13)$$

$$v_{ij} - v_{i-1j+1} = -\frac{1}{\rho C_V}(\sigma_{ij} - \sigma_{i-1j+1}) + \left(\frac{\sigma_{ij} - \sigma_e}{E_0 + E_1 + 2\alpha\varepsilon_{i-1j+1} + 3\beta\varepsilon_{i-1j+1}{}^2}\right)\frac{(X_{ij} - X_{i-1j+1})}{\theta_1} \quad (14)$$

$$\varepsilon_{ij}-\varepsilon_{i-1j}=\frac{\sigma_{ij}-\sigma_{i-1j}}{\frac{d\sigma_e}{d\varepsilon}+E_1}+\frac{\sigma_{i-1j}-\sigma_e}{\frac{d\sigma_e}{d\varepsilon}+E_1}\cdot\frac{t_{ij}-t_{i-1j}}{\theta_1} \tag{15}$$

Equations (13), (14) and (15) are the differential equations of nonlinear viscoelastic materials, which are the display formats. All the stress states in the computational domain can be solved, including stress, strain and particle velocity.

4. Simulation calculation and result analysis

Based on the above discrete solution, the stress state of the subregion of the target depends on the stress state of its circumferential region, and the stress state of any subregion will only affect the four adjacent subregions. In such a mechanical property, the continuous equation is discretized by building differential equations (13), (14) and (15). And the numerical calculation program based on Lagrange coordinates is compiled with FORTRAN90. The initial value of the stress state parameters is 0, the boundary condition is set as the free end boundary, and the appropriate time and space step are selected, and stress σ, strain ε, and particle velocity v can be obtained by the recursive calculation.

Combined with the actual situation and material size, the length of the calculation area is set to 77.5mm, the computation time is 50 μs, the space step length is 0.5mm, and the time step is 0.5 μs. The waveforms of the input stress are simplified to rectangular waves and are solved in equations (13), (14) and (15). The comparison between theoretical calculation and experimental results is shown in Table 1.

Table 1. Comparison of output stress experiment and theoretical calculation.

Test number	Input stress /*MPa*	Output stress /*MPa*		Absolute value of error
		Experimental value	Calculated value	
1	165.3	49.9	51.7	3.6%
2	179.5	50.7	52.8	4.1%
3	153.6	48.7	51.1	4.9%
4	149.9	38.8	40.7	4.9%
5	166.1	46.7	48.9	4.7%
6	194.6	65.0	68.1	4.8%

From table 1, it can be seen that the output stress calculated theoretically and the output stress error measured experimentally are within 5 %, indicating that the model solution method is feasible.

5. Conclusion

Based on the propagation characteristics of stress waves, a three-group independent characteristic line equation is constructed. Because the characteristic line equation is nonlinear, the analytical solution cannot be derived. For this reason, the differential equations of the characteristic line compatible relation are constructed, the initial value conditions and the boundary value conditions are determined, and the dynamic state of the entire computational domain is gradually deduced by selecting the appropriate time step length and space step length. The theoretical results are in good agreement with the experimental results, which verify the correctness of the nonlinear dynamic model solving method based on the characteristic line method. This method provides an effective theoretical analysis and calculation method for the study of dynamic characteristics of structures under impact.

Acknowledgment
The project was supported by the National Natural Science Foundation of China (11172331).

References
[1] Wang L L, Ren H P, Yu J L, Zhou F H, Wu X Y, Tang Z P, Hu S S, Yang L M and Dong X L 2013 *Chin. J. Solid Mechanics*, **34**:217-40.
[2] Pang B J, Yang Z Q, Wang L W and Chi R Q 2011 *Chin. J. High Pressure Physics*, **25**:407-15.
[3] Zhou X R, Wang Q and Wang B Z 2007 *Chin. J. Vibration and Shock*, **26**:33-7.
[4] Ruwan J, David P T, Nimal J P and Vladis K 2011 *Int. J. Engineering Structures*, **33**:3483-95.
[5] Robin O 2010 *Int. J. Solids and Structures*, **47**:2884-92.
[6] Qian L Z, Li J and Ning Q L 2011 *Chin. J. Science & Technology Review*, **29**: 40-3.
[7] Qian L Z, Ning Q L, Li J and Jiang B A 2015 *Chin. J. Acta Armamentarii*, **36**: 1875-81.
[8] Wang L L, Shi S Q, Chen J and Huang D J 2000 *Chin. J. Journal of Ningbo University (NSEE)*, **13(Supplement)**: 141-9.
[9] Wang L L 2005 *Foundation of Stress Waves* (Beijing: National Defense Industry Press): 166-72
[10] Wang L L and Wang Y G 2005 *Chin. J. Explosion and Shock Waves*, **24**: 17-25.

IOP Conf. Series: Materials Science and Engineering **474** (2019) 012034 doi:10.1088/1757-899X/474/1/012034

Optimizing of hardness, microstructure and adhesive strength of plasma-sprayed FeSiAl coatings using orthogonal analysis

Y Guo[1], X Q Feng[1], S L Jiang[1], X Jian[2,*], L Zhang[1], X Wang[1] and L J Deng[1]

[1] National Engineering Researching Centre of Electromagnetic Radiation Control Materials, Key Laboratory of Multi-spectral Absorbing Materials and Structures of Ministry of Education, State Key Laboratory of Electronic Thin Films and Integrated Devices, School of Electronic Science and Engineering, University of Electronic Science and Technology of China, Chengdu, 611731, China

[2] School of Materials and Energy, University of Electronic Science and Technology of China, Chengdu, 611731, China

*E-mail: jianxian@uestc.edu.cn

Abstract. The traditional high-efficiency absorber FeSiAl has gradually achieved the attention at high temperature. The paper focuses mainly on the preparation and characterization of plasma-sprayed FeSiAl coatings. The plasma-sprayed FeSiAl coatings were prepared through the Praxair 100HE high enthalpy plasma spray gun on stainless steel 316 substrate by orthogonal experiment. The cross-section and properties of sprayed coatings were analyzed. The results reveal that the best parameters of sprayed coatings can be represented by $C_3B_2A_3D_2$ for microhardness, i.e., 30 SCFH, 140 mm, 50 KW, 3 PRM. The optimal parameters can be expressed as $C_3D_1B_3A_3$ for adhesive strength, i.e., 30 SCFH, 2 PRM, 200 mm, 50KW. The cross-section of sprayed coatings is composed of fattening FeSiAl particles (pale gray) and oxidation phase of FeSiAl (dark). The percentage of disorder structure reduces in FeSiAl coatings.

1. Introduction

An electromagnetic shielding and microwave absorbing materials have been urgently needed to reduce electromagnetic radiation and radar reflection in some extreme environment, like high temperature [1]. FeSiAl plays a vital role among high temperature absorbing coating due to high permeability and a saturation magnetization in recent years [2-3]. As of now, the focus of attention is on the preparation process of excellent absorbing coatings. The preparation technics for absorbing coatings includes mainly chemical vapor deposition (CVD) [4], sol-gel method [5], plasma spraying method [2] and so on. Plasma spraying process gradually gains the public' favor because of numerous spraying materials and high deposition efficiency. However, seldom systematic research concerning for FeSiAl coating is reported using the plasma spraying process.

The paper explored the influence the parameter of plasma spraying on hardness and adhesive strength of plasma-sprayed FeSiAl coatings using orthogonal analysis. Meanwhile, the orthogonal analysis was employed to calculate each control factors of spraying parameters on hardness and adhesive Strength of FeSiAl coatings. The similar investigation conclusions can be considered as guidance in industrial production for a mixture of FeSiAl and any other materials (for example, ceramic and alloy).

2. Experimental procedure
FeSiAl powder (9.6 wt. % Si, 5.4wt. % Al, and balance Fe) was used as raw material with an average particle size of 45 μm. Stainless steel 316 was regarded as the base material.

2.1. Preparation of coatings
The Praxair 100HE high enthalpy plasma spray gun of Progressive Technologies Inc was applied to produce FeSiAl coatings. The moving speed and step distance of the plasma gun was 150mm/s and 3mm respectively. Argon was used as the plasma carrier gas. The substrates were sandblasted with corundum sand before plasma-sprayed. FeSiAl coatings were deposited on Stainless steel 316 with respective serial numbers and different processing parameters in an L_9 (3^4) orthogonal experiment. The as-sprayed coatings were controlled with a thickness of 0.15-0.30 mm.

2.2. Microstructure and properties
The size distribution of the FeSiAl powder was measured using a Malvern MS3000. SEM images were collected on a JEOL SEM 7600F electron microscope, with the accelerating voltage of 15 kV. Microhardness of these samples was obtained by using Digital Microhardness tester. Bonding strength was measured according to the national standard of "Thermal Spraying Determination of Tensile Adhesive Strength [6]".

3. Results and Discussion
Various factors affect the plasma spraying coatings prepared by the different equipment, such as spraying power, spraying distance, carrier gas flow, power feeding rate, cooling gas flow, sandblasting distance, sandblasting pressure, sandblasting angle [7] and so on. Based on our preliminary investigation and survey for the Praxair 100HE high enthalpy plasma spray gun, the microstructure and properties of FeSiAl coatings can be controlled by spraying power, spraying distance, carrier gas flow, power feeding rate with the same sandblasting conditions and the others. To further explore the effect of these four significant factors on the microstructure and properties of FeSiAl coatings, three levels of each factor were selected to establish the orthogonal test, that is, *L9* (3^4) four factors and three levels orthogonal arrangement test, as can be seen in tables 1 and 2.

To analyze the effect of these four significant factors on the microstructure and properties of FeSiAl coatings, range factor R was calculated, as shown in table 3. Figure 1 shows the response curve of control factors for microhardness and adhesive strength of P_1-P_9. As shown in table 3, four factors including carrier gas flow, spraying distance, spraying power, power feeding rate when microhardness was considered as the primary index. The optimal parameters for microhardness can be expressed as C_3B_2 A_3D_2, which is 30 SCFH, 140 mm, 50 KW, 3 PRM. Meanwhile, it can be concluded that primary and secondary sequences of spraying parameters for adhesive strength of coatings are carrier gas flow, power feeding rate, spraying distance, spraying power (see chart 3). The best parameters of spraying coatings can be represented by $C_3D_1B_3A_3$, in the case, 30 SCFH, 2 PRM, 200 mm, 50KW.

Table 1. Control factors and levels of L_9 (3^4) orthogonal.

Level	A [Spraying Power (KW)]	B [Spraying Distance (mm)]	C [Carrier gas flow (SCFH)]	D [Power feeding rate (PRM)]
1	40	80	10	2
2	45	140	20	3
3	50	200	30	4

Table 2. L_9(3^4) orthogonal experimental layout with microhardness and adhesive strength.

Experiment number	A	B	C	D	Microhardness (HV)	Adhesive strength (MPa)

P_1	40	80	30	3	737.2	22.395
P_2	40	140	10	4	500.8	7.037
P_3	40	200	20	2	492.2	25.956
P_4	45	80	10	2	467.9	13.044
P_5	45	140	20	3	738.1	13.812
P_6	45	200	30	4	589.9	29.442
P_7	50	80	20	4	746.5	10.140
P_8	50	140	30	2	825.8	56.576
P_9	50	200	10	3	413.0	15.091

Table 3. Range analysis of $L_9(3^4)$ orthogonal experimental.

Control factors	A	B	C	D
K'_1	576.73	650.53	460.57	593.30
K'_2	598.63	688.23	658.93	629.43
K'_3	661.77	498.37	717.63	612.4
R'	85.03	189.87	257.07	34.13
The sequence of the factors		C>B>A>D		
The optimum scheme		$C_3B_2A_3D_2$		
K''_1	18.46	15.19	11.72	25.19
K''_2	18.77	19.14	16.64	17.10
K''_3	20.60	23.50	29.47	15.54
R''	2.14	8.30	17.75	9.65
The sequence of the factors		C>D>B>A		
The optimum scheme		$C_3D_1B_3A_3$		

As shown in figure 1a, the value of microhardness increases with spraying power increasing to 50KW, the reason lies in the higher-power accelerates the melting degree of FeSiAl powder, which improving the density of coatings. The microhardness values firstly increase and then declines as the spraying distance changing. According to the gas cooling theory, the spraying powder has longer flights under the cooling gas, the containing lower heat energy decreases the melting degree of FeSiAl powder, finally, reducing the compactness of coatings. The phenomenon can be explained by the spraying powder has the best melting condition during the spray process. The heat energy of spraying powder for 200 mm keeps a lower value comparing with 140 mm and 80 mm spraying distance. The increasing carrier gas flow upgrades the values the microhardness, possible because the increasing carrier gas flow reduces the porosity and microcracks of coatings and the increasing the amplitude of carrier gas flow greater than the decrease of temperature. The microhardness values increase and then decrease upon the power and feeding rate increasing. It is probably that the added spraying powder cannot be timely incomplete fusion reducing the compactness of the coatings. As shown in figure 1b, the adhesive strength of coatings appears to be rising as the spraying power increase due to the higher-power promoting the compactness of coating. The adhesive strength of coatings gradually increases with the increasing spraying distance. It is highly possible that thermal failure has little influence on the coating. Meanwhile, a larger distance gives plenty of time to release the residual stress of the coating. The increasing carrier gas flow upgrades the adhesive strength because of the increasing carrier gas flow reducing the porosity and micro cracks of coatings. Interestingly, the adhesive strength of coatings is progressively lowering as the rising power feeding rate. It might be due to that the larger power feeding rate, the greater residual stress. The larger power feeding rate leads to the number of unmelted particles, which improving the residual stress of coatings.

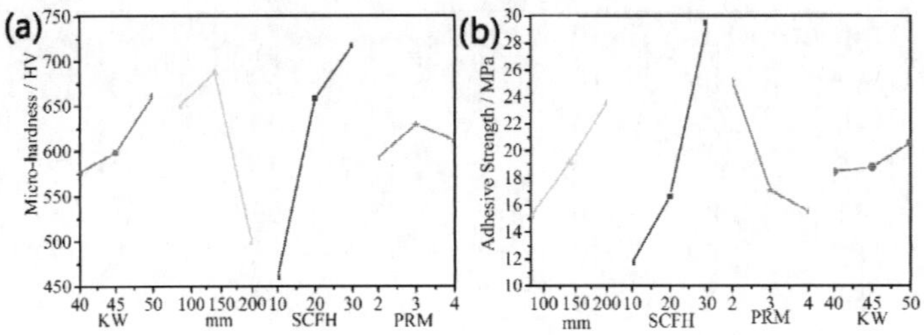

Figure 1. Response curve of control factors: (a) microhardness, (b) adhesive strength.

SEM and EDS analysis for the P_1-P_9 coatings are shown in figure 2 and table 4. As can be seen from figure 2, it can be found that the coatings are composed of fattening FeSiAl particles (pale gray) and oxidation phase of FeSiAl (dark color). Flattening indicates FeSiAl particles fully melt in the spray process. The spray powder hit the surface of the substrate and lamellar cambium structure experiencing full deformation and stacking during deposition. Furthermore, the cross-sectional morphology emerges the feature of uniform, compact, less porosity. Selected areas A and C are the highly oxidized phases of FeSiAl. Selected areas B and D are flatting FeSiAl. Selected area E located in the substrate is found to contain Mn, Ni and Cr element besides iron. The highly oxidized phase formed due to taking place the inflight oxidation and the post-impact oxidation.

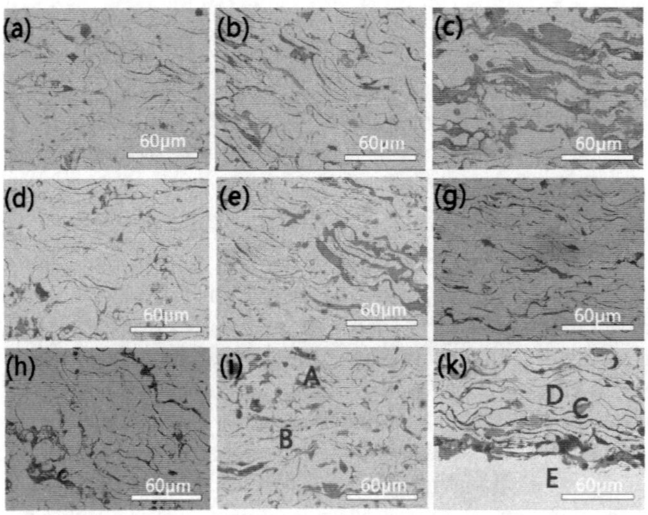

Figure 2. Secondary electron images of cross-sectional for P_1-P_9 samples.

Table 4. EDS results corresponding to figure 2 (i) and (k) (normalized, in wt. %).

Figure Number	Selected area	Fe	Si	Al	O	Cr	Mn	Ni	Total
3 (i)	A	70.86	12.00	1.98	15.16				100.00
	B	86.62	9.07	4.31					100.00
3 (k)	C	73.57	8.31	2.81	15.32				100.00
	D	85.39	8.97	5.64					100.00
	E	50.16	1.20		3.02	25.44	1.57	18.60	100.00

Figure 3 shows the SEM-mapping photograph of Fe, Si, Al, O, Cr, Mn, and Ni in P_9. Fe is in all zones of P_9, and its content is the highest. Si and Al are mainly distributed in the coating region. Cr, Mn, and Ni are primarily distributed in the substrate region. O is randomly dispersed in the matrix.

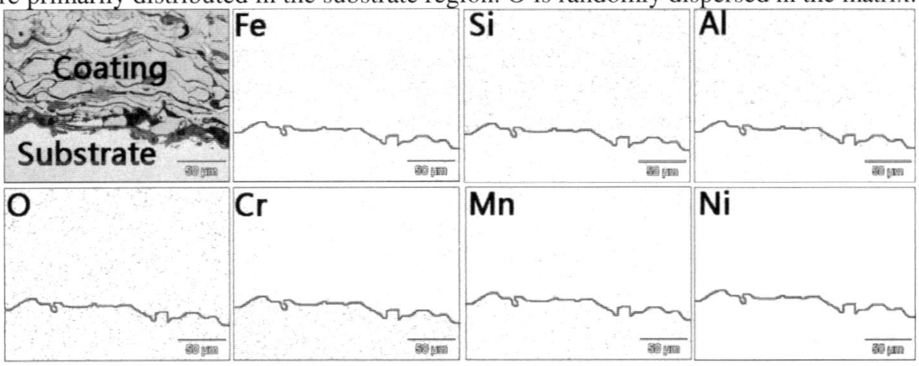

Figure 3. SEM Mapping of Fe, Si, Al, O, Cr, Mn, and Ni in P_9.

Figure 4 exhibits the X-ray diffraction (XRD) spectra of P_1-P_9 coatings deposited by plasma spraying and as-received. The as-received XRD pattern presents seven peaks at about 27.3, 31.2, 44.8, 53.29, 55.7, 65.4, and 82.9°, corresponding to (111), (200), (220), (311), (222), (400) and (422) of FeSiAl, respectively. It indicates that the as-received is nearly single phase and structure of the major phase includes A2, B2, and D03 [8-10]. According to the results of calculation and simulation by using Materials Studio, the disordered A2 structure consists of (200), (400) and (422) diffraction peak, corresponding to 44.8, 65.1, and 82.5°, respectively. B2 is defined as the intermediate state between A2 and D03. D03 contains a (111) diffraction peak at 27.3° and a (200) diffraction peak at 31.2°. (200) and (111) diffraction peak appear with the increasing degree of order for FeSiAl phase structure, also known as characteristic peaks of hyperlattice. For the P_1-P_9 coatings, the XRD patterns show the major diffraction peaks contain (200), (220), (400) and (422). Comparing with the XRD pattern of as-received, the intensity of diffraction peak (111) and (222) decrease, which means the percentage of the disordered structure reduces in FeSiAl coating. Maybe because the phase structure of FeSiAl powder gradually become disordered in the process of striking forward the substrate after melting.

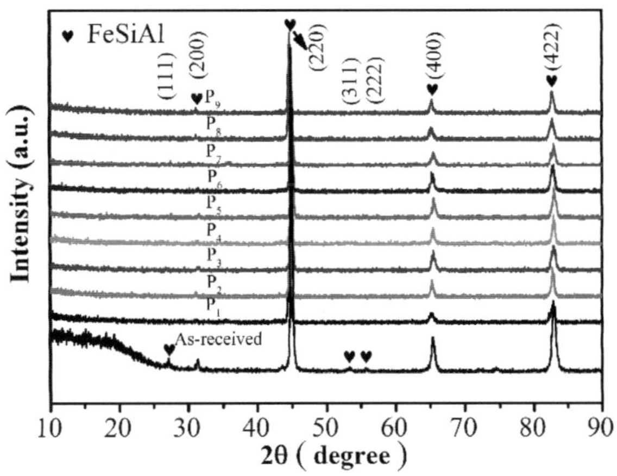

Figure 4. XRD patterns of P_1-P_9 samples.

4. Conclusion

The plasma-sprayed FeSiAl coatings were prepared through the Praxair 100HE high enthalpy plasma spray gun on stainless steel 316 substrate by orthogonal experiment. The main conclusions in this paper are as follows:

(i) The best parameters of sprayed coatings can be represented by $C_3D_1B_3A_3$ for adhesive strength, i.e., 30 SCFH, 2 PRM, 200 mm, 50KW. The optimal parameters can be expressed as $C_3B_2A_3D_2$ for microhardness, i.e., 30 SCFH, 140 mm, 50 KW, 3 PRM.

(ii) The cross-sectional of sprayed coatings are composed of fattening FeSiAl particles (pale gray) and oxidation phase of FeSiAl (dark).

(iii) Comparing the original FeSiAl particles, the percentage of disorder structure is reduced in FeSiAl coatings.

Acknowledgment
This work was financially supported by the Open Foundation of Key Laboratory of Multi-spectral Absorbing Materials and Structures of Ministry of Education (A03017023801009).

References

[1] Guo Y, Zhang X Z, Feng X Q, Jian X, Zhang L and Deng L J, 2018, *Composites Part B: Engineering,* **155**, 282-287.
[2] Qing Y, Su J, Wen Q, Luo F, Zhu D and Zhou W, 2015, *J. Alloys Comp.* **651**, 259-265.
[3] Qing Y, Zhou W, Luo F and Zhu D, 2017, *Ceramics International,* **43**, 870-874.
[4] Jian X, Xiao X, Deng L J and Tian W, 2018, *ACS Applied Materials Interfaces,* **10**, 9369-9378.
[5] Tian W, Liu S, Deng L J and Jian X. 2018, *Composites Part B: Engineering,* **149**, 92-98.
[6] Lu S, Chen Y, Chen J and Fu L, 2011, *Adv. Materials Res.,* **189-193**, 1292-1295.
[7] Zou H and Zhang H, 2012, *Key Engineering Materials,* **531-532**, 290-294.
[8] Schwindt C, Sandoval M, Ardenghi J S and Bechthold P, 2015, *J. Magnetism Magnetic Materials,* **389**, 73-76.
[9] Legarra E, Apiñaniz E, Plazaola F, 2012, *J. Alloys Compounds,* **536**, 282-286.
[10] Zhang N, Li G, Wang X, Liu T and Xie J, 2016, *J. Alloys Compounds,* **672**, 176-181.

IOP Conf. Series: Materials Science and Engineering **474** (2019) 012035 doi:10.1088/1757-899X/474/1/012035

The effect of Sn-Al-C composite powders on the electromagnetic interference shielding of capsule building materials

F S Hung*

Department of Leisure, Recreation and Tourism Management, Southern Taiwan University of Science and Technology, Tainan, Taiwan

E-mail: fshung@stust.edu.tw; Tel: +886-6-253-3131 ext. 8192

Abstract. The Sn substrate composite powder technology for capsule building was tested in this study, and the trend of adding graphite was studied. Then, a series of materials with higher resistance to electromagnetic waves were investigated. Based on low-cost advantages and particular categories of architectural application, the Sn-Al-C (SAC) ternary system powder was studied using the coating method, which can be applied to cement or plastic surfaces. The effects of particle size, content ratio of C and Sn-Al additions, stacking effect, and influence of porosity and thickness on the electromagnetic shielding mechanism were analyzed and discussed in detail.

1. Introduction

Earth's energy reduction and stricter environmental requirements increase the demand for green and environmental-friendly building materials. This provides a lot of opportunities to nanotechnology development, especially in nanomaterials, new structures, properties and the nanomaterial function adjunct on known materials, as well as addresses innovative issues [1-3]. For current capsule buildings or hotels with small internal space and high-density arrangement, the material shielding from electromagnetic wave interference is essential. Furthermore, the outdoor capsule materials with low cost, high-temperature resistance, rust resistance, excellent electrical conductivity, and thermal conductivity, etc., are required to cope with the trend of environmental-friendly materials [4-6].

Based on the above considerations, such elements as Sn, Al and C have not only excellent EMI shielding properties but are also competitive in price [7]. A low-cost silver-white powder of Sn-Al- C (SAC) was prepared by a simple technique, and its electromagnetic wave interference shielding effect versus particle size, C content ratio, and thickness of coating were examined by the coating method [8-9].

2. Experimental method

Improved powder microstructure by used powder test and coasting test. First, discussed the correlation between Sn-Al-C (SAC) and the influence of shielding electromagnetic interference (EMI) based on the ternary system powder in the architectural application category. Besides, the content of graphite (C) in building materials was clarified, and related characteristics were analyzed to propose the optimum ratio of Sn alloy powder coating (film) mechanism for construction using [10-12]. A construction uses a ternary system as the main body to develop shielded electromagnetic wave powder (film) material for building construction. Through added graphite (C) to review the electromagnetic wave shielding effect of SAC before and after the heat treatment and magnetic evolution changed to the comprehensive mechanism and further clarified the characteristics of the composition, light transmittance, and conductivity, etc. Finally, integrity review and analysis were based on the EMI mechanism and the shielded electromagnetic wave film system to suggest the healthy building materials application reference [13-15].

In this study, system powders such as Sn-xAl (x=40 wt.%)-yC (x=5, 10.15 wt.%) were selected for EMI-related experiments. A suitable amount of Polypropylene and Sn-Al-C (SAC) and system powder mix well (Polypropylene selected 15 wt.% ratio), let SAC powder was completely uniformly dispersed in Polypropylene. The SAC was mixed well with a Polypropylene by used blade forming apparatus, then composite colloid was coated on a glass substrate (glass substrate size 2 cm x 2 cm x 2 cm, total one piece), and solidify at room temperature for 48 h [16-17]. Estimated number of coatings in the experiment is two layers and three layers coating; each coasting thickness was controlled at 25 ± 10 μm. To understand the characteristics of the microstructure. High-order instruments such as SEM were used to analyze the particle size of SAX powder and the appearance characteristics of the coating, to review the evolutionary correlation of shielding effect on electromagnetic waves between the Sn-Al-C and the system.

This study used a coaxial electromagnetic wave shielding effect tester with a scan frequency range from 300K to 3G Hz with an accuracy of ±10 ppm (250 ± 50). The measurement method of this experiment is used the plane wavelength of normal incidence to test; the measured frequency range is controlled from 50MHz to 3000MHz, all the data average values were 3~7 points. The shielding efficiency value was obtained by calculation [18-20].

3. Results and Discussion

3.1. Effect on EMI shielding mechanism characteristics of Sn-Al-xC powder coating structure and film structure

Sn-xAl-C was coated on a glass substrate in this experiment. After the film was prepared, the material properties (microstructure, interface) of the initial coating and powder were analyzed, and the electromagnetic wave shielding experiment was also performed. Then the properties of the sample microstructure were analyzed to compare the changes of the film before and after heat treatment. After powder coating experiment, when the Al content in the Sn-xAl coating was increased, the electrical conductivity was lowered, but the stacking effect could improve the electromagnetic wave shielding property under low-frequency conditions. The volume fraction of the intermetallic compound in the powder is reduced by decreased the Al content to increase the powder conductivity and improve the high-frequency electromagnetic wave shielding property. Besides, the channel effect of the Polypropylene was remarkable but single coating without electromagnetic shielding. The three-layered Sn-xAl coating used in increased porosity solidifying (up to ~15 vol.%), so the electromagnetic shielding was caused deterioration (confirmation: the more the number of layers is, the better the shielding effect is not). On the other hand, more importantly, in the Sn-Al film experiment increased film thickness and increased conductivity was an important finding in increasing the efficiency of EMI electromagnetic shielding.

According to reasonable inference, increasing the Al content in Sn-40Al-10 C powder can improve then shielding characteristics of low-frequency electromagnetic waves based on electromagnetic wave shielding characteristics of Sn-xAl powder coatings study. On the other hand, after annealing test, high-temperature crystallization process (crystallization phase homogenization) and high conductivity can be improved (as shown in table 1) EMI shielding property of the Sn-Al nanofilm (increased high-frequency shielding efficiency), as shown in table 2. Sn-40Al-10C (330nm) thin film EMI shielding experiment, Figure 1(a) was the electromagnetic wave shadowing effect evolution map (from 330nm, 560nm, and 780nm) by different sputtering film thicknesses. Similarly, table 2 also showed electromagnetic shielding testing result. Increased the thickness of the sputter film not only improved crystallinity (IOC) but also enhanced electromagnetic shielding efficiency. To avoid affecting particle size, crystallization rate, and coating formation, 330nm was annealed in a vacuum environment (200°C/1 h) and then cooled at room temperature (0.5°C/min.). And a control EMI masking experiment was performed for the 330 nm-H coating and the 330 nm coating. According to the result of EMI shielding effect of 330nm-H and 330nm, the EMI shielding efficiency of the (330nm-H) film after annealing was better than that of the 330nm film (fig. 1(b); table 2). From the above result, annealing experiment (200°C/1h) not only improved the film conductivity but also had better crystallization rate and lower resistivity. So that could enhance the high-frequency electromagnetic shielding effect. However, for building materials application, cost of the Sn-40Al-10C film had high competitiveness, and low-frequency shielding of the Sn-40Al coating was superior to high-frequency conditions. Considering the electromagnetic space environment has mostly low frequency the application of the Sn-40Al-10C film to the interface layer of building materials should be helpful for EMI shielding.

Table 1. Four-point probe result of Sn-alloy.

	Sheet resistance
Sn-40Al-5C	1.59×10^{-4} Ω-cm
Sn-40Al-10C	2.87×10^{-4} Ω-cm

Table 2. The average dB value of coatings at different frequencies.

Specimen	300 MHz (-dB)	900 MHz (-dB)	1.80 GHz (-dB)	3 GHz (-dB)
330nm	-11.6737	-13.8131	-14.9646	-21.1212
560nm	-15.3353	-19.2621	-25.8131	-21.2161
780nm	-21.1113	-22.1671	-20.2212	-24.2235
330nm-H	-15.2259	-16.2852	-17.7233	-24.3989

Table 3. Four-point probe result of Sn-Al-xC.

	SAC	Sn-40Al-5C	Sn-40Al-10C	Sn-40Al-15C
SAC	As-sputtered	5.44×10^{-5}	6.22×10^{-5}	3.39×10^{-4}
	Annealed	1.09×10^{-5}	3.36×10^{-5}	3.21×10^{-4}

Table 4. The average dB value of the Sn-40Al-C thin film.

Specimen	300 MHz(-dB)	900MHz(-dB)	1.80 GHz(-dB)	3 GHz(-dB)
Sn-40Al-10C -780nm	-13.19914	-16.3332	-22.0809	-14.8712
Sn-40Al-10C -780nm-H	-7.3321	-8.2356	-15.1399	-13.0101

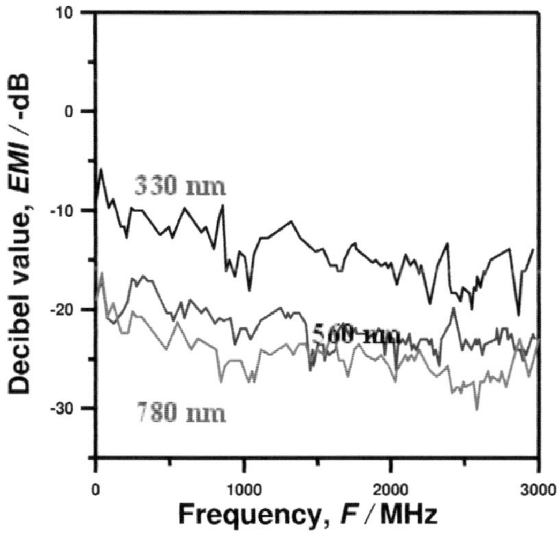

(a)

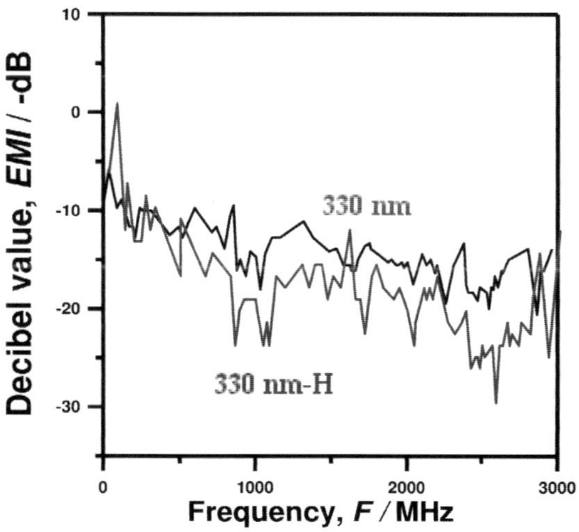

(b)

Figure 1. EMI shielding of Sn-Al sputtering thin films: (a) different thickness (b) before and after annealing.

3.2. Sn-40Al-xC coated film microstructure characteristic and EMI shielding mechanism

Sn-Al systems had excellent EMI electromagnetic shielding performance and had an advantage of price competitiveness. According to the literature and the above experimental results, Al content can affect the EMI shielding efficiency at different frequencies. Comparing the Sn-40Al-x C film, as shown in figure 1(a) (select x=10 and 40wt.% ratio), the Nanofilm was sputtered on the glass substrate (a total of one glass substrate, the substrate size is 2 cm x 2 cm per each glass, as shown in figure 2(a)), and the Sn-40Al film thickness is controlled to be 330 nm to 780 nm. It is worth noting that the Sn-40Al (330nm) film was annealed in a vacuum environment (330nm-H, 200°C/1 hr), the results show that the EMI shielding effect of Sn-40Al (330 nm) is better and the grain surface features are more detailed (as shown in figure 2(b), the particle size is about 20~25nm), as shown in figure 1(b). This is the reason for further planning will leading-in heat treatment to improve the uniformity. Since target material of Sn-Al-C had a more excellent high electrical conductivity (1.09 x 10-4 (-cm) and reasonable price competitiveness as shown in table 3. Therefore, the experiment of the electromagnetic wave shielding effect was used SAC. According to previous studies, the effect of 300k ~ 3 GHz electromagnetic shielding in high-purity Sn substrate materials is optimum. Figure 3 shows a comparison of the coated Sn-40Al-10 C (780nm) film. After the experiment of the electromagnetic shielding effect, the EMI shielding effect of Sn-40Al-10 C (780nm) film is better than Sn- 40Al-10C (780nm) film (table 4). But it is worth noting that the Sn-40Al-10C (780nm) film does have a partial frequency shielding effect, which is another discovery in this experiment (figure 3).

IOP Conf. Series: Materials Science and Engineering **474** (2019) 012035 doi:10.1088/1757-899X/474/1/012035

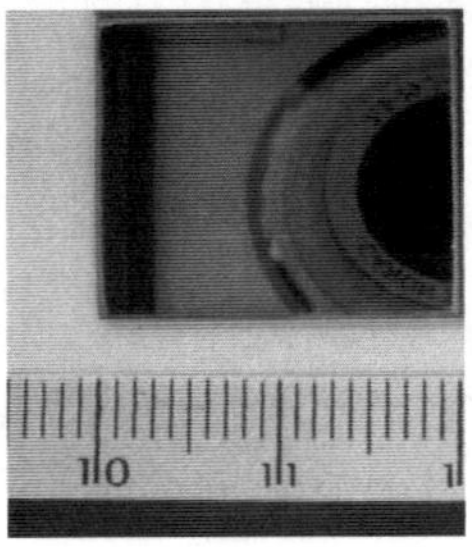

(a) Appearances

(b) The surface characteristic

Figure 2. Sn-Al sputtering thin films (330 nm):

According to the experiment, the film after annealing experiment can not only increase the conductivity but also improve the excellent EMI shielding effect. However, it is worth noting that the Sn-40Al-10C film has been deteriorated after annealing experiments. (figure 4 and table 4). More importantly is the thickness of the Sn-40Al-10C film is only half of Sn-40Al-10C (this result is different from the previous Sn-xAl film shielding effect). Besides, from SAC (780nm) film experiment, the EMI of the high-frequency shielding effect was degraded due to the increase in IMCs plot ratio and the double cause after the annealing process. The experimental results show that Sn-40Al is significantly improved in overall electromagnetic shielding after annealing treatment. It is noteworthy that low-frequency electromagnetic shielding is better than high frequency when C content is increased. Since Sn-Al-C has higher conductivity and magnetic strength, the electromagnetic wave shielding full-frequency shielding can be improved.

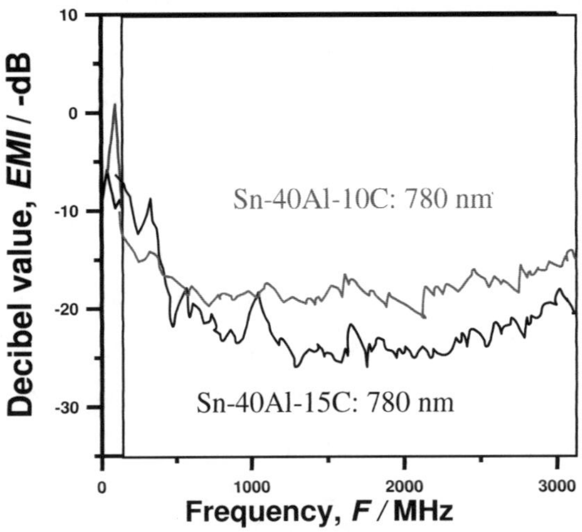

Figure 3. EMI shielding of Sn-40Al-10C (780nm) and Sn-40Al-15C (780nm).

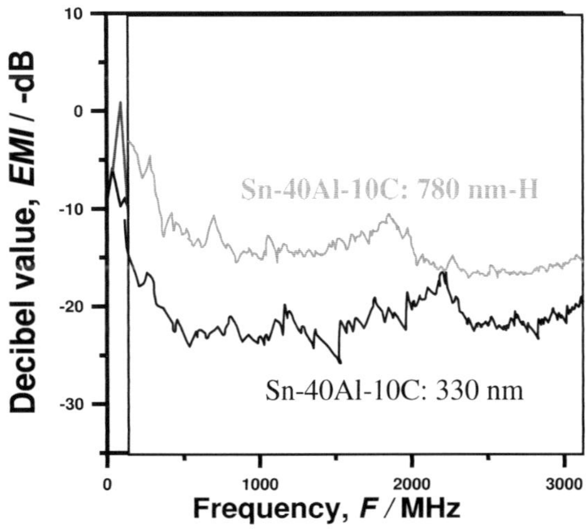

Figure 4. EMI shielding of Sn-40Al-10C (annealed) and Sn-40Al-10C (un-annealed)

4. Conclusions

- ✓ The conductivity is decreased when Al content in Sn-xAl-C coating is increased, but the stacking effect can be improved the electromagnetic wave shielding under low-frequency conditions. And the volume rate of intermetallic compound powder is reduced when Al content decreased so that the electrical conductivity is improved and the high-frequency electromagnetic wave shielding property can be effectively improved.
- ✓ The effect of conductivity and magnetic strength is significantly improved, the low-frequency electromagnetic shielding effect is better than high-frequency one when C content is increased.

Acknowledgment

The authors are grateful to the Department of Leisure, Recreation and Tourism Management, Southern Taiwan University of Science and Technology for the support of this research.

References

[1] Dawould M, Taha I, Ebeid S J and Wiss U M 2015 *Werkstofftech*, **46** (12) 1185.

[2] Carutasu N L, Simion I, Carutasu G, Jiga G and Arion A F 2015 *Material Plastic*. **52** (3) 397.

[3] Castles F, Isakov D, Lui A, et al. Grant 2016 *Scientific Reports*. 1-8.

[4] Seoung Y H 2015 *Journal of the Korea Academia-Industrial Cooperation Society*. **16** (9) 6019.

[5] Zhang C S, Ni Q Q, Fu S Y and Kurshiki K 2007 *Composites Science and Technology*. **67** 2793–980.

[6] Wang T, Chen G, Wu C and Wu D 2007 *Progress in Organic Coatings*. **59** 101–5.

[7] Han G, Kim E A and Oh K W 2001 *Synthetic Metals* **123** 469–76.

[8] White G E 2007 *U.S. Patent No. 20070085108*.

[9] Liu D and Suqiao D 2000. *Asia-Pacific Conference on Environmental Electromagnetics* 326–32.

[10] Chiang W Y and Cheng K Y 1997. *Polymer Composites.* **18** (6) 748.

[11] Park J T and Park J S 2006 *The Institute of Electronics Engineering of Korea.* **43** (4) 122.

[12] Rocha C R, Perze A R T, Roberson D A, Shemelya C M, Macdonald E and Wicker R B 2014 *Journal of materials research.* **29** (17) 1859.

[13] Hung F Y, Lui T S and Liao H C (2007) *Applied Surface Science, Sci. Eng. A*, **253**, 7443-8.

[14] Zenha A, Faria M and Figueiras R 2011 *Construction and Building Materials.* **25**, (11) 108-16.

[15] Perini K, Ottele M, Fraaij A L A, Haas E M and Raiteri R 2011 *Building and Environment*, **46** (11), 2287-94.

[16] Barroca N, Borges L M, Velez F J, Monteiro F and Castro-Gomes J 2013 *Construction and Building Materials.* **40** 1156-66.

[17] Sham J F C, Lo T Y and Memon S A 2012 *Energy and Buildings.* **53** 108-16.

[18] Tan A Y K and Wong N H 2012. *Energy and Buildings.* **53** 29-7.

[19] Xie J L, Zhu Q Y, Xu X H 2012 *Journal of Central South University.* **19** (6) 1663-7.

[20] Juan P R, Camacho N, Barberis E and Palma S D 2018 *Journal of 3D Printing in Medicine.* **2**(**3**).

Study on the Mg-Zn binary alloys fabricated by gas-phase alloying

Y T Yan*, L B Niu, Y X Long and J Ma
College of Materials Science and Engineering, Xi´an University of Science and Technology, China
*E-mail:18829345717@163.com

Abstract. By adjusting the evaporation temperatures and changing the Mg/Zn atomic ratios (in the range of 0.5 to 4), various Mg-Zn binary alloys were fabricated usofing a novel technique named as gas-phase alloying. The theoretical calculation by the formula of saturated vapor pressure is to determine the evaporation temperature, which is 900-1200K for Mg and 700-1000K for Zn. The specimens were subjected to SEM, EDS, and Brinell hardness examinations. The results obtained show that, regardless of atomic ratios, microstructures include Mg_7Zn_3 particles and α-Mg matrix, the former is uniformly distributed into the latter. However, with increasing Mg/Zn atomic ratios, the microstructural transformation occurs from columnar precipitation to granular one.

1. Introduction

With the significant advantages of low density, high specific strength, Mg-based alloys are desirable materials for the transport industry and replace the aluminum-based alloys in some particular cases [1-2]. It is well known that zinc is a preferable alloying element and can form the different intermetallic phases, especially Mg_7Zn_3 [3-4]. However, Mg/Zn atomic ratio can dramatically change the microstructure and influence the properties of Mg-Zn alloy.

Because vapor condensation is the effective and simple technology for fabricating new materials or coatings, Xi et al. [5] prepared the Zn-Mg alloy through the chemical vapor deposition (CVD) method and observed the self-corrosion current density and impedance resistance of the alloy. In the paper, the authors aimed to produce Mg-Zn alloys with indirectly adjustable Mg/Zn ratios under the vacuum conditions and to investigate the effect of Zn content on the microstructure and formation of Mg_7Zn_3 intermetallic particles. Besides, the expected findings will be used to substantiate the idea of fabricating a novel composite reinforced by intermetallic Mg_7Zn_3 particles.

2. Experimental processing

The Mg-Zn binary alloys were fabricated by the evaporation-mixture-condensation method (nominated as gas-phase alloying technique). Figure 1 shows a conceptual diagram of the experimental device. The crude Mg (99.96wt.% purity) was evaporated in chamber 1, and the crude Zn (99.98wt.% purity) was evaporated in chamber 2. Two kinds of metal vapors entered into chamber 3 (700±10K), simultaneously, and were mixed under the function of the baffles inside chamber 3. Subsequently, the mixture of metal vapors was condensed inside chamber 4 (600±10K). The whole device was made up of stainless steel. There are two similar graphite crucibles fixed inside chambers 1 and 2, respectively, the crude Mg and Zn (similar in size and shape) were loaded into the graphite crucibles. During every experiment, the system vacuum degree was 30±5Pa and the soaking time was 0.5h at a certain temperature. Microstructure and elemental analysis were conductedby using a TESCAN VEGA 3 scanning electron microscope (SEM) and energy dispersive X-ray spectroscopy (EDS).

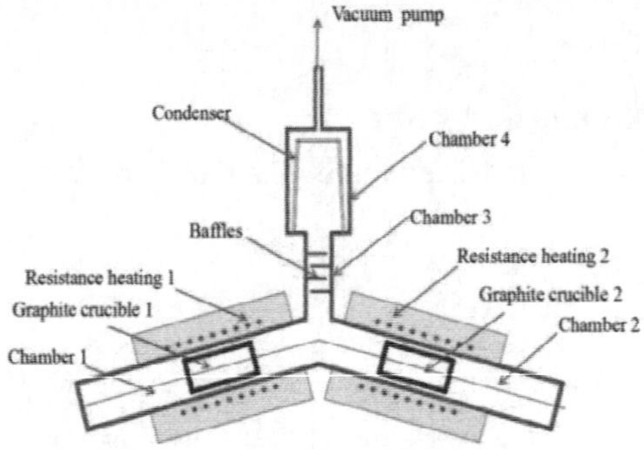

Figure 1. Conceptual diagram of the gas-phase alloying device.

To ensure the formation of uniform Mg/Zn binary alloys, the preferable mixture degree of two metal vapors is very critical; therefore, several baffles were set up in chamber 3. These baffles not only are beneficial to improve the mixture degree but also can reduce the kinetic energies of two metal vapor atoms and favor the condensation [6]. However, the more baffles cannot ensure the mixture vapors tosuccessfully condensate inside chamber 4 successfully. According to the preliminary experiments, the appropriate number of baffles is four-layer.

3. Theoretical analysis and calculation

3.1. Determination of the ratios of Mg/Zn

In the study, the Mg/Zn atomic ratio is the most important factor to control the microstructural characteristic and related to the evaporation rate. According to the literature [6-7], the vapor rate of certain metal can be calculated by formula (1), and the saturated vapor pressure of Mg and Zn metals at certain temperature can be computed through equations (2) and (3), respectively, and the calculated results were listed in table.1.

$$w_i = 1.574\alpha P\left(\frac{M}{2\pi R T}\right)^{\frac{1}{2}} \quad (1)$$

$$\lg P^*_{Mg} = -\frac{7550}{T} - 1.41 \lg T + 14.90 \quad (2)$$

$$\lg P^*_{Zn} = -\frac{6620}{T} - 1.255 \lg T + 14.465 \quad (3)$$

where w_i is evaporation rate for certain metal, α is the condensation constant for most crystalline metals approaches unity, M is the molar mass of metal (g/mol), P_i^* is the saturated vapor pressure (Pa), T is the temperature in Kelvin scale (K).

Table 1. The saturated vapor pressures of Mg and Zn at different temperatures.

T(K)	700	800	900	1000	1100	1200
P^*_{Mg}	-	-	220.08	1309	5557.8	18348
P^*_{Zn}	27.384	352	2522.3	120226	-	-

Then, the vapor ratio for Mg/Zn can be determined using the following equation (4).

$$w\frac{T_{Mg}}{Mg}/w\frac{T_{Zn}}{Zn} = P_{Mg}/P_{Zn}\left(\frac{M_{Mg}T_{Zn}}{T_{Mg}M_{Zn}}\right)^{\frac{1}{2}} \quad (4)$$

Therefore, $w_{Mg}^{1200}/w_{Zn}^{1000} = 0.4695$, $w_{Mg}^{1100}/w_{Zn}^{900} = 0.94438$, $w_{Mg}^{1000}/w_{Zn}^{800} = 2.021$,

$w_{Mg}^{900}/w_{Zn}^{700} = 4.32$

where superscript and subscript in $w_i^{T_i}$ indicate the evaporation temperature and certain metal, respectively. To simple observation, the ratios of Mg/Zn for various specimens are approximate to 0.5, 1.0, 2.0 and 4.0, the corresponding evaporation temperatures for the chamber 1 and 2 are shown in table.2.

Table 2. The different evaporation temperatures for various specimens.

Vapor ratio for Mg/Zn (at.%)	0.5	1	2	4
Mg evaporation temperature (K)	900	1000	1100	1200
Zn evaporation temperature (K)	700	800	900	1000

4. Formation mechanism

When two kinds of the vapors were mixed in chamber 3, which temperature is lower than the dew point of Mg vapor and higher than the dew point of Zn vapor [7], Mg vapor will transform into Mg tiny droplets following equation (5), and Zn vapor cannot be converted into droplets and still keep as a state of steam flow under the vacuum function.

$$Mg_{(g)} = Mg_{(1)} \quad (5)$$

Some tiny Mg droplets can collide with the Zn vapor atoms, some reactions between them will take place via equation (6), and the intermetallic compound Mg_xZn_y will be formed, these particles are expected to act as the reinforcement of novel composite [8].

$$xMg_{(g)}+yZn_{(g)}=Mg_xZn_y \quad (6)$$

But a large portion of tiny Mg droplets cannot collide with the Zn vapor atoms which directly reach condensing wall and rapidly solidify at orientated growth. This part of condensation production has less Zn content and can be expected to act as the matrix of novel composites reinforced by the intermetallic compound.

5. Results and Discussion

Figure 2 gives the microstructures of alloy under the different atomic ratios of Mg/Zn. When the atomic ratio is 0.5, namely, the Zn vapor content is much higher than that of the Mg content. The microstructure is comprised of sizeable white flake precipitation and granular production, the size of them is less than 5 micrometer in width, as seen in figure 2(a). Based on the EDS results, Mg content in the black area (point A in figure 2(a) and table 3) is rich**,** and Zn content in the white area (point B in figure 2(a) and table 3) is rich. The white phase is determined **by** Mg_7Zn_3**,** and the black area belong**s** to the α-Mg matrix, in which the Zn content is less**.** This indicate**s** that some Mg droplets collided with Zn vapor, **the** peritectic reaction can be carried out, and the reactive products Mg_7Zn_3 can directly grow perpendicular to the cooling surface. At the same time, non-collided Mg droplets were directly cooled as the matrix. These findings are agreeable to thefront mechanism interpretation.

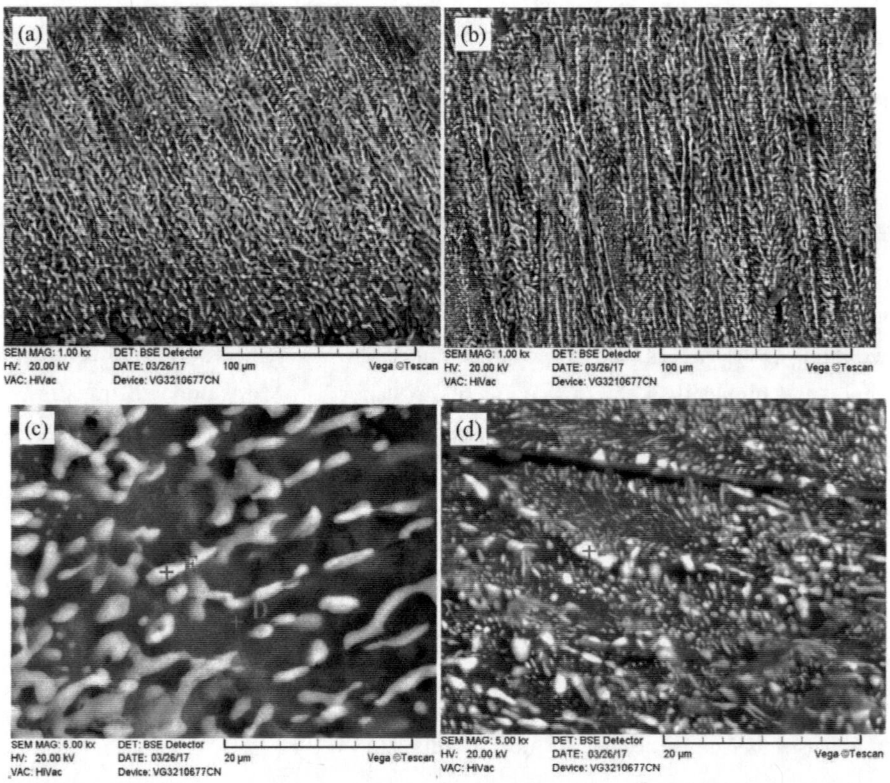

Figure 2. Microstructures of Mg-Zn alloys with different vapor ratios: (a) 0.5, (b) 1, (c) 2, (d) 4.

Table 3. EDS for different points in figure 2.

Point	A	B	C	D	E	F	G
Mg(at.%)	71.41	95.02	68.30	94.14	73.52	75.75	92.02
Zn(at.%)	28.59	4.98	31.70	5.86	26.48	24.25	7.98

When the atomic ratio of Mg/Zn is added up to 1, the greater flake precipitation can be avoided, the more obvious orientated growth and the column distribution can be found, as well as some finer granular precipitation can be detected between column structure. According to the EDS results (point C in figure 2(b) and table.3), the finer particles are deemed to Mg_7Zn_3 and not other intermetallics, such as $MgZn_2$, MgZn, etc. These phenomena adequately indicate that the trendy for the formation of Mg_7Zn_3 is superior to other intermetallic compounds.

When the atomic ratio of Mg/Zn is up to 2, the white and black structure is still the main characteristic as shown in figure.2(c). Due to the solidification of a larger of Mg droplets, releasing latent heat can melt a part of the inter-metallic compound and leads to the occurrence of rounded shape production (point E in figure 2(c) and table 3). Meanwhile, Zn atoms from the dissolution of Mg_7Zn_3 diffuse into the black area and increase Zn content in the dark area (point D in figure 2(c) and table.3).

When the maximization of the atomic ratio is 4, the quantity of Zn vapor is too less to ensure the formation of the continuous columnar production, instead of the granular precipitation. From the Figure 2(d), it can be seen that the initial reactive product present very fine particles which will grow gradually and may provide a more substantial granulate precipitation. These particles are expected to improve the properties of Mg-based composite (point F in Figure. 2(d) and Table 3).

From above analysis, it is concluded that, regardless of the Mg/Zn ratios (in the range of 0.5 to 4), the Mg_7Zn_3 is the only kind of intermetallic compound, and presents a change from a columnar structure to a granular one with increasing the Mg/Zn atomic ratios.

ofSince the size of intermetallic particles is very small, the pressure head of microhardness is not able to sufficiently load on the particle. Therefore, Brinell hardness is selected to compare the difference in term of hardness. Figure 3 gives the decreasing trendy of hardness with increasing the Mg/Zn atomic ratios, the maximum value (127.2±0.5HB for the ratio of 0.5) is nearly 1.5 times higher than the minimum value (82.3±0.5HB for the ratio of 4), which exceeds the hardness of AZ31 in a forging state [9]. It can be interpreted by the fact that abundant Zn vapor can react with the Mg droplets and contributes to the formation of plenty of Mg_7Zn_3 compounds which has higher hardness and expected to act as the reinforcement of composite.

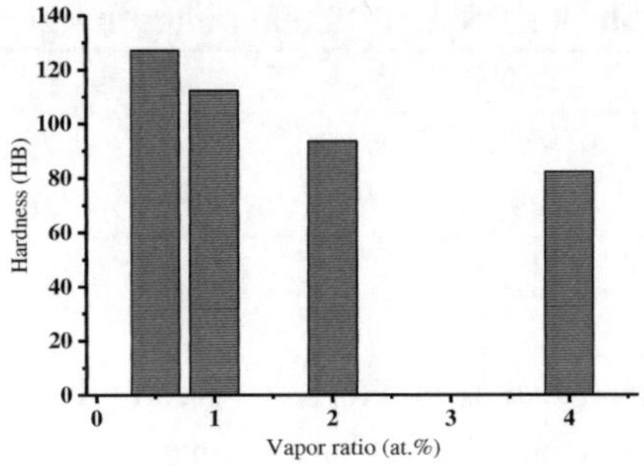

Figure 3. Hardness variation with the Mg/Zn ratios.

6. Conclusions

Mg-Zn binary alloys were prepared by gas-phase alloying technique under the vacuum condition. The ratios of Mg/Zn vapor atoms can be explained and effectively controlled through the difference of evaporation temperatures. When theof Mg/Zn is higher, it is advantageous to form Mg_7Zn_3 particles. When the ratio of Mg/Zn is lower, Mg_7Zn_3 particles present orientated growth and columnar structure. Regardless of the Mg/Zn ratio, all condensations have high hardness. These findings substantiate the idea to fabricate a novel composite reinforced by intermetallic Mg_7Zn_3 particles.

References

[1] Cao F, Shi Z and Hofstetter J 2013 *Corros. Sci.* **75** 78.
[2] Mordike B L and Ebert T 2001 *Mater. Sci. Eng.* A **302** 37.
[3] Maeng D Y, Kim TS and Lee J H 2000 *Scripta Mater.* **43** 385
[4] Kaplan H I, Hryn J N and Clow B B 2013 *Magnesium Technol.* **75** 243.
[5] Xi Z G, Zhang Q F and Jiang S M 2016 *Surf. Coat. Tech.* **306** 418.
[6] Zhang R, Khalizov A and Wang L 2012 *Chem. Rev.* **112** 1957.
[7] Koo A, Brooks G A and Nagle 2008 M *J. Cryst. Growth* **310** 2659.
[8] Deng C J, Wong M W, Ho M W, Yu P and Dickon H L 2005 *Compos. Part A-Appl.* **36** 551.
[9] Dziubinska A, Gontarz A and Horzelska K 2015 *Procedia Manufacturing* **2** 337.

IOP Conf. Series: Materials Science and Engineering **474** (2019) 012037 doi:10.1088/1757-899X/474/1/012037

Study on mechanical properties of FGD matrix composites

T W Liu, Z Wang*, G Z Li, G P Shi and X Zhao

School of Material Science and Engineering, University of Jinan, Jinan, China

E-mail: wangzhi@ujn.edu.cn Tel: +86- 0531-82767636

Abstract. Flue gas desulphurization gypsum (FGD), fly ash (FA) and steel slag (SS) were by-products during the industrial production process, and comprehensive utilization of these by-products could both protect the environment and save resources. The effects of FA and SS on the mechanical properties of FGD and different activating agents on the mechanical properties of FGD matrix composites were studied. The results showed that mechanical properties gradually decreased with the increase of the content of FA and SS which was principally attributed to the poor hydration activity of FA and SS. The flexural strength and compressive strength of FGD matrix composites were significantly improved by adding NaOH, $Al_2(SO_4)_3$, and compound activating agent respectively and the compound activating agent had a more positive effect than NaOH and $Al_2(SO_4)_3$.

1. Introduction

In recent years, flue gas desulphurization gypsum (FGD) has attracted an increasing interest of researchers due to its many excellent properties such as high early strength, low density, nontoxicity and low energy consumption [1-3]. Besides, FGD is a kind of industrial waste solid produced in desulfurization of flue gases in power stations and heating plants, and its output is enormous every year in all the world [4]. As one of the most critical cementitious materials, its energy consumption is lower than that of ordinary Portland cement during the production process which results in the decrease of greenhouse gas emissions [5]. Therefore, FGD is a promising material that can be used for the construction industry. However, its application is limited by some shortcomings such as massive water consumption and water absorption [6]. These problems can be solved by adding siliceous materials such as FA and SS into FGD. Moreover, FA that has pozzolanic activity due to the existence of Al_2O_3 and SiO_2 and SS that has cementing action caused by C_3S and C_2S are also industrial waste solid [7-8].

Nevertheless, the poor activity of FA and SS hinders their application. One of the most effective methods used to improve the hydration activity is additions of activating agents such as sodium hydroxide and sodium silicate [9-10]. Although FGD with FA and SS addition have enormous potential to be a promising material for applications in renewable energy buildings, few pieces of research about mechanical properties of ternary cementitious materials have been reported.

In this paper, the effects of FA and SS on the mechanical properties of FGD and different activating agents on the mechanical properties of FGD matrix composites were studied.

2. Experimental

2.1. Raw materials

FGD and FA were supplied from Shandong Laiwu Power Plant, and SS was supplied from Laiwu Iron and Steel Group. The chemical compositions of these materials were listed in Table 1. The commercially available powder NaOH (purity≥96.0%) and $Al_2(SO_4)_3$ (purity≥98.0%) were used as activating agent to improve the hydration rate of SS and FA in this study. A compound activating agent (made up of Portland cement clinker, $Al_2(SO_4)_3$ and Na_2SiO_3) was also used to stimulate the activity of FA and SS.

Table 1. The chemical composition of raw materials (wt%).

	SO_3	MgO	SiO_2	CaO	Al_2O_3	Fe_2O_3
FGD	43.11	2.01	1.45	31.80	0.53	0.21
FA	0.64	0.93	47.13	4.13	40.33	-
SS	1.14	4.83	15.44	46.01	5.57	18.25

2.2. *Sample preparation and test methods*
The FGD was calcined at 155 °C for 200 min and placed at room temperature for 7 days before using it to improve it cementitious performance. The FA and SS were ball-milled using a planetary ball milling machine at a speed of 300 rpm for 6 h before the experiment to enhance hydration activity of FA and SS. The proportions of FGD, FA and SS were given in Table 2. The above material mixtures were mixed using a paste mixer at a speed of 140 rpm for 150 s. After milling, the mixtures were cast into a steel mold of size 16 × 4 × 4 cm and demolded after 24 h. The specimens were cured in air at room temperature until the testing days. The compressive and flexural strengths were measured by an electromechanical universal testing machine (CMT5105, China) and the values reported were obtained by the average of six strength tests.

Table 2. Mixture proportions of samples (wt%).

Sample	A1	A2	A3	A4	A5
FGD	100	90	80	70	60
FA	0	5	10	15	20
SS	0	5	10	15	20

3. Results and Discussion

3.1. *The effect of FA and SS addition on mechanical properties of FGD*
The flexural strength and compressive strength of FGD matrix composites with different mass percentages of FA and SS added are showed in Figure 1. It can be seen from figure 1 that flexural and compressive strength values of samples after 7 and 28 days show a decreasing trend with the increase of content of FA and SS. When the additions of FA and SS are both 20wt%, the flexural and compressive strength values after 7 days reach a value of 2.12 and 8.76 MPa, respectively, which implies the decline by 34.37 and 31.62%, as compared to blank samples (A1). The drop in mechanical properties is principally attributed to the poor activity of FA and SS which causes the low speed of hydration reaction. Besides, flexural and compressive strength values of samples after 7 days have a steeper trend than those of samples after 28 days. Flexural strength and compressive strength at 28 days reach a value of 2.67 and 10.78 MPa, respectively dropping by 18.10 and 16.17%, as compared to samples without FA and SS. It can be observed that the addition of FA and SS significantly reduces the mechanical properties of FGD after both 7 and 28 days. Therefore, it is essential to improve the poor activity of FA and SS using the addition of activating agents which are propitious to the increase of mechanical properties.

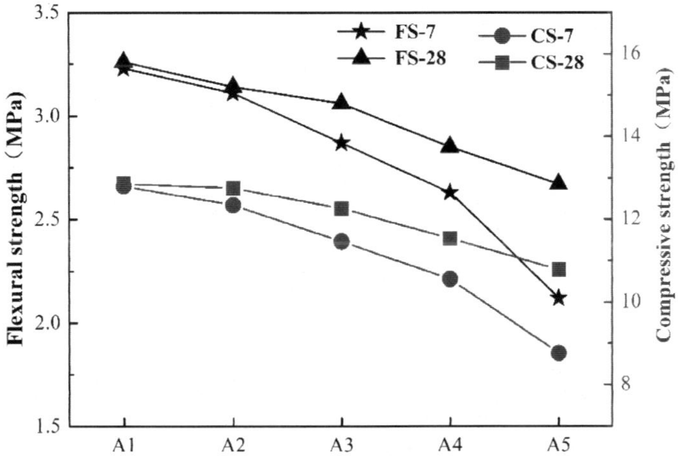

Figure 1. The flexural strength and compressive strength of FGD matrix composites with different mass percentages of FA and SS added.

3.2. The effect of NaOH addition on mechanical properties of FGD matrix composites

Figure 2 presents that the flexural strength and compressive strength of FGD matrix composites with NaOH addition (adding 0-2.5wt% NaOH into A5 sample). It can be observed that the flexural strength and compressive strength after 7 and 28 days of FGD matrix composites have mostly the same variation tendency. These mechanical properties are enhanced with the increase of NaOH at the initial stage (0-2.0wt%) and then reach the maximum values of 2.59 MPa, 10.86 MPa, 3.41 MPa and 14.13 MPa when the content of NaOH is 2.0wt% which increase by 22.17%, 23.97%, 27.72%, and 31.08% respectively compared with the samples without NaOH addition. NaOH can improve the basicity of FGD matrix composites and destroys the glass structure of FA and SS resulting in the release of silicon-oxygen bond and aluminum-oxygen bond which can react with Ca^{2+} ions SO_4^{2-} ions to produce more hydration products such as ettringite and hydrated calcium silicate. While continually increasing NaOH from 2.0wt% to 2.5wt%, these mechanical properties exhibit a downward trend which reveals that excessive NaOH addition has an adverse effect on FGD matrix composites.

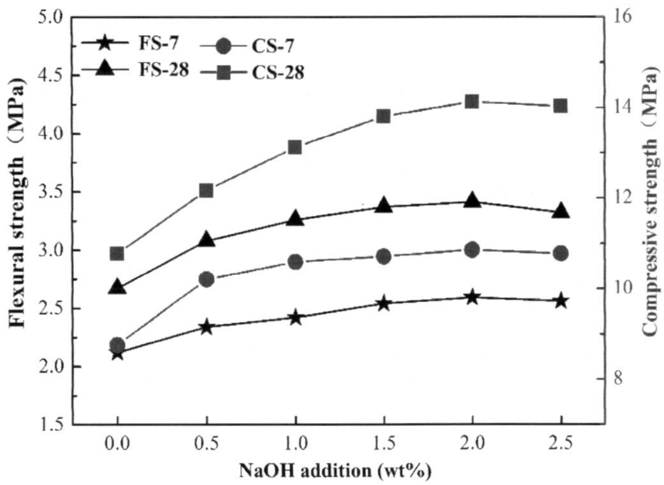

Figure 2. The flexural strength and compressive strength of FGD matrix composites with NaOH addition (0-2.5wt%).

3.3. The effect of $Al_2(SO_4)_3$ addition on mechanical properties of FGD matrix composites

Figure 3 exhibits the influence of $Al_2(SO_4)_3$ addition on flexural strength and compressive strength of FGD matrix composites (adding 0-2.5wt% $Al_2(SO_4)_3$ into A5 sample). It is evident that flexural strength and compressive strength after 7 and 28 days gradually increase when the content of $Al_2(SO_4)_3$ ranges from 0 to 1.5%. The samples with 1.5wt% $Al_2(SO_4)_3$ addition possess the highest values of 2.63, 11.02, 3.48, and 14.56 MPa, which are higher by 24.06, 25.80, 30.34, and 35.06%, respectively, as compared to samples without $Al_2(SO_4)_3$ addition. The release of Al^{3+} ions from $Al_2(SO_4)_3$ is the primary factor leading to the improvement of mechanical properties. Al $(OH)^{2+}$ is generated by the reaction of Al^{3+} and OH^- that originates from FGD matrix composites system and then converted into zeolite hydration product through reacting with Ca^{2+} and Na^+ ions to enhance the mechanical properties of samples. The slight drop in the mechanical properties emerges when the content of $Al_2(SO_4)_3$ is over 1.5wt% which illustrates the counterproductive effect of excessive $Al_2(SO_4)_3$ addition.

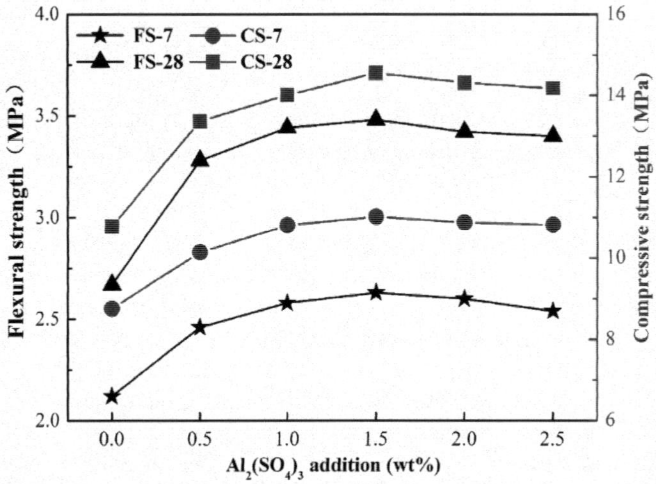

Figure 3. The flexural strength and compressive strength of FGD matrix composites with $Al_2(SO_4)_3$ addition (0-2.5wt%).

3.4. The effect of compound activating agent addition on mechanical properties of FGD matrix composites

The flexural strength and compressive strength of FGD matrix composites with compound activating agent addition are presented in Figure 4. As indicated in the curve graph, compound activating agent can significantly improve mechanical properties of FGD matrix composites. It can be clearly seen that the flexural compressive strengths after 7 days and 28 days increase gradually from 2.12, 8.76, 2.67, 10.78 MPa to 2.86, 11.94, 3.83, and 15.12 MPa with the increase of compound activating agent addition from 0 to 1.5wt%, which is attributed to the rise of hydration products caused by combined effect of various components of compound activating agent such as Portland cement clinker, $Al_2(SO_4)_3$ and Na_2SiO_3. It can be concluded by calculation that the flexural strength and compressive strength of samples with 1.5wt% compound activating agent added after 7 and 28 days is 34.90, 36.30, 43.46, and 40.26%, respectively, higher than those of samples without activating the added agent [11]. When the content of compound activating agent increase from 1.5 to 2.5wt%, the mechanical properties of samples remain essentially unchanged which discloses that more beneficial effects cannot be produced through adding overmuch compound activating agent into FGD matrix composites. Moreover, comparing compound activating agent with NaOH and $Al_2(SO_4)_3$, the activated effect of the former on FGD matrix composites is better than that of the latter two.

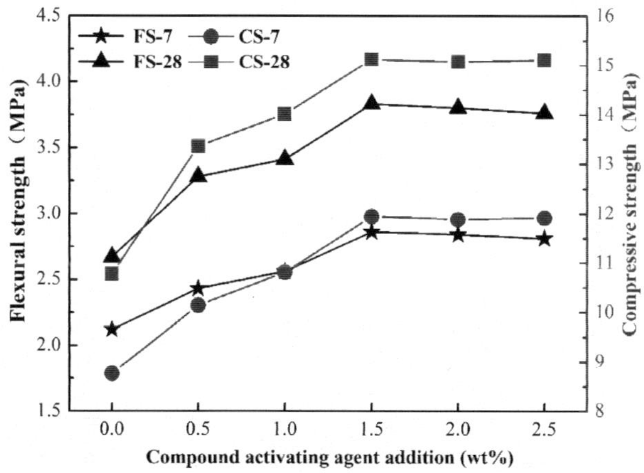

Figure 4. The flexural strength and compressive strength of FGD matrix composites with compound-activating agent addition (0-2.5wt%).

3.5. The XRD analysis

XRD patterns of FGD and FGD matrix composites are presented in Figure 5. It can be seen from pattern X that the main hydration product of pure desulphurization gypsum is made up of dihydrate gypsum, while peaks of ettringite emerge in the pattern of the FGD with FA and SS. However, the peak intensity of ettringite is very weak due to the poor activity of FA and SS which causes the low speed of hydration reaction. As demonstrated in the XRD patterns of samples with compound activating agent addition (pattern Z), the peak intensity of ettringite is strengthened compared to the peak intensity of ettringite in the samples without compound activating agent addition. It can be considered that the activation of steel slag and fly ash is stimulated after adding compound activating agent which is propitious to produce hydration products through the reaction between FGD, FA, and SS, resulting in improvement of mechanical properties of FGD matrix composites. Besides, dihydrate gypsum peak intensity of samples with FA and SS (pattern Y and Z) is lower than that of samples with FA and SS addition (pattern X) which also indicates that the reaction between FGD, FA and SS causes the reduction of dihydrate gypsum.

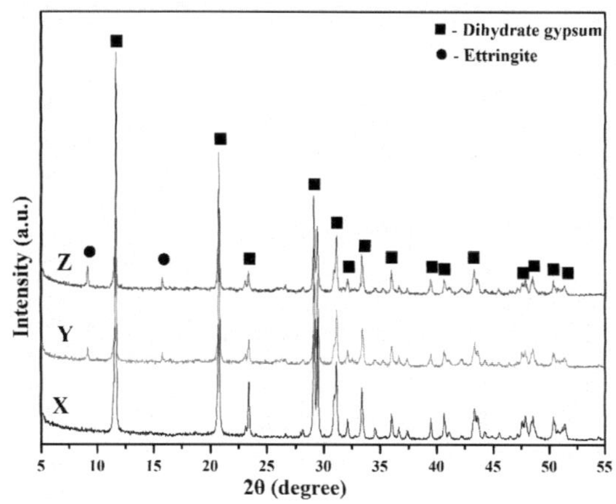

Figure 5. XRD patterns of FGD and FGD matrix composites (X: pure FGD gypsum; Y: FGD with 15wt% FA and 15wt% SS; Z: sample Y with 1.5wt% compound activating agent).

4. Conclusions

In general, the mechanical properties of FGD matrix composites gradually decrease with the increase of the proportion of FA and SS in the composite material system which is principally attributed to low hydration ability of FA and SS in the early stage.

The flexural strength and compressive strength of FGD matrix composites are improved by adding NaOH and $Al_2(SO_4)_3$ respectively which reveals that NaOH and $Al_2(SO_4)_3$ have a positive effect on hydration process of FA and SS.

The activated effect of the compound activating agent on FGD matrix composites is higher than that of NaOH and $Al_2(SO_4)_3$ due to the combined impact of various components of the compound activating agent. When the content of the compound activating agent is 1.5wt%, the mechanical properties of FGD matrix composites reach the maximum values.

Acknowledgments

This work has been financially supported by the Major Program of Shandong Province Natural Science Foundation (Grant No. ZR2017ZC0736). The authors would also like to appreciate the financial supported by the National Natural Science Foundation of China (Grant No. 51872118).

References

[1] Tesárek P, Drchalová J, Kolísko J, Rovnaníková P and Cerny R 2007 *Constr. Build. Mater* **21** 1500–9
[2] Xu L, Wu K, Li N, Zhou X and Wang P 2017 *J. Clean. Prod* **161** 803–11
[3] Zhang Y, Pan F and Wu R 2016 Study on the performance of FGD gypsum-metakaolin-cement composite cementitious system *Constr. Build. Mater* **128** 1–11
[4] Shi W, Lin C, Chen W, Hong J, Chang J, Dong Y and Zhang Y 2017 *Renew. Sust. Energy. Rev* **72** 1–9
[5] Zhu Z W, Zhuo Y Q, Fan Y M and Wang Z P 2017 *Fuel. Process. Tech* **156** 485–90
[6] Zhao F Q, Liu H J, Hao L X and Li Q 2012 *Constr. Build. Mater* **27** 531–3
[7] Yang T, Zhu H, Zhang Z, Gao X, Zhang C and Wu Q 2018 *Cement. Concrete. Res* **109** 198–207
[8] Jiang Y, Ling T C, Shi C and Pan S Y 2018 Resources *Resour. Conserv. Recy* **136** 187–97
[9] Hadi M N S, Mustafa A A and Yu T 2018 *Constr. Build. Mater* **175** 41–54
[10] Furlani E, Maschio S, Magnan M, Aneggi E, Andreatta F, Lekka M, Lanzutti A and Fedrizzi L 2018 *Ceram. Int* **44** 5226–36
[11] Duan S, Liao H, Cheng F, Song H and Yang H 2018 *Constr. Build. Mater* **187** 1113–20

Mechanical properties and thermal conductivity of lightweight thermal insulation composites

T W Liu, Z Wang*, G Z Li, G P Shi and X Zhao

School of Material Science and Engineering, University of Jinan, Jinan, China

*E-mail: wangzhi@ujn.edu.cn Tel: +86- 0531-82767636

Abstract. Lightweight thermal insulation composites with different mass percentages of expanded polystyrene (EPS) were prepared, and the effects of Polypropylene fiber (PPF) and redispersible polymer powder (RPP) addition on mechanical properties of lightweight thermal insulation composites were investigated. The results indicated that the density, thermal conductivity, flexural and compressive strength of lightweight thermal insulation composites significantly decreased with the increase of EPS. Besides, PPF and RPP both had a positive effect on the mechanical properties of lightweight thermal insulation composites.

1. Introduction

In recent years, EPS has been used in thermal insulation applications such as building wall insulation material due to its low bulk density, small thermal conductivity, excellent durability and low cost [1-2]. However, just like other organic thermal insulation materials such as polyurethane foam and phenolic foam, it tends to aggravate the fire and cause loss of life and properties in case of fire due to its inherent flammability [3-4]. Therefore, it is significant to develop thermal insulation material with superior fire retardant performance.

The inorganic cementitious materials such as ordinary Portland cement, gypsum and composite cementitious materials are non-combustible, but they also possess some defects in density and thermal insulation properties compared to organic thermal insulation materials [5]. Based on the above view, lightweight thermal insulation composites composed of EPS and the inorganic cementitious materials have given rise to broad interests of many pieces of research owing to their advantages of non-combustible, low bulk density and thermal conductivity [6-7]. Besides, industrial solid wastes such as flue gas desulphurization gypsum (FGD), silica fume (SF), fly ash (FA) and steel slag (SS) have been used to prepare inorganic composite cementitious materials, which solves the problems of occupying land resources and polluting the environment caused by the industrial solid wastes [8-9]. Nevertheless, further application of lightweight thermal insulation composites made up of EPS, and industrial solid wastes is limited by poor mechanical properties. To improve the mechanical properties, PPF [10] and RPP [11] can be added as additives.

In this paper, lightweight thermal insulation composites were prepared by mixing EPS used as lightweight aggregate and composite cementitious materials consisting of FGD, FA, and SS. The effects of PPF and RPP on mechanical properties of lightweight thermal insulation composites were researched.

2. Experimental

2.1. Raw materials

FGD and FA were supplied from Shandong Laiwu Power Plant, and SS was supplied from Laiwu Iron and Steel Group. The chemical compositions of these materials were listed in Table 1. EPS (thermal conductivity ≤ 0.041 W·(M·K)$^{-1}$, packing density: 15 kg/m^3) was used as lightweight aggregate mainly designed to reduce the density and thermal conductivity of the material. NaOH was used as activators to improve the activity of fly ash and steel slag. PPF with 850 kg/m^3 of bulk density and 25 mm in length and RPP mainly composed of Ethylene–vinyl acetate (EVA) were used as additives to improve the mechanical performance of lightweight thermal insulation composites.

Table 1. The chemical composition of raw materials (wt%).

	SO_3	MgO	SiO_2	CaO	Al_2O_3	Fe_2O_3
FGD	43.11	2.01	1.45	31.80	0.53	0.21
FA	0.64	0.93	47.13	4.13	40.33	-
SS	1.14	4.83	15.44	46.01	5.57	18.25

2.2. Sample preparation and test methods

The FGD was calcined at 155 °C for 200 min and placed at room temperature for 7 days before using it to transform it into hemihydrate gypsum. The FA and SS were ball-milled using a planetary ball milling machine at a speed of 300 rpm for 6 h before the experiment to enhance hydration activity of FA and SS. The ternary cementitious composites (TCC) were made of FGD, FA, and SS with a mass ratio of 7:2:1. The ratio of water-to-binder (TCC) in this study was 0.51. EPS was added as a lightweight aggregate at different mass percentages (0-8wt%), as shown in Table 2. PPF and RPP were combined with varying proportions of mass (0-1.0wt% and 0-5.0wt%, respectively) into lightweight thermal insulation composites (Sample A5). The above material mixtures were mixed using a mortar mixing machine (UJZ-15) at a speed of 180 rpm for 200 s. After mixing, the mixtures were placed into a steel mold and cured at room temperature in a sealed condition for 7 days. Finally, the samples were demolded (as shown in Figure 1) and tested.

Table 2. Mixture proportions of samples (wt%).

Sample	TCC	Water	NaOH	EPS
A1	100	51	1.2	0
A2	100	51	1.2	2
A3	100	51	1.2	4
A4	100	51	1.2	6
A5	100	51	1.2	8

Figure 1. Samples demolded.

According to the standard GB/T 5486-2008, the apparent density of samples with a size of 300 mm × 100 mm × 40 mm was determined as the ratio of the mass to a volume by weighing using an electronic balance (JM-1001). The flexural strength with of samples with a size of 300 mm × 100 mm × 40 mm and compressive strength with a size of 100 mm × 100 mm × 40 mm were evaluated by an electromechanical universal testing machine (CMT5105, China) and the values reported were acquired by the average of three strength tests. According to the standard GB/T 10294-2008, the samples with a size of 300 mm × 300 mm × 25 mm was prepared, and the thermal conductivity was measured by a double-plate thermal conductivity tester (IM-DRY3001).

3. Results and Discussion

3.1. The effect of EPS on properties of lightweight thermal insulation composites

The density, thermal conductivity, flexural and compressive strength of lightweight thermal insulation composites with different mass percentages of EPS added are presented in Figures 2 and 3. It can be seen that the density, thermal conductivity, flexural and compressive strength of samples significantly are affected by the addition of EPS. According to the figures, the density, thermal conductivity, flexural and compressive strength of lightweight thermal insulation composites have substantially the same variation tendency. These properties decrease with the increase of EPS from 0 to 8wt%. When the content of EPS is 8wt%, the density, thermal conductivity, flexural and compressive strength of lightweight thermal insulation composites reach a value of 213 $kg \cdot m^{-3}$, 0.053 $W \cdot (m \cdot K)^{-1}$, 0.11 MPa and 0.26 MPa which decline by 83.96, 93.23, 96.37, and 97.86%, respectively, as compared to the samples without EPS addition. EPS is a porous material that is used widely for thermal insulation applications due to its excellent performance such as low density and superior thermal insulation ability [1-2]. Therefore, the samples with a high proportion of EPS addition possess the advantages of low density and thermal conductivity. However, it can also be found that the addition of EPS has a disadvantageous effect on flexural and compressive strength. Given the above-mentioned defects caused by EPS addition, PPF and RPP are added to improve flexural and compressive strengths of lightweight thermal insulation composites, as discussed in Sections 3.2 and 3.3.

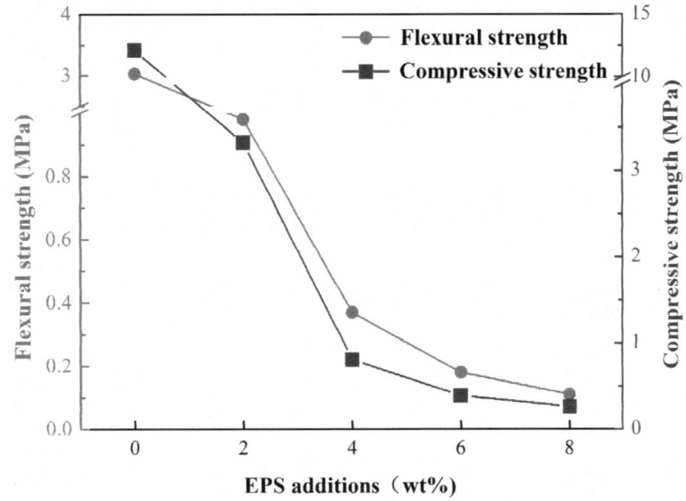

Figure 2. The flexural and compressive strength of lightweight thermal insulation composites with different mass percentages of EPS added.

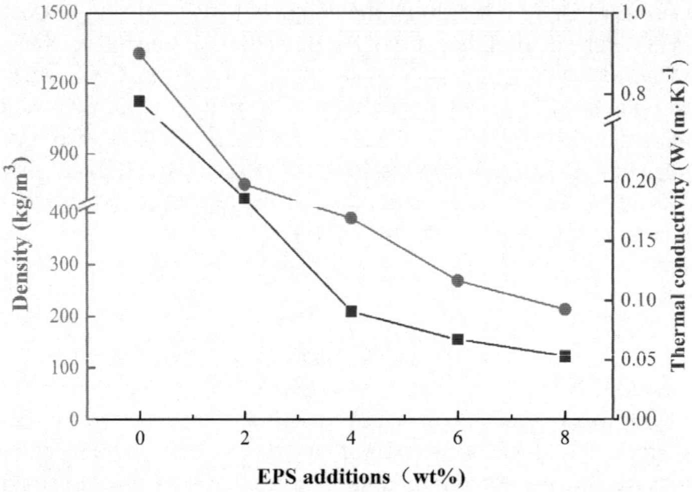

Figure 3. Density, thermal conductivity of lightweight thermal insulation composites with different mass percentages of EPS added.

3.2. The effect of PPF on mechanical properties of lightweight thermal insulation composites

Figure 4 shows the impact of different mass percentages of PPF addition on mechanical properties of lightweight thermal insulation composites. It is apparent that the flexural strength and compressive strength of the lightweight insulation material increase at the initial stage (0-0.8wt%) and then decrease (0.8wt%-1.0wt%) with the increase of polypropylene fiber. When the content of PPF is 0.8wt%, the flexural strength and compressive strength of the sample reach the maximum values of 0.18 MPa and 0.27 MPa which are 63.64 and 7.69%, respectively, higher than those of the samples without PPF. When lightweight insulation materials are subjected to external forces, cracks will produce and then propagate quickly in the matrix of composites which cause a rapid loss in mechanical properties. However, PPF distributed among the matrix will intercept the development of crack or make the crack propagation path longer resulting in the improvement in flexural strength and compressive strength [10]. Besides, another reason why PPF can improve the mechanical properties of composites is that PPF is a material with excellent mechanical properties such as high strength and outstanding wear resistance. Nevertheless, when the content of fibers is too high, the fibers are difficult to disperse uniformly and aggregate together inside the material matrix which leads to the decline of mechanical properties.

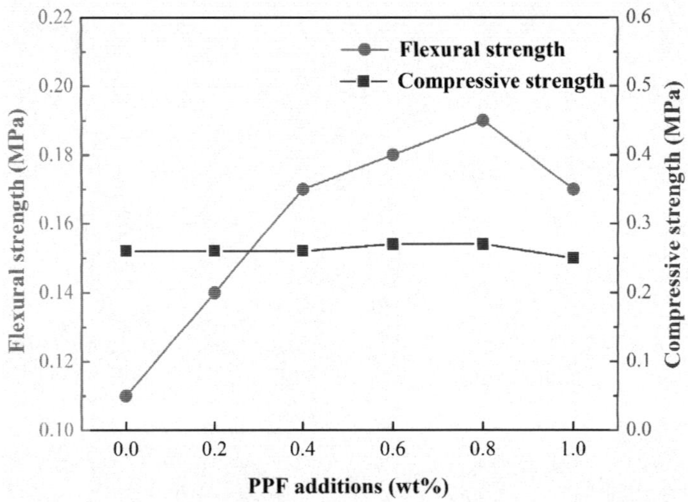

Figure 4. The flexural and compressive strength of lightweight thermal insulation composites with different mass percentages of PPF added.

3.3. The effect of RPP on mechanical properties of lightweight thermal insulation composites

The impact of RPP addition on flexural strength and compressive strength of lightweight thermal insulation composites (adding 0-5wt% RPP into A5 sample) are presented in Figure 5. It is seen that the flexural strength and compressive strength of lightweight thermal insulation composites gradually increase when the content of RPP ranges from 0 to 3wt%. While continually increasing from 3wt% to 5wt%, it remains the peak value. When the RPP addition is 3wt%, flexural strength and compressive strength reach a maximum value of 0.16 and 0.35 MPa, which increase by 45.45 and 34.62%, respectively, as compared to the samples without RPP addition. On the one hand, the increase of mechanical properties is attributed to the improvement of pore structure by the addition of RPP [11]. On the other side, the interface combining effect of TCC and EPS is not stable due to the poor wettability between them which leads to a decline in mechanical properties. It can be seen from Figure 6 that the EPS surface of samples without RPP added is smooth, while the EPS surface of samples with RPP added adheres in no small amount of cementitious materials, which illustrates that RPP can improve interface combining effect increasing mechanical properties.

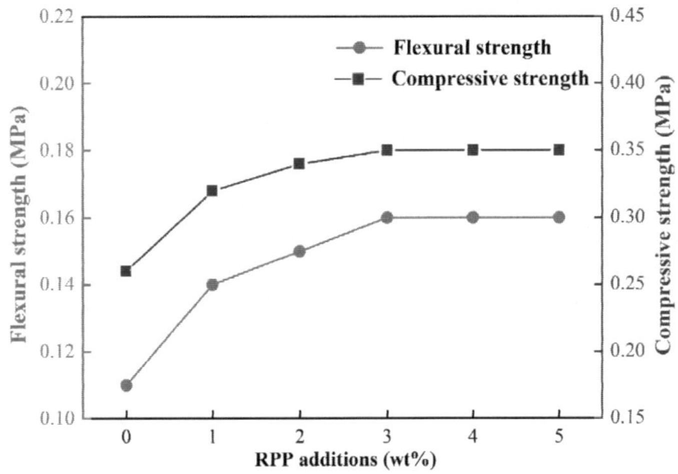

Figure 5. The flexural and compressive strength of lightweight thermal insulation composites with different mass percentages of RPP added.

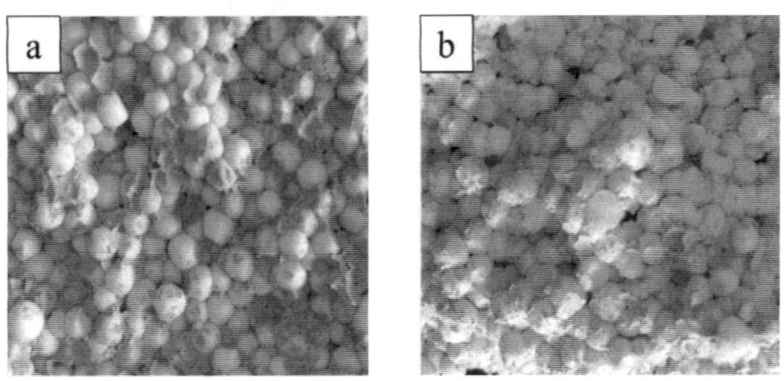

Figure 6. The images of fracture surface for lightweight thermal insulation composites with (a) 0wt%, (b)3wt% RPP added.

4. Conclusion

The density, thermal conductivity, flexural and compressive strengths of lightweight thermal insulation composites sharply decreased with the increase of EPS which was mainly attributed to low bulk density, small thermal conductivity and poor mechanical strength of EPS. PPF could improve the mechanical properties of lightweight thermal insulation composites to a certain extent. When the content of PPF exceeded 0.8wt%, it was proven to be counterproductive due to the aggregation effect of PPF in the composites. The flexural and compressive strength of lightweight thermal insulation composites gradually increased with the increase of RPP and then attained the peak value with the continual rise of RPP from 3wt% to 5wt%.

Acknowledgments

This work has been financially supported by the Major Program of Shandong Province Natural Science Foundation (Grant No. ZR2017ZC0736). The authors would also like to appreciate the financial supported by the National Natural Science Foundation of China (Grant No. 51872118).

References

[1] Brooks A L, Zhou H and Hanna D 2018 *Constr. Build. Mater* **159** 316–328
[2] Miskinis K, Dikavicius V, Buska A and Banionis K 2018 *Appl. Acoust* **137** 62–68
[3] Hu X M and Wang D M 2013 *J. Appl. Polym. Sci* **129** 238–246
[4] Zhou L, Chen A, Gao L and Pei Z 2017 *Fire. Safety. J* **91** 155–164
[5] Zhang Y, Pan F and Wu R 2016 *Constr. Build. Mater* **128** 1–11
[6] Colangelo F, Roviello G, Ricciotti L, Ferrándiz-Mas V, et al. *Concrete. Comp* **86** 66–272
[7] Sayadi A A, Tapia J V, Neitzert T R and Clifton G C 2016 *Constr. Build. Mater* **112** 716–724
[8] Duan S, Liao H, Cheng F, Song H and Yang H 2018 *Constr. Build. Mater* **187** 1113–20
[9] Jiang Y, Ling T C, Shi C and Pan S Y 2018 *Resour. Conserv. Recy* **136** 187–197
[10] Li L G, Chu S H, Zeng K L, Zhu J and Kwan A K H 2018 *Concrete. Comp* **93** 196–204
[11] Ma C and Chen B 2016 *Constr. Build. Mater* **113** 255–263

Water resistance of flue gas desulphurization gypsum-fly ash-steel slag composites

T W Liu, Z Wang*, G Z Li, G P Shi and X Zhao

School of Material Science and Engineering, University of Jinan, Jinan, China

E-mail: wangzhi@ujn.edu.cn Tel: +86- 0531-82767636

Abstract. Flue gas desulphurization gypsum-fly ash-steel slag (FGD-FA-SS) composites with different mass percentages of NaOH and redispersible polymer powder (RPP) were prepared, and the effects of NaOH and RPP added on the water resistance of composites were researched. The results demonstrated that the addition of NaOH improved the water absorption and softening coefficient of composites due to the production of more hydration products. And RPP could modify the pore structure and decrease the gypsum solubility resulting in the improvement of water resistance. Besides, NaOH and RPP co-doped had a more positive effect on water absorption and softening coefficient of composites compared to NaOH or RPP addition separately which was mainly attributed to the synergistic effect of NaOH and RPP on FGD-FA-SS composites.

1. Introduction

In recent years, a large number of industrial solid wastes such as FGD, FA, and SS have been produced with the rapid development of industrialization which occupy land resources and cause environmental pollutions [1-2]. Therefore, measures must be taken on the utilization of industrial waste solid in large scale. FGD, as a cementitious material, has attracted the attention of researchers owing to its high early strength, low energy consumption, nontoxicity and low density [3-5]. However, it is hard for pure FGD to meet the requirements in building materials application due to its weakness such as low late strength and poor water resistance [6]. FA and SS are also industrial solid wastes that can be added into FGD to improve its weakness owing to pozzolanic activity of FA and potential cementitious action of SS which lead to the production of various hydration products such as the hydrated calcium silicate and hydrated calcium aluminate that can fill pore spaces and reduce the water permeation into the hardened cementitious composites [7-8]. Besides, hydration activity of FA and SS is improved by the addition of NaOH that can destroy the glass phase by increasing the alkalinity of the composites to generate more hydration products, which are beneficial to the improvement of water resistance [9-10]. RRP is an organic copolymer material that is usually produced by spray-drying polymer emulsion containing an anti-caking agent and a colloidal stabilizer. Ma *et al.* [11] investigated the influence of RPP addition on the waterproof performance of cement-based composites and concluded that adding RPP could significantly improve the water resistance by modifying the pore structure and reducing the water permeation.

Moreover, Pagona et al. [12-13] also drew the same conclusion by testing on the OPC mortar. Hence, it is believed that water resistance of FGD-FA-SS composites has excellent potential to be strengthened by the addition of RPP. Although NaOH and RPP have a beneficial effect on water resistance, few pieces of research on the impact of NaOH and RPP addition on water resistance properties of FGD-FA-SS composites has been reported.

This paper aimed to enhance the waterproof performance of FGD-FA-SS composites by adding NaOH and RPP and the water resistance was evaluated by water absorption and softening coefficient.

2. Experimental

2.1. Raw materials and treatment

FGD and FA used in this work were collected from Shandong Laiwu Power Plant in China and SS was supplied from Laiwu Iron and Steel Group in China. The chemical compositions of FGD, FA and SS were shown in table 1. The commercially available NaOH and RPP were used in this study to improve the waterproof performance of FGD-FA-SS composites. Before the experiment, the FGD was calcined at 155 °C for 200 min in a high temperature industrial furnace and placed at room temperature for 7 days to transform it into hemihydrate gypsum. Besides, the FA and SS were ball-milled using a planetary ball milling machine at a speed of 300 rpm for 6 h before using them to excite hydration activity of FA and SS.

Table 1. The chemical composition of raw materials (wt%).

	SO_3	MgO	SiO_2	CaO	Al_2O_3	Fe_2O_3
FGD	43.11	2.01	1.45	31.80	0.53	0.21
FA	0.64	0.93	47.13	4.13	40.33	-
SS	1.14	4.83	15.44	46.01	5.57	18.25

2.2. Samples preparation and test methods

In this study, the ratio of water-to-binder (FGD-FA-SS composites) was 0.51, and the proportions of FGD, FA, and SS in FGD-FA-SS composites were 7:2:1. Moreover, NaOH and RPP were added with different mass percentages, as shown in table 2. The above raw materials were mixed in a paste mixer at a speed of 140 rpm for 150 s. The samples were formed in a steel mold and demolded after 2 hours. The specimens were cured at room temperature for 28 days.

Table 2. The mixed proportions of FGD-FA-SS composites (wt%).

Sample	FGD	FA	SS	Water	NaOH	RPP
A0	70	20	10	0.51	0	0
AI	70	20	10	0.51	0.5	0
A2	70	20	10	0.51	1.0	0
A3	70	20	10	0.51	1.5	0
A4	70	20	10	0.51	2.0	0
A5	70	20	10	0.51	2.5	0
B1	70	20	10	0.51	0	1
B2	70	20	10	0.51	0	2
B3	70	20	10	0.51	0	3
B4	70	20	10	0.51	0	4
B5	70	20	10	0.51	0	5
C1	70	20	10	0.51	0.5	1
C2	70	20	10	0.51	1.0	2
C3	70	20	10	0.51	1.5	3
C4	70	20	10	0.51	2.0	4
C5	70	20	10	0.51	2.5	5

The water resistance of composites was evaluated by water absorption at 2 hours and 24 hours and softening coefficient in this paper. The water absorption of samples was calculated using the following equation:

$$W = (G_2 - G_1)/G_1 \times 100\% \tag{1}$$

where W is water absorption of samples at 2 or 24 hours, G_1 is mass of the completely dried samples before putting in water, G_2 is mass of the samples after absorbing water for 2 or 24 hours. The softening coefficient was measured according to equation (2) as follows:

$$f = R_2/R_1 \tag{2}$$

where f is softening coefficient of samples, R_1 is compressive strength of the completely dried samples before putting in water, R_2 is compressive strength of samples whose water content reaches the saturation point.

3. Results and Discussion

3.1. The effect of NaOH on water absorption and softening coefficient of FGD-FA-SS composites

The water absorption and softening coefficient of samples with different mass percentages of NaOH added are presented in figure 1. It can be seen that the addition of NaOH reduces water absorption at 2 hours and 24 hours while it increases softening coefficient of samples, which illustrates that the water resistance of composites is improved by the addition of NaOH. With regard to the water absorption at 2 hours and 24 hours, they initially decrease with the increase of NaOH addition from 0 to 1.5wt% and then basically remain the minimum value of 18.36 and 19.93%, respectively, when the content of NaOH is over 1.5wt%, which decreased by 40.52 and 37.80%, respectively, as compared to samples without NaOH addition. As for the softening coefficient, it exhibits an increasing trend at the beginning stages and then remains mainly unchanged. When the content of NaOH is 1.5%, softening coefficient reaches a maximum value of 0.53 increasing by 51.43%, as compared to blank samples. The alkalinity of FGD-FA-SS composites is improved by the addition of NaOH which decomposes the glass structure of FA and SS and promotes the release of silicon-oxygen bond and aluminum-oxygen bond that trend to react with Ca^{2+} and SO_4^{2-} ions to produce more hydration products such as ettringite and hydrated calcium silicate. The holes between dihydrate gypsum crystals can be filled with these products resulting in densification of FGD-FA-SS composites. Besides, the products are adhered to the surface of dihydrate gypsum crystals which hinders dissolution of desulfurized gypsum and improve the water resistance of FGD-FA-SS composites.

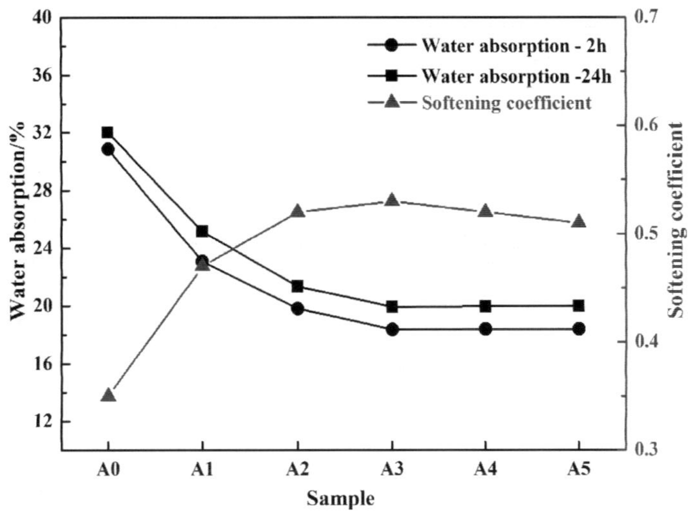

Figure 1. The water absorption and softening coefficient of samples with different mass percentages of NaOH added.

3.2. The effect of RPP on water absorption and softening coefficient of FGD-FA-SS composites

Figure 2 illustrates the impact of RPP on water absorption and softening coefficient of FGD-FA-SS composites. As shown in the curve graph, the water absorptions at 2 and 24 hours gradually decrease, while softening coefficient gradually increases with RPP from 0 to 5wt%. It is evident that the tendency of descent or increase is sharp when the content of RPP is below 3wt% and the curves of water absorption and softening coefficient both become smooth when the content of RPP is over 3wt%. According to the values of water absorption and softening coefficient in figures 1 and 2, water absorption at 2 and 24 hours and softening coefficient of composites with 3wt% RPP addition are 16.36%, 18.93% and 0.55 slightly better than those of the samples with NaOH addition. The increase of water resistance for samples with RPP addition is chiefly attributed to the improvement of the pore structure which is propitious to reduce the water permeation into FGD-FA-SS composites [12]. Besides, RPP tends to generate polymer film on the surface of dihydrate gypsum crystals which reduces the solubility of gypsum resulting in improvement of water resistance.

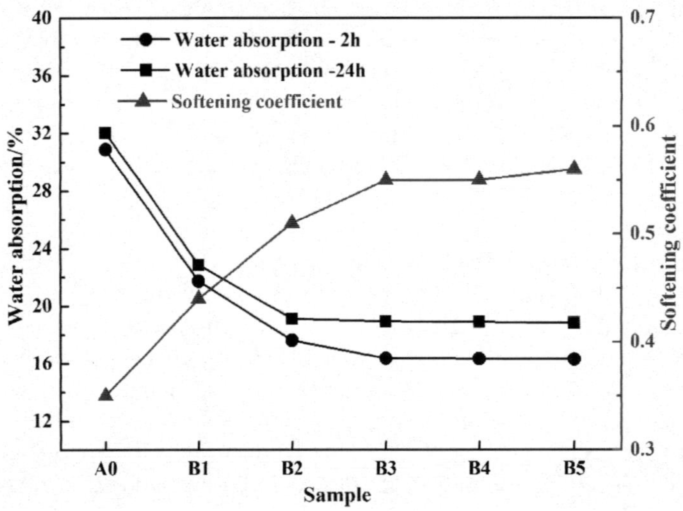

Figure 2. The water absorption and softening coefficient of samples with different mass percentages of RPP added.

3.3. The effect of NaOH and RPP co-doped on water absorption and softening coefficient of FGD-FA-SS composites

The results of water absorption and softening coefficient of FGD-FA-SS composites with NaOH and RPP co-doped are presented in figure 3. As shown in figure 3, water absorptions at 2 hours and 24 hours exhibit a decreasing trend and softening coefficient shows an increasing trend at the beginning stage, while they remain substantially unchanged when the content of NaOH exceeds 1.5wt% (or the content of NaOH exceeds 3wt%). It can be found that water absorptions at 2 hours and 24 hours and softening coefficient of FGD-FA-SS composites with NaOH and RPP co-doped (1.5wt% NaOH and 3wt% RPP) basically reach an optimal value of 12.15%, 13.68% and 0.62 which are better than water resistance of the blank samples remarkably and similar to water resistance of other cementing materials [14-15]. Moreover, NaOH and RPP co-doped have a more positive effect on water absorption and softening coefficient of composites compared to NaOH or RPP addition separately which is mainly attributed to the synergistic effect of NaOH and RPP on FGD-FA-SS composites. It can also be found from Sections

3.1 and 3.2, NaOH and RPP both improve the pore structure and decrease the gypsum solubility in different ways.

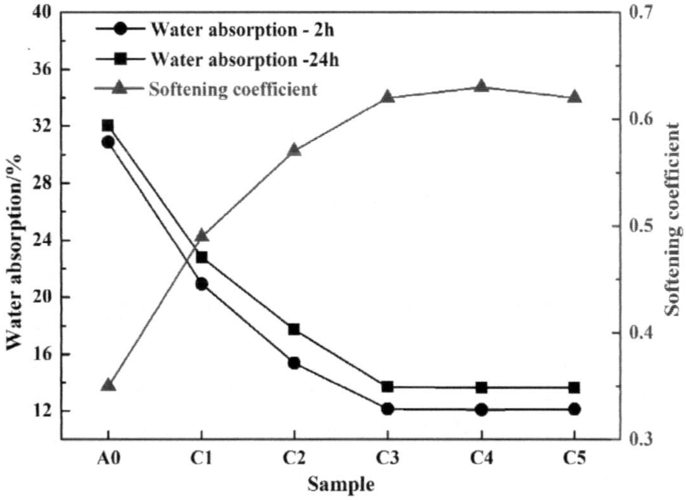

Figure 3. The water absorption and softening coefficient of samples with different mass percentages of NaOH and RPP co-doped.

3.4. The SEM analysis

Figure 4. The SEM micrographs of fracture surface for FGD-FA-SS composites with (a) 0wt%, (b)1.5wt% NaOH, (c) 3wt% RPP, (d) 1.5wt% NaOH and 3wt% RPP added.

The SEM micrographs of the fracture surface for FGD-FA-SS composites are shown in figure 4. It can be seen that the pore contents decrease with the addition of NaOH and more hydration products can be observed on the surface of FA and SS in the samples with NaOH addition which gives rise to the improvement of water resistance (figure 4(a) and (b)). Besides, the samples with RPP addition are denser than those without RPP addition, which is mainly attributed to the improvement of the pore structure caused by RPP addition [12]. It can also be observed from figure 4(a) and (d) that the decreasing trend of pore contents is caught sight and the more hydration products are adhering to the surface of FA and SS in the FGD-FA-SS composites with NaOH and RPP addition due to combined effects of NaOH and RPP resulting in the improvement of waterproof performance (discussed in Section 3.3).

4. Conclusion

NaOH could promote the production of more hydration products such as ettringite and hydrated calcium silicate that led to the improvement of water resistance of FGD-FA-SS composites. When the content of NaOH was 1.5wt%, water absorptions at 2 and 24 hours and the softening coefficient reached an optimum value of 18.36%, 19.93%, and 0.53, respectively.

The addition of RPP could remarkably improve the water resistance performance due to the improvement of the pore structure and decrease of the gypsum solubility. Besides, the water absorption at 2 and 24 hours and softening coefficient of composites with 3wt% RPP addition were 16.36%, 18.93% and 0.55 slightly better than those of the samples with 1.5wt% NaOH addition.

The FGD-FA-SS composites with NaOH and RPP co-doped had a better water resistance performance compared to the samples with NaOH, or RPP added separately which was mainly attributed to the synergistic effect of NaOH and RPP on composites.

Acknowledgments

This work has been financially supported by the Major Program of Shandong Province Natural Science Foundation (Grant No. ZR2017ZC0736). The authors would also like to appreciate the financial supported by the National Natural Science Foundation of China (Grant No. 51872118).

References

[1] Yan Y, Li Q, Sun X, Ren Z, He F, Wang Y and Wang L 2015 *J. Colloid. Interf. Sci* **457** 1–11
[2] Ren C, Wang W, Mao Y, Yuan X, Song Z, Sun J and Zhao X 2017 *J. Clean. Prod* **167** 1314–24
[3] Zhang Y, Pan F and Wu R 2016 *Constr. Build. Mater* **128** 1–1
[4] Xu L, Wu K, Li N, Zhou X and Wang P 2017 *J. Clean. Prod* **161** 803–811
[5] Tesárek P, Drchalová J, Kolísko J, Rovnaníková P and Cerny R 2007 *Constr. Build. Mater* **21** 1500–09
[6] Zhao F Q, Liu H J, Hao L X and Li Q 2012 *Constr. Build. Mater* **27** 531–533
[7] Jiang Y, Ling T C, Shi C and Pan S Y 2018 *Resour. Conserv. Recy* **136** 187–197
[8] Duan S, Liao H, Cheng F, Song H and Yang H 2018 *Constr. Build. Mater* **187** 1113–20
[9] Hefni Y, Zaher Y A E and Wahab M A 2018 *Constr. Build. Mater* **172** 728–734
[10] Gu X, Yu B, Dong Q and Deng Y 2018 *J. Clean. Prod* **181** 102–108
[11] Ma C and Chen B 2016 *Constr. Build. Mater* **113** 255–263
[12] Maravelaki-Kalaitzaki P 2007 *Cement. Concrete. Res* **37** 283–290
[13] Collepardi M 1999 *Cement. Concrete. Comp* **21** 147–154
[14] Zhang X, Li G X, Niu M D and Song Z P 2018 *Constr. Build. Mater* **175** 768–776
[15] He P P, Poon C S and Tsang D C W 2017 *Constr. Build. Mater* **147** 519–524

IOP Conf. Series: Materials Science and Engineering **474** (2019) 012040 doi:10.1088/1757-899X/474/1/012040

Effects and mechanisms of slag reinforced coal gangue geopolymers

H Q Ma*, H G Zhu, H Y Chen, Y D Ni, T Wang and S Yang

School of Mechanics and Civil Engineering, China University of Mining and Technology (Beijing), Beijing 100083, China

*E-mail: mahongqiang1992@163.com

Abstract. This paper uses alkali activators with the modulus of 1.30 to prepare coal gangue geopolymers, thereby studying the reinforcement effects of slag content on coal gangue geopolymers. Besides, XRD, TG-DTG, and SEM are introduced to analyze changes in the microstructure and further explore the change mechanisms. The experimental results show that the compressive strength of coal gangue geopolymers increases with the increase in slag content. Compared to the P.O 42.5 pure cement paste, coal gangue geopolymers can obtain higher early strength. Besides, the 28d compressive strength of coal gangue geopolymers is greater than P.O 42.5 pure cement specimens when the slag content is 20%. The geopolymer products of slag-containing coal gangue geopolymers are C-(A)-S-H and N-A-S-H gels. The two kinds of gels, which are highly compatible, jointly constitute the network-like disordered structure of polymers. The XRD, TG-DTG and SEM results are consistent with the strength analysis results.

1. Introduction

The cement industry has always been a heavy polluter among the various building materials industries. About 1.6 billion tons of cement is consumed annually, emitting 150.4 million tons of CO_2 [1-2]. Cement accounts for 5% to 8% of global CO_2 emissions [3]. Therefore, it is the development direction of modern cement industry to develop green ecological cement. Geopolymers mainly refer to inorganic polymeric aluminosilicate materials that have similar properties to cement. Studies have shown that geopolymer cement has high strength, corrosion resistance, fire resistance and better durability [4-7]. Geopolymers have become the most promising cementitious materials to replace cement due to their low carbon footprint.

Coal gangue coexists with coal, accounting for 15%-20% of coal production. Coal gangue has become one of the most significant industrial solid wastes in coal-rich countries such as China, India, Australia, and South Africa [8-10]. Currently, industrial solid wastes containing silicon-aluminum materials, including fly ash, slag, kiln dust, steel slag, and red mud, can all be used to prepare geopolymers. So far there has been little research on the coal gangue geopolymer (alkali-activated coal gangue), hydration characteristics and microstructure have also not been studied. Geng *et al.* [11] mixed 20% coal gangue into red mud to prepare high-strength red mud-coal gangue geopolymers but did not determine its optimal application parameters. Zhang [12] mixed 30% coal gangue into cement and found that the main hydration products of burned coal gangue-cement were C-S-H, C-A-H, CH, and AFM. Zhang *et al.* [13] revealed that the activation product of alkali-activated coal gangue is an amorphous alkali-aluminosilicate gel having a structure similar to that of zeolitic precursors. Yi *et al.* [14] prepared coal gangue geopolymers with sodium hydroxide and sodium silicate and found that coal gangue geopolymers can obtain higher early strength, but the 28d strength is lower than P.O. 42.5 pure cement specimens. Through the analysis of alkali-activated slag, steel slag, and red mud, alkali-activated silicon-aluminum materials containing calcium can achieve higher strength. The differences in precursor reactants eventually result in the differences in geopolymerization products. Bignozzi *et al.* discovered that calcium could improve the mechanical properties and the microstructure of silicon-aluminum geopolymers [15]. In the process of geological polymerization, the alkali metal cation Na+ is used to charge-balance, but it is possible to balance the charge by Ca^{2+} ion rather than Na^{+} in the presence of calcium source [16].

In summary, this paper uses sodium hydroxide and sodium silicate as the alkali activators to study the impact of slag as a calcium source on the strength of coal gangue geopolymers. Besides, XRD, TG-DTG, and SEM are introduced to explore the geopolymerization mechanisms further. The primary purpose of this study is to investigate the reinforcing effect of slag on coal gangue geopolymer and analysis the reaction mechanism of alkali activator slag-coal gangue.

2. Experimental procedures

2.1. Materials

The coal gangue selected (clay rock coal gangue) was calcined at a constant temperature of 700°C for 2 hours. Table 1 shows that the primary chemical compositions of calcined coal gangue and slag. In the experiment, the alkali activators were used 96% sodium hydroxide(NS) and sodium silicate(NS) with a modulus of 3.22 (26.5% SiO_2, 8.5% Na_2O, 65% H_2O).

Figure 1 demonstrates the XRD patterns of original and 700°C calcined coal gangue. The principal mineral components of coal gangue are kaolinite, muscovite and quartz. After high-temperature calcination, the diffraction peak of kaolinite in coal gangue decreases, and the diffraction peak of quartz increases.

Table1. The primary chemical composition of coal gangue and slag.

Materials	SiO_2	Al_2O_3	CaO	Fe_2O_3	MgO	Na_2O	TiO_2
Coal gangue	56.56	36.78	0.62	1.95	0.22	0.42	2.10
Slag	30.58	14.04	38.43	0.35	10.57	0.57	1.93.

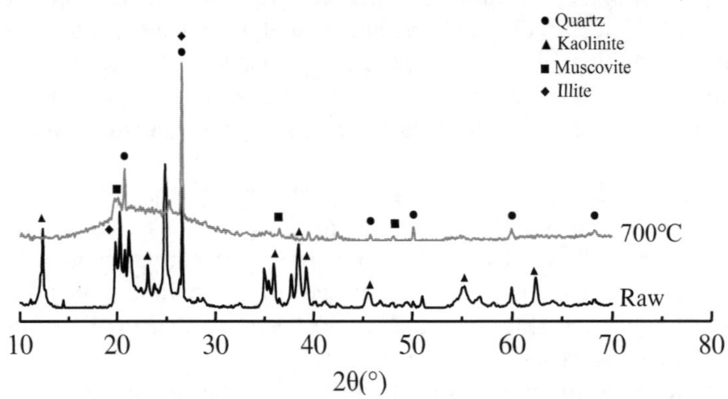

Figure 1. XRD patterns of original and 700°C calcined coal gangue.

2.2. Specimens preparation

The modulus of alkali activators and the liquid-solid ratio were set to 1.30 and 0.36 separately. In addition, the coal gangue was 0, 10, 20, 30, 40, and 50% replaced by slag, respectively. The geopolymer mix proportion is shown in table 2. The alkali activation solutions were prepared 24 hours before use, allowed to cool to room temperature, ensuring uniform mixing of the activator components and improve the polymerization process. After the coal gangue and slag were uniformly mixed, we added alkali activation solutions to the mixture and mixed agitation for 3 minutes, then quickly pour the fresh paste into the steel mold (40mm×40mm×40 mm) and vibrate on the electric vibration table to remove residual air. Then covered with a thin polyethylene film, demolded after 24h of curing under standard curing conditions(curing temperature 20±2°C, humidity≥95%), the test block was moved to standard curing room until test age(1d, 3d, 7d, and 28d).

Table 2. Mix proportion of geopolymer.

Specimens	Coal gangue: slag	Liquid-solid ratio	Alkali-activator modulus
Cement	—	0.36	—
S0	100:0	0.36	1.30
S10	90:10	0.36	1.30
S20	80:20	0.36	1.30
S30	70:30	0.36	1.30
S40	60:40	0.36	1.30
S50	50:50	0.36	1.30

*a**: mass ratio

2.3. Specimens and test methods

2.3.1. Compressive strength. The compressive strength of the paste specimens was tested following the Standard for Test Method of Performance on Building Mortar. The 1d, 3d, 7d and 28d compressive strengths of each set of specimens were tested, and three specimens of each set were selected to determine the average compressive strength.

2.3.2. Microstructure. Microscopic tests for this test included X-ray diffraction (XRD), Thermogravimetric analysis (TG-DTG), scanning electron microscopy analysis (SEM). The XRD test used the German Bruker D8 X-ray diffractometer, and the specimen was continuously scanned at 5° to 70° (2θ°) with a step size of 0.02°. The TG-DTG test used the German Netzsch STA 449 F3 thermal analyzer, the experiment was carried out under a nitrogen atmosphere to 600°C, and the heating rate was 10°C/min. The SEM test used the field emission scanning electron microscope (SU8010; Hitachi; Japan) to observe surface morphology of the broken specimens.

3. Results and Discussion

3.1. Compressive strength

Figure 2 exhibits the compressive strength of pure cement specimens and specimens with different slag contents. It can be seen that the compressive strength of coal gangue geopolymers increases with the increase of slag content. In comparison with S0 specimens, specimens with the slag content of 10, 20, 30, 40 and 50% exhibited an increase in compressive strength by 18.78, 23.17, 29.22, 32.93, and 55.98%, respectively. Besides, coal gangue geopolymers achieved higher initial strength, thereby enjoying higher 1d, 3d, and 7d compressive strengths. The compressive strength of geopolymers increased slowly from the 7th day, and their 7d compressive strength reached 95% of the 28d compressive strength. However, the 28d compressive strength value of P.O 42.5 cement specimens was 48.42% greater than their 7d compressive strength value. Besides, geopolymers had a greater 28 compressive strength than P.O 42.5 cement specimens when the slag content was 20%. Therefore, coal gangue geopolymers are more suitable for projects that require high early strength. The increase in slag content tends to bring a more substantial increase in the compressive strength.

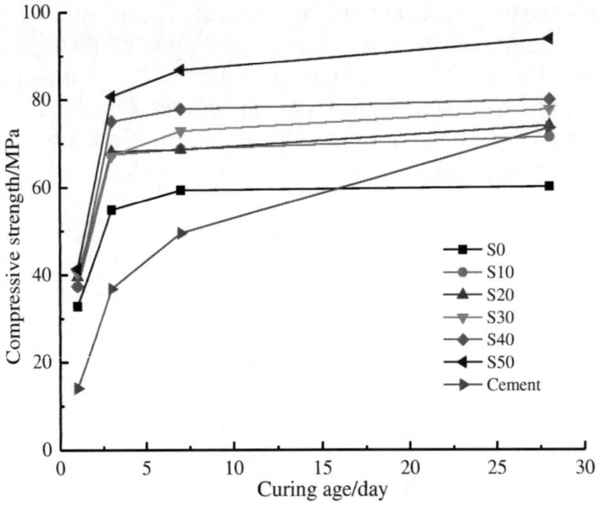

Figure 2. The compressive strength of specimens with different slag content and cement.

3.2. XRD analysis

Figure 3 demonstrates the XRD patterns of specimens with different slag contents. In the figure, sharp and broad peaks between 20° and 30° are essentially the characteristic reflections of amorphous geopolymers, indicating the dominant position of SiO_2 and Al_2O_3-containing components. Besides, the diffraction peaks of quartz in the original coal gangue still exist, there is also the formation of new minerals such as $CaAl_5Si_{20}O_{10}\cdot 4H_2O$, $NaAlSiO_2\cdot H_2O$, and AlSi2O5·xH_2O. It can be seen from the figure that the diffraction peak of zeolite phases enhances while that of quartz weakens with the increase in slag content. According to the analysis of XRD patterns by Jade, there are diffraction peaks of calcium zeolite ($Ca[Al_2Si_3O_{10}]\cdot 3H_2O$, PDF#29-0809). Namely, calcium alumino-silicate is generated in this process. The availability of additional reactive calcium would have increased the amount of reactive calcium favorable for the formation of C–S–H gel [17]. As the slag content increases, the amount of active substances involved in the alkali-activated reaction increases, thereby improving the chemical reactivity. It not only reduces the content of inert quartz in the mixture but also increases the amount of calcium alumino-silicate produced. The experimental results are consistent with the strength analysis results.

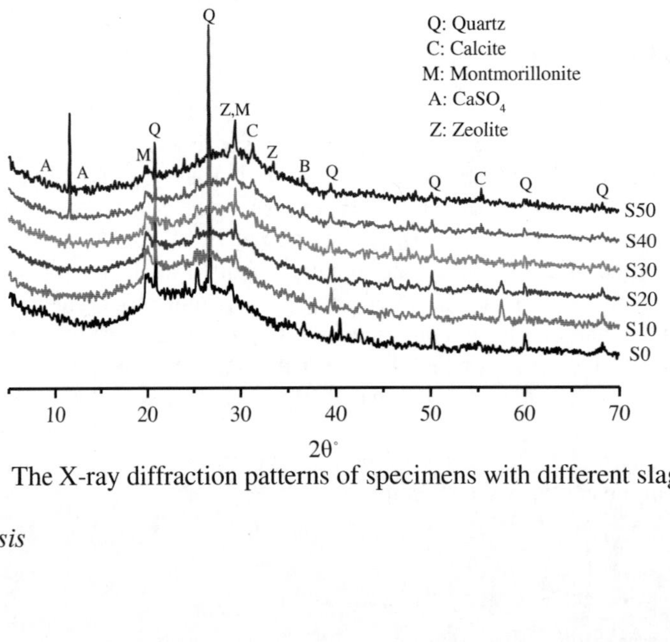

Figure 3. The X-ray diffraction patterns of specimens with different slag content

3.3. TG-DTG analysis

Figure 4 shows the TG-DTG curves (40°C to 600°C) of specimens with the slag content of 0%, 30% and 50% separately. Weight loss of the specimens, which is mainly manifested as the water loss of zeolite minerals and the binding gel. Specifically, specimens with the slag content of 0%, 30%, and 50% reached the weight loss peaks at 116.28°C, 119.82°C and 134.77°C respectively, which indicates that the weight loss peak temperature increases with the increase in slag content, due to modification of the structure of coal gangue geopolymer gel by incorporation of slag and the formation of C–A–S–H gel whose free water is more tight to the structure [16]. The peaks at 110°C-250°C are related to the dehydration of the (C, N)-A-S-H gel and ettringite. Weight loss of the three specimens below 250°C was 7.50%, 8.43%, and 8.83% respectively while the total loss of the three specimens was 8.39%, 9.83%, and 10.98% respectively. As the slag content increases, the amount of hydrous aluminosilicate gels (mainly calcium aluminum silicate gels) increases accordingly. The growth in hydrous aluminosilicate gels densifies the structure of geopolymers and improves the compressive strength, which is consistent with the strength analysis results.

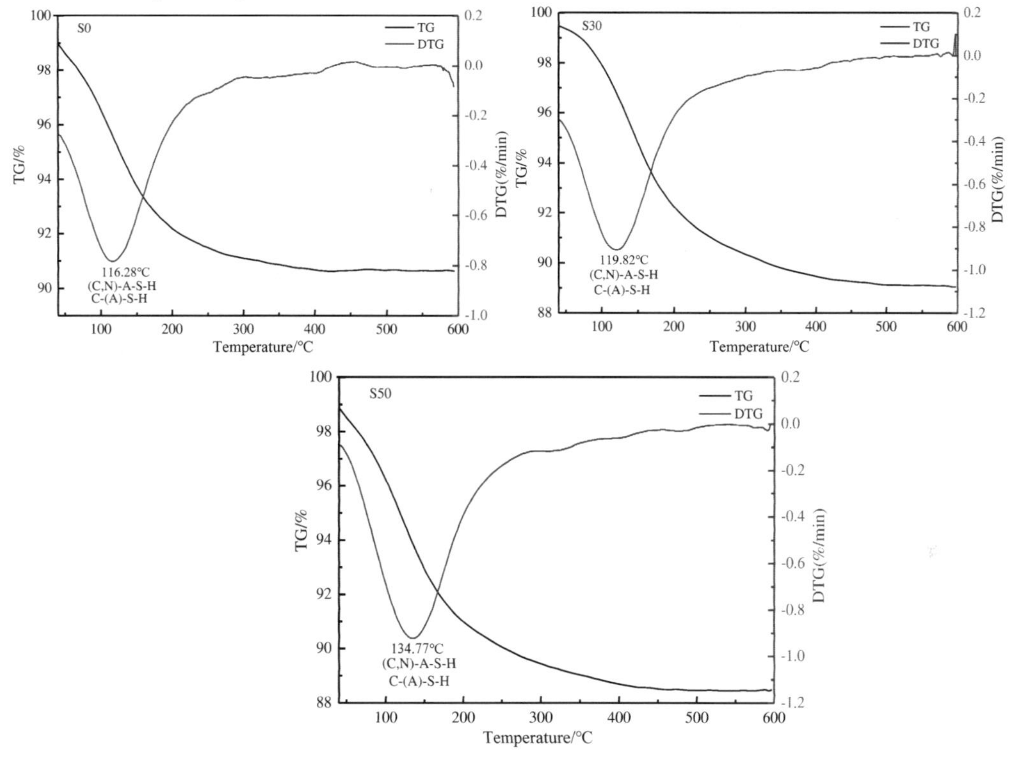

Figure 4. Thermogravimetric analysis of S0, S30, and S50 specimens (40°C~600°C).

3.4. SEM analysis

Figure 5 displays the SEM patterns of specimens with the slag content of 0%, 30% and 50% separately. It can be seen from the figure that there are significant differences in the microstructure of geopolymer specimens. Besides, the amount of gelatinous and flocculent products significantly increases with the increase in slag content. As the slag content increases, a large amount of CaO is introduced into the mixture. Ca in the slag promotes the exchange with sodium ions. The products are therefore transformed from N-A-S-H gels into C-A-S-H gels, and the original three-dimensional disordered structure is retained [18]. Some CaO reacts with the depolymerized SiO_4 and AlO_4 tetrahedrons to form C-S-H gels while some Ca replaces Na to form C-S-H and C-A-S-H gels. The geopolymerization process is necessarily the polycondensation of SiO_4 and AlO_4 tetrahedrons. Na^+ and Ca^{2+} are used to balance the charges. The hydration products of slag-containing coal gangue geopolymers are mainly C-(A)-S-H and N-A-S-H gels. The two kinds of gels, which are highly compatible, jointly constitute the network-like disordered structure of polymers. The high strength development of geopolymers cementitious materials is attributed to the stable, cross-linked and three-dimensional structure of aluminosilicate polymers. As the figure shows, the disordered network structure is enhanced with the increase in slag content, which is consistent with the strength analysis results.

Figure 5. SEM analysis of the CGG specimens of slag contents 0%, 10%, 30%, and 50%.

4. Conclusions

The paper studies the impact of slag content on the strength of coal gangue geopolymers and explores the geopolymerization mechanisms by using XRD, TG-DTG, and SEM. The concrete conclusions are as follows:

(1) Coal gangue can be used to prepare geopolymers. As the slag content increases, the compressive strength of coal gangue geopolymers increases. Compared with P.O 42.5 pure cement specimens, coal gangue geopolymers can obtain higher early strength. Besides, the rise in slag content tends to bring a more substantial increase in the compressive strength.

(2) With the rise in slag content, the amount of calcium-containing zeolite phases and the surface roughness of coal gangue geopolymers increase accordingly. The addition of Ca leads the C-(A)-S-H and N-A-S-H gels to interweave with each other and jointly constitute the network-like disordered structure of polymers.

(3) The XRD, TG-DTG and SEM results are consistent with the strength analysis results. Slag can significantly improve the compressive strength of coal gangue geopolymers and generate compatible products when added as a calcium source.

Acknowledgments

This research was sponsored by the National Natural Science Foundation of China (NSFC, NO.51578539), Beijing Natural Science Foundation projects (NO.8164061), and the Open Research Project of State Key Laboratory of Coal Resources and Safe Mining, China University of Mining and Technology, (SKLCRSM16KFD07). The authors gratefully acknowledge test support from Materials and Durability Laboratory of China University of Mining and Technology (Beijing).

IOP Conf. Series: Materials Science and Engineering **474** (2019) 012040 doi:10.1088/1757-899X/474/1/012040

References

[1] Gartner E 2004 *Cem. Concr. Res.* **34**(9) 1489-98.
[2] Rashad A M 2013 *Constr. Build. Mater* **41**(2) 751-65.
[3] Huntzinger D N and Eatmon T D 2009 *J. Clean. Prod.* **17**(7) 668-75.
[4] Thokchom S, Ghosh P and Ghosh S 2010 *J.Eng. Sci. .Technol. Rev* **3**(1) 31–8.
[5] Zhang J, Shi C, Zhang Z and Ou Z 2017 *Constr. Build. Mater.* **152** 598-613.
[6] Rangan B V 2014 *Indian Concr. J.* **88**(4) 41-59.
[7] W D A Rickard, L Vickers and A Van Riessen 2013 *Appl. Clay Sci.* **73** 71-7.
[8] Yi C, Ma H Q, Zhu H G, Dong Z C, Su Z J, Zhang Y T and Chu Z 2017 *J. Build. Mater.* **20**(5) 787-93.
[9] Li Y J, Xing Y, Zhang X and Yan P X 2013 *J. China Coal Soc.* **38**(7) 1215-19.
[10] Yi C, Ma H Q, Zhu H G, Li W J, Xin M L, Liu Y L and Guo Y D 2018 *Constr. Build. Mater.* **167** 649-56.
[11] Geng J, Zhou M, Zhang T, Wang W, Wang T and Zhou X 2017 *Mater. Struct.* **50**(2) 109.
[12] Zhang C S 2006 *J. Wuhan Univ. Technol.* **21**(4) 150-153.
[13] Zhang C S and Fang L M 2004 *Mater. Sci. Technol.* **12**(6) 597:601.
[14] Yi C, Ma H Q, Chen H Y, Wang J X, Shi J, Li Z H and Yu M K 2018 *Constr. Build. Mater.* **187** 318-26.
[15] Bignozzi M C, Manzi S, Lancellotti I, Kamseu E, Barbieri L and Leonelli C 2013 *Appl. Clay Sci.* **73**(1) 78-85.
[16] Djobo J N Y, Tchakouté H K, Ranjbar N, Elimbi A, Tchadjié L N and Njopwouo D 2016 *J. Am. Ceram. Soc.* **99**(9) 3159-3166.
[17] Ranjbar N, Kashefi A and Maheri M R 2018 *Cem. Concr. Compos.* **86** 1-8.
[18] Garcia-Lodeiro I, Fernández-Jiménez A, Palomo A and Macphee D E 2011 *Cem. Concr. Res.* **41**(9) 923-31.

Effect of pre-annealing heat treatment on destabilization behavior of 28 wt. % Cr-2.6 wt. % C high-chromium cast iron

A Wiengmoon[1,*], J Khantee[1], J T H Pearce[2] and T Chairuangsri[3]

[1] Department of Physics, Faculty of Science, Naresuan University, Phitsanulok 65000, Thailand

[2] Panyapiwat Institute of Management, Nonthaburi 11120, Thailand

[3] Department of Industrial Chemistry, Faculty of Science, Chiang Mai University, Chiang Mai 50200, Thailand

*E-mail: ampornw@nu.ac.th

Abstract. In the present work, the effects of pre-annealing heat treatment on microstructure and hardness in destabilized 28 wt. % Cr-2.6 wt. % C high-chromium irons have been investigated by X-ray diffraction, light microscopy, and scanning electron microscopy. As-cast specimens were annealed at 800°C for 4 h, and then furnace cooled. Destabilization after annealing was carried out at 1000°C for 2 and 4 h, followed by forced air-cooling to room temperature. Vickers macro-hardness was measured on specimens from each condition. The results revealed that the as-cast microstructure consisted of primary austenite dendrites with an interdendritic eutectic structure of M_7C_3 carbide and eutectic austenite. The eutectic austenite had partially transformed to martensite during cooling in the mold. During annealing at 800°C, austenite converted to carbides + ferrite and some pearlite. The hardness decreased from 507 to 425 HV30. Destabilization led to precipitation of secondary carbides and transformation of dendritic and eutectic austenite to martensite with some retained austenite. After destabilization, the macro-hardness increased from 736 HV30 in the iron without prior annealing up to 843 HV30 in the iron with annealing plus destabilization. For maximum hardness, prior annealing reduced the destabilization time from 4 to 2 h.

1. Introduction

High-chromium cast irons are widely used in applications in which high abrasive wear resistance is required such as wear parts in mining, mineral and cement industries. The performance of these irons can be improved by heat treatment, and alloying element [1-7] and their wear resistance depends on the type and the volumetric fraction of matrix and carbides. Although un-machined high-chromium cast iron casting can be used for abrasion resistance. There are some applications of these irons where machining operations become necessary such as the mating faces between pump bodies and their end liners and the faces of roll segments used in crushing rolls. Machined fitting faces ensure that correct assembly with other parts can be achieved during the production or maintenance of crushing and pumping equipment. High-chromium cast irons are difficult to machine due to the presence of large amounts of hard eutectic carbides coupled with work hardening austenitic or hard martensitic matrices in their microstructures. Machining such structures can cause rapid wear damage to cutting tools. To allow conventional tool machining in place of grinding, annealing heat treatments at 700-950°C can be used to reduce hardness levels to below 400 HV [8-12]. During annealing, austenite decomposes into ferrite and pearlitic and/or secondary carbides [8, 10-12]. The problems involved in the machining of white alloy irons, in general, have been considerably reduced by the use of cubic boron nitride (CBN) cutting tool tips, which can replace conventional carbide or alumina ceramic tooling and grinding with silicon carbide wheels [13-14]. CBN tooling enables cost-effective machining of the as-cast austenitic matrix or hardened irons since it avoids the need to use grinding, which is not only slow but may also introduce grinding cracks. Even when CBN tooling is available annealing to produce lower hardness levels is often required notably when integral screw threads have to be provided to avoid the use of cast-in inserts in slurry pump parts [15]. After annealing treatment and machining, irons can then be hardened by re-austenitization and destabilization following by air hardening heat treatment to provide a martensite matrix with higher hardness and wear resistance. Destabilization involves heating to and holding at 900-1025°C for up to 6 h to destabilize the austenite matrix by encouraging precipitation of secondary carbides. After destabilization and air hardening, the microstructure consists of eutectic carbide (M_7C_3) and secondary carbides (normally M_7C_3 and/or $M_{23}C_6$) in a matrix of martensite and some retained austenite [1-4]. The hardness can be increased up to 800 HV30 after destabilization [3-4]. Irons are typically tempered after air hardening to transform retained austenite, reduce internal stress, and control final hardness to provide an optimum balance of wear and impact resistance for a given application [9-12, 16-17].

Apart from reducing the hardness of the irons to accommodate tool machining as mentioned earlier [8-12], pre-annealing was done before destabilization for homogenizing [17]. It was found that pre-annealing did no effect to the hardness of the irons after hardening, but it reduced the destabilization time to develop full hardness [17]. However, higher total hardness and lower destabilization time to establish full hardness were reported for the irons after destabilization with pre-annealing [12]. In the present work, the effects of pre-annealing at 800°C on the microstructure and hardness after destabilization of an iron-containing 28 wt. % Cr-2.6 wt. % C were investigated.

2. Experimental method

2.1. Material preparation and heat treatment

Table 1 shows the chemical composition for high-chromium cast iron prepared by melting in a magnesia crucible, using an electric resistance furnace. The metal was cast into a sand mold to produce cylindrical test bars with a diameter of 25 mm and length of 300 mm.

The as-cast samples were annealed at 800°C for 4 h in an electric furnace with a heating rate of 5°/min and then cooled in the furnace with a cooling rate of 40°/hr. Destabilization after annealing was performed at 1000°C for 2 and 4 h, followed by air-cooling to room temperature. This iron is air-hardenable.

Table 1. The chemical composition of iron in the experiment.

Element (wt.%)									
Cr	C	Mo	Si	Mn	S	Ni	P	W	Fe
27.86	2.63	0.03	0.20	0.11	0.01	0.20	0.02	0.01	Bal.

2.2. Characterization and hardness testing

For microstructural examination, the samples were cut from the as-cast and heat-treated materials. Samples for X-ray diffraction technique (XRD) and scanning electron microscopy (SEM) were ground on silicon carbide papers and then polished with diamond paste. Specimens were etched in 50 ml of HCl and 10 g of $Na_2S_2O_5$ in 100 ml distilled water. The microstructures were studied using an LEO 1455VP SEM operated at 20 kV. The phases present were determined using a Philips X'pert diffractometer with a Cu Kα X-ray source. Vickers macro-hardness testing was performed on un-etched specimens using 30 kgf load and 15 s indenting time. The mean values are based on eight different areas in each specimen.

3. Results and Discussion

3.1. Microstructures

The as-cast microstructures shown in figure 1(a), consisted of primary austenite (γ) dendrites with an interdendritic eutectic structure of M_7C_3 carbide (M = Fe, Cr, Mo) + eutectic austenite which was partially transformed to martensite (α′) during cooling in the sand mold. After annealing at 800°C for 4 h, as seen in figure 1(b), the ferrite (α) matrix + secondary carbides were observed, due to decomposition of austenite during annealing. Small areas of pearlite were also present in the central regions of dendrite arms due to microsegregation effects. The presence of pearlite indicated that the transformation of austenite to ferrite and secondary carbides was incomplete [8].

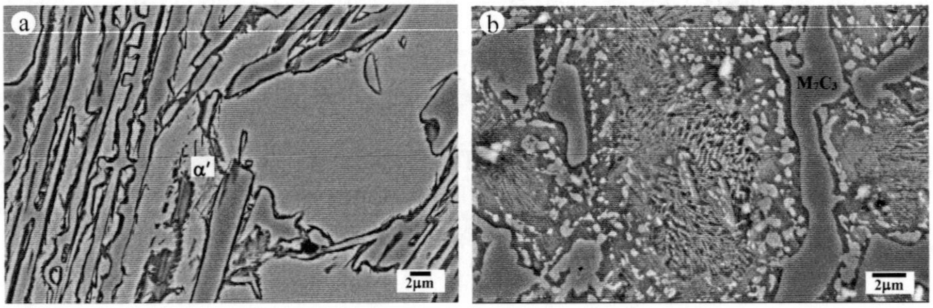

Figure 1. The microstructure of 28 wt. % Cr-2.6 wt. % C iron: (a) as-cast condition (b) annealing at 800°C for 4 h.

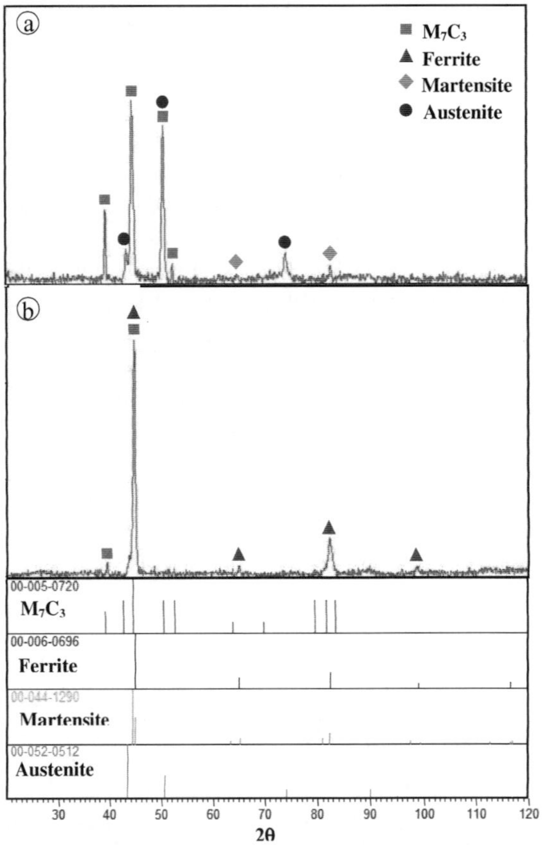

Figure 2. XRD traces obtained from 28 wt. % Cr-2.6 wt. % C iron: (a) as-cast condition (b) annealing at 800°C for 4 h.

Figure 2 shows the XRD patterns of the as-cast iron and after annealing at 800°C for 4 h as compared with the XRD lines of austenite, ferrite, martensite, and M_7C_3 carbide. XRD results confirmed the presence of austenite, martensite and M_7C_3 carbide in the as-cast condition. Peaks from austenite at 2θ = 43.279°, 50.408°, 74.054°, martensite at 2θ = 65.542°, 81.943°, and M_7C_3 at 2θ = 39.135 °, 44.370°, 50.375°, 52.553°, were observed. After annealing, M_7C_3 carbide and ferrite formed. Peaks from ferrite at 2θ = 44.674°, 65.023°, 82.335°, and 98.949°, were observed. The absence of some M_7C_3 peaks in the annealed iron can be attributed to texture effects.

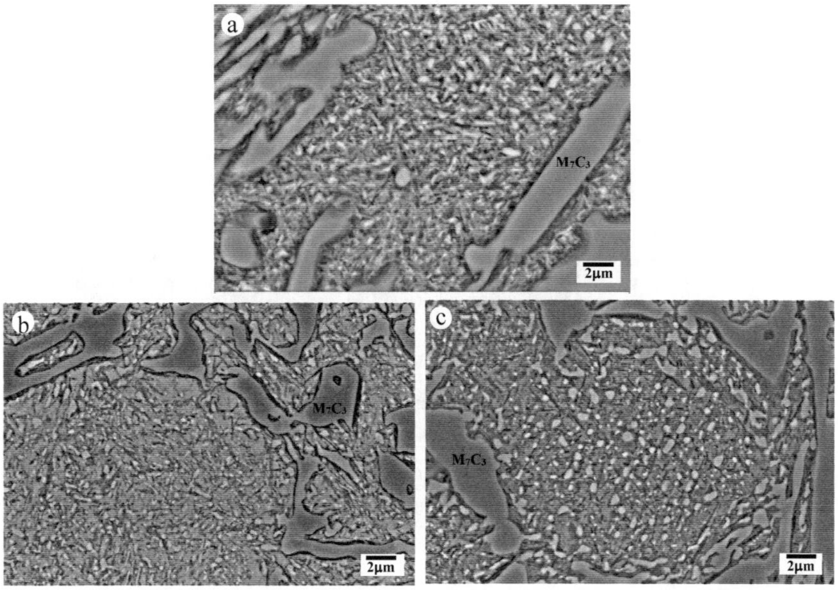

figure 3. The microstructure of 28 wt. % Cr-2.6 wt. % C iron: (a) after destabilization for 4 h, (b) pre-annealing plus destabilization for 2 h and (c) pre-annealing plus destabilization for 4 h.

The microstructures after destabilization for 4 h and annealing plus destabilization for 2 and 4 h, as shown in figure 3, consisted of secondary carbides in a martensite matrix. Pre-annealing increased the amount and sizes of secondary carbides after destabilization. At longer destabilization time of 4 h, as shown in figure 3(c), the secondary carbides were observed to be coarser and fewer in number. This is in agreement to previous work [3-4, 12] and can result from coarsening or the Ostwald ripening of the secondary carbide particles [3].

3.2. Macro-hardness

Figure 4 shows the effect of annealing and destabilization on the macro-hardness. After annealing at 800°C for 4 h, the macro-hardness decreased from 507 HV30 in the as-cast condition to 425 HV30. During annealing at the 800°C transformation of the austenite matrix to ferrite plus secondary and pearlitic carbides is expected to be almost complete such that the furnace cooling rate used in work gave an acceptable low hardness of 425 HV30 since no martensite was formed.

The macro-hardness after destabilization at 1000°C for 4 h in the iron without pre-annealing was 736 HV30. Pre-annealing increased the macro-hardness after destabilization at 1000°C for 2 and 4 h up to 843 and 830 HV30, respectively. This is most likely due to more even distribution or higher degree of homogenization of C and Cr as suggested by Inthidech *et al.* [17], and hence more uniform secondary carbide precipitation. Therefore, it can be suggested that pre-annealing affects increasing the hardness of the irons after destabilization hardening as reported by Cubillos *et al.* [12] and also reduces the destabilization time to develop full hardness as reported by Cubillos *et al.* [12] and Inthidech *et al.* [17].

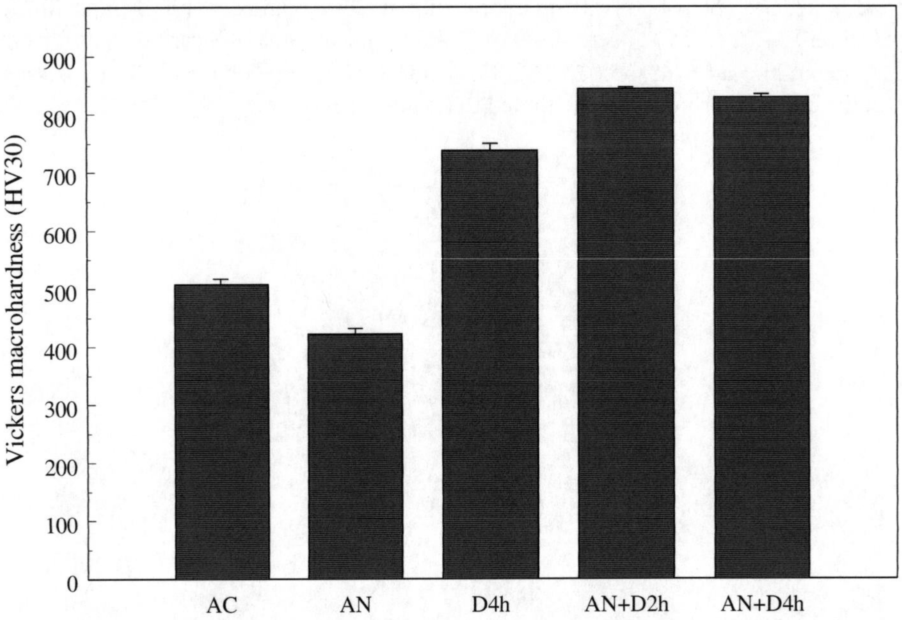

Figure 4. Effect of annealing and annealing plus destabilization on the macro-hardness (AC = as-cast, AN = annealing, D4h = destabilization for 4 h, AN+D2h = annealing plus destabilization for 2 h, AN+D2h = annealing plus destabilization for 4 h).

4. Conclusions

The results obtained and their discussion make it possible to draw the following conclusions:

- ✓ The microstructure of the as-cast 28 wt. % Cr-2.6 wt. % C iron consisted of austenite dendrites, eutectic carbide and some martensite adjacent to carbides.
- ✓ The average as-cast hardness is 500 HV30. A hardness of 425 HV30 can be obtained after annealing at 800°C for 4 h.
- ✓ Pre-annealing before destabilization increased the amount and sizes of secondary carbides in the iron after destabilization.
- ✓ Pre-annealing as an effect on increasing the hardness of the irons after destabilization hardening. The maximum hardness of 843 HV30 was obtained in the iron with pre-annealing at 800°C for 4 h plus destabilization at 1000°C for 2 h.
- ✓ Annealing is used to facilitate machining. This study suggests that pre-annealing can also reduce the destabilization time needed to obtain the maximum hardness.

Acknowledgments

The Naresuan University, Thailand, is acknowledged for the financial support of this work under Grant No. R2562B078.

References

[1] Tabrett C P, Sare I R and Ghomashchi M R 1996 *Int Mater Rev* **41** 59.
[2] Pearce J T H 2002 *Foundryman* **95** 156.
[3] Wiengmoon A, Chairuangsri T and Pearce J T H 2004 *ISIJ Int 44* No. **2** 396.
[4] Wiengmoon A, Pearce J T H and Chairuangsri T 2011 *Mater. Chem. Phys.* **125** 739.
[5] Imurai S., Thanachayanont C, Pearce J T H, Tsuda K and Chairuangsri T 2014 *Mat Char* **90** 99.

[6] Carrillo E C, Jacuinde A B, Mejía I, Zepeda C M, Silva J Z and Lopez F V G 2017 *Wear* **376-377** 77.
[7] Ma Y, Li X, Liu Y, Zhou S and Dang X 2013 *Bull. Mater. Sci* **36** 839.
[8] Kibble K A and Pearce J T H 1995 *Cast Met* **8** 123.
[9] Amorim P, Santos H, Santos J, Coimbra S and Sá C 2004 *Mat Sci Forum* **455-456** 290.
[10] Berkun M N, Volchok I P, Zhivitsa I V and Topal V I 1971 *Met Sci Heat Treat* **13** 69.
[11] Chen L, Iyengar S, Zhou J, Turba K and Ståhl J E 2015 *J Mater Eng Perform* **24** 98.
[12] Cubillos P O, Bernardini, P A N, Fredel M Cand Campos, R A 2015 *Revista Facultad de Ingeniería* **76** 134.
[13] Blackman T N 1990 *The Foundryman* **1** 17.
[14] Chen L, Zhou J, Bushlya V and Stahlet J E 2015 *Procedia CIRP* **31** 172.
[15] Herbert S 1986 *Industrial Diamond Review* **46** 512.
[16] Lv Y, Sun Y, Zhao J, Yu G, Shen J and Hu S 2012 *Mat Des* **39** 303.
[17] Inthidech S, Sricharoenchai P and Matsubara Y 2006 *Mater Trans* **47** 72.

Experimental study on the utilization of waste foundry sand as embankment and structural fill

M F Iqbal[1, 2, *], Q F Liu[1, 2, **] and I Azim[1, 2]

[1] State Key Laboratory of Ocean Engineering, School of Naval Architecture, Ocean & Civil Engineering, Shanghai Jiao Tong University, Shanghai, China

[2] Collaborative Innovation Center for Advanced Ship and Deep-Sea Exploration (CISSE), Shanghai, China

*E-mail: farjad51@sjtu.edu.cn; **E-mail: liuqf@sjtu.edu.cn

Abstract. Metal casting industries, either ferrous or non-ferrous, produce a large quantity of the waste foundry sand (WFS). Currently, it is either discarded to landfill or collected by waste organizations to extract iron remnants after washing. This research evaluates the reuse of WFS as a replacement of sand in structural fill and embankments. Sand was used to replace WFS at a replacement level of 0, 4, 6, 8 and 10% by weight. Geotechnical and environmental tests were performed to assess the durability and physical properties of WFS samples. Features like the gradation and friction angle were similar for the samples having varying sand percentages. Moreover, the California bearing ratio indicated that WFS having 10% sand replacement will make the hardest fill material. However, it is recommended to use WFS with 6% sand replacement, since it best meets the criteria of specific gravity, optimum moisture content, maximum dry density, permeability and the California bearing ratio for structural fill, embankment and road sub-base material. The chemical composition of WFS indicated the absence of any hazardous material, which may hinder its use as fill material.

1. Introduction

Ferrous and non-ferrous metal casting foundries generate large quantities of by-products [1]. These foundries utilize silica sand mixed with some binder for their molding and casting process [2]. The repeated utilization of high-quality silica sand for casting and molding results in the production of waste foundry sand (WFS) [3, 4]. The annual generation of WFS is approximately 9 and 12 million metric tons in Europe and the United States, respectively [5, 6].
Moreover, countries such as Australia, China, and Taiwan are also generating millions of tons of WFS every year which are posing an environmental threat, since they are dumped into landfills [3, 4]. The landfilling cost of WFS is US$135-675 per ton which includes transportation, storage, and labor costs [7]. The considerable disposal cost has made the current practice of WFS disposal unsustainable for foundries [7]. Due to the lack of information and regulations on the beneficial reuse of WFS, only 32% of WFS is utilized for construction practice.

The consumption of river sand, in the construction works, is high. Therefore its demand is also very high [8]. Several countries are facing problems in meeting the demand for river sand for construction [8, 9]. Multiple pieces of research have been conducted to study the use of WFS in construction as a replacement of river sand.

The application of WFS can be subdivided into three classes namely concrete, highways and geotechnical or soil based applications [8]. The California bearing ratio(CBR), compaction, resilient modulus and hydraulic conductivity tests on WFS and its blends with geosynthetics [4] indicated that it is suitable to be used as highway sub-base material [5, 10]. Arulrajah *et al.* [3] determined the geotechnical properties (optimum moisture content (OMC), CBR, maximum dry density (MDD), permeability) of WFS and compared them with those of a standard recycled glass waste material. It was concluded that WFS could be used as fill material in embankment and pipe bedding.

Researchers have concluded that replacing fine aggregates with a certain percentage of WFS has improved the compressive and split tensile strength, modulus of elasticity (MOE), abrasion resistance [1, 2, 11-13] and durability properties, i.e. resistance to carbonation and chloride penetration of concrete [14]. The percentage of WFS successfully replaced in concrete ranges from 5% to as high as 30%. Siddique *et al.* concluded that there was an increase of 9.8% and 9% in the compressive strength and splitting tensile strength, respectively for the concrete mix containing 30% WFS compared to the standard mix (0% WFS) [11]. At 28-days, the flexural strength of control mix (0% WFS) was 3.41 MPa and increased by 9% (4.18MPa) for the concrete mix (30% WFS) [12]. The mechanical properties of concrete containing WFS also improve with age. The modulus of elasticity of concrete mixtures (containing 0-20% WFS) increased by 6.02-12.37% at 91 days compared to 28 days average MOE of 29.9 GPa [13]. The chloride ion permeability of concrete decreases with the increase in WFS content. At 28 days, the charges passed were 1368, 1250, 1150 and 1060 coulombs at 0, 5, 10, 15% WFS content [1].The replacement percentage proposed to achieve the desired strength varies depending on the type of WFS, additives and testing conditions.

A problem with using foundry sand is the potential for environmental impact caused by the leaching of heavy metals present in the foundry sand [15]. Therefore, past researchers have studied the environmental characteristics such as chemical composition and leachate analysis of WFS and rendered that if water comes in contact with WFS during drainage, the quality of water will not be affected [3, 5].

The majority of the past research work focused on the properties of blends in which WFS was used as a secondary component, rather than sole properties of WFS [3]. If WFS fulfills the desired requirement, it can be used without mixing any other material, consequently saving a lot of cost and time required for the mix design and blending [3]. However, it has been observed that properties of pure WFS may deviate from standard criteria because of the presence of impurities and various binders used by different sources.

In this research, geotechnical investigations, including specific gravity, size gradation, permeability, direct shear, compaction and CBR tests were carried out on WFS samples containing 0, 4, 6, 8 and 10% by weight sand replacement. The chemical composition of WFS sample was also determined to ensure its sustainable use. The results were benchmarked against the standard parameters cited in the previous researches or codes. Finally, the WFS mixture having a minimum sand percentage, which can be utilized as fill material, is proposed.

2. Experimental

2.1. Materials

2.1.1. Waste foundry sand. WFS used in this study was delivered in bags from a metal casting foundry. It was black due to the presence of contaminant and burning during the metal casting process. It was chemically bounded using phenolic urethanes before using for casting process by industry. It was washed with fresh water twice and placed in sunlight for two days before testing [8].

2.1.2. Fine aggregate. Locally available river sand was used as fine aggregate, having a maximum size of 4.75 mm, confronting to the ASTM C33 requirement. The particles of the sand were round in shape and had a smooth texture. The clay and silt content was low due to excessive washing in the river. It was tested based on ASTM C128 and C136. The physical properties and mineralogy (obtained by petrographic modal analysis) of the aggregates are listed in tTable 1 and 2, respectively.

Table 1. Physical properties of the fine aggregates.

Property	Specific gravity	Fineness modulus	Water absorption %	Moisture content %	Maximum size (mm)
Values	2.65	3.06	1.48	0.14	4.75

Table 2. Petrographic modal analysis of the fine aggregates.

Mineral	Quartz	Amphibole	Quartzite + polygrain quartz	Granite	Feldspar	Biotite	Slate/ argillite
Percentage (%)	63.0	6.0	5.0	3.5	8.5	3.5	2.0
Mineral	Carbonate	Magnetite	Acid to intermediate volcanic	Muscovite	Chert	Lithic arenite	Epidote
Percentage (%)	3.0	1.0	1.0	1.0	0.5	1.0	0.5

2.2. *Preparation of specimens*
Five WFS specimens (GS-1, GS-2, GS-3, GS-4, and GS-5) were prepared at room temperature having 0, 4, 6, 8 and 10% sand by weight as replacement of WFS, respectively. Special care was taken for handling, transportation and placing during the testing procedure since the samples were very fragile.

2.3. *Testing methods*
An X-ray diffraction analysis was performed to determine the chemical composition of WFS. The hazard category of WFS was defined based on the Environmental Protection Authority standards and ASTM D3987. The environmental properties of the filtrate were analyzed for different metals.

The particle size distribution of the specimen was obtained using the ASTM D6913. A 500 g sample was used so that the overloading limits for each sieve were satisfied. The specific gravity was determined according to ASTM C128. 100 g oven dry samples were used along with a 500 mL pycnometer.

To estimate the MDD and OMC, the standard Proctor test following ASTM D698 was performed. For each specimen, GS-1, GS-2, GS-3, GS-4, and GS-5, four different sub-samples having a moisture content of 8, 10, 12 and 14% were prepared.

Shear strength is an important parameter required for the determination of the bearing capacity of the soil. The direct shear strength test was performed on the specimen following ASTM D3080. Two shear tests were performed on each of the five samples at 10 and 20 lbs standard loads. The friction angle and cohesion of the samples were also calculated.

The constant head permeability test was conducted to determine the hydraulic conductivity of the samples according to ASTM D2434. Sand was compacted in three layers in a 152 mm-diameter mold by applying a standard compaction effort.

The CBR test was carried out following ASTM D1883. All the samples were tested at their OMC. The CBR values of 2.54 and 5.08 mm were recorded using stress-penetration curves, and the highest value is reported in the study.

3. Results and Discussion

3.1. Geotechnical properties

The geotechnical properties of WFS and mixtures with sand were determined to evaluate its use as structural and embankment fill. The results are discussed below.

3.1.1. Grain size distribution. Figure 1 represents the grain size distribution of the five test samples and ASTM C33 limits for the fine aggregate. Approximately, 80% of the particles, for each of the five specimens, lie between 0.08-0.8mm. WFS used as an embankment, and structural fill has the particle size ranging from 0.2-1mm [16]. Percentage of the fines lies between 1.8-2.2%. The gradation curves for the samples are consistent with the curves reported for the 17 WFS samples tested by Deng *et al.* [7]. However, GS-3 best meets the gradation criteria published in the literature [7].

The coefficient of uniformity (Cu) and coefficient of curvature (Cc) range between 1.5-2.8 and 0.6-1.2, respectively. Hence, the samples are classified as poorly graded. A similar classification was reported and compared with recycled glass which is well accepted as a waste material by Arulrajah *et al.* [3]. Owing to the low percentage of fines, the Atterberg limits cannot be applied to the samples and are termed as non-plastic [3].

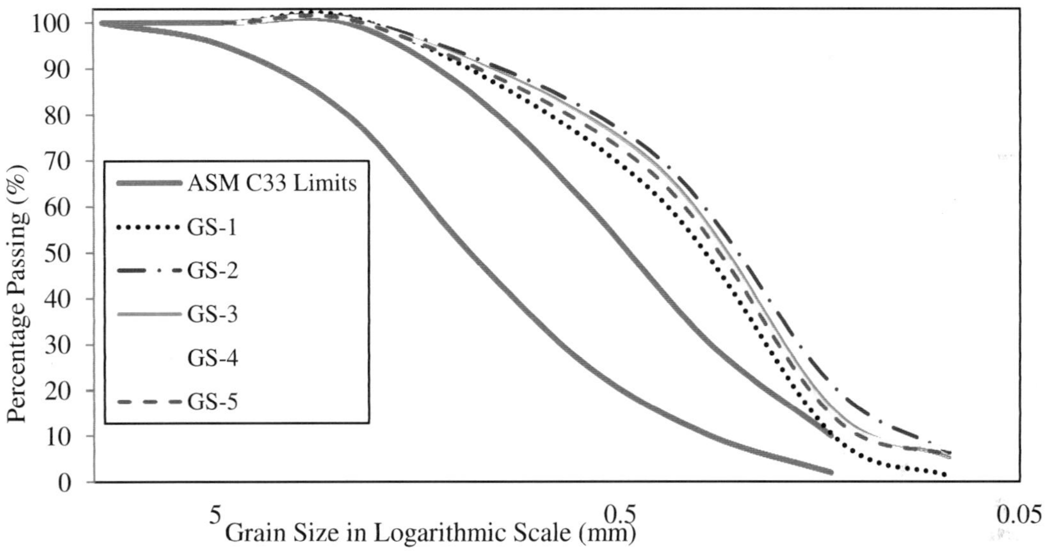

Figure 1. Gradation curves of the WFS specimens and standard fine aggregates.

3.1.2. Specific gravity and compaction characteristics. The specific gravity of the specimens lies in the range of 2.5-2.66. Sand commonly used as structural and embankment fill has a specific gravity of 2.50-2.80 [16]. The variation in the specific gravity and MDD with varying sand content is due to the variation in sand mineralogy, particle gradation, grain shape and fine content [7].

Table 3. Optimum moisture content and maximum dry density results.

Designation	GS-1	GS-2	GS-3	GS-4	GS-5	Benchmarked standard [17]
Optimum moisture content (%)	10.86	12.45	11.92	12.06	12.85	10
Maximum dry density (kg/m^3)	1810	1790	1786	1800	1820	1763

The OMC shows a slight increase with an increase in sand content, which may be due to the presence of moisture and 2% fine content in the sand used (table 3). The curves shown in figure 2 are relatively flat compared to the plastic soils and similar to the curves for WFS (without clay) used as fill material [16]. The OMC and MDD of GS-3 lie in the standard range, i.e. approx. 10% and 1763 kg/m^3, respectively [16] and, therefore. GS-3 is termed as the most suitable structural and embankment fill material.

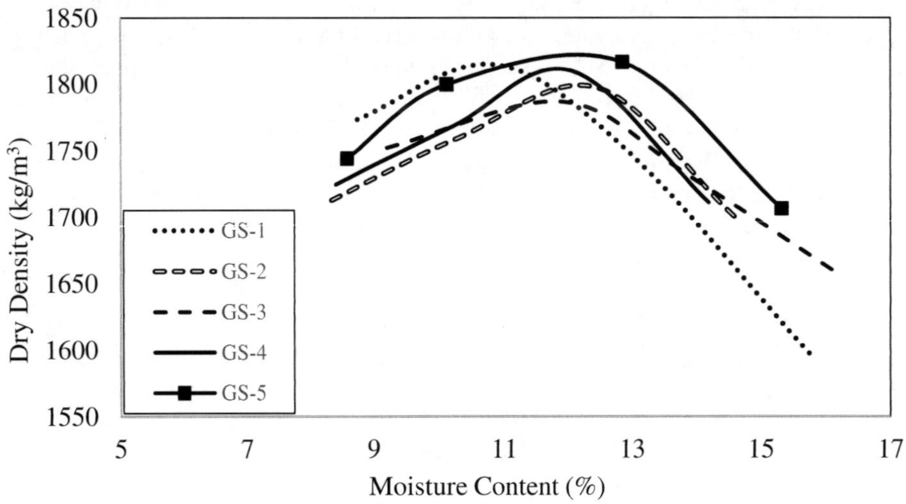

Figure 2. Compaction curves of the tested samples.

3.1.3. Shear strength. The cohesion, as shown in Figure 3, is very small as compared to the cohesion of WFS, i.e. 177 kN/m^2 (3700 psf) due to the negligible clay content in the samples [16]. The cohesion increased with the increase in sand content. This could be due to the improved density and binding with the increase in sand percentage [17].

Since the clay content is negligent, WFS used for embankment/structural fill was characterized based on the friction angle. The slight increasing trend (Figure 4) shows that as the sand content in the specimens increases, the friction angle approaches to that of the standard sand. The average friction angle is 34° for the standard sand used as a fill material [16]. The GS-3 having a friction angle of 34.2° fulfills the specified criteria.

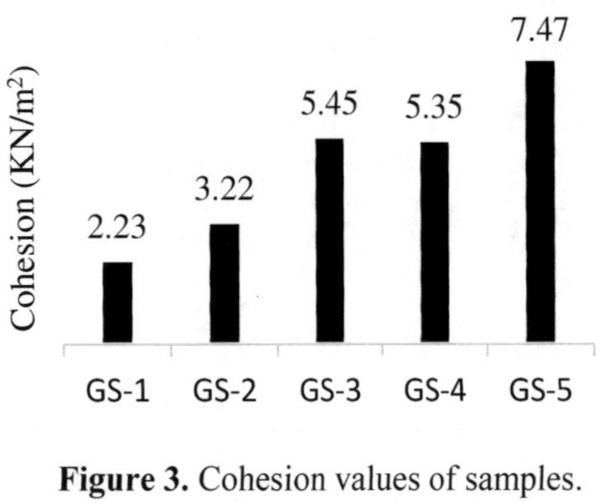

Figure 3. Cohesion values of samples.

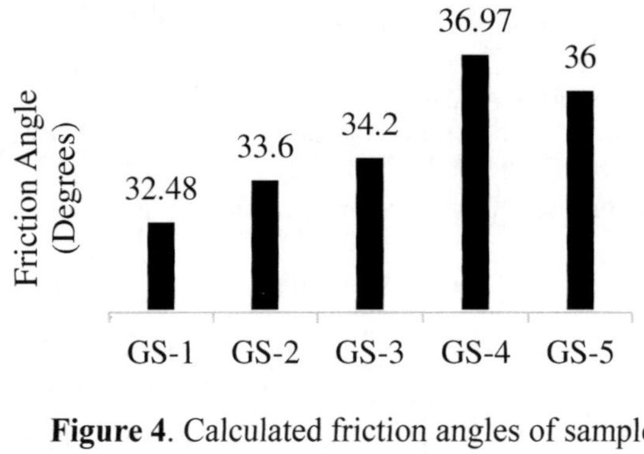

Figure 4. Calculated friction angles of samples.

3.1.4. Permeability. The permeability of WFS is lower than typical sand since it is not considered as a free draining material [3]. Table 4 shows the permeability results of the specimen. The results lie well within the range of permeability of sand used as fill material, i.e. $6x10^{-7}$ - $5x10^{-6}$ m/s [16].

Table 4. The permeability of the samples.

Designation	GS-1	GS-2	GS-3	GS-4	GS-5
Permeability $x10^{-6}$ (m/s)	5.26	5.2	5.24	4.88	4.7

The blend of WFS with sand indicates that a high WFS content increases the water absorption due to the high fineness, surface area and porosity [18, 19]. The porosity increase results in the expansion of the pore diameter and an increase in permeability [20]. Therefore, it can be concluded from the results in Table 4 that GS-5 has the lowest WFS content. Therefore permeability is minimum. The low permeability indicates that penetration of chlorides, sulfates or alkali ions will be hindered when used as fill material and can be termed as a durable material [21].

3.1.5. California bearing ratio. The CBR values of the tested specimens lie in the range of 15-22% as shown in Table 5. For the fill material, the standard CBR is 11-30% whereas for sub-base application the requirement is 4-20 % [16].

It can be seen from Figure 5 that as the sand content increases, the stress required to achieve maximum penetration increases with GS-5 having the highest stress at maximum penetration. Thus it will make the hardest fill for roads. It can also be concluded, based on the CBR results, that WFS can be used as an alternative material for road embankment, structural fill, and sub-base material.

Table 5. California bearing ratio (CBR) of the investigated samples.

Designation	GS-1	GS-2	GS-3	GS-4	GS-5
CBR (%)	15.1	17.15	19.37	22.08	20.40

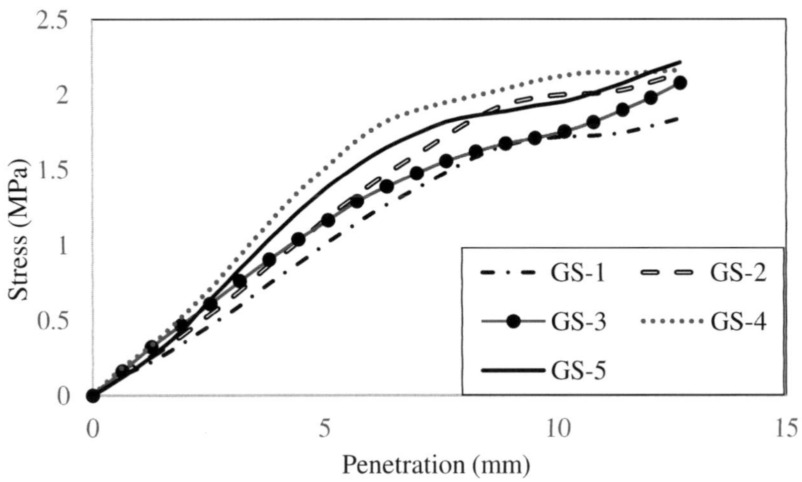

Figure 5. Stress penetration curves for the samples.

3.1.6. The chemical composition of WFS. During the casting process, WFS was exposed to molten metals at high temperatures several times. Therefore, an X-ray diffraction analysis was carried out to determine the chemical composition of WFS (Table 6). Moreover, the results were compared with similar samples reported by Arulrajah *et al.* and Singh *et al.* [3, 13]. Since the origin of the materials is sand, therefore it contains a high content of silica which improves hardness [3]. All the metals, other than iron oxide, are at a very low percentage. The presence of excess iron remnants could hinder the binding of WFS which can cause leaking problems, particularly when used as fill material. The result shows that the specimen utilized in this study has the lowest percentage of iron oxide (0.831%) among others reported in literature thereby mitigating the leakage issue. The leachate obtained from WFS does not contain any harmful metals.

Table 6. The chemical composition of WFS and its comparison with the results from literature [3, 13].

Constituent (%)	Silica (SiO_2)	Alumina (Al_2O_3)	Calcium oxide (CaO)	Magnesium oxide (MgO)	Ferric oxide (Fe_2O_3)	Chromium oxide (Cr_2O_7)	Sulfur trioxide (SO_3)	Titanium dioxide (TiO_2)
WFS	87.220	3.352	1.831	0.316	0.831	0.004	0.813	0.23
Arulrajah *et al.* [3]	84.145	11.817	1.507	-	1.533	-	0.453	0.257
Singh *et al.* [13]	83.8	0.81	1.420	0.86	5.39	-	0.21	0.22

4. Conclusions and recommendations

From the experimental observations, the following conclusions and recommendations can be made.

- ✓ The gradation results indicated that the samples are poorly graded and non-plastic. The maximum dry density (MDD) and optimum moisture content (OMC) are slightly higher than the standard fill due to variations in the sand mineralogy, particles size, shape, and fine contents.
- ✓ The cohesion of the samples is negligible due to the absence of clay content. The friction angle increased with the increase in sand content in the samples. The waster foundry sand (WFS) sample having a 6% sand has a friction angle (34.2°) similar to the average friction angle of sand used as fill material.
- ✓ The increase in sand content in WFS increased the density, thereby improving the binding and durability properties. It is owing to the same reason that WFS having 10% sand would make the hardest road embankment based on the CBR values.
- ✓ Based on the experimental evaluations, it is recommended that the WFS mixture with 6% sand can be used as structural, embankment fill and sub-base material, since its geotechnical properties (specific gravity, OMC, MDD, permeability, and CBR) are comparable to that of standard fill material.

Acknowledgments

The work was supported by the Natural Science Foundation of China (51508324), the Shanghai Chenguang Program, China (16CG06), the State Key Laboratory of High Performance Civil Engineering Materials (2018CEM006) and the Open Research Fund of State Key Laboratory of Simulation and Regulation of Water Cycle in River Basin, China Institute of Water Resources and Hydropower Research (IWHR-SKL-201705).

References

[1] Singh G and Siddique R 2012 *Construction and Building Materials* **26** 416-22
[2] Siddique R, Singh G, Belarbi R, Ait-Mokhtar K and Kunal 2015 *Construction and Building Materials* **83** 216-22
[3] Arulrajah A, Yaghoubi E, Imteaz M and Horpibulsuk S 2017 material: *Sustainable Cities and Society* **28** 343-9

[4] Lin K L, Cheng C J, Cheng A and Chao S J 2012 *Sustainable Environment Research* **22** 91-7
[5] Guney Y, Aydilek A H and Demirkan M M 2006 *Waste Management* **26** 932-45
[6] Abichou T, Edil T B, Benson C H and Tawfiq K 2004 *Conf. on Recycled Materials in Geotechnics, (Baltimore, Maryland, USA/ASCE, 2004)* pp 186-200
[7] Deng A and Tikalsky P J 2008 *Waste Management* **28** 2161-70
[8] Manoharan T, Laksmanan D, Mylsamy K, Sivakumar P and Sircar A 2018 *Waste Management* **71** 454-60
[9] Bahoria B V, Parbat D K, Nagarnaik P B and Waghe U P 2003 *International Journal of Application or Innovation in Engineering & Management* **2** 387-94
[10] Kleven J, Edil T and Benson C 2000 *Transportation Research Record: Journal of the Transportation Research Board* 40-8
[11] Siddique R, Schutter G D and Noumowe A 2009 *Construction and Building Materials* **23** 976-80
[12] Siddique R and Singh G 2011 *Resources, Conservation & Recycling* **55** 885-92
[13] Singh G and Siddique R 2012 *Construction and Building Materials* **28** 421-6
[14] Siddique R, Aggarwal Y, Aggarwal P, Kadri E-h and Bennacer R 2011 *Construction and Building Materials* **25** 1916-25
[15] Siddique R, Kaur G and Rajor A 2010 *Resources, Conservation and Recycling* **54** 1027-36
[16] Foundry-Industry F I R S T 2004 *Foundry Sand Facts for Civil Engineers* Report No. FHWA-IF-04-004 (Washington DC, USA: American Foundrymen's Society Inc. for Federal Highway Administration Environmental Protection Agency.) pp 7-33
[17] Basar H M and Deveci N 2012 *Construction and Building Materials* **35** 508-15
[18] Şahmaran M, Lachemi M, Erdem T K and Yücel H E 2011 *Materials and Structures/Materiaux et Constructions* **44** 1193-204
[19] Khatib J M, Herki B A and Kenai S 2013 *Construction and Building Materials* **47** 867-71
[20] Pathak N and Siddique R 2012 *Construction and Building Materials* **34** 512-21
[21] Zhang S P and Zong L 2014 *Advances in Material Science and Engineering* **2014**

A novel approach to assessing the anisotropic permeability parameters of fractured media

X Y Song[1,2], Y T Liu[1,2], X E Jiang[1,2], Z P Ding[3] and L Xue[1,2,*]

[1] State Key Laboratory of Petroleum Resources and Prospecting, China University of Petroleum, Beijing 102249, China

[2] College of Petroleum Engineering, China University of Petroleum, Beijing102249, China.

[3] CNOOC Research Center, Beijing 100028, China

*E-mail: xueliang@pku.edu.cn.

Abstract. The fracture permeability in is very high and directional, which brings a series of problems for the petroleum reservoir development, such as earlier water breakthrough and low sweep efficiency. It is vital to accurately characterize fractures and acquire the parameters of fracture permeability to alleviate these issues, which are favorable for well pattern design and flow modeling. Permeability tensor is employed to comprehensively and expediently reflect the effect of fracture anisotropy on the fluid diversion. Based on information of fracture log and production, the complete permeability tensors of single group fractures and multiple group fractures are identified, where the coordinate transformation theory and tensor property are used. With the simplified form of the derived full permeability tensor, the anisotropic parameters, principal values and principal directions of permeability for the fractured reservoir, can then be calculated. The proposed method integrates the static characterization and percolation feature of fractures and can be applied merely into field practices of naturally fractured reservoir. This work is beneficial for the management of fractured reservoir and optimization of field development.

1. Introduction

About 50% of the world's remaining conventional hydrocarbons are stored in the naturally fractured reservoirs [1]. But due to the presence of a large number of natural fractures, these reservoirs are always anisotropic and heterogeneous, which brings a lot of difficulties to the development of oil and gas fields. In such reservoirs, the anisotropic fracture permeability plays a significant role. For example, the anisotropic permeability has a destructive impact on the general patterns of wells [2]. Therefore, a precise characterization of anisotropic fracture permeability is the key to make reasonable decisions on the development of the fractured reservoir. The permeability via the conventional Darcy rule fails to properly characterize the percolation features of the fractured reservoir, considering that full tensor permeability in modeling fracture reservoir is a necessity even for reservoirs with initial isotropic permeability [3]. Thus, a nine-component tensor is introduced to capture the anisotropy of the naturally fractured reservoir, where the orientation and magnitude of permeability may vary in the same position. The full permeability tensor is commonly symmetric [4] and its geometric form behavior is similar to that of an ellipsoid [5].

Experimental approaches are generally utilized to measure the anisotropic permeability. After substituting permeability tensor into the conventional Darcy rule, the permeability can be

IOP Conf. Series: Materials Science and Engineering **474** (2019) 012043 doi:10.1088/1757-899X/474/1/012043

assessed by measuring the flow rate and pressure gradient of different direction [6]. Similarly, the full hydraulic conductivity tensor is computed with a linear least squares algorithm by measuring the average velocity and gradient [7]. However, this approach is not conclusive because the accuracy is not satisfied. Since the sample used for the experiments is heterogeneous, the cubic sample is used to eliminate the effect of heterogeneity when measuring the anisotropic permeability in a tri-axial cell [8]. For the experimental methods, the end effect of the sample always affects the result of the measurement. Besides, the results obtaining from cores cannot characterize the dynamic changes of anisotropic permeability.

Analytical models are also employed to determine the anisotropic permeability. Snow [9] estimated anisotropic permeability by applying means of superposition of parallel plate flow under the assumption of an impermeable rock matrix. An approach incorporating Snow's tensor and permeability tensor calibration to assess permeability tensor is proposed [10]. The models based on material balance [11] and the lowest order finite formula [12] are used to investigate the anisotropic permeability of cross-bedded reservoir. A consistent rock physics model for effective permeability and stiffness tensors of the fractured reservoir are employed to estimate the anisotropic permeability [13]. But this method can deal with the reservoir containing a single set of vertical fractures. From the extended Poiseuille flow approximation of the velocity and pressure field, the analytical model for effective permeability was developed [14]. However, the reliability depends on the fiber's direction, and the anisotropy of the horizontal plane cannot be identified. Chi and Heidari [15] integrated a directional pore connectivity factor into a nuclear-magnetic-resonance-based permeability model to determine directional permeability where the lattice Boltzmann method was used to verify the results. This approach requires micro-CT scans on the rock sample, which prevents the real-time application of this model.

The geophysical method is also applicable for determining anisotropic permeability. The resistivity anisotropy and conventional log derived permeability are used to interpret permeability anisotropy [16]. But it assumes that the formation is transversely isotropic. Frequency-dependent seismic amplitude versus angle and azimuth data and Monte Carlo method performing a Bayesian inversion are utilized to obtain the information of fractures including permeability [17].

The presented numerical simulator can handle the anisotropic issues, but it is assumed that the axial directions of the coordinate system of the girds are aligned with the principal directions of permeability. Under this assumption, the off-diagonal terms of permeability tensor are neglected. The principal directions of permeability do not coincide with the axial direction of simulators' coordinate system which causes significant errors calculating the magnitude and orientation of the Darcy flow rate [18] and the magnitude and direction of velocity [11].

This work focuses on the anisotropic permeability of fractured reservoir. The complete permeability tensor of fractures is first determined by combing the log information and production information. Next, the approach to obtain parameters of anisotropic permeability by the derived anisotropic permeability tensor of fracture is presented. In the final part, an example of the computation of the principal permeability values and principal permeability directions of a practical fractured reservoir is illustrated.

2. Methodology

2.1 The calculation method of anisotropic permeability tensor of fractures

The single group fracture permeability is firstly obtained. Then the permeability tensor of multiple groups of fractures is derived by a simple summation. Next, the approach for calculating the significant parameter average permeability K is introduced.

2.1.1. The anisotropic permeability tensor of single group fractures.

Assume that the azimuth angles are β and the dip angles are α of an arbitrary set of parallel fractures in the reservoir, of which the average permeability is K. At first, the Cartesian coordinate system is established by taking the earth as the reference. As shown in Figure 1, the north, east, and the directions perpendicular to them are considered as the coordinate lines,

IOP Conf. Series: Materials Science and Engineering **474** (2019) 012043 doi:10.1088/1757-899X/474/1/012043

which correspond to the three unit vectors e_1, e_2, and e_3, respectively. Next, a Cartesian coordinate system is established by taking the fracture as the reference. One of the coordinate lines is the intersection line of the fracture face and horizontal plane, which corresponds to the unit vector f_2. Another coordinate line is perpendicular to f_2 within the fracture surface and corresponds to the unit vector f_1. The direction perpendicular to f_1 and f_2 is taken as the third coordinate line and corresponds to a unit vector f_3.

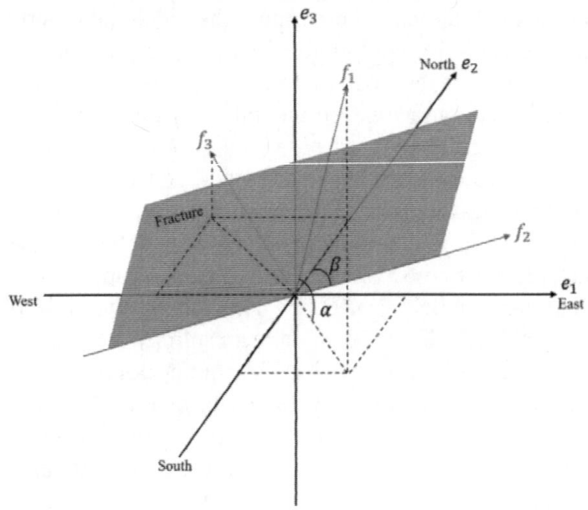

Figure 1. The stereo azimuth of fracture and the coordinate transformation relation.

The expression of permeability tensor $\boldsymbol{K}$ of an arbitrary set of fractures in the coordinate system (e_1,e_2,e_3) will be derived below.

Assume that the expression of permeability tensor $\boldsymbol{K}$ of fractures in the coordinate system (e_1,e_2,e_3) in figure 1 is [19]:

$$\boldsymbol{K}=(e_1,e_2,e_3)\cdot\overline{K}_{\mathrm{e}}\cdot\begin{pmatrix}e_1\\e_2\\e_3\end{pmatrix}=(e_1,e_2,e_3)\cdot\begin{pmatrix}k_{\mathrm{e}11}&k_{\mathrm{e}12}&k_{\mathrm{e}13}\\k_{\mathrm{e}21}&k_{\mathrm{e}22}&k_{\mathrm{e}23}\\k_{\mathrm{e}31}&k_{\mathrm{e}32}&k_{\mathrm{e}33}\end{pmatrix}\cdot\begin{pmatrix}e_1\\e_2\\e_3\end{pmatrix}=\sum_{i=1}^{3}\sum_{j=1}^{3}K_{eij}\,e_ie_j \tag{1}$$

where $\overline{K}_{\mathrm{e}}$ is the fracture permeability tensor in the coordinate system (e_1,e_2,e_3).

Assuming the expression of permeability tensor $\boldsymbol{K}$ of fractures in the coordinate (f_1,f_2,f_3) in figure 1 is:

$$\boldsymbol{K}=(f_1,f_2,f_3)\cdot\overline{K}_{\mathrm{f}}\cdot\begin{pmatrix}f_1\\f_2\\f_3\end{pmatrix}=(f_1,f_2,f_3)\cdot\begin{pmatrix}k_{\mathrm{f}11}&k_{\mathrm{f}12}&k_{\mathrm{f}13}\\k_{\mathrm{f}21}&k_{\mathrm{f}22}&k_{\mathrm{f}23}\\k_{\mathrm{f}31}&k_{\mathrm{f}32}&k_{\mathrm{f}33}\end{pmatrix}\cdot\begin{pmatrix}f_1\\f_2\\f_3\end{pmatrix}=\sum_{i=1}^{3}\sum_{j=1}^{3}K_{\mathrm{f}ij}\,f_if_j \tag{2}$$

where $\overline{K}_{\mathrm{f}}$ is the fracture permeability tensor in the coordinate system (f_1,f_2,f_3). Based on the known condition, we get:

$$\overline{K}_{\mathrm{f}}=\begin{pmatrix}K&0&0\\0&K&0\\0&0&0\end{pmatrix} \tag{3}$$

According to the three-dimensional position of the fracture and the spatial relationship between the coordinates (e_1,e_2,e_3) and (f_1,f_2,f_3) in figure 1, the transformation method of the coordinate vector can be expressed as follows:

$$(\boldsymbol{f}_1,\boldsymbol{f}_2,\boldsymbol{f}_3)=(\boldsymbol{e}_1,\boldsymbol{e}_2,\boldsymbol{e}_3)\cdot\boldsymbol{T} \tag{4}$$

where

$$\boldsymbol{T}=\begin{pmatrix} cos\alpha\cdot cos\beta & sin\beta & -sin\alpha\cdot cos\beta \\ -cos\alpha\cdot sin\beta & cos\beta & sin\alpha\cdot sin\beta \\ sin\alpha & 0 & cos\alpha \end{pmatrix}$$

Due to the essential property of tensor, equation (5) can be derived from equations (1) and (2).

$$(\boldsymbol{e}_1,\boldsymbol{e}_2,\boldsymbol{e}_3)\cdot\overline{K}_e\cdot\begin{pmatrix}\boldsymbol{e}_1\\ \boldsymbol{e}_2\\ \boldsymbol{e}_3\end{pmatrix}=(\boldsymbol{f}_1,\boldsymbol{f}_2,\boldsymbol{f}_3)\cdot\overline{K}_f\cdot\begin{pmatrix}\boldsymbol{f}_1\\ \boldsymbol{f}_2\\ \boldsymbol{f}_3\end{pmatrix} \tag{5}$$

Substituting equation (4) into (5), the following result can be obtained:

$$(\boldsymbol{e}_1,\boldsymbol{e}_2,\boldsymbol{e}_3)\cdot\overline{K}_e\cdot\begin{pmatrix}\boldsymbol{e}_1\\ \boldsymbol{e}_2\\ \boldsymbol{e}_3\end{pmatrix}=(\boldsymbol{e}_1,\boldsymbol{e}_2,\boldsymbol{e}_3)\cdot\boldsymbol{T}\cdot\overline{K}_f\cdot\boldsymbol{T}^{\mathrm{T}}\cdot\begin{pmatrix}\boldsymbol{e}_1\\ \boldsymbol{e}_2\\ \boldsymbol{e}_3\end{pmatrix} \tag{6}$$

Apparently,

$$\overline{K}_e=\boldsymbol{T}\cdot\overline{K}_f\cdot\boldsymbol{T}^{\mathrm{T}} \tag{7}$$

Substituting equation (3) into (7), the final expression can be derived:

$$\overline{K}_e=K\begin{pmatrix} cos^2\alpha\cdot cos^2\beta+sin^2\beta & sin^2\alpha\cdot cos\beta\cdot sin\beta & cos\alpha\cdot sin\alpha\cdot cos\beta \\ sin^2\alpha\cdot cos\beta\cdot sin\beta & cos^2\alpha\cdot sin^2\beta+cos^2\beta & -cos\alpha\cdot sin\alpha\cdot sin\beta \\ cos\alpha\cdot sin\alpha\cdot cos\beta & -cos\alpha\cdot sin\alpha\cdot sin\beta & sin^2\alpha \end{pmatrix} \tag{8}$$

2.1.2. The anisotropic permeability tensor of multiple groups of fractures.

The azimuth of the fractures is β_i and the dip angle is α_i ,respectively, i=1, 2, …, N. The density and aperture of every single group of fracture are d_i and b_i respectively and setting $c_i=b_i\cdot d_i$. The average permeability of each group of fractures is K_i . Assume that the permeability of fracture is proportional to c_i and the permeability is K_0 per unit c_i for each group of fractures. The average permeability of an arbitrary single group among the N groups of fractures is $K_i=c_i\cdot K_0$

According to equation (8), the permeability tensor of the i*th* group in the earth coordinate system is:

$$\overline{K}_{ei}=K_0\,\overline{K}_i \tag{9}$$

where,

$$\overline{K}_i=c_i\begin{pmatrix} cos^2\alpha_i\cdot cos^2\beta_i+sin^2\beta_i & sin^2\alpha_i\cdot cos\beta_i\cdot sin\beta_i & cos\alpha_i\cdot sin\alpha_i\cdot cos\beta_i \\ sin^2\alpha_i\cdot cos\beta_i\cdot sin\beta_i & cos^2\alpha_i\cdot sin^2\beta_i+cos^2\beta_i & -cos\alpha_i\cdot sin\alpha_i\cdot sin\beta_i \\ cos\alpha_i\cdot sin\alpha_i\cdot cos\beta_i & -cos\alpha_i\cdot sin\alpha_i\cdot sin\beta_i & sin^2\alpha_i \end{pmatrix}$$

The summation of permeability tensor of N groups is:

$$\overline{K}_{eN}=\sum_{i=1}^{N}\overline{K}_{ei}=K_0\sum_{i=1}^{N}\overline{K}_i \tag{10}$$

2.1.3 Approach to calculate the average permeability.

It can be found that the average permeability value K is crucial for assessing the full permeability tensor for fractured reservoirs. The average permeability K is the mean value of the permeability of different directions, which can be obtained by using production data.

For vertical wells, K can be derived from the production formula of plane radial flow:

$$Q = \frac{2\pi Kh(P_e - P_w)}{\mu B_O (ln\frac{R_e}{R_w} + S)} \tag{11}$$

For horizontal wells, K can be obtained according to production formula:

$$Q = \frac{2\pi Kh(P_e - P_w)}{\mu B_O (ln(\frac{a + \sqrt{a^2 - (\frac{L}{2})^2}}{\frac{L}{2}}) + \frac{h}{L}(\ln(\frac{h}{2\pi R_w}) + S))} \tag{12}$$

where Q is the liquid production, m^3/d; μ is the viscosity of fluid, mPa·s ; B_O is the oil volume factor, m^3/ m^3; P_e is initial pressure of reservoir, MPa; P_w is bottom hole pressure, MPa; h is the height of the reservoir, m; R_e is supply radius, m; R_w is the radius of wellbore, m; S is the skin factor; L is the length of the horizontal well, m; a is the semi-major axis of the elliptical oil drainage region , m.

Log data gives the information of fracture orientation including azimuth, dip angle, density, and aperture. Production data can determine the average permeability K. Combing the two, anisotropic permeability tensor of an arbitrary set of fractures $\overline{K}_e$ and that of multiple groups of fractures $\overline{K}_{eN}$ can be identified.

2.2 Approach to calculate the anisotropic permeability parameters

During the field practice, the most concerned issue is the anisotropic permeability parameters, which are the principal directions and principal values of permeability for fractured media. For the principal direction, it can be determined by the fracture orientation data from logging. The method to obtain the three principal values in the permeability tensor will be discussed below.

2.2.1. A simplified form of permeability tensor.

Based on the identified principal directions, the axis direction of the global coordinate is set to be aligned with the principal direction. In this case, off-diagonal terms of the permeability tensor can be eliminated.

Accordingly, when the principal directions are east, north and vertical direction in the coordinate system (Sec. 2.1.1) taking earth as a reference, equation (8) for single group fractures can be simplified into:

$$\overline{K}_e = K\begin{pmatrix} cos^2\alpha \cdot cos^2\beta + sin^2\beta & 0 & 0 \\ 0 & cos^2\alpha \cdot sin^2\beta + cos^2\beta & 0 \\ 0 & 0 & sin^2\alpha \end{pmatrix} \tag{13}$$

Equation (9) for multiple groups of fractures can be reduced to the following form:

$$\overline{K}_{eN} = \sum_{i=1}^{N}\overline{K}_{ei} = K_0\sum_{i=1}^{N}\overline{K}_i$$

$$= K_0\sum_{i=1}^{N}c_i\begin{pmatrix} cos^2\alpha_i \cdot cos^2\beta_i + sin^2\beta_i & 0 & 0 \\ 0 & cos^2\alpha_i \cdot sin^2\beta_i + cos^2\beta_i & 0 \\ 0 & 0 & sin^2\alpha_i \end{pmatrix} \tag{14}$$

According to equation (14), it is evident that the ratio of the three principal permeability values can be written as:

$$k_{xx} : k_{yy} : k_{zz} = S_x : S_y : S_z \tag{15}$$

where

$$\begin{cases} S_x = \sum_{i=1}^{N} c_i(cos^2 \alpha_i \cdot cos^2 \beta_i + sin^2 \beta_i) \\ S_y = \sum_{i=1}^{N} c_i(cos^2 \alpha_i \cdot sin^2 \beta_i + cos^2 \beta_i) \\ S_z = \sum_{i=1}^{N} c_i(sin^2 \alpha_i) \end{cases}$$

2.2.2. Approach to calculate the principal permeability values.

After collecting log data and getting the average permeability K, the permeability tensor of each fracture can be calculated by equation (8). The macroscopic permeability tensor of the fractured reservoir can be computed by equation (9). The eigenvalue and eigenvector of the tensor are the principal permeability values and principal permeability directions of fractured media respectively. For the scenario where the principal directions have been confirmed, the three principal values can be calculated by the following method.

The average permeability K can be derived via the anisotropic seepage theory [20]:

$$K = (k_{xx} \cdot k_{yy} \cdot k_{zz})^{1/3} \tag{16}$$

Substituting (15) into (16), the three principal permeability values can be obtained as:

$$\begin{cases} k_{xx} = K / (\frac{S_y \cdot S_z}{S_x^2})^{1/3} \\ k_{yy} = K / (\frac{S_x \cdot S_z}{S_y^2})^{1/3} \\ k_{zz} = K / (\frac{S_x \cdot S_y}{S_z^2})^{1/3} \end{cases} \tag{17}$$

Therefore, the anisotropic permeability parameters, principal directions and principal values of permeability, are identified, which can be employed to optimize the pattern design and infilling of the well.

3. Results and Discussion

The target area is a volcanic reservoir in Liaohe Oilfield. Natural fractures are well developed in this block, which leads to a strong heterogeneity and anisotropy. The objective is to obtain the principal permeability directions and the three principal permeability values of the area. Sixteen wells are drilled, and the pay zone is divided into three layers in the vertical direction.

The flow chart of the solution is devised as Figure 2. Taking well 18-30 as an example, the detailed procedures are given as follows.

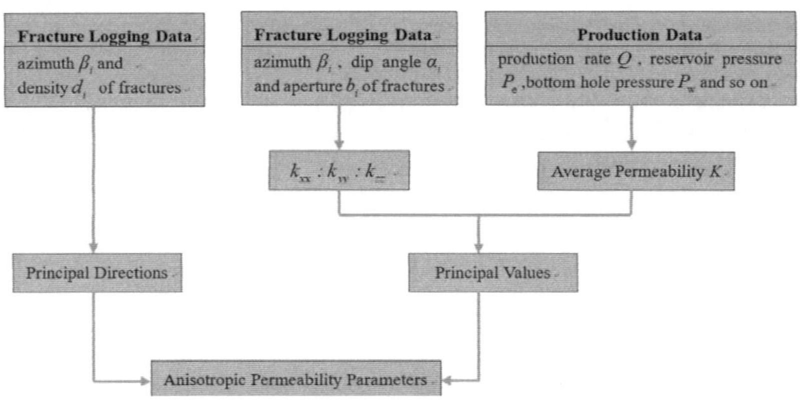

Figure 2. Illustration of workflow to compute the anisotropic parameters.

Step 1: The identification of principal directions. The azimuth of 1969 fractures are shown in Table 1 and statistically analyzed in Figure 3, where the length of the radius represents the number of fractures of different directions. And an adjacent ellipse of data points is drawn in Figure 3. The directions of the major axis and minor axis of the ellipse indicate that the fractures distribute mostly in 25° north by the west but least in 65° north by east. Since the fracture permeability is dominated by the orientation and density of fractures, 65° north by east, 25° north by west, and the direction perpendicular to the two are used as the three principal permeability directions, and they are taken as x-, y-, and z-axis of the Cartesian coordinate system, respectively.

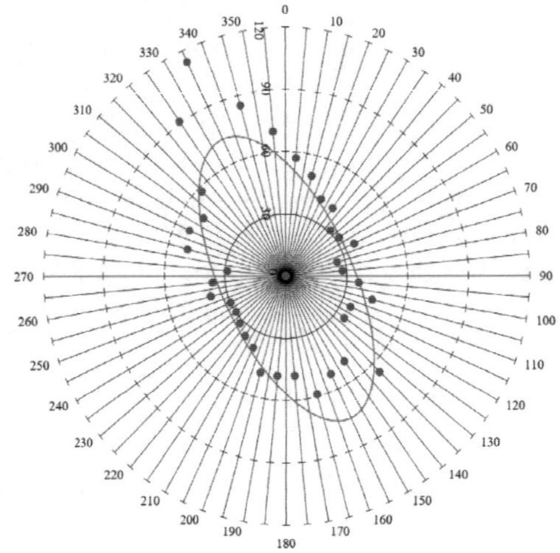

Figure 3. Fracture azimuth distribution circle diagram.

Step 2: Calculating the ratio of the three principal permeability values. As shown in Figure 4, the established system $(\boldsymbol{x}, \boldsymbol{y}, z)$ having an $\theta = 25^{\circ}$ included angle with the coordinate system $(\boldsymbol{e}_1, \boldsymbol{e}_2, \boldsymbol{e}_3)$ mentioned, equation (15) should be transformed into the present coordinate system. Thus, the ratio of the three principal permeability value should be written as:

$$k_{xx} : k_{yy} : k_{zz} = S_x^{'} : S_y^{'} : S_z^{'} \tag{18}$$

where

$$\begin{cases} S_x^{'} = cos\theta(\sum_{i=1}^{N} c_i(cos^2 \alpha_i \cdot cos^2 \beta_i + sin^2 \beta_i)) - sin\theta(\sum_{i=1}^{N} c_i(sin^2 \alpha_i \cdot cos\, \beta_i \cdot sin\, \beta_i)) \\ S_y^{'} = sin\theta(\sum_{i=1}^{N} c_i(sin^2 \alpha_i \cdot cos\, \beta_i \cdot sin\, \beta_i)) + cos\theta(\sum_{i=1}^{N} c_i(cos^2 \alpha_i \cdot sin^2 \beta_i + cos^2 \beta_i)) \\ S_z^{'} = \sum_{i=1}^{N} c_i(sin^2 \alpha_i) \end{cases}$$

It is assumed that the azimuth of fractures is the primary direction 25° north by west, and the secondary direction 65° north by east. The weights of the primary and secondary directions account for 66.67 and 33.33%, respectively, according to the number of fractures. The inclination angles of fractures are summarized in Table 2. Then, substituting the related data into equation (18), the result can be obtained as:

$$k_{xx} : k_{yy} : k_{zz} = 0.5712 : 0.6295 : 0.7348 \tag{19}$$

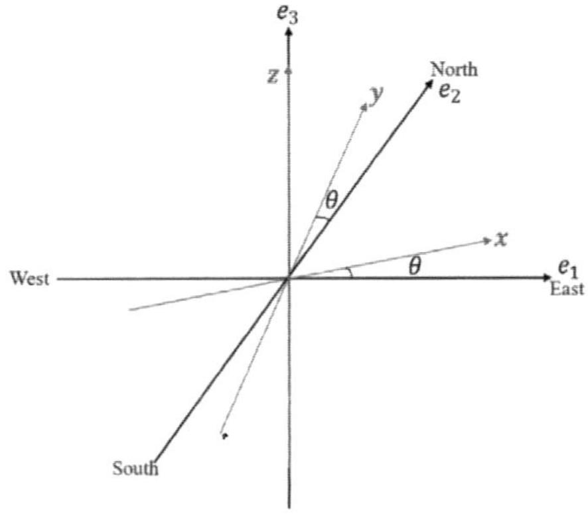

Figure 4. The coordinate system of this field problem.

Table 1. The distribution of fracture azimuth of well 18-30.

Angle (°)	Number	PCT (%)	Angle (°)	Number	PCT (%)	Angle (°)	Number	PCT (%)	Angle (°)	Number	PCT (%)
0-10	2	2.47	90-100	0	0	180-190	3	3.70	270-280	2	2.47
10-20	6	7.41	100-110	0	0	190-200	2	2.47	280-290	2	2.47
20-30	4	4.94	110-120	3	3.70	200-210	1	1.23	290-300	4	4.94
30-40	4	4.94	120-130	1	1.23	210-220	2	2.47	300-310	6	7.41
40-50	1	1.23	130-140	0	0	220-230	1	1.23	310-320	3	3.70
50-60	1	1.23	140-150	1	1.23	230-240	2	2.47	320-330	2	2.47
60-70	3	3.70	150-160	2	2.47	240-250	1	1.23	330-340	3	3.70
70-80	1	1.23	160-170	0	0	250-260	0	0	340-350	5	6.17
80-90	0	0	170-180	6	7.41	260-270	2	2.47	350-360	5	6.17

Table 2. The distribution of fracture dip angle of well 18-30.

Angle (°)	Number	PCT (%)	Angle (°)	Number	PCT (%)	Angle (°)	Number	PCT (%)	Angle (°)	Number	PCT (%)
10-15	2	2.47	30-35	4	4.94	50-55	9	11.11	70-75	11	13.58
15-20	1	1.23	35-40	4	4.94	55-60	3	3.70	75-80	7	8.64
20-25	2	2.47	40-45	5	6.17	60-65	10	12.35	80-85	4	4.94
25-30	4	4.94	45-50	3	3.70	65-70	10	12.35	85-90	2	2.47

Step 3: Calculating the average permeability. According to equation (11), the average permeability $K = 57.03\times10^{-3}\mu m^2$. The used production data are listed in Table 3.

Table 3. The relevant production data to calculate average permeability of well 18-30.

Parameters	Liquid production (Q) t/d	Reservoir pressure (P_e) MPa	BHFP (P_w) MPa	Area of reservoir (A) Km^2	Radius of well (R_w) m	Skin factor (S)	Perforation height (H) m	Oil viscosity (μ) mPa.s	Volume factor (B_o)	Oil density (ρ_o) g/cm^3
Value	63.2	29.33	28.08	7.40	0.12	68.55	75	0.50	1.47	0.84

Since the permeability is proportional to the fracture density, the average permeability of each layer can be obtained, as is shown in Table 4.

Table 4. The fracture density and average permeability of well 18-30.

Layer	II layer	II-1 Layer	II-2 Layer	II-3 Layer
Density (1/m)	0.3122	0.4000	0.2485	0.1765
Average permeability ($10^{-3}\mu m^2$)	57.03	73.07	45.39	32.24

Step 4: Calculating principal permeability. Substituting (19) and the average permeability value *K* of each layer into (17), the corresponding principal permeability values of well 18-30 can be identified. Similarly, the principal permeability values of other 15 wells are calculated. The results are shown in
Table **5**, which lists single wells' parameters of anisotropic permeability of three layers. It can

be concluded that principal values vary with different positions due to the non-uniform distribution of fractures.

Table 5. Principal permeability values of different wells of three layers ($10^{-3}\mu m^2$)

Well number	II-1 layer			II-2 layer			II-3 layer		
	x	y	z	x	y	z	x	y	z
18-30	65.05	71.68	83.67	40.41	44.53	51.97	28.70	31.63	36.92
22	1.49	2.04	1.81	1.50	2.02	1.83	1.49	1.90	1.95
23	19.66	27.04	23.99	15.96	21.48	19.45	8.26	10.54	10.80
24	0.95	0.84	0.54	1.44	1.38	0.52	0.34	0.36	0.21
25	0.61	0.79	0.52	1.02	0.92	0.36	2.56	2.41	2.28
14-14	3.28	3.68	3.02	2.69	3.07	2.54	2.92	4.54	3.62
14-18	8.00	16.86	17.23	6.46	14.28	12.63	2.89	4.25	4.70
14-26	1.44	1.56	0.87	1.52	2.17	1.89	0.81	1.74	1.32
16-22	4.20	3.06	3.45	4.17	3.10	3.42	4.20	3.29	3.21
16-24	9.36	23.86	19.34	11.50	20.81	16.09	13.47	17.20	17.62
16-26	17.26	23.73	21.06	17.38	23.40	21.20	17.29	22.07	22.61
16-32	0.87	0.91	0.41	1.40	1.42	0.86	0.98	2.33	2.28
18-24	3.98	5.79	4.58	1.80	3.15	4.19	2.33	2.06	1.22
18-34	3.73	4.97	3.37	6.42	7.57	2.19	3.49	3.41	2.29
18-36	2.26	2.03	1.22	0.29	0.24	0.22	1.76	1.88	1.11
20-38	3.58	3.06	1.98	3.99	4.45	1.65	0.57	4.00	3.54

In this study, the derived complete tensor permeability calculated using the static log data and dynamic production data not only characterizes the fracture permeability quite accurately but also provides reliable simulation results. On the other hand, after the identification of the principal direction with log data, the simplified form also gives a good performance. With the calculation results, the principal permeability direction of fractures helps one to design the direction of the array of the vertical wells and the wellbore direction of horizontal wells. The principal value of the permeability gives a guidance to determine the well spacing. These designs, decisions, and adjustments based on the calculation results are of great significance and are instrumental in retarding the water breakthrough time, improving the displacement effect, and increasing the ultimate oil recovery.

4. Conclusions

The results obtained made it possible to draw the following major conclusions.

- ✓ The expression of full permeability tensor of fractures is derived by converting the fracture permeability values from different local coordinate systems into one global coordinate system based on the tensor theory, which offers a comprehensive and accurate description of the flowability. Upon this approach incorporation into simulators, the precision of the simulation is significantly improved.
- ✓ The proposed method for assessing the parameter of anisotropic permeability is expedient and applicable. Principal directions are derived based on the log data. Principal values are identified using log data and production data. This approach can be used for adjusting well patterns properly and precisely.
- ✓ The field application strongly indicates that the methodology proposed in this paper is effective in assessing the anisotropic permeability parameters. Thus, it can facilitate the production system management and identification of the optimal well pattern layout.

Acknowledgments

The paper is sponsored by National Natural Science Foundation Project (No. 51374222), the National Key Basic Research & Development Program (No. 2015CB250905), National Major Project (No. 2017ZX05032004-002 and 2016ZX05037003), and the Science Foundation of China University of Petroleum - Beijing (No. 2462014YJRC038).

References

[1] Ahmed E M and Geiger S 2012 *EAGE Annual Conf. and Exhibition* (Denmark, Society of Petroleum Engineers)
[2] Liu Y 2007 *Int. Oil Conf. and Exhibition* (Mexico, Society of Petroleum Engineers) pp27-30
[3] Bagheri M and Settari A 2008 *EAGE Annual Conf. and Exhibiti*on. (Rome, Society of Petroleum Engineers)
[4] Wang H F 1983 *Journal of Jianghan Petroleum Institute* **2** (90), pp 90-6
[5] Liakopoulos A C 1965 *Int. Association of Scientific Hydrology Bulletin* pp 41-48
[6] Asadi M, Ali G I, Rose W D and Shirazi M K 2000 *SPE Asian Pacific Conf.* (Japan, Society of Petroleum Engineers)
[7] Renard P 2001 *Journal of Geophysical Research* **106** (B11) pp 443-52
[8] Pan Z, Ma Y, Connell L D, Down D I and Camilleri M 2015 *Journal of Natural Gas Science and Engineering* pp 336-44
[9] Snow D T 1969 *Water Resources Research* **5** (6) pp 1273-89
[10] R E Avila, Gupta A and Penuela G 2001 *Journal of Canadian Petroleum Technology* **40** (12)
[11] Kasap E and Lake L W 1990 *SPE Formation Evaluation* pp192-200
[12] King M J 1993 *Western Regional Meeting* (Alaska, Society of Petroleum Engineers) pp 745-6
[13] Jakobsen M, Liu E and Chapman M 2007 *SEG/San Antonio Annual Meeting* (San Antonio, Society of Exploration Geophysicists) pp 99-103
[14] Nordlund M, Lopez P D J, Stolz S, Kuczaj A, Winkelmann C, and Geurts B J 2013 *Int. Journal of Engineering Science* **68** pp 38-60
[15] Chi L and Heidari Z 2016 *SPE Journal*. 21. 10.2118/179734-PA
[16] Hou J, Mekic N, Quirein J, Donderici B and Torres D 2016 *SPE Annual Technical and Exhibition* (Dubai, Society of Petroleum Engineers) pp 1436-49
[17] Ali A and Jakobsen M 2014 *Geophysical Prospecting* **62** pp 293-314
[18] Fanchi J R 2005 Directional Permeability Presented at the SPE Annual Technical *Conf.* and Exhibition (San Antonio, Society of Petroleum Engineers) pp 24-7
[19] Liu Y T, Ding Z P, Qu Y G and Zhao C J 2011 *Acta Petrolei Sinica* **32** (5) pp 842-6
[20] Muskat M 1982 *IHRDC Boston Mass* pp 225-7

Testing of glutathione contents by calcein-Cu(II) fluorescence spectra

L Liu*, W Wu, Y Zhang, W Chen, J Y Guo and S Z Li

School of Electronic and Electrical Engineering, Wuhan Textile University, No.1 Yangguang Street, Wuhan, China
*E-mail: liuli@wtu.edu.cn

Abstract. Glutathione (GSH) is a natural active tripeptide formed by glutamic acid, glycine, and cysteine, which plays a vital role in many physiological processes of the living body. The GSH contents of the sample were detected by using calcein-Cu(II) fluorescence system in our work. In a Na_2HPO_4-NaH_2PO_4 buffer with a pH of 8.0, Cu (II) could coordinate with calcein. Calcein -Cu(II) complex was not fluorescent which led to a reduction of calcein fluorescence intensity. For the strong affinity of GSH with copper (II), the addition of GSH solution would make calcein free from calcein-Cu(II) complex and recover fluorescence. Excitation spectra and emission spectra of calcein and calcein-Cu(II) system were researched. And fluorescence spectra of calcein-Cu(II) system with different GSH contents were investigated. The influence of experimental parameters such as reaction time and the addition sequence of the solution on results was also studied. We got the following conclusions: first, Cu(II) can make the calcein fluorescence be quenched, but the wavelength at the peak of fluorescence is unchanged. Second, the addition of Glutathione can make the fluorescence of calcein - Cu (II) system recover. And the change of the fluorescence intensity has a linear relationship with the concentration of GSH in the range from 2×10^{-6} to 1.0×10^{-5} mol/l. Third, response time may be extended appropriately for complete reaction in the system. At last, the experimental results have shown that calcein - GSH - Cu(II) system has higher sensitivity and better stability than calcein - Cu(II)- GSH system. This test method is of great importance in practical application.

1. Introduction

Glutathione (GSH) is a kind of important tripeptide containing sulfhydryl group, amino group, and gamma amide bond, which is formed by condensation of glutamic acid, glycine, and cysteine. It is widely distributed in body fluid and tissue fluid of human and animals and is a kind of metabolic regulators in cells [1]. It not only removes peroxide and other free radicals in the body but also promotes the activity of the liver enzyme and maintains the integrity of the erythrocyte membrane. At the same time, it also has many physiological functions such as maintaining DNA's biosynthesis and cellular immunity. Therefore, rapid, efficient and accurate detection on GSH is vital to the application of prevention, investigation, and treatment of some diseases.

In recent years, many test methods have been used, such as spectrophotometry, chromatography, capillary electrophoresis, an electrochemical method, surface enhanced Raman scattering (SERS), fluorescent probe and so on [2]. Fluorescence analysis has the advantages of a simple test, high selectivity and short response time [3, 4], and more importantly, it can also be utilized in real-time monitoring and imaging of organisms and has been widely applied in biomedicine, analytical chemistry, chemical biology, and many other fields.

GSH could be tested usually when it was combined with a fluorescence system. Cao et al. determined glutathione in cucumber by using orthophthalaldehyde as a complexing agent, the determining limit was $0.1\mu gml^{-1}$ [5]. Xu et al. synthesized water-soluble ZnS quantum dot by using mercaptopropanoic acid (MPA) as the capping ligand for the detection of GSH [6]. The fluorescence intensity of the ZnS QDs was significantly quenched in the presence of Ni^{2+} ions. In the presence of GSH, the fluorescence of ZnS QDs was recovered because Ni^{2+} preferably reacted with GSH. The degree of recovery of the fluorescence intensity of ZnS QDs was proportional to the GSH concentration. Huang reported a method to realize precise and rapid GSH detection by using silver-nanoclusters as a fluorescent probe [7]. Yang designed and synthesized a fluorescent probe for the detection of biothiols and investigated its properties for labeling glutathione [8]. After reaction with glutathione in 2-[4-(2-hydroxyethyl)-piperazinyl] ethanesulfonic acid (HEPES) buffer, the color of the solution changed from light yellow to pink, which could be detected by naked eyes.

Contents of acetylcysteine, traces of sulfide and GSH were tested by fluorescence spectroscopy using calcein-Cu(II) system [9-10]. In a buffer with a suitable pH, copper would coordinate with calcein, which caused the quenching of fluorescence. Due to strong affinity of sulfide with copper (II), copper (II) would be seized from the calcein-Cu (II) complex and the released calcein leading to the recovery of fluorescence. Recovery degree of the fluorescence was linear with a concentration of sulfide ion.

Based on the principle of fluorescence analysis, a method of quantitative determination of GSH was established according to the variation of fluorescence intensity with different concentrations of GSH standard solution in calcein-Cu(II) system. The comparison of our results with Jing et al. [10] is provided.

2. Experimental

2.1. Reagents and instruments

The following reagents were used: Glutathione (Shanghai Aladdin Bio-Chem Technology Co., Ltd), calcein (Sinopharm Chemical Reagent Co., Ltd), $CuSO_4$ (Sinopharm Chemical Reagent Co., Ltd), NaOH (Sinopharm Chemical Reagent Co., Ltd), $Na_2HPO_4 \cdot 12H_2O$ (Sinopharm Chemical Reagent Co., Ltd), $NaH_2PO_4 \cdot 2H_2O$ (Sinopharm Chemical Reagent Co., Ltd). All reagents used in the experiment were AR.

Spectropro® 500 fluorescence spectroscopy (Acton Company, USA), FA2004 analytical balance (Shanghai Precision Instrument Co., Ltd), SZ-96 automatic water distiller (Shanghai Yarong Biochemical Instrument Factory).

2.2. Method

2.2.1. Selection of concentration ratio of calcein to Cu(II) and optimum concentration. It was reported that the relative contents of copper(II) and calcein would influence the intensity of the background signal and the linear range of detection results. When the content ratio of copper(II) to calcein was less than 1:1, the fluorescence quenching of calcein-Cu(II) system increased with the addition of copper (II). When the content ratio of copper(II) to calcein was over than 1:1, the degree of fluorescence quenching was tending to stable. Compared with calcein, sulfide had a stronger affinity with copper(II), if there were excessive copper(II) in the solution, sulfide ion would will first react with the excess of copper(II) instead of seizing the copper (II) ions in calcein-Cu(II) complex, which would result in the lower sensitivity, and destroyed the linear relationship of fluorescence intensity with calcein concentrations in low concentration. Therefore, the concentration ratio of copper (II) and calcein was 1:1 in their experiments. Because GSH molecules contain active sulfhydryl groups, which chemical properties are similar to sulfur ions, we consulted the reference [9] and chose the concentration ratio of copper(II) and calcein as 1:1 and the optimum concentration of calcein was 1×10^{-5} mol/l. The concentrations of GSH were 2×10^{-6}, 4×10^{-6}, 6×10^{-6}, 8×10^{-6}, and 1.0×10^{-5} mol/l, respectively.

2.2.2. Preparation of stock solution. Calcein standard solution (1.0×10^{-3}mol/l): 0.0623g calcein solid was accurately weighed by analytical balance and 25ml NaOH solution of 0.1mol/l concentration was added into it. After the calcein solid was dissolved completely, the solution was transferred into a 100ml volumetric flask, and distilled water was added to the volume. The calcein standard solution should be stored at 4°C, which was diluted with distilled water to the desired concentration step by step when used.

Copper(II) ion standard solution(1.0×10^{-3}mol/l): 0.0160g $CuSO_4$ was accurately weighed by analytical balance. After $CuSO_4$ was dissolved completely by little-distilled water, the solution was transferred into a 100ml volumetric flask, and distilled water was added to the final volume. The calcein standard solution was diluted with distilled water to the desired concentration step-by-step when used.

The GSH standard solution (1.0x10-3mol/l): 0.0307g GSH was accurately weighed by analytical balance, which was dissolved completely by little-distilled water. The solution was transferred into a 100ml volumetric flask, and distilled water was added to the final volume. The GSH standard solution was diluted with distilled water to the desired concentration step by step when used. Because the reduced GSH could be easily oxidized, GSH standard solution should be prepared before detection.

Buffer solution (0.2mol/l): 7.1600g $Na_2HPO_4\cdot 12H_2O$ was accurately weighed by analytical balance, which was dissolved completely by little-distilled water and cooled down to room temperature. The solution was transferred into a 100ml volumetric flask, and distilled water was added to scale line, and 0.2mol/l Na_2HPO_4 solution was obtained. 3.1200g $NaH_2PO_4\cdot 2H_2O$ was accurately weighed by analytical balance, which was dissolved completely by little-distilled water. According to the above steps, 0.2mol/l NaH_2PO_4 solution was got. The mixed 94.7ml Na_2HPO_4 solution with a 5.3ml NaH_2PO_4 solution, the 100ml buffer with a pH of 8 was prepared.

2.2.3. Detection. Calcein standard solution (1×10^{-3} mol/l), GSH standard solution (1.0×10^{-3} mol/l) and Copper(II) ion standard solution (1.0×10^{-3} mol/l) was diluted by distilled water to the standard solution with 1×10^{-4} mol/l concentration respectively. 2.5ml of 0.2mol/l buffer solution and 2.5ml of 1×10^{-4} mol/l calcein standard solution were transferred into five 50ml volumetric flasks , then 1×10^{-4} mol/l GSH standard solution of 1.0, 2.0, 3.0, 4.0, and 5.0 ml were added, respectively, 2.5ml 1×10^{-4}mol/l Copper(II) standard solution was added subsequently and distilled water was dropped to final volume. Five samples were prepared.

These samples were heated in water bath at 40°C for 20 min, and their excitation spectra and emission spectra were obtained. The standard curve of concentration of GSH with the peak intensity of fluorescence was constructed.

3. Results and Discussion

3.1. Excitation spectra and emission spectra of calcein and calcein-Cu (II)

From figure 1, it can be seen when the concentration of calcein solution was 1×10^{-5}mol/l, and pH value of buffer solution equals to 8.0, strong fluorescence of calcein appeared. The wavelength at excitation peak and emission peak was 497 nm and 518nm. After adding copper (II) to calcein solution, the quenching of fluorescence occurred, and the wavelength at excitation peak and emission peak had no apparent change. In a Na_2HPO_4-NaH_2PO_4 buffer with pH of 8.0, copper (II) could coordinate with calcein. Calcein -Cu(II) complex was not fluorescent which led to the reduction of calcein fluorescence intensity.

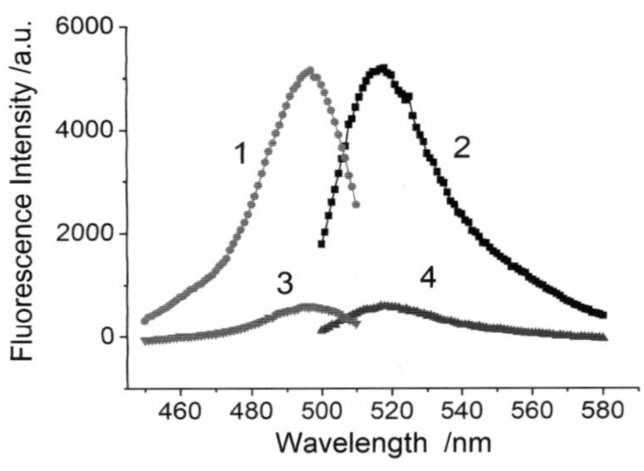

Figure 1. Excitation spectra and emission spectra of calcein and calcein-Cu(II) curves labeled 1, 2 corresponding to excitation and emission spectra of calcein and curves labeled 3, 4 corresponding to excitation and emission spectra of calcein-Cu(II).

3.2. Fluorescence spectra of calcein-GSH-Cu(II) system

For simplicity, according to the sequence of adding reagents, the solution obtained by the above steps were referred to as calcein-GSH-Cu (II) system. Figure 2 depicts its fluorescence spectra with different concentration of GSH. For a strong affinity of GSH with copper (II), the addition of GSH solution would make calcein free from the calcein-Cu(II) complex and recover the fluorescence intensity. With an increase in GSH concentration, the fluorescence intensity enhanced and the wavelength at peak intensity remained unchanged.

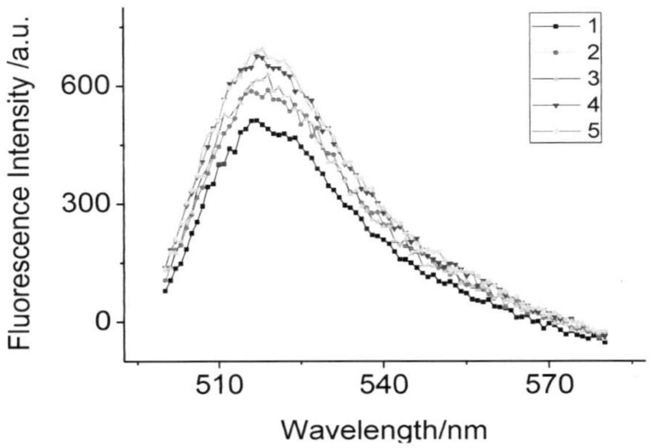

Figure 2. Fluorescence spectra of calcein-GSH-Cu (II) system, curves labeled 1, 2, 3, 4, 5 corresponding to 2×10^{-6}, 4×10^{-6}, 6×10^{-6}, 8×10^{-6}, and 1.0×10^{-5} mol/l concentration of GSH, respectively.

3.3. Comparative analysis of experimental results of calcein-GSH- Cu(II)with calcein-Cu(II) -GSH
To study the influence of the sequence of adding GSH and copper, the sequence of reagents was changed. For the calcein-Cu(II)-GSH system, after 1×10^{-4}mol/l GSH standard solution of 1.0, 2.0, 3.0, 4.0, and 5.0 ml was added, and samples were heated in water bath at 40°C for 20 minutes, the fluorescence spectra were determined, as shown in figure 3. With an increase of GSH concentration, fluorescence intensity enhanced and the wavelength at peak intensity had no change, which was similar to calcein-GSH-Cu(II) system.

*3.3.1. Comparison of sensitivity of calcein-GSH- Cu(II) with calcein-Cu(II) -GSH.*The difference between the maximum and minimum of fluorescence peak intensity was written as ΔF, while ΔF_1 corresponding to calcein-GSH-Cu(II) system and ΔF_2 corresponding to calcein-Cu(II) -GSH, which is shown in figure 4 by a bar graph. ΔF_1= 176.34, ΔF_2= 99.33, ΔF_1 was much larger than ΔF_2, so calcein-GSH-Cu (II) system was more sensitive than calcein-Cu (II) -GSH system.

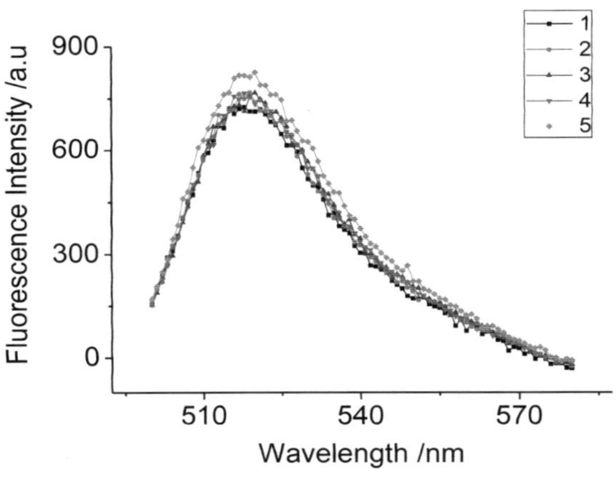

Figure 3. Fluorescence spectra of calcein-Cu(II)-GSH system, curves labeled 1, 2, 3, 4, 5 corresponding to 2×10^{-6}, 4×10^{-6}, 6×10^{-6}, 8×10^{-6}, and 1.0×10^{-5} mol/l concentrations of GSH.

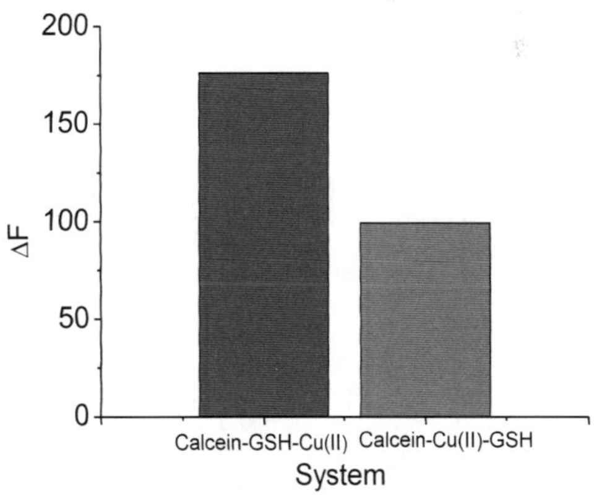

Figure 4. Comparison of sensitivity of calcein-GSH-Cu(II) with calcein-Cu(II)-GSH

3.3.2. Comparison of stability of calcein-GSH-Cu (II) with calcein-Cu(II)-GSH. For calcein-GSH-Cu(II) system, there were five sets of data corresponding to different concentrations of GSH, and the time required for stability was 20, 20, 35, 40, and 60 min, respectively. The highe the concentration of GSH, the longer the time required for stability. For calcein-Cu(II)-GSH system, the time required for stability was 20, 20, 35, 100, and 115 min, respectively. As compared to the time necessary for stability in two fluorescence systems, calcein-GSH-Cu(II) system required less time to stabilize, and its stability was better than calcein-C(II)-GSH. calcein-GSH-Cu(II) system should be chosen in practical application.

3.3.3. Influence of reaction time on the results. The reaction time of the reagents had significant impact on the results. When the concentrations of calcein, copper (II) ion and GSH were all 5×10^{-6} mol/l and the pH was 8.0 in PBS buffer, the reaction time and fluorescence intensity values are shown in table 1 and figure 5.

In both calcein-GSH-Cu(II) and calcein-Cu-GSH(II) systems, the longer the reaction time, the better the stability. Under the same concentration of GSH, calcein-GSH-Cu(II) system would need less time for the reaction and stabilization.

Table 1. Reaction time and fluorescence intensity of two systems

Serial number	Reaction time /min	Fluorescence intensity /a.u. calcein-GSH-Cu^{2+}	Reaction time /min	Fluorescence intensity /a.u. calcein-Cu^{2+}-GSH
1	20	497.00	20	583.67
2	25	548.67	35	676.83
3	35	612.33	55	744.67
4	55	656.67	70	783.00
5	80	683.00	100	809.67
6	100	687.67	130	812.33

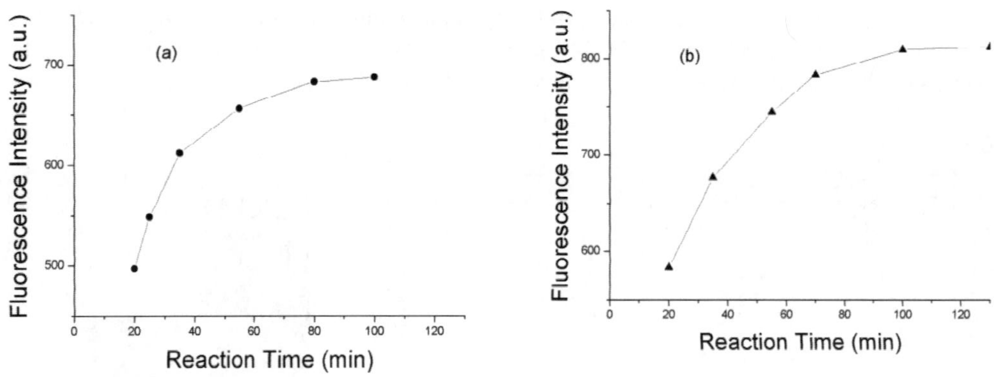

Figure 5. Relation of fluorescence intensity with reaction time. (a) calcein-GSH-Cu(II) system, (b)calcein-Cu-GSH(II) system

3.3.4. Standard curve and linear equation. GSH concentration with relative fluorescence intensity was plotted, and standard curves of calcein-GSH-Cu(II) system and calcein-Cu-GSH(II) system were given by using the linear fitting method in figure 6. Concentration of GSH was used as X-axis, and fluorescence intensity was as Y-axis, the linear equation was F=22.35C+480.43 for calcein-GSH-Cu(II) system and F=11.183C+690.37 for calcein-Cu-GSH(II) system, the correlation coefficient was 0.991 and 0.972 respectively, the linear range of these two equations was from 2×10^{-6} mol/l to 1.0×10^{-5} mol/l. Compared with calcein-Cu-GSH(II) system, the standard curve for calcein-GSH-Cu(II) system had a larger slope and more sensitivity. The standard glutathione solution with a content of 2.0×10^{-5} mol·L^{-1} was tested ten times and a relative standard deviation of 1.2%.

3.3.5. Effects of other ions on results. The impact of some common ions on the determination of glutathione was investigated. Interference experiments were performed on 2.0×10^{-5} mol/l glutathione under the selected experimental conditions. The results showed that in the range of ± 5% of the interference level, 200 times of the amount of PO_4^{3-}, NO_3^-, Na^+, K^+ and ten times of the amount of Al^{3+} had no interference, and ten times of the amount of Fe^{2+} would interfere the results.

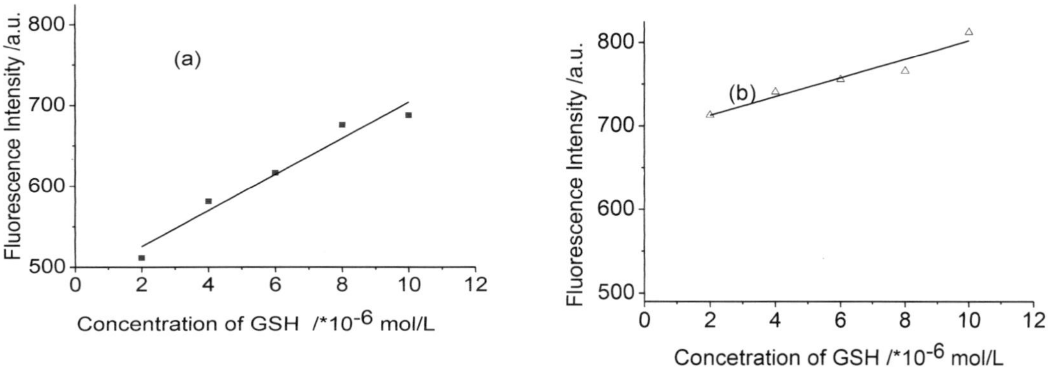

Figure 6. Standard curves of (a) calcein-GSH-Cu(II) system, (b) calcein-Cu-GSH(II) system

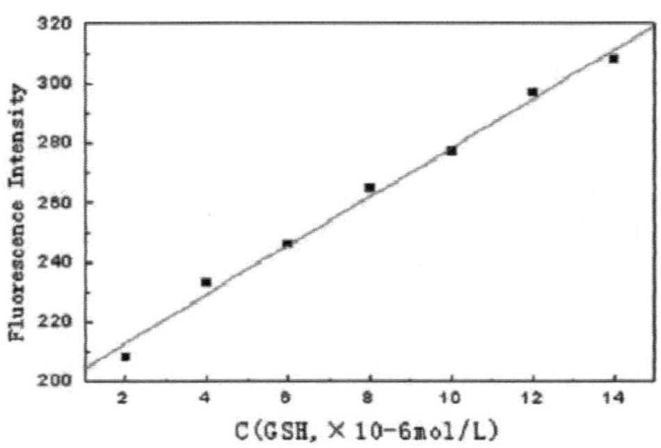

Figure 7. A standard curve of GSH reported by Jing et al. [10]

3.3.6. Comparison of our results with reference [10]. Jing B.B et al. had done the similar work on testing the GSH concentration. The wavelength at excitation peak and emission peak was 492nm and 519nm. In the sodium borate buffer solution with pH of 9.18, according to the dropping step of buffer solution→calcein→GSH→Cu^{2+}, the standard curve equation was F=8.1964C+196.43, the linear range is in 2.0×10^{-6}-1.4×10^{-5} mol/l, and the detecting limit is 1.0×10^{-6}mol L^{-1}. We adopted Na_2HPO_4-NaH_2PO_4 buffer with pH of 8.0, while wavelength at excitation peak and emission peak was 497 and 518nm in our experiment. The wavelength at fluorescence peak has no noticeable difference with Jing's report, but excitation wavelength is red-shifted. From the linear equations and figures 6 and 7, because of having a larger slope, our standard curve has more sensitivity. Therefore, it can be considered that the pH value and the type of buffer solution would also influence the results.

4. Conclusions

The GSH content was determined by fluorescence spectra of calcein-Cu(II) system in our work. In a Na_2HPO_4-NaH_2PO_4 buffer with pH of 8.0, copper(II) can coordinate with calcein which led to the reduction of fluorescence intensity of calcein. The addition of GSH solution would make calcein free from the calcein-Cu(II) complex and recover the fluorescence intensity. The change of the fluorescence intensity had a linear relationship with the concentration of GSH in the range of 2×10^{-6}mol/l - 1.0×10^{-5} mol/l, the linear equation was $F=22.35C+480.43$. We also researched the influence of reaction time and the sequence of adding reagents on the experimental results. It was shown that the more the concentration of GSH, the longer the time needing for stability. We could appropriately extend reaction time for more stability. Calcein - GSH- Cu(II) system has a higher sensitivity and better stability than the calcein - Cu(II)- GSH system. This test method is of great importance in practical applications.

Acknowledgment

This work was supported by the National Natural Science Fund of China (61108033).

References

[1] Zhong H, Zhang H and Xu H 2014, *Amino Acids & Biotic Resources*, vol. **36**, 23-6.
[2] Shi L, Hung L and Gong S 2018, *Guangzhou Chem. Industry*, vol. **46**, 23-7,107.
[3] Zhu Q, He S and Dou Y 2016, *Fermen. Tech. Comm.*, vol. **45**, 170-4.
[4] Wu H and Gu Y 2017 *Spectrosc. Spectral Analy.*, vol. **31**, 2100-4.
[5] CAO X, Chen Y 2000, *Resource Develop. & Market,* vol. **16**, 272-3.
[6] Xu Q, Wei H, Hu X 2013, *Chin. J. Analy. Chem.,* vol. **41**, 1102-6.
[7] Huang K, Qin C, Cao X, Yang T, Chen Y, Zhang S, Pan H and Xu J 2016, *Spectrosc. Spectral Analy.*, vol. **36**, 3973-7.
[8] Yang R, Tang Y and Zhu W 2016, *Chem. J. Chin. Universities*, vol. **37**, 643-7.
[9] Zhang S, Nie Y, Hu D, Zhang P, Zhu Y 2011, *Chin. J. Analy. Lab.*, vol. **30**, 103-6.
[10] Jing B, Yang X, Zhang S, Zhang S H, Da M, Zhang P and Zhu Y 2011 *Progress in Modern Biomed.*, vol. **11**, 1359-62.

IOP Conf. Series: Materials Science and Engineering **474** (2019) 012045 doi:10.1088/1757-899X/474/1/012045

Characterization of hot deformation behavior of TC32 titanium alloy

J Li*, L W Zhu, M B Li, X N Wang, G Q Shang and Z S Zhu

Beijing Institute of Aeronautical Materials, Beijing 100095, China

*E-mail: tjulijing@126.com

Abstract. The hot deformation behavior of TC32 alloy is investigated in the temperature range of 873~948°C and strain rate range of 0.01, 0.1, 1, and $10s^{-1}$ by hot compression experiments, and the constitutive relationship of TC32 alloy is established on the base of the Arrhenius equations. It is found that, with an increase in deformation temperature and the strain rate reduction, the flow stress decreases, and the deformation mechanisms of TC32 alloy exhibit dynamic recovery feature within the high-temperature region and recrystallization feature within the low-temperature one. Deformation activation energy values are 404.7 kJ/mol in the alpha-beta region and 526.6 kJ/mol in the beta region. Finally, the constitutive equation of TC32 titanium was established.

1. Introduction

With the rapid development of the new generation of the aeronautic and astronautic vehicle, tank and switch antitank vehicle, etc. the design and application demand for excellent performance, high weight loss, long-lived and low-cost increase [1, 2]. The innovative TC32 titanium alloy was developed by alloying, comprehensive strengthening and toughening technologies in Beijing Institute of Aeronautical Materials. The alloy is a new kind of Ti-Al-Mo-Cr-Zr system titanium alloy with low cost and excellent performance of low density, good strength-plastic-toughness-fatigue properties which could match different microstructure types [3-7].

The mechanical behavior in thermal deformation is related to chemical composition, microstructure and deformation temperature, deformation degree, strain rate of the alloy, etc. [8-12]. High temperature working above the beta transus (beta phase field) is an essential step for homogenization of the cast structure while deformation below the transus (alpha-beta phase field) converts the transformed beta structure to equiaxed alpha-beta structure [13].

2. Experimental procedures

The TC32 (Ti-5Al-3Mo-3Cr-1Zr-0.15Si) titanium alloy which was three-time melted by vacuum self-consuming arc melting, forged in the alpha-beta region and duplex-annealed in the investigation. A bimodal microstructure containing 40% primary alpha phase was obtained, which is characterized that discontinuous primary alpha phase equiaxed or elliptical-shaped were distributed in the transfer beta matrix with lamellar microstructure, as depicted in figure 1. And the mechanical properties of the original condition of TC32 alloy are shown in table 1. The phase transition temperature measured was 918°C. Cylindrical specimens of TC32 alloy machined to Φ (8±0.5) mm × (15±0.1) mm were prepared. The hot compression tests were carried out on the Gleeble 3500 thermal simulation testing machine. The specimens were heated to the testing temperature for 5 mins with the heating rate of 10°C/min, then deformed. The as-deformed samples were immediately quenched in water for retaining high temperature structure. The temperature of 873, 888, 903, 933, and 948°C and strain rates of 0.01, 0.1, 1, and10s^{-1} were carried out in the experiments, the true stress-strain curves were obtained according to data collected automatically by a Gleeble 3500 thermal simulation machine, including load, displacement, temperature, and time, etc. The stress-strain curves in different conditions were analyzed, and the constitutive equation was established.

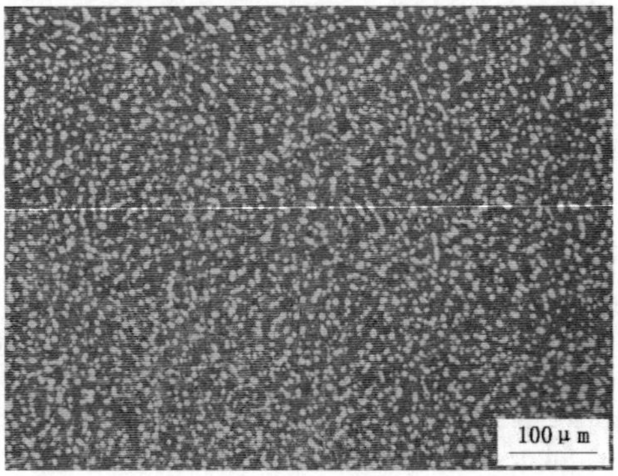

Figure 1. The original microstructure of TC32 titanium alloy

Table 1. Mechanical properties of the original condition of TC32 alloy at room temperature.

Mechanical properties	σ_b (MPa)	$\sigma_{p0.2}$(MPa)	$\delta_{0.5}$ (%)	Ψ (%)	α_{ku}(J·cm^{-2})	K_{IC}(MPa·m$^{1/2}$)
Tensile properties	1021	960	15.9	52.6	/	/
Impact properties	/	/	/	/	98.9	/
Fracture toughness	/	/	/	/	/	119

3. Results and Discussion

3.1. *True stress-true strain curves*

The true stress-true strain curves obtained at different deformation temperature are shown in figure 2, the hot deformation behavior of TC32 titanium alloy is sensitive with temperature and strain rate. The stress decline with increasing temperature and decreasing strain rate. The curves exhibit an obvious peak which is attributed to dynamic recrystallization at the beginning of hot deformation, dynamic recovery and working-hardening occur concomitantly, the dislocation density increase until a critical value, new grains nucleate and lead to flow-softening. A balance is established, and the phenomenon presents a stable level in the curves with a slight fluctuation because of work-hardening and flow-softening processing alternately in the steady-state stage, especially at a higher temperature.

The peak flow stress (σ_p) of TC32 titanium alloy with different conditions of strain rate and temperature are listed in table 2. It can be found that when temperature increases, the peak flow stress decrease especially under a higher strain rate of 10s^{-1}, σ_p have little change when 0.1s^{-1}. At high strain amplitude, dislocation multiplication occurred, and plastic deformation cannot be fully completed. As a result, the degree of elastic deformation increase and lead to the rise in the peak of flow stress. Strain rate has a significant impact on the properties of titanium alloy; the larger strain rate makes a more considerable deformation resistance and higher energy loss. On the contrary, higher energy loss caused by longer deformation time makes a failure of hot deformation.

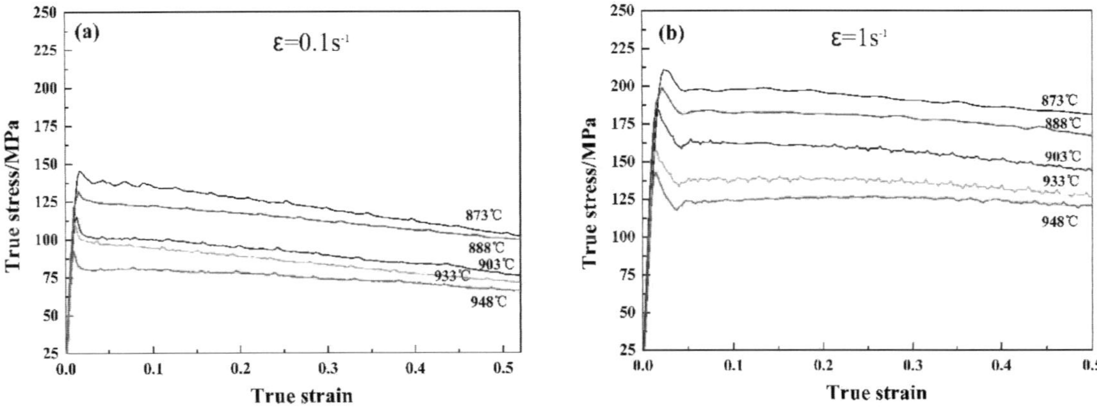

Figure 2. Flow curves obtained under different strain rate: (a) $0.1s^{-1}$ and (b) $1s^{-1}$.

Table 2. Measured values of the peak flow stress (σ_p/MPa) of titanium alloy with different conditions of strain rate and temperature.

Strain rate(s^{-1})	Temperature (℃)				
	873	888	903	933	948
0.01	81	71	65	58	56
0.1	145	132	115	111	92
1	211	199	182	157	143
10	305	285	249	215	190

3.2. Kinetic analysis

The flow stress depends on temperature and strain rate during the high temperature deformation. To understand the rate-controlling step during the deformation, the flow stress data are analyzed using the kinetic rate equation given by:

$$\sigma = \sigma(Z, \varepsilon) \tag{1}$$

$$Z = \dot{\varepsilon} exp[Q(RT)^{-1}] \tag{2}$$

where $\dot{\varepsilon}$ is strain rate, σ is flow stress, E is elastic modulus, Q is apparent activation energy, R is ideal gas constant, T is the temperature in Kelvin, and Z is the Zener-Hollomon parameter.

The deformation mechanism of titanium alloy is related to strain rate. The relationship between flow stress and strain rate is described by the equation reported in [14-16] as follows:

$$\dot{\varepsilon} = A_1 \sigma^{n_1} \exp[-Q(RT)^{-1}] \tag{3}$$

$$\dot{\varepsilon} = A_2\, exp(\beta\sigma)\, exp[-Q(RT)^{-1}] \tag{4}$$

$$\dot{\varepsilon} = A[sin(\alpha\sigma)^n exp[-Q(RT)^{-1}] \tag{5}$$

where $A1$, $A2$, A, $n1$, n, β, α are frequency factors, and $\alpha = \beta/n_1$. Equations (3-5) are generally used to describe the low-, high-stress, and entire stress regions, respectively. The natural logarithm of equation (3-4) can be obtained as equations (6) and (7):

$$In\dot{\varepsilon} = InA_1 + n_1 In\sigma - Q(RT)^{-1} \tag{6}$$

$$In\dot{\varepsilon} = InA_2 + \beta\sigma - Q(RT)^{-1} \tag{7}$$

Taking partial derivative for equations (8) and (9), the values of n_1 and β can be obtained as follows:

$$n_1 = \frac{\partial In\dot{\varepsilon}}{\partial In\sigma} \tag{6}$$

$$\beta = \frac{\partial In\dot{\varepsilon}}{\partial \sigma} \tag{7}$$

A plot of $In\dot{\varepsilon}$ vs. σ is shown in figure 3, while $In\dot{\varepsilon}$ vs. $In\sigma$ is depicted in figure 4, while n_1 and β have been estimated of 5.1877 and 0.0391, respectively, then α is obtained by calculation of 0.00753.

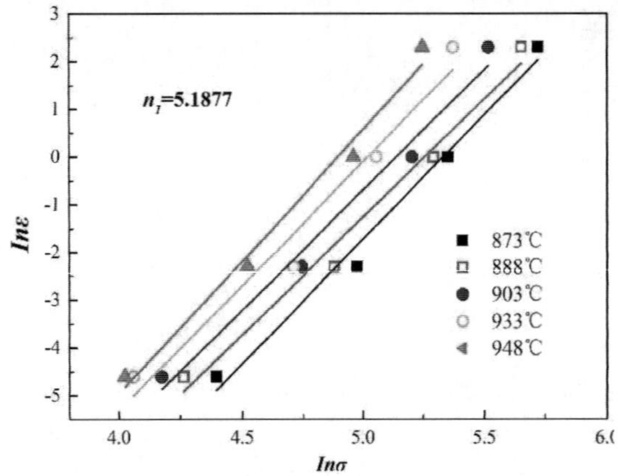

Figure 3. The interaction between $In\dot{\varepsilon}$ and $In\sigma$

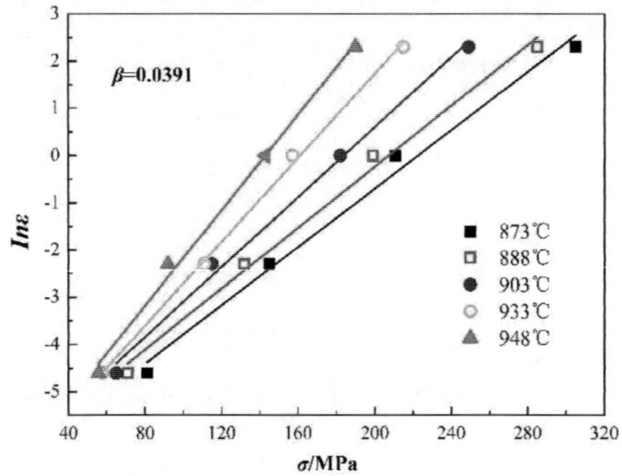

Figure 4. The interaction between $In\dot{\varepsilon}$ and σ

The natural logarithm of equation (5) can be obtained as:

$$In\dot{\varepsilon} = InA + nIn[sin(\alpha\sigma)] - Q(RT)^{-1} \tag{10}$$

Then the value of n_2 can be obtained as:

$$n = \frac{\partial In\dot{\varepsilon}}{\partial In[sinh(\alpha\sigma)]} \tag{11}$$

The hot deformation activation energy Q (J/mol) can be calculated by the following equation [14]:

$$Q = R\frac{\partial In\dot{\varepsilon}}{\partial In[sinh(\alpha\sigma)]}\bigg|_T \frac{\partial In[sinh(\alpha\sigma)]}{\partial(1000T^{-1})}\bigg|_\varepsilon = Rn_2n_3 \tag{12}$$

A plot of $In\dot{\varepsilon}$ vs. *Insin[h(ασ)]* is given in figure 5, *Insin[h(ασ)]* vs. $1000T^{-1}$ in figure 6, n and n_3 have been estimated of 3.55327/4.32130 and 13.6976/14.6564 in alpha-beta region/beta region, respectively.

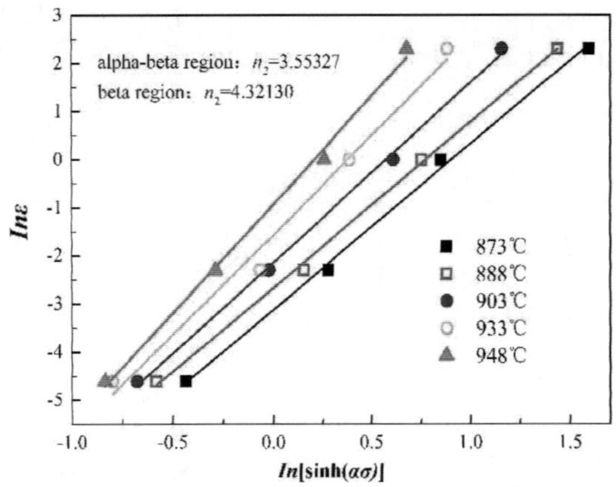

Figure 5. The interaction between $In\dot{\varepsilon}$ and *In[sinh(ασ)]*

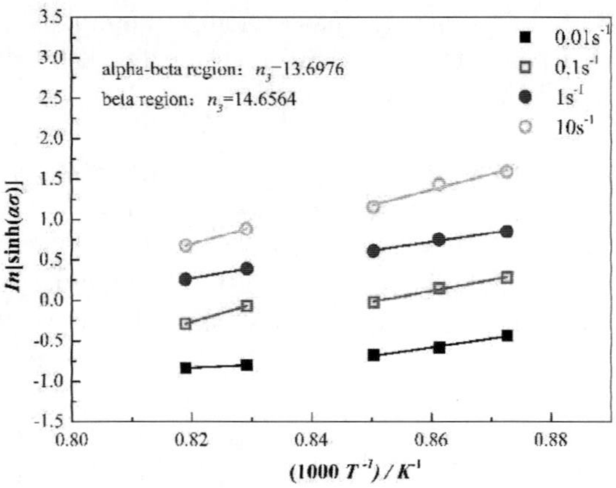

Figure 6. Interaction between *In[sinh(ασ)]* and $1000T^{-1}$

According to equation (11), the hot deformation activation energy Q can be calculated of 404.7kJ/mol and 526.6kJ/mol in the alpha-beta region and beta region respectively.

Substitution of equation (5) into equation (2) yields

$$Z = Asinh(\alpha\sigma)^n \tag{13}$$

Take the natural logarithm for equation (14) is obtained as follows:

$$InZ = InA + nIn[sinh(\alpha\sigma)] \tag{14}$$

A plot of *InZ* vs. *In[sinh(ασ)]* is depicted in figure 7. It yields that the value of *n* is 3.54235/4.30593 and *InA* is 39.26174/50.51993 in alpha-beta region/beta region. Taking *A, α, n, Q* which are obtained from the above calculation into equation (5) and the constitutive equation for TC32 titanium alloy can be expressed as equations (15) and (16).

In the alpha-beta region:

$$\dot{\varepsilon} = [sinh(0.00753\sigma)]^{3.54235}\exp[-404700(RT)^{-1} + 39.26174] \tag{15}$$

In the beta region:

$$\dot{\varepsilon} = [sinh(0.00753\sigma)]^{4.30593}\exp[-526600(RT)^{-1} + 50.51993] \tag{16}$$

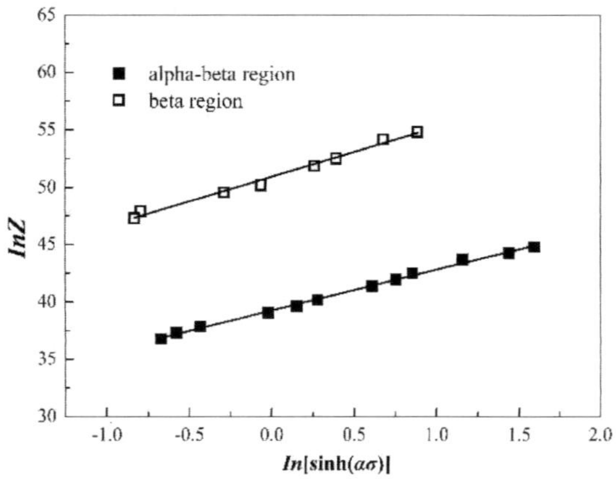

Figure 7. Interaction between *InZ* and *In[sinh(ασ)]*.

4. Conclusions

In this study, high-temperature deformation tests with different temperatures of 873~943°C and strain rate of 0.01~10s^{-1} were carried out. The conclusions obtained from the experimental results are as follows:

(1) With an increase in temperature and decrease in strain rate, the reduction of flow stress will occur.

(2) The deformation activation energy of TC32 alloy is 404.7kJ/mol and 526.6kJ/mol in the alpha-beta and beta regions, respectively. The constitutive equations were obtained.

Acknowledgments

This study was financially supported by the National Defence Scientific Research Project of China (Nos. A0520110059).

References

[1] Zhu Z S 2014 *J Aeronaut Mater* **34** 44-50
[2] Brewer D.W.D, Brid R.K and Wallace T A 1998 *Mater. Sci. Eng.* A. **243** 299-304
[3] Fei Y, Zhu Z S, Wang X N, Li J, Shang G Q and Zhu L W 2013 *Rare metals* **37** 186-91
[4] Wang X N, Fei Y, Liu Z, Shang G Q, Li J, Zhu L W and Zhu Z S 2013 *Titanium* **30** 7-10
[5] Shang G Q, Wang X N, Fei Y, Li J, Zhu L W and Zhu Z S 2013 *Failure analysis and Prevention.* **8** 74-78
[6] Li M B, Zhu Z S, Wang X N, Fei Y, Zhu L W, Shang G Q and Li J. 2016 *J Aeronaut Mater.* **36** 7-13
[7] Zhu Z S, Shang G Q, Wang X N, Fei Y and Li J 2012 *Titanium* **29** 1-5
[8] Li J, Zhu L W, Wang X N, Fei Y, Shang G Q and Zhu Z S 2016 *Mater Sci Forum* **849** 332-39
[9] Yao P P, Li P, Li C M, Xue K M and Gan G Q 2015 *Rare metals* **11** 967-74
[10] Cui X F, Mi X J, Luo Z, Tao H M and Lin C G 2015 *JMEP* 24 67-79

[11] Mao B P, Guo S L and Shen J 2008 *Rare Metals* **32** 674-78
[12] Tang Z, Yang H, Sun Z C, Li Z Y and Duan H 2008 *Chin J Nonferrous Met* **18** 722-27.
[13] Seshacharyulu T, Medeiros S C, Morgan J T, Malas J C, Frazier W G and Prasad Y V 1999 *Scripta Mater* **3** 283-88
[14] Zhu L W, Wang X N, Fei Y, Li J and Zhu Z S 2016 *Mater Sci Forum* **849** 309-16.
[15] Feng L, Qu H L, Zhao Y Q, Li H, Zhang Y N and Zeng W D 2004 *J Aeronaut Mater* **24** 11-13
[16] Liu Q, Xue X Y, Fu B Q and Wang Y C 2009 *Mater Heat Treat.* **38** 43-47

IOP Conf. Series: Materials Science and Engineering **474** (2019) 012046 doi:10.1088/1757-899X/474/1/012046

The safety thickness of blast furnace hearth lining and simulation analysis for different interventions

J F Chen*, **L Y Chen and Y Li**

School of Mechanical Engineering & Automation of Northeastern University, Shenyang, China

*E-mail: chenjf1993@yeah.net

Abstract. In the late period of the blast furnace operation, the residual thickness of the hearth lining is insufficient. To avoid the burning through of hearth, intervention measures with variable operating conditions are needed. Plugging wind gap, adding Titanium ore and controlling production are frequently used at the scene of the project. A 1-D method is used to determine the safe thickness of the hearth lining. A general CFD software FLUENT is used to simulate the iron tapping of the hearth model under different interference conditions. The influence of different conditions on erosion state of the hearth is explored.

1. Introduction

During the lifecycle of blast furnace hearth, physical and chemical erosion occurs when the hot surface of the lining contacts with molten iron directly. With the development of erosion, the thickness of the hearth lining decreases gradually. The erosion of hearth lining is affected by operating conditions to a great extent. Because the life of hearth is closely related to the residual thickness of lining, it is crucial to adjust operating conditions reasonably to delay the erosion of hearth.

Currently, the lining erosion diagnosis technology based on heat transfer theory is ripe at domestic and foreign countries. Zhang *et al.* [1] considered four possible situations and calculated the hot face position in the side wall by using a 1-D heat transfer model. A 2-D inverse method for identifying erosion boundary was proposed [2].

The previous method for calculating safe thickness was based on the thermal stability strength of cooling stave [3]. This paper puts forward a new calculation method of the safe thickness based on the strength theory.

In recent years, as Tachimon [4] and Ohno *et al.* [5] attempted to study the relationship between molten iron flow and lining heat transfer, various comprehensive models have been developed to simulate the flow of molten iron in the hearth and lining heat transfer. Kouji *et al.* [6] established a mathematical model for estimating the transient erosion process of blast furnace lining, which can simulate the flow and heat transfer of molten iron and the erosion of carbon brick. Chenn *et al.* [7] proposed a comprehensive computational fluid dynamics model to simulate the flow of hot metal in blast furnace hearth and heat transfer in the lining. Sun [8] discussed the influence of different dead porosity of dead coke column, convective heat transfer coefficient and volume utilization factor on hearth tapping and theoretical operation furnace type. Previous studies mainly focused on modeling the original model of blast furnace hearth, but few on modeling the hearth model under erosion condition.

Gas flow is weak, and there is no dripping slag iron when the wind gap is plugging. Titanium compounds with a high melting point are efficiently accumulating at the bottom of hearth and can form a viscous substance with molten iron. This substance can repair the corroded hearth to some degree [9]. High productivity is very adverse to the stability and smooth operation of the furnace. Controlling productivity can delay the erosion of the lining. Plugging wind gap, adding titanium ore and controlling production are commonly used for furnace protection in the late period of blast furnace [10].

Using the 1-D theory to calculate the safe thickness of hearth, and taking measures to intervene after the hearth is eroded to the safe thickness. This is of great significance for realizing the safety and longevity of the blast furnace.

2. The calculation model of safety thickness

The hearth structure is generally composed of lining, cooling stave, filler, and shell. When there is no gap between the layers and the shell has a good tighten action, the circumferential stress of the lining shows compressive stress. Take the single layer cylinder structure as the object, shown in Figure 1. The inner and outer radius of the structure are respectively r_1 and r_E. The lining hot surface temperature takes the iron equilibrium solidification temperature 1150° C. The outer surface of the lining is convective heat transfer boundary. The convective heat transfer coefficient and temperature are respectively h_C and T_C. The thermal conductivity, linear expansion coefficient and elastic modulus of the lining are respectively k_1, α_1, E_1. The reference temperature of thermal strain is T_0 the interface pressure of lining structure are respectively p and p_k.

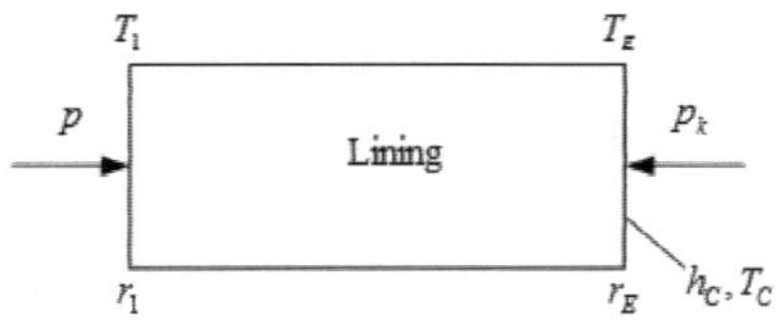

Figure 1. Single layer lining structure.

According to the calculation method of the heat transfer of the cylinder, the temperature of the outer surface of the lining can be calculated by the following formula

$$T_E = \frac{T_1 + r_E R_{E1} h_C T_C}{1 + r_E R_{E1} h_C} \quad (1)$$

where R_{E1} is thermal resistance of lining

$$R_{E1} = \frac{1}{k_1} \ln \frac{r_E}{r_1} \quad (2)$$

The circumferential stress of the inner surface of the lining is

$$\sigma_\theta = \frac{(1+\omega_1^2)p - 2p_k + 2p_{T1}}{1-\omega_1^2} - E_1 \alpha_1 \Delta T_1 \quad (3)$$

where $p_{T1} = \frac{\theta_1}{2}(\Delta T_E - \omega_1^2 \Delta T_1) - \frac{\theta_1}{4}(1-\omega_1^2)\frac{T_1 - T_E}{\ln \omega_1}$, $\omega_1 = \frac{r_1}{r_E}$, $\Delta T_1 = T_1 - T_0$.

According to the strength condition of lining,

$$|\sigma_\theta| \le \sigma_y, \quad (4)$$

where σ_y is the allowable compressive strength of lining.

Given the radius of the outer edge and the compressive strength, the radius r_1 of the inner edge of the liner can be obtained by iteration calculation from the formulas (3) and (4). The safety thickness of lining is calculated by Equation (5).

$$L_S = r_E - r_1 + L_0 \quad (5)$$

where L_0 is the deterioration thickness of lining.

3. The numerical simulation model

The erosion model of the hearth is studied, because of the symmetry of the model, taking half of the hearth model as the calculation model. The erosion model is established as shown in Figure 2. Furnace hearth is round-shaped erosion as a whole, and local mushroom-shaped erosion occurs at -45° azimuth. The most severe erosion has reachedto a safety thickness of 0.3m. The dimensions of the hearth are as follows: the diameter of the hearth is 6m, the height of the hearth is 5.2m, the average thickness of the side wall of the hearth is 0.65m, the average thickness of the hearth bottom is 1.3m, the height of the iron tap is 4.1m, and the diameter of the iron tap is 0.1m. The hearth is composed of a mixture of iron and slag and a central dead coke zone, and they are porous media with a porosity of 0.8 and 0.3 respectively. The combination of iron and slag is mixed by the mass ratio of 3:1. The properties of carbon brick are shown in table 1.

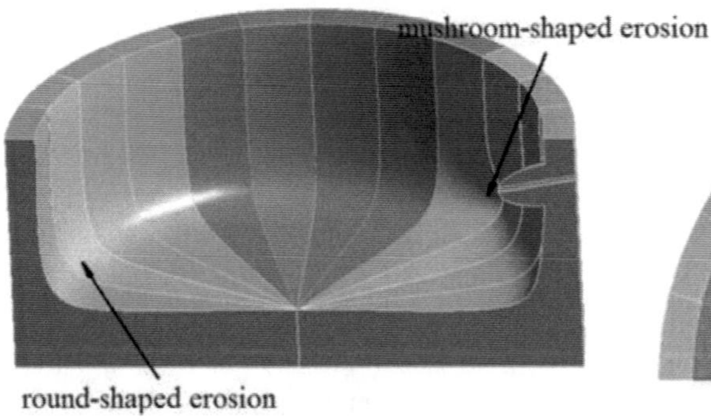

Figure 2. The hearth erosion model.

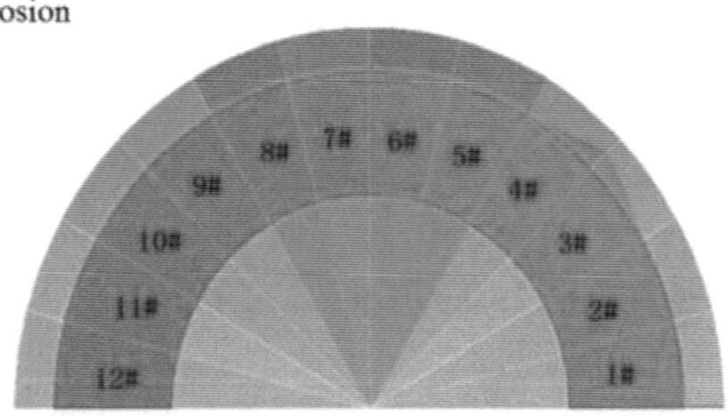

Figure 3. Sketch of wind gap number.

Table 1. Physical properties of carbon brick.

Density / kgm^{-3}	Heat capacity / $kg^{-1}K^{-1}$	Thermal conductivity / $Wm^{-1}K^{-1}$
1600	700	14

The simulation of plugging different wind gap above the hearth is realized by changing the boundary conditions. Under the condition of supplying air normally, the top of the hearth is the velocity boundary. Under the condition of blocking wind gap, the area without blocking wind gap is velocity boundary; and the area of blocking wind gap becomes wall boundary. As shown in figure 3, the wind gaps are numbered from 1 to 12, the site of occurring local depression is under the 3# and 4# wind gap. The tapping process of supplying air normally and blocking single wind gap are simulated respectively.

The simulation of adding titanium ore is realized by changing the initial erosion morphology of the lining. The residual thickness of the lining is 0.3m on the area of occurring mushroom erosion without adding titanium ore, according to the amount of titanium ore, the minimum thickness values of lining are 0.4, 0.5, and 0.6m. The simulation of the hearth model with different thickness is carried out respectively.

The simulation of controlling production is realized by changing the volumetric utilization coefficient [11]. According to the daily tapping quantity, the inlet speed above the hearth can be calculated according to Equation (6). In this paper, the volume utilization coefficients are equal to 2.4, 2.6, 2.8, 3.0, 3.2, and 3.4 $t{\cdot}m^{-3}d^{-1}$, respectively. The inlet velocities of different volume utilization coefficients are listed in table 2.

$$v = \frac{1000(M_{iron} + M_{slag})}{3600 \times 24\pi r^2 \rho} \quad (6)$$

where M_{iron} is the daily iron quantity, M_{slag} is the daily slag quantity, r is the inner radius of lining, ρ is the density of the iron and slag mixture.

The general CFD software FLUENT is used to simulate the iron tapping of the hearth model, assuming that the iron-slag is a viscous incompressible liquid, the k-ε model is used to simulate the turbulence. The solver settings of FLUENT: the inlet is velocity boundary, and keeps a constant temperature of 1500° C. The outlet is a pressure boundary, and the pressure is 1 atmospheric, the outlet temperature is 1420° C. The wall-boundary is non-slip boundary condition. The side wall convection coefficient of the hearth is 60W/(m^{2}K), and the bottom convection coefficient is 40W/(m^{2}K).

Table 2. The inlet velocity of different volume utilization coefficients.

Volume utilization coefficient, tm^{-3}d^{-1}	Inlet velocity, ms^{-1}
2.4	0.00024047
2.6	0.00026051
2.8	0.00028055
3.0	0.00030059
3.2	0.00032063
3.4	0.00034066

4. The numerical simulation model

The mesh element size is 0.15m for the total hearth model, the mesh at tap-hole is locally refined, and the mesh size is 0.02m. Usually, the 1150°C isotherm of the axial section of the hearth is taken as the erosion boundary of the lining. If the erosion line of 1150°C does not enter the inner of carbon brick, the slagging occurs at this position. Because the lining thickness of the model at -45° axial section is minimal, there will be a certain thickness of slag skin attached to the lining surface under the condition of the external cooling condition. The thicker the slag skin thickness, the smaller the erosion radius of the position, and the safer the hearth. The erosion radius is taken as an indicator of the erosion degree.

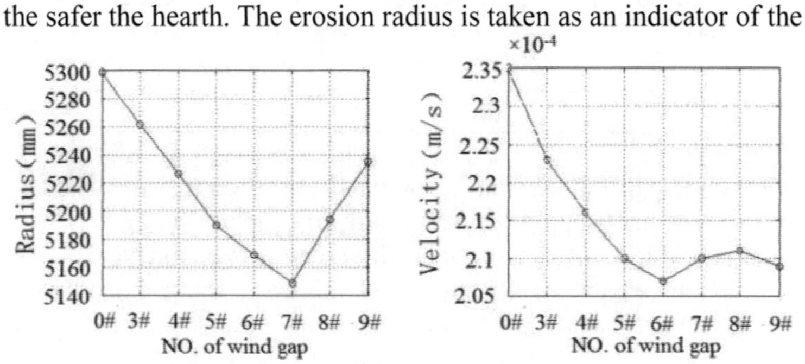

Figure 4. The result of plugging wind gap.

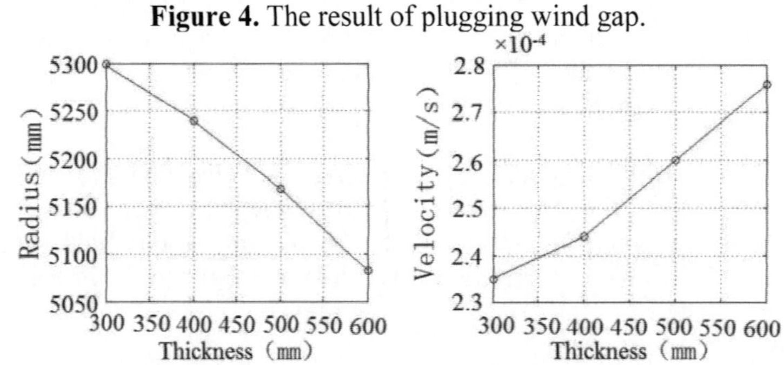

Figure 5. The result of adding Titanium ore.

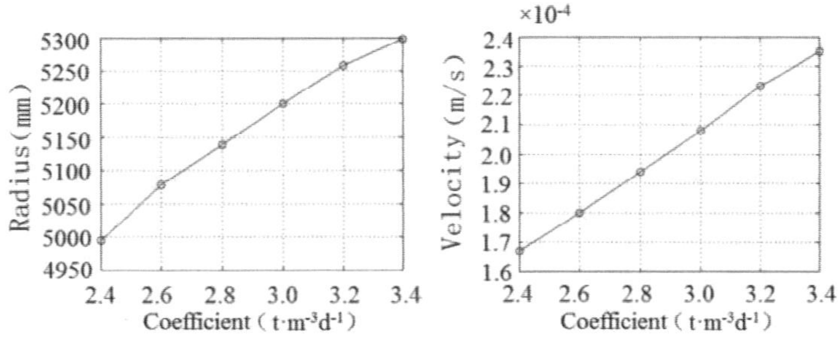

Figure 6. The result of controlling production.

The velocity of iron-slag at -45° azimuth is smaller than that of other places because of mushroom erosion. But the erosion at this position is more dangerous, the velocity of iron-slag is taken as another indicator of the erosion degree. The calculation results of FLUENT extract the erosion radius at a 1.8m attitude of -45° axis section and iron-slag velocity at the evaluation point (-3.5m, 1.8m, 3.5m).

Figure 4 shows the relationship between the erosion radius of the mushroom-shaped erosion position and the velocity of iron-slag with the serial number of wind gap. Blocking 0# wind gap represents the situation of supplying air usually. It can be seen that the erosion radius of the hot lining surface is the smallest at the 1.8m elevation of the -45 degree axis section when the 7# wind gap is blocked. And the velocity of hot metal at the evaluation point is the smallest when the 6# wind gap is blocked. The velocity of iron-slag for plugging 7#, 8# and 9# wind gap is not much different than that of 6# wind gap. By comparison, plugging 7# wind gap is the best choice when plugging single outlet.

Figure 5 shows the relationship between the erosion radius of the mushroom-shaped erosion position and the velocity of iron-slag with the minimum residual thickness of the lining. The larger the residual thickness of lining is, the smaller the erosion radius is, and the higher the flow velocity of iron-slag is. Because adding titanium ore is directly repairing the lining from the root and slagging is helpful to protect the hearth lining, but adding titanium ore will strengthen the scour of the lining.

Figure 6 shows the relationship between the erosion radius of the mushroom-shaped erosion position and the velocity of iron-slag with the volume utilization coefficient. It can be seen that both erosion radius and iron-slag velocity decrease with the decrease of volume utilization coefficient. With the reduction of smelting strength, the volume utilization coefficient decreases by 0.2 and the erosion radius decrease by about 50-80 mm. Since after the production control, all kinds of indices will be reduced, the production control is an effective measure to protect the lining.

5. Conclusions

Through the 1-D theoretical prediction, a calculation method for determining the safe thickness of lining based on strength theory is deduced. The theoretical prediction has practical significance, since its results indicate whether the respective interventions are required or not.

In this paper, the hearth tapping process under the three interventions is studied. The study on the measures of blocking wind gap has explored the sequence number of the best blocking wind gap. The intervention measure of adding titanium ore was realized by changing the hearth model. The study on the intervention measures of controlling the production capacity revealed that the smaller the production capacity, the lower the erosion rate.

References

[1] Zhang Y, Rohit D and Huang D 2008 *International Journal of Heat and Mass Transfer* 51 186-197

[2] Chen L Y, Li Y and Wu H L 2008 *J. Iron and Steel* **43** 12-15

[3] Chen L Y, Jiang H and Sun J D 2013 *Proc. 14th Annual National Conf. on Blast Furnace Ironmaking* pp 23-30

[4] Tachimon M 1984 *J. Tetsu to Hagane* 2224.

[5] Ohno J, Tachimori M and Nakamura M 1985 *J. Tetsu To Hagane* **71** 34.

[6] Takatani K, Inada T and Takata T 2001 *J. ISIJ International* **41** 1139-1145

[7] Chenn Q Z and Huang D(Frank) 2010 *J. Thermal Science and Engineering Application* **2**

[8] Sun J D 2013 *Numerical Calculation And Thermal Design of Iron Casting Model for Blast Furnace Hearth* (Shenyang Northeastern University)

[9] Zhang F M and Cheng S S 2012 *Modern Blast Furnace Longevity Technology* (Metallurgical Industry Press)

[10] Ren Z D and Yu H X 2017 *J. Ironmaking* **36** 53-55

[11] Li Y 2008 *Thermal State Model and Lining Erosion Identification of Blast Furnace Hearth* (Shenyang: Northeastern University)

The durability of reactive powder concrete: a review

S G Liu[1], Z Xiang [1], R H Huang[1], D H Wang[2,*] and Y Z Ju[2]

[1] Shenzhen Power Supply Co., Ltd, Shenzhen, 518001, China;
[2] School of Civil Engineering and Architecture, Northeast Electric Power University, Jilin 132012, China.

*E-mail: hitwdh@126.com.

Abstract: Reactive powder concrete (RPC) is a new cement-based composite material with ultra-high strength, high ductility and good toughness, which has an extensive application prospects. Based on the domestic and foreign research achievements, the durability of reactive powder concrete is integrally discussed from the standpoints of freezing-thawing resistance chloride ion penetration, carbonization, erosion, and abrasion. Under freeze-thaw cycle, the mass loss of RPC is minimal, chloride ion permeability of RPC is almost negligible, active admixtures can improve the sulfate corrosion resistance of RPC significantly. Moreover, RPC has excellent carbonation resistance and abrasion resistance. The mechanisms of the supreme durability of RPC are analysed in detail. The outlook on topical problems and promising research directions in this field is presented.

1. Introduction

Reactive powder concrete (RPC) is a new type of ultra-high strength and ultra-high performance cement-based composites developed by Richard [1-2] of Bouygues Company of France according to the linear packing density model (LPDM) and the compressible packing model (CPM) and fiber reinforced material technology. RPC has the characteristics of high strength, high toughness, high durability, and high volume stability. Dugat et al. [3] carried out the experiment on the mechanical properties of RPC with a cylinder specimen diameter of 70 mm, height of 180 mm and steel fiber content is 146 kg/m^3. The RPC compressive strength varied from 194 to 203 MPa and its elastic modulus ranged from 62 to 66 GPa. Tai [4] investigated the dynamic compression mechanical properties of RPC under different strain rates and reported that the static strength of RPC was significantly higher than that of normal concrete, while the dynamic energy absorption of RPC increased with the steel fiber content and was proportional to the impact energy. Zheng [5] conducted quasi-static tests of RPC beam-column joints, and test results indicated the seismic performance and shear strength of RPC beam-column joints were better than that of reinforced concrete joints. RPC with ultra-high performance has broad application prospects in civil engineering, mining, nuclear power, municipal, marine, and military engineering. The engineering application of RPC is becoming more and more common. For example, the Jean Bouin stadium in France, which was completed in August 2013 and featured a double curved surface structure, had the shell consisting of 3600 pieces of precast RPC sheets.

China will be at the peak of its construction in the future for quite some time. The durability of concrete materials determines the durability of reinforced concrete structures. Therefore, it is of great practical significance to study the durability of RPC materials. At present, scholars in China and abroad have carried out extensive research on the durability of RPC materials, including frost resistance, carbonation resistance, chloride ion resistance, sulfate resistance, chemical solution corrosion resistance and wear resistance. From the research on the durability of RPC at home and abroad, this paper makes a comprehensive discussion on the aspects of frost resistance, chloride ion penetration resistance, carbonation resistance, corrosion resistance and wear resistance and so on.

2. Review of the durability of reactive powder concrete

2.1 Freezing-thawing resistance

Using the orthogonal test method, nine groups of RPC specimens were designed and made by Ju [6-7]. According to the fast freezing test method of the long-term performance and durability test method of GBJ82285 normal concrete, 100 freeze-thaw cycle tests were carried out on RPC specimens. The effects of water binder ratio, silica fume cement ratio and steel fiber content on freeze-thaw properties of ordinary concrete and reactive powder concrete were explored. The results indicated that RPC had good freeze-thaw resistance. The water-binder ratio was the most critical factor, followed by the ratio of silica fume and cement, and the content of steel fiber.

Liu [8] performed freeze-thaw cycle tests on RPC prism specimens according to ASTMC666 standard (US rapid freeze-thaw test standard). The durability coefficient and mass loss rate are used to evaluate the frost resistance of concrete. After freeze-thaw cycles, the mass loss was about 0.3%, which was close to 0, and the durability coefficient exceeded 100.

Cwirzen [9] placed the RPC cube with a side length of 100mm and compressive strength of 130~202MPa into the 3% NaCl solution and conducted 56 freeze-thaw cycles. The study showed that the smaller the relative humidity inside the RPC specimens, the smaller the loss of the dynamic elastic modulus. The mass loss rate of the surface of the RPC specimens did not exceed 300g/m^2, some were lower than 200g/m^2. Thus, the requirement of the 200 years design life in the environment of freeze-thaw in Finland was satisfied.

Graybeal [10] conducted 690 freeze-thaw cycles of RPC, according to the “Standard test method for resistance of concrete to rapid freezing and thawing” (ASTM C 666). The mass loss rate of RPC after 28d standard conservation was 1.60~2.56%, the mass loss rate of RPC after 44h steam curing of 60~90°C was 0.07%~0.41%; the maximum dynamic elastic modulus of all the RPC specimens were not more than 4%, the maximum decrease of primary vibration frequency was not more than 5%.

2.2 Resistance chloride ion penetration

Roux [11] put the concrete with the thickness of 5mm(the cylinder compressive strength of the specimens with 100mm diameter and 200mm diameter were 35MPa and 90MPa respectively) and the disk-shaped RPC specimens into the electrochemical solution (distilled water for the anode, and the 0.5 Mol/L NaCl solution for the cathode), and applied the voltage of 12V, measured the migration power, respectively. The chloride ion diffusion coefficients of the three materials were 11×10^{-13}, 6×10^{-13}, and 0.2×10^{-13} m^2/s, respectively.

Graybeal [10], in accordance with the ASTM C1202, contacted one side of the cylindrical specimens of RPC with a diameter of 100mm, a height of 50mm to NaOH solution, and the other side to NaCl solution, and applied the voltage of 60V, measured the power through the specimen in 6 h was 18°C, then judged that the chloride ion permeability of UHPC can be ignored.

IOP Conf. Series: Materials Science and Engineering **474** (2019) 012047 doi:10.1088/1757-899X/474/1/012047

Peng [12] used the RCM method, tested the chlorine ion penetration of steel slag RPC and the compressive strength of cement mortar specimen was 120MPa, and compared with the C50 concrete with the water-binder ratio of 0.33, and found that: the chloride ion diffusion coefficient of steel slag RPC was 0.51×10^{-13}~$0.58\times10^{-13}m^2/s$, and the chloride ion diffusion coefficient of C50 concrete was 7×10^{-13}~$9\times10^{-13}m^2/s$.

Yu [13] test the electric flux of cylinder RPC with the compressive strength of 110.3~120.9MPa, according to ASTM C1202 "Standard test method for electrical indication of concrete s ability to resist chloride ion penetration", the results showed that: the permeability of RPC increase with the decrease of the content of silica fume. The power through the RPC specimens in 6 h are all below 60C in different groups, lower than the concrete of 1~2 orders of magnitude, while chloride ion permeability can be ignored.

Wang [14] reported the durability of reactive powder concrete with the compressive strength of 150MPa under chloride-salt freeze-thaw cycling. The mass-loss rate was 0.62 % after 1500 freeze-thaw cycles in a 5.0 wt% NaCl solution and the relative dynamic-elastic modulus varied from 98 to 102%. The compressive strength of reactive powder concrete after 1000 and 1500 chloride-salt freeze-thaw cycles decrease by 17.3 and 41.6%, respectively.

2.3 Corrosion resistance

Liu [8] immersed RPC specimens, whose compressive strength of cement mortar was about 200MPa, into the mixed solution of 17.62% NaCl, 0.17% $CaCl_2$, 1.07% $MgCl_2\cdot6H_2O$ and 7.32% $MgSO_4\cdot7H_2O$. After 90 days, the mass loss of RPC was 0, the loss of dynamic elastic modulus was only 0.5%. This is because the hydration reaction of mineral admixtures created a dense RPC structure and the crack resistance effect of steel fiber.

Li [15] loaded the RPC blocks whose size were 100mm×100mm×400mm to 60, 70, and 80% of the compressive strength and then unloaded, then put them into 5% Na_2SO_4 solution and conducted 60 times of the wet-dry cycle. The test results showed that: the basic natural frequency of all the RPC blocks was increased at first and then decreased. But all of them were higher than the values when the test started in the end; the most significant decrease in the mess of the blocks was no more than 0.3%.

Song [16] conducted 20 dry-wet cycles of the RPC blocks whose side length was 100mm, and the cube compressive strength was 125MPa (a cycle means soaked in the Na_2SO_4 for 24h and drying for 24h in the drying oven of 80℃), the studies showed that: the mass loss of RPC blocks were around 1%, but the strength increased, indicating that the active admixtures can improve the sulfate corrosion resistance of RPC significantly.

2.4 Carbonation resistance

Wei [15] conducted the accelerated carbonation test of RPC whose fly ash content accounted for 48%. When the concentration of CO_2 was 60 %, the temperature was 20℃, and the relative humidity was 70%, the carbonation depth of RPC at 28d is 0, The carbonation depth of C80 concrete was 1.3~1.5mm in the same test conditions.

Roux [11] tested the natural carbonation and accelerated carbonation of normal concrete with the diameter of 110mm, the height of 220mm and the cylinder compressive strength of 35MPa and the RPC with the diameter of 70mm, the height of 140mm and the cylinder compressive strength of 170~230MPa. The results showed that: the carbonation coefficient of normal concrete is $50mm/yr^{0.5}$, but the RPC placed in the 100% CO_2 after 90d, the depth of carbonation is 0.

2.5 Abrasion resistance

Roux [11] obtained the wear resistance coefficient of C30, C80, and RPC200 concrete by a wear resistance test. The coefficient of wear resistance is characterized by the ratio of the wear rate of the specimen to the wear rate of the glass. The results showed that the wear resistance coefficient of RPC200 was only 1.3, which was comparable to that of cement mortar made of Emery (wear resistance coefficient is 1.2).

The wear coefficient of regular concrete (RC), high strength mortar (HSM), and reactive powder concrete (RPC) was measured by Lee [17]. The wear coefficient is obtained under 500 and 1000 wear cycles. The results indicated that the wear coefficients of RC, HSM and RPC were 0.68, 0.86 and 0.95, respectively, under 500 wear cycles, and 0.33, 0.67 and 0.92, respectively, under 1000 wear cycles.

3. Durability mechanism of reactive powder concrete

Although the above studies used different tests and indicators for the investigation of RPC durability, however, all of the results show that the durability of RPC is superior to that of high strength concrete and ordinary concrete. The reasons are as follows:

1) The low water-cement ratio of RPC makes the number of gel pores, capillaries and connected pores in concrete very small. No coarse aggregate is added to improve the fineness of the components, and many kinds of active admixtures with different particle sizes and graded sand are added to optimize the particle gradation, which makes the concrete structure compact, uniform in texture and reduces the internal defects (pores and micro-cracks) of the concrete. Test results using mercury injection method indicated the porosity of RPC is very low, and most porosities are internal and harmless porosity [13]. For the reasons mentioned above, RPC has good impermeability. Capillary water and toxic ions are difficult to enter the concrete, and the channel of corroded concrete is blocked.

2) The pozzolanic effect of silica fume, slag, fly ash, and other active mineral admixtures is also one of the main reasons for the high durability of RPC. Because silica fume and other mineral admixtures are rich in SiO_2 or Al_2O_3, the reaction of Ca $(OH)_2$ with SiO_2 and Al_2O_3 reduce the content of $Ca(OH)_2$ and form stable hydrated calcium silicate with high strength. Especially under the condition of heat curing, most of Ca $(OH)_2$ is converted into hydrated calcium silicate. On the other hand, calcium silicate hydrated gel forms a protective layer on the surface of aluminum-containing compounds, which reduces the RPC erosion caused by harmful ions.

3) The excellent wear resistance of RPC can be attributed to two aspects. For one thing, RPC is usually made of quartz sand. The content of SiO_2 in quartz sand is more, the strength and hardness of quartz sand are higher than that of ordinary sand. Quartz sand is the most significant component in particles of RPC; its excellent wear resistance helps to improve the wear resistance of RPC. Besides, the high compactness caused by the extremely low water-cement ratio of RPC and the microaggregate filling effect and the pozzolanic effect, which improve the internal pore structure, interface structure of concrete, and the wear resistance of RPC.

4. Conclusions

From the review of the above studies, it can be seen that extensive research on both durability tests and mechanism analysis at home and abroad in recent years, and some achievements have been achieved. However, there are still some problems to be solved.

(1) The quantitative relationship between high-performance durability, mix ratio parameters and curing conditions need to be further investigated. The strength and fluidity of RPC are affected by water-cement ratio, sand content, admixture amount of active minerals, a variety of components, curing system, and so on. Whether the change of these conditions will affect the durability of RPC, if so, the degree of influence of various factors on RPC durability needs to be studied.

(2) Study on durability under multi-factor coupling. In the real environment, multiple erosion factors exist

at the same time. For example, in the splash zone and tidal range in marine environment, there are three kinds of erosion, chloride ion erosion, sulfate ion erosion, and dry-wet cycle. There is a freeze-thaw cycle along the northern coast of China. Therefore, multiple erosion factors should also be considered in the durability study of RPC.

(3) The problems of microstructure and mechanisms. The microstructure of RPC remains to be thoroughly studied, and the formation mechanism of its structure and strength is not fully understood.

(4) There is a lack of research on the durability of RPC components. The durability design theory and method of RPC structure or component need to be developed and perfected.

Acknowledgments
The authors gratefully acknowledge Shenzhen Power supply Bureau Co., Ltd. (project No.: 090000KK52170163) and the National Natural Science Foundation of China (Project No. 51878128 and No. 51378095).

References

[1] Richard P.and Cheyrezy M 1995 *Cement Concrete Res* **25** (7) 1501-11.
[2] Richard P.and Cheyrezy M 1994 *ACI Special Publication* **114** 507-18.
[3] Dugat J, Roux N and Bernier G 1996 *Mater & Struct* **29** (4) 233-40.
[4] Tai Y S 2009 *Theor Appl Fract Mec* **52** (1) 14-21.
[5] Zheng W, Wang D and Ju Y 2018 *Adv Civ Eng* 2018 1-12.
[6] Ju Y Z, Wang D H and Shan M 2012 *Experimental Mechanics* **27** (02) 214-220.
[7] Ju Y Z, Wang F and Wang D H 2011 *Advanced Materials Research* 168-70 1742-48.
[8] Liu S, Wei S, Wei L and et al. 2003 *Journal of the Chinese Ceramic Society* **31** (11) 1080-1085.
[9] Cwirzen A, Penttala V and Vornanen C 2008 *Cement Concrete Res* **38** (10) 1217-1226.
[10] Graybeal B and Tanesi J 2007 *J Mater Civil Eng* **19** (10) 848-854.
[11] Roux N 1996 *J Mater Civil Eng* **8** (1) 1-6.
[12] Peng Y 2009 *Study on Composition, Structure, and Properties of Reactive Powder Concrete Containing Steel Slag Powder* Wuhan: Wuhan University of Technology 19-116.
[13] Yu Z R, Gao K, An M Z et. al. 2013 *J. Xi'an Univ. of Arch* (Natural Science Edition) **45** (1) 31-37.
[14] Wang Y, An M Z, Yu Z R and et al. 2016 *Mater Struct* **50** (1).
[15] Li T L 2011 *Study on Durability of Reactive Powder Concrete after Damage* Beijing: Beijing Jiaotong University 70-84.
[16] Song S M and Wei C X 2006 *Concrete* (2) 72-73.
[17] Lee M G, Wang Y C and Chiu C T 2007 *Constr Build Mater* **21** (1) 182-189.

Oxidation resistance improvement of ZrB_2 powders by the deposition of $Al_2O_3/Y_2O_3/ZrO_2$ coatings via chemical co-precipitation method

X Q Feng, X Wang*, Y Liu, Y Guo, M Zhang, L Zhang, J L Xie and L J Deng

National Engineering Research Center of Electromagnetic Radiation Control Materials, State Key Laboratory of Electronic Thin Film and Integrated Devices, Key Laboratory of Multi-spectral Absorbing Materials and Structures of Ministry of Education, School of Electronic Science and Engineering, University of Electronic Science and Technology of China, 2006 Xiyuan Road, Chengdu 611731, P.R. China

*Email: xinwang@uestc.edu.cn

Abstract. Zirconium diboride (ZrB_2) is an important class of advanced ceramic material with excellent performance and good application prospect, but the oxidation restricts its application greatly. In order to improve the oxidation resistance of ZrB_2 ceramics at high temperature, we coat Al_2O_3, Y_2O_3 and ZrO_2 layers on the surface of ZrB_2 particles to prepare the ZrB_2@Al_2O_3, ZrB_2@Y_2O_3 and ZrB_2@ZrO_2 composite particles via the chemical co-precipitation method. The results of XRD SEM and EDS strongly suggest that the coated composite particles are prepared successfully, while TG and DSC curves prove that the oxidation resistance of ZrB_2 particles can be improved significantly by the proposed method.

1. Introduction

With the development of space and energy technologies, the needs for materials that can withstand the extreme work environments are becoming more urgent. Therefore, the ultra-high temperature ceramics (UHTCs) are widely researched recently due to their unique properties of mechanical and chemical resistance and configuration stability. ZrB_2 has become one of the most promising candidate materials among UHTCs, because of its high melting temperature (> 3000 °C), low theoretical density (6.09 $g\cdot cm^{-3}$), high hardness, high thermal conductivity (58.2 $W\cdot(mK)^{-1}$) and good thermal shock resistance. [1-8]

ZrB_2 has many advantages, in comparison with other high temperature materials, but it is easily oxidized at high temperature. This poor oxidation resistance of ZrB_2 ceramics at high temperature is one of the major problems to retard its widespread application.

Noteworthy is that ZrB_2 is oxidized into porous ZrO_2 via reaction (1) and volatile B_2O_3 via reaction (2):

$$ZrB_2(c)+\frac{5}{2}O_2(g)\rightarrow ZrO_2(c)+B_2O_3(l) \quad (1)$$

$$B_2O_3(l)\rightarrow B_2O_3(g) \quad (2)$$

The reaction products provide little protective effect against oxidation, and the oxidation rate increases with the oxidation temperature [9-10]. Multiple researchers adopt lots of methods to improve the oxidation resistance of ZrB_2 ceramics [11-15]. Adding Si-containing additives is the most common method, which can form a borosilicate protective layer to isolate oxygen [16-19]. Nonetheless, the borosilicate protective layer is dissociative at high temperature, and can be removed and disrupted in the high temperature airflow, exposing the surface below to continued oxidation [2,12,20]. However, the refractory oxides would not evaporate and run off in the faster-moving airflow, insofar as they are solid-state and quite stable at high temperature.

Thus, we propose an alternative method, which does not rely on the SiO_2 protective layer to improve the oxidation resistance of ZrB_2. We aim to deposit a refractory oxide protective coating on the surface of ZrB_2 particles to prepare the coated composite particles (ZrB_2@Al_2O_3, ZrB_2@Y_2O_3 and ZrB_2@ZrO_2). The protective coating can inhibit the oxygen diffusion and improve the oxidation resistance of ZrB_2.

2. Experimental procedure

2.1. Sample preparation

Raw ZrB_2 (10-15 μm, 99.5 %) powders are commercial chemicals and obtained from Eno Material, China. Analytic grade $Al(NO_3)_3 \cdot 9H_2O$, $ZrOCl_2$, HNO_3, $NH_3 \cdot H_2O$ and chemical grade Y_2O_3 are purchased from Kelong Chemical Reagent Factory, China.

Y_2O_3 and HNO_3 are added to deionized water to prepare $Y(NO_3)_3$ solution (equation 3). The ZrB_2 powders are ultrasonically dispersed in deionized water to prepare the ZrB_2 suspension with concentration of 25 g/L. The ZrB_2 suspensions is stirred and $Al(NO_3)_3$/$Y(NO_3)_3$/$ZrOCl_2$ solution (0.2 mol/L) is dripped slowly into ZrB_2 suspension by peristaltic pump (BT100F, Lead Fluid) with dripping speed of 3 mL/min. At the same time, ammonias with corresponding concentration of 0.6 mol/L, 0.6 mol/L and 0.8 mol/L are added slowly as the precipitator to synthesize the precursors (ZrB_2@$Al(OH)_3$, ZrB_2@$Y(OH)_3$ and ZrB_2@$Zr(OH)_4$) by chemical co-precipitation method (equation 4, 5, 6). Then the precursors are filtered from the suspensions and washed by deionized water and anhydrous ethanol for three times. Then these precursors are dried for 24 h at 80 °C. Finally, these dry precursors are calcined (via equations (7- 9)) in a tabular furnace with molybdenum disilicide ($MoSi_2$) as a heating element at 600°C for 3 h to obtain the final particles with the coating structure (ZrB_2@Al_2O_3, ZrB_2@Y_2O_3 and ZrB_2@ZrO_2).

$$Y_2O_3 + 6HNO_3 \rightarrow 2Y(NO_3)_3 + 3H_2O \quad (3)$$

$$Al(NO_3)_3 + 3NH_3 \cdot H_2O \rightarrow Al(OH)_3 \downarrow + 3NH_4NO_3 \quad (4)$$

$$Y(NO_3)_3 + 3NH_3 \cdot H_2O \rightarrow Y(OH)_3 \downarrow + 3NH_4NO_3 \quad (5)$$

$$ZrOCl_2 + 4NH_3 \cdot H_2O \rightarrow Zr(OH)_4 \downarrow + (NH_4)_4OCl \quad (6)$$

$$2Al(OH)_3 \rightarrow Al_2O_3 + 3H_2O \uparrow \quad (7)$$

$$2Y(OH)_3 \rightarrow Y_2O_3 + 3H_2O \uparrow \quad (8)$$

$$Zr(OH)_4 \rightarrow ZrO_2 + 2H_2O \uparrow \quad (9)$$

2.2. Sample characterization

Microstructure of samples was observed by the scanning electron microscopy (SEM, JSM-7600F, JEOL). Crystalline phases of samples were identified by X-ray diffraction (XRD, XRD-7000, Shimadzu) using Step Scan Mode with a scan rate of 6 $° \cdot min^{-1}$ in the range from 5 to 95°. The chemical composition of samples was examined by the energy dispersive spectroscopy (EDS, NORAN SYSTEM 7, Thermo Scientific). The thermal analysis of samples was investigated by differential scanning calorimetry (DSC) using a differential scanning calorimeter (STA 449C, Netzsch). Samples were filled into alundum crucible and heated up to 1450 °C with the heating rate of $10°C \cdot min^{-1}$ in the dry airflow (20 % oxygen and 80 % nitrogen) with the flow rate of $20\ ml \cdot min^{-1}$ to get the thermal analysis curve.

3. Results and Discussion

3.1. Microstructure of the particles of precursors

The preparation of $ZrB_2@Al(OH)_3$, $ZrB_2@Y(OH)_3$ and $ZrB_2@Zr(OH)_4$ precursors was the first step for the synthesis of $ZrB_2@Al_2O_3$, $ZrB_2@Y_2O_3$ and $ZrB_2@ZrO_2$ composite particles. The addition of ammonias could urge the metal cations (Al^{+3}, Y^{+3} and Zr^{+4}) in ZrB_2 suspensions to combine the hydroxyl ions (OH^-) to separate out as precipitated hydroxides. Pure ZrB_2 particles, which act as the sites of nucleation, were coated with $Al(OH)_3$, $Y(OH)_3$ and $Zr(OH)_4$ coatings to form precursors through heterogeneous nucleation[21-24], as shown in figure 1. The Al $(OH)_3$ coatings on the surface of ZrB_2 particles were inconspicuous in secondary electron images (figure 1-a), while the coatings could be observed obviously in backscattered electron images (figure 1-b). It can be seen that $Al(OH)_3$ in darker color were coated on the surface of ZrB_2 particles. However, some of ZrB_2 particles had a clean and smooth surface with lighter color, which indicated that ZrB_2 particles were not well wrapped by $Al(OH)_3$. The $Y(OH)_3$ coatings had a similar color with ZrB_2 particles in backscattered electron images (figure 1-d), hence it was difficult to distinguish $Y(OH)_3$ coatings from ZrB_2 particles by the colors. But the flocculent structure of $Y(OH)_3$ coatings could be observed clearly in secondary electron images (figure 1-c). There were also some ZrB_2 particles being not well wrapped by $Y(OH)_3$ coatings. The same Zr element led to the same color of $Zr(OH)_4$ coatings and ZrB_2 particles in backscattered electron images (figure 1-f), and $Zr(OH)_4$ coatings did not have a flocculent structure. But it was still sure that ZrB_2 particles were coated with $Zr(OH)_4$ coatings, because of the transformation of the surface condition of ZrB_2 particles from clean and smooth to uneven and rough. The $Zr(OH)_4$ coatings still had the same problem, that some of the exposed surfaces of ZrB_2 particles were not well wrapped.

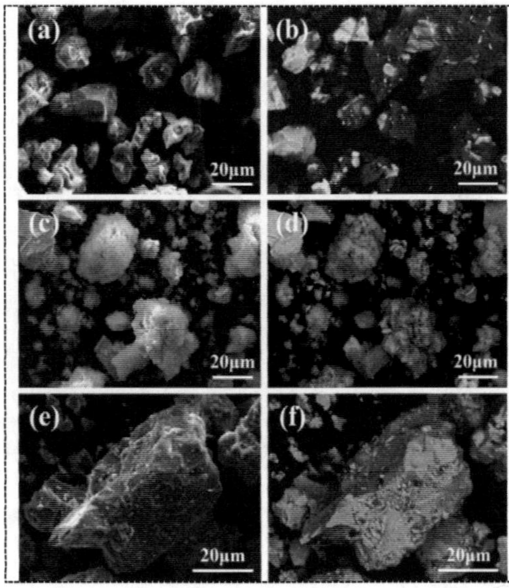

Figure 1. SEM images of precursor particles: (a) secondary electron image of $ZrB_2@Al(OH)_3$, (b) backscattered electron image of $ZrB_2@Al(OH)_3$, (c) secondary electron image of $ZrB_2@Y(OH)_3$, (d) backscattered electron image of $ZrB_2@Y(OH)_3$, (e) secondary electron image of $ZrB_2@Zr(OH)_4$, (f) backscattered electron image of $ZrB_2@Zr(OH)_4$.

The XRD patterns of $ZrB_2@Al(OH)_3$, $ZrB_2@Y(OH)_3$ and $ZrB_2@Zr(OH)_4$ precursors are shown in figure 2. Only the hexagonal phase of ZrB_2 (PDF #34-0432) was detected in the three precursors, and there were no other characteristic peaks observed. This can be explained by the fact that the hydroxides produced in the precipitation reaction in these three precursors were amorphous.

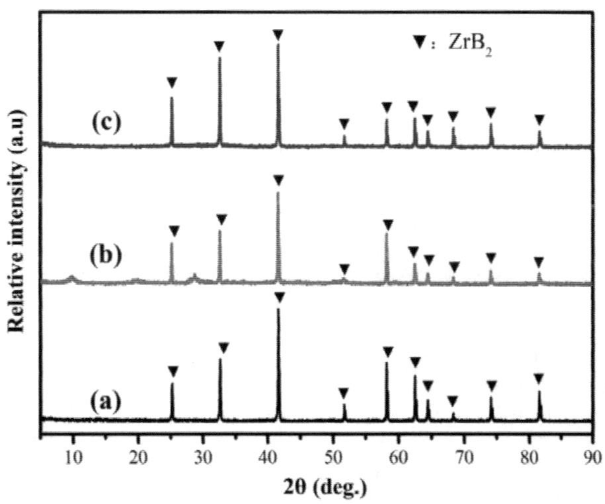

Figure 2. XRD patterns of the precursors: (a) ZrB_2@$Al(OH)_3$, (b) ZrB_2@$Y(OH)_3$, (c) ZrB_2@$Zr(OH)_4$.

3.2. Microstructure of the composite particles

The target products ZrB_2@Al_2O_3, ZrB_2@Y_2O_3 and ZrB_2@ZrO_2 composite particles were obtained from ZrB_2@$Al(OH)_3$, ZrB_2@$Y(OH)_3$ and ZrB_2@$Zr(OH)_4$ precursors by calcination. The hydroxides of the precursors were dehydrated to form the corresponding oxides during the calcination process. The oxides were coated on the surface of ZrB_2 particles to form the coating structured ZrB_2@Al_2O_3, ZrB_2@Y_2O_3 and ZrB_2@ZrO_2 particles, as shown in figure 3. The Al_2O_3 coatings on the surface of ZrB_2 particles were in lighter color in secondary electron images (figure 3-a), due to the enrichment of secondary electron in non-conductive Al_2O_3 coatings. Meanwhile, Al_2O_3 coatings were in darker color in backscattered electron images (figure 3-b), because of the smaller atomic number of Al in comparison with Zr. The Y_2O_3 coatings were similar to $Y(OH)_3$ coatings in SEM images, they both had a flocculent structure (figure 3-c,d). The ZrO_2 coatings were looser than $Zr(OH)_4$ coatings (figure 3-e,f) because of the evaporation of water generated from the calcination of $Zr(OH)_4$. There was a common problem in these three coatings: some of the exposed surfaces of ZrB2 particles were not wrapped completely.

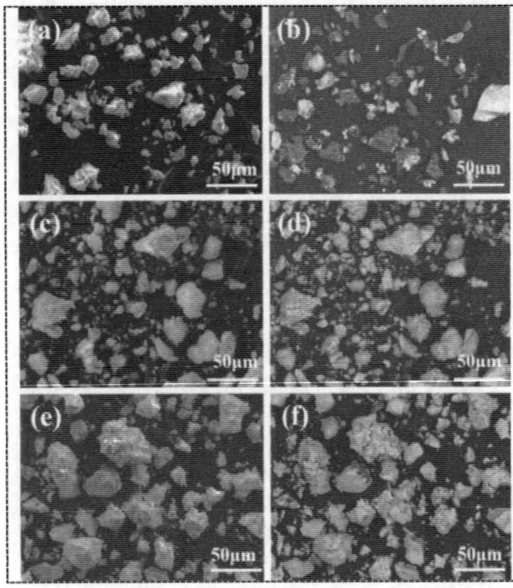

Figure 3. SEM images of target products: (a) secondary electron image of ZrB_2@Al_2O_3, (b)backscattered electron image of ZrB_2@Al_2O_3, (c) secondary electron image of ZrB_2@Y_2O_3, (d) backscattered electron image of ZrB_2@Y_2O_3, (e) secondary electron image of Z ZrB_2@ZrO_2, (f) backscattered electron image of ZrB_2@ZrO_2.

Table 1 lists the chemical composition of the oxide coatings. The major elements in the coating of ZrB_2@Al_2O_3 particle are Al and O, which proves that the ZrB_2 particles are coated by Al_2O_3 successfully. It is evident that the coating of ZrB_2@Y_2O_3 is Y_2O_3, according to revealed Y and O elements in the particle surface. Moreover, Zr and O elements were detected in the coatings of ZrB_2@ZrO_2 particles, which indicated the successful synthesis of ZrO_2 coatings.

Table 1. Elemental analysis results to the surface of target products.

Sample	Compositions			
	O (weight %)	Al (weight %)	Y (weight %)	Zr (weight %)
ZrB_2@Al_2O_3	48.34	51.66	0	0
ZrB_2@Y_2O_3	22.19	0	77.81	0
ZrB_2@ZrO_2	38.32	0	0	61.68

The XRD patterns of ZrB_2@Al_2O_3, ZrB_2@Y_2O_3 and ZrB_2@ZrO_2 composite particles are shown in figure 4. It was found that the major phase was ZrB_2 (PDF #34-0432) for these particles. There were no other characteristic peaks being observed in the ZrB_2@Al_2O_3 particles. YBO_3 (PDF #16-0277) and Y_2O_3 (PDF #41-1105) were indexed in ZrB_2@Y_2O_3 particles, which indicated the reaction of Y_2O_3 coating and ZrB_2 particles. Some monoclinic ZrO_2 peaks (m-ZrO_2; PDF #37-1484) and tetragonal ZrO_2 peaks (t-ZrO_2; PDF #50-1089) appeared simultaneously in ZrB_2@ZrO_2 particles.

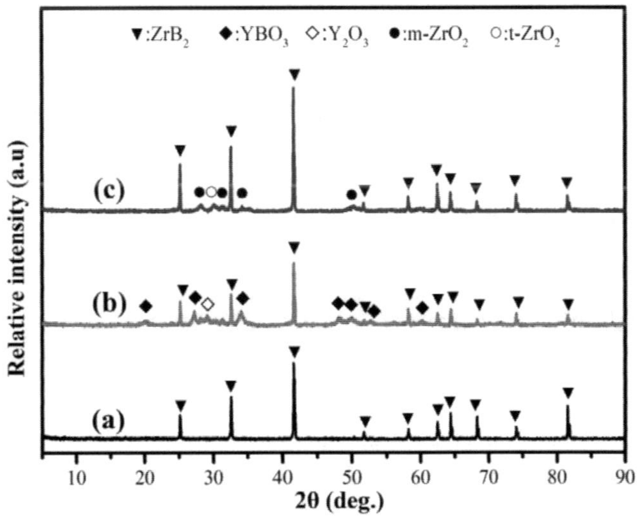

Figure 4. XRD patterns of the target products: (a) $ZrB_2@Al_2O_3$, (b) $ZrB_2@Y_2O_3$, (c) $ZrB_2@ZrO_2$.

3.3. Thermal analysis

Figure 5 illustrates the thermal analysis (TG and DSC) curves of $ZrB_2@Al_2O_3$, $ZrB_2@Y_2O_3$, $ZrB_2@ZrO_2$ and pure ZrB_2 particles. The starting temperature of weight gain for $ZrB_2@Al_2O_3$, $ZrB_2@Y_2O_3$ and $ZrB_2@ZrO_2$ particles were higher than that of pure ZrB_2 particles. The peak value of heat release of $ZrB_2@Al_2O_3$, $ZrB_2@Y_2O_3$ and $ZrB_2@ZrO_2$ particles were all decreased and moved toward higher temperature in comparison with pure ZrB_2 particles. The slope of TG curve for pure ZrB_2 particles was larger than all slopes of TG curves for the other three composite particles in the range of 700~1000°C, while the final weight of pure ZrB_2 particles after oxidation was the largest. The reduction of the peak value of heat release and the slope of TG curve strongly indicate that the oxidation rate of the composite particles is reduced. This is ascribed to the fact that the oxides' protective layer inhibits the oxygen diffusion and drops this diffusion rate. The latter can slow down the oxidation because the oxidation behavior is controlled by the diffusion step [25-27]. Therefore, the Al_2O_3, Y_2O_3 and ZrO_2 protective shell on the surface of ZrB_2 particles can significantly improve the oxidation resistance of ZrB_2 ceramics.

Besides, the variation of TG curves is terminated, and the heat release values drop to nearly zero at the temperatures above 1400 °C, which indicates the complete oxidation of ZrB_2 in the samples to ZrO_2. The phase composition of the oxidized sample (after temperature runs) must be the phase of ZrO_2 and the phase of corresponding oxide coatings.

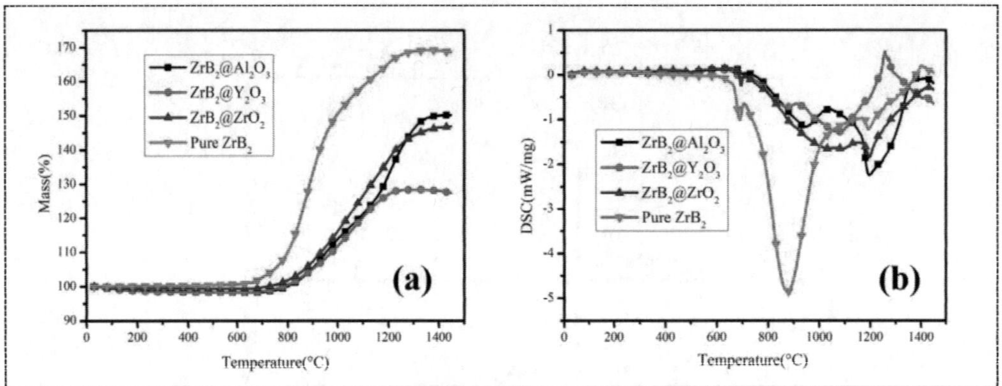

Figure 5. Thermal analysis curves of $ZrB_2@Al_2O_3$, $ZrB_2@Y_2O_3$, $ZrB_2@ZrO_2$ and pure ZrB_2 particles: (a) TG curves, (b) DSC curves.

4. Conclusions

In this work, the precursors ($ZrB_2@Al(OH)_3$, $ZrB_2@Y(OH)_3$ and $ZrB_2@Zr(OH)_4$) were prepared by the chemical co-precipitation method, and the composite particles ($ZrB_2@Al_2O_3$, $ZrB_2@Y_2O_3$ and $ZrB_2@ZrO_2$) with coating structure were obtained by the calcinations of corresponding precursors at 600 °C for 3 h. The Al_2O_3, Y_2O_3 and ZrO_2 coatings acted as a barrier of oxygen diffusion to improve the oxidation resistance of ZrB_2 particles, which was proved by the thermal analysis results of $ZrB_2@Al_2O_3$, $ZrB_2@Y_2O_3$, $ZrB_2@ZrO_2$ and pure ZrB_2 particles. Coating the Al_2O_3, Y_2O_3 and ZrO_2 protective layers on the surface of ZrB_2 particles was an effective method to improve the oxidation resistance of ZrB_2 particles.

Acknowledgment

This work was supported by the National Natural Science Foundation of China (No. 51772042).

References

[1] Hu P, Wang G L and Wang Z 2009 *Corros. Sci.*, **51** 2724-32.
[2] Gasch M J, Ellerby D T and Johnson S M 2005 *Handbook of Ceramic Composites*, Springer US, Boston, MA, 197-224.
[3] Opeka M M, Talmy I G and Zaykoski J A 2004 *J. Mater. Sci.*, **39** 5887-904.
[4] Opeka M M, Talmy I G, Wuchina E J, Zaykoski J A and Causey S J 1999 *J. Eur. Ceram. Soc.*, **19** 2405-14.
[5] Guo S Q 2009 *J. Eur. Ceram. Soc.*, **29** 995-1011.
[6] Guo S Q Yang J M, Tanaka H and Kagawa Y 2008 *Compos. Sci. Technol.*, **68** 3033-40.
[7] Li W, Zhang Y, Zhang X, Hong C and Han W 2009 *J. Alloy Compd.*, **478** 386-91.
[8] Monteverde F, Guicciardi S and Bellosi A 2003 *Mater. Sci. Eng. A*, **346** 310-9.
[9] Monteverde F and Bellosi A 2003 *J. Electrochem. Soc.*, **150** B552-9.
[10] Fahrenholtz W G, Hilmas G E, Talmy I G and Zaykoski J A 2007 *J. Am. Ceram. Soc.*, **90** 1347-64.
[11] Zhang S C, Hilmas G E and Fahrenholtz W G 2011 *J. Am. Ceram. Soc.*, **94** 1198-205.
[12] Zhang S C, Hilmas G E and Fahrenholtz W G 2008 *J. Am. Ceram. Soc.*, **91** 3530-5.
[13] Kazemzadeh Dehdashti M, Fahrenholtz W G and Hilmas G E 2015 *Corros. Sci.*, **91** 224-31.
[14] Silvestroni L, Meriggi G and Sciti D 2014 *Corros. Sci.*, **83** 281-91.
[15] Sciti D, Brach M and Bellosi A 2005 *Scr. Mater.*, **53** 1297-302.
[16] Silvestroni L, Landi E, Bejtka K, Chiodoni A and Sciti D 2015 *J. Eur. Ceram. Soc.*, **35** 4377-87.
[17] Williams P A, Sakidja R, Perepezko J H and Ritt P 2012 *J. Eur. Ceram. Soc.*, **32** 3875-83.
[18] Gao D, Zhang Y, Xu C, Song Y and Shi X 2013 *Ceram. Int.*, **39** 3113-9.

[19] Hu P, Wang G L and Wang Z 2009 *Corros. Sci.*, **51** 2724-32.
[20] Han J, Hu P, Zhang X, Meng S and Han W 2008 *Compos. Sci. Technol.*, **68** 799-806.
[21] Song J G, Zhang L M, Li J G and Song J R 2007 *J. Reinf. Plast. Comp.*, **26** 139-45.
[22] Song J G, Zhang L M, Li J G and Song J R 2007 *Surf. Rev. Lett.*, **14** 117-22.
[23] Song J G, Zhang L M, Li J G and Song J R 2007 *Surf. Rev. Lett.*, **14** 445-9.
[24] Song J G, Li J G, Song J R and Zhang L M 2007 *Surf. Rev. Lett.*, **14** 945-50.
[25] Sharp J H, Brindley G W and Achar B N N 1966 *J. Am. Ceram. Soc.*, **49** 379-82.
[26] Sharp J H and Wentworth S A 1969 *Anal. Chem.*, **41** 2060-2.
[27] Chou K C and Hou X M 2009 *J. Am. Ceram. Soc.*, **92** 585-94.

Synthesis of α-cordierite based glass-ceramic from Bayan Obo tailing and fly ash through volume crystallization

B W Li, Z Y Zhang, W L Wu, M Zhao* and Y Shi

Key Laboratory of Integrated Exploitation of Bayan Obo Multi-Metal Resources, Inner Mongolia University of Science and Technology, Baotou, Inner Mongolia Autonomous Region, P.R. China

*E-mail: philip@imust.edu.cn

Abstract. The α-cordierite glass-ceramics samples were synthesized from Bayan Obo mine tailing and fly ash through volume crystallization as a result of using $(NH_4)\ H_2PO_4$ and MgF_2 as the nucleating and glass network relaxing agents, respectively. The application of some natural impurities within above wastes, i.e., CaO, Na_2O, and K_2O, as the flux to lower the viscosity of the base glass, also contributed to this successful synthesis. The density, bending strength, Vickers hardness, dielectric constant and loss within the 1-10 MHz testing frequency range, and the average coefficient of thermal expansion of the glass-ceramic were characterized at 2.56 g cm^{-3}, 113.5±15.87 MPa, 7.04±0.26 GPa, ~7.3, ~10^{-2} and 1.86×10^{-6} $^{o}C^{-1}$ respectively. These properties make this solid waste-derived material a lucrative candidate for replacing the industrial cordierite in metallurgy and construction. The results of this study may serve as a useful example of the re-utilization of solid wastes similar to Bayan Obo tailing with high added value.

1. Introduction

The α-cordierite ($Mg_2Al_4Si_5O_{18}$) glass-ceramic is a material of outstanding commercial and technical importance because of its low dielectric permittivity and loss, high mechanical strength, and excellent thermal stability and strong resistance to thermal shock [1-2]. However, it contains more than 50 wt% of SiO_2 in its starting materials [3]. This characteristic makes its parent glass very resistant to crystallization. Therefore, α-cordierite glass-ceramics are not easy to produce by the mechanism of volume-crystallization [4]. As one of the counteracting measures to this problem, the sintering fabrication process has widely been used, in which surface defects of the starting material powders can be employed as the hetero-nucleating sites to enhance the crystallization of α-cordierite grains[3, 5-6]. The glass-ceramics fabricated by this process have inherent pores that may inevitably deteriorate both the mechanical and dielectric property of the final product. Alternatively, the high-temperature (e.g., 1150°C) volume crystallization method can also be used to produce α-cordierite glass-ceramics, wherein some expensive nucleating agents [7], such as TiO_2, ZrO_2, and flux materials, for example, CeO_2, were used. Recently, the ever-escalating scarcity of resources has been calling for better use of different natural resources and industrial solid wastes. The R&D of the cordierite based material should not be an exception. As an answer to this calling, drift sand [8], kaolin [9], fly ash [10], gold tailing [11] and ferrochromium slag [12] were successfully adapted as the starting materials to fabricate cordierite glass-ceramics. However, to the best of our knowledge, no attempt has ever been made on Bayan Obo tailing. The accumulation of this specific industrial solid waste has reached 170 million tons in the dam for its reservation at Bao Tou and posed as a constant threat to the environment and people nearby.

In this paper, we report the successful fabrication of an $(NH_4)H_2PO_4$ and MgF_2 co-doped α-cordierite glass-ceramic from Bayan Obo tailing and fly ash through the volume crystallization process without using TiO_2 or ZrO_2, the more commonly applied nucleating agents with high values. The factors affecting the fabrication was analyzed using techniques, such as DTA, XRD, and SEM orientation imaging techniques and discussed with the relevant works of peers.

2. Experimental

The compositions of Bayan Obo tailing and the fly ash were decided using the XRF and the traditional chemical analysis method, respectively, and their results are listed in table 1. Referencing to this table and the composition of cordierite glass, the starting constituents of our cordierite glass-ceramic were designed as (all in weight ratios): 13.57% Bayan Obo tailing, 29.32% fly ash, 25% industrial grade silica sand, 11.58% Al_2O_3, 7.48% MgO, 6.81% borax, 4.84 % MgF_2, and 0.81% $(NH_4)H_2PO_4$.

Table 1. Constituents of Bayan Obo tailing and the fly ash used in this study.

Constituents	MgO	Al_2O_3	SiO_2	CaO	Fe_2O_3	R_2O*	P_2O_5	Loss	Other
Tailings	11.76	1.89	14.21	15.87	16.56	2.71	1.65	25.59	9.76
Fly ash	0.78	30.71	51.68	3.90	8.60	2.28	-	-	-

* R_2O: Na_2O and K_2O in total

The well-mixed raw materials (200g) were melted in a corundum crucible in air at 1580 °C for four hours. The obtained glass melt was cast in a preheated (600 °C) steel mold. The obtained glass plate was transferred immediately to an electric furnace also preheated to 600 °C, wherein it was further annealed for three hours to release the internal stress generated in the process of casting. A small fraction of the aforementioned melt was directly quenched into cold water to create some frit particles. Then those particles were ground and measured in the differential thermal analysis (DSC- NETZSCH STA 449C) with pure Al_2O_3 as the reference standard within the temperature range of the ambient air to 1200 °C at a heating rate of 10 °C min^{-1}. Based on the endothermic inflection and the exothermic peak temperature on the DSC curve, the glass transition point (T_g) and crystallization point (T_c) were assessed and used as the reference to determine the temperatures for the final nucleating and crystallizing heat-treatments respectively. The soaking time for these two heat-treatments was equally set at 2.5 hours with a heating rate of 2.5 K min^{-1}. The crystalline phase assemblage of the heat-treated sample was identified by a Panalytical X'pert powder XRD. Based on this result, the crystallinity of the sample was calculated according to the Rietveld method. The surface morphology was examined on a field emission scanning electron microscope (FESEM-Zeiss Supra 55) operating in both second electron (SE) and orientation imaging (OIM) modes. According to the standard JT/T 263-93, bending strength was determined by the three-point method with spans of 15 mm at a cross-head speed of 0.5 mm min^{-1} on a Universal Testing Machine (CSS-88000) using samples with dimensions of 3×4×40 mm. For each measurement, the reported value was the average of six samples. The hardness was determined on a Vickers hardness tester (HV-SOA) with a load of 3 N and a dwell time of 10 s. Testing results of six samples were averaged and reported as the Vickers hardness. The coefficient of thermal expansion (CTE) was measured by a thermal expansion tester (ZRPY-1400) within the temperature range of 100-600 °C using 5×5×50 mm samples. The heating rate of the CTE test was 10 K min^{-1}. The classic parallel plate method was used to determine the dielectric constant and loss tangent of the final material on a Keysight 4294 A RLC meter at the frequency range of 1-10 MHz.

3. Results

DSC results on the cordierite melt frit powder are depicted in figure 1. It shows T_g =720°C and T_c =1000.4°C. The only exothermal peak marked by T_c is apparently a result of the formation of α-cordierite, the single main crystalline phase confirmed by of the XRD pattern of the obtained cordierite glass-ceramic depicted in figure 2.

Referencing to the as measured T_g, T_c and also our experience, the temperature for the nucleating and crystallization heat-treatment were set at 760 and 1050 °C respectively to finalize the fabrication of our cordierite glass-ceramic. As shown in figure 2, α-cordierite (JCPDS: 01-082-1884) is the main crystalline phase. Besides, figure 2 also confirms the formation of just a small amount of $CaFeO_4$ (JCPDS: 00-003-0804) as the secondary phase. The crystallinity of the whole sample was calculated at around 85% with a residual error value of <14.89%, while the crystallinity of α-cordierite alone was determined at over 83%.

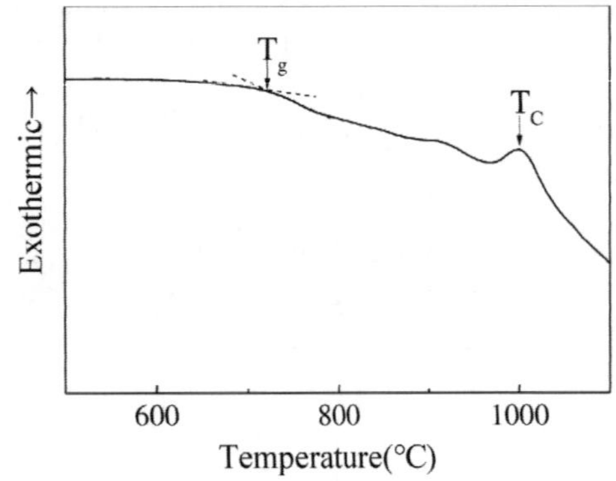

Figure 1. DSC result of the frit powder obtained by pouring a fraction of the melt into water.

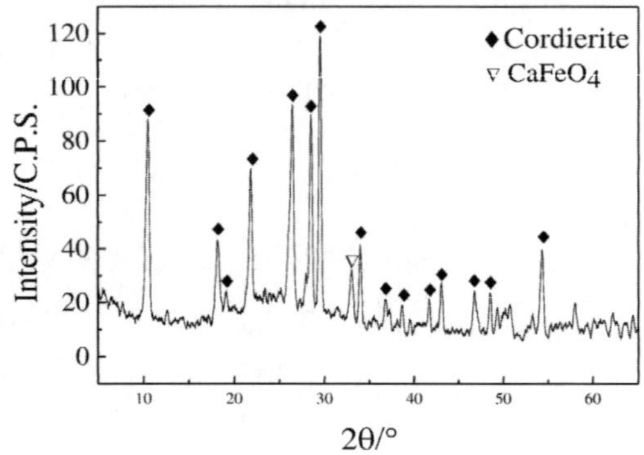

Figure 2. XRD pattern of the cordierite glass-ceramic fabricated mainly from Bayan Obo mine and fly ash.

The SEM photo shown in figure 3 exhibits that there are some darker blocks embedded within the light gray matrix of the sample. These blocks resemble those α-cordierite crystals closely in the cordierite based glaze for conventional floor tile [13]. Therefore, they should be the α-cordierite main phase identified in our XRD result.

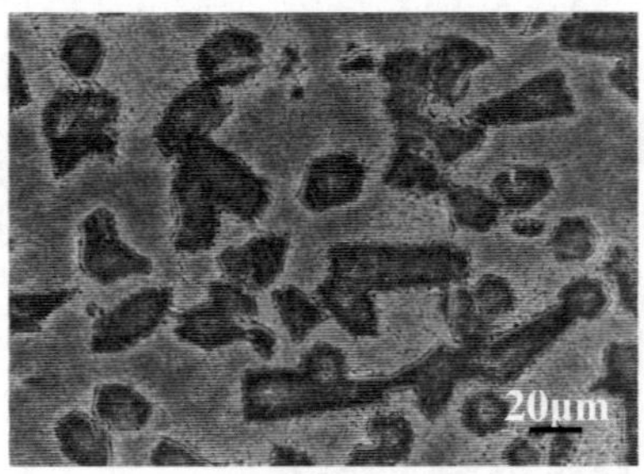

Figure 3. SEM photo of the cross-section surface of the as studied cordierite glass-ceramic.

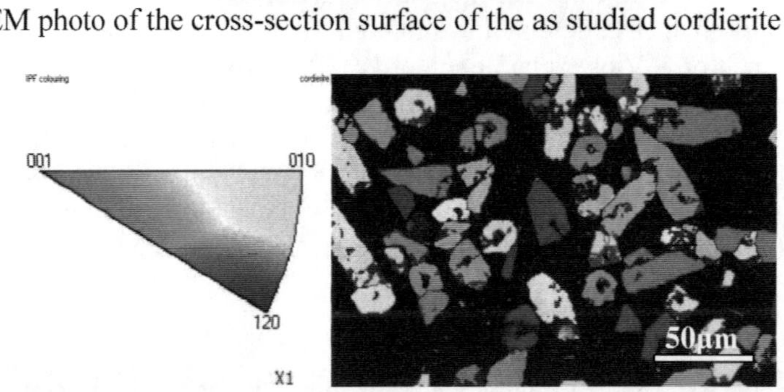

Figure 4. Orientation imaging (OIM) photo of the as studied cordierite glass-ceramic.

The OIM-SEM photo of the sample material using the lattice parameters of α-cordierite as the precondition is presented in figure 4. The color contrast in such picture was generated by the electrons diffracted by different crystalline planes of the crystals on the sample surface. Thus the shape of a colored area represents the shape of a crystal. It is obvious the colored regions in figure 4 strongly resemble the α-cordierite crystals in figure 3. Therefore, this result further confirms the above identification of those darker blocks as α-cordierite crystals, and their different color in figure 4 suggests a random distribution of those α-cordierite crystals within the matrix.

Another thing worth noticing here is the surface of the gray matrix area in figure 3 should be much rougher than the surface of α-cordierite crystals, since only rougher surface can generate more SE signals, thereby resulting in a stronger contrast (being whiter in color) in an SEM picture. This fact suggests the matrix should also contain some cordierite crystals, whose size was too small to generate detectable diffracted electron signals in the OIM photo shown in figure 4. Because the phase boundaries between small cordierite crystals and residual glass were less resistant to acid erosion and could have been etched out by light HF solution in the sample preparation process for SEM observation, yielding a rough surface as observed in the aforementioned matrix areas. This deduction is following the above XRD result which shows the crystallinity of this sample is more than 80%.

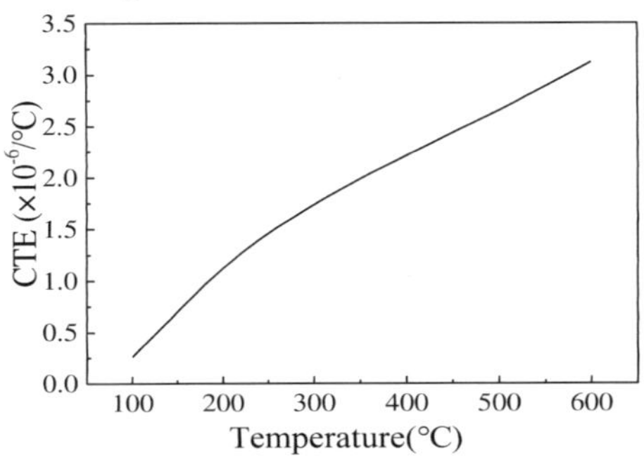

Figure 5. Variation of the Coefficient of thermal expansion (CTE) of the as studied cordierite glass ceramic.

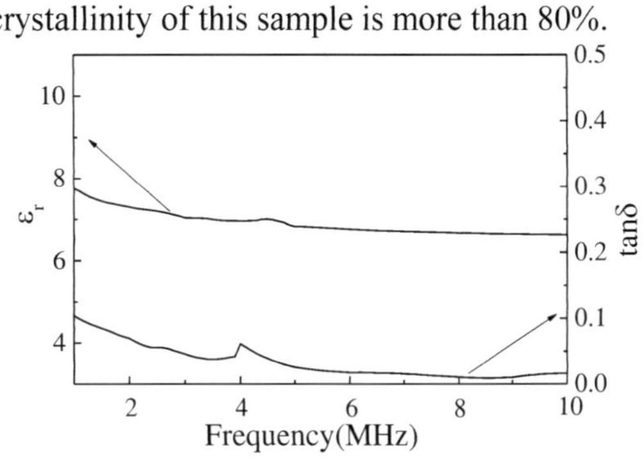

Figure 6. Variations of the dielectric constant and loss of the as studied cordierite glass.

The variation of the coefficient of thermal expansion (CTE) of the as-prepared cordierite glass over the temperature range of 100-600 °C is illustrated by figure 5. The recorded CTE value increased continuously from 0.25×10^{-6} to 3×10^{-6} °C^{-1} and averaged at 1.86×10^{-6} °C^{-1}. This result is comparable with the earlier reported CTE values for the cordierite glass-ceramics fabricated from pure chemicals [14]. An opposite trend can be observed in the variation of the dielectric constant with the increase of frequency in figure 6, which exhibits the dielectric constant decreased slowly from about 7.8 to slightly over 7 within the tested frequency range of 1 to 10 MHz. Meanwhile, the dielectric loss varies between 0.1 and 0.0024. Both dielectric parameters were somewhat larger than the commonly reported ones [5]. This is a result of the decrease in the network connectivity introduced by those natural impurities within the raw materials, so ions within the network of the final material can be more flexible under the alternative electrical field, thereby yielding a higher dielectric constant and loss [15].

Table 2. Summary table of some physical properties of the as-fabricated α-cordierite glass-ceramic (GC) and those reported in peer study.

Reference	Density (g/cm^3)	Bending strength (MPa)	Vickers hardness (GPa)	Averaged CTE over 100~600℃ ($\times10^{-6}$/ °C)
α-cordierite GC	2.56	113.5±15.87	7.04±0.26	1.86
Industrial cordierite [16]	2.5	110	-	2.5

Moreover, the additional tests showed that the density of this solid-waste derived material was 2.56 g/cm^{-3}, the bending strength was about 113MPa, and the Vickers hardness was about 7 GPa. These results, together with the obtained averaged CTE values, are listed in table 2. The corresponding values of the industrial cordierite are also listed in this table for comparison. It shows the mechanical and thermal properties of our waste derived α-cordierite glass-ceramic are generally comparable to those of industrial cordierite. Therefore, this glass-ceramic can be a feasible candidate to replace the industrial cordierite.

4. Discussion

As it has been revealed, the T_c in this study was 1000.4 °C, considerably lower than those commonly reported ones at around 1150 °C [4], suggesting the crystallization of α-cordierite from this waste-derived material has become relatively easier. The reason should be related to those chemical ingredients within the raw materials other than forming the basic crystalline structure of cordierite. Numerous studies have shown CaO [17], Fe_2O_3[18] and R_2O [19] (Na_2O and K_2O), the natural residing impurities within Baiyan Obo tailing and the locally collected fly ash can all be used as a flux to reduce the viscosity of cordierite melt, therefore benefiting the crystallization of α-cordierite. Meanwhile, P_2O_5, one of the decomposing products of $(NH_4)H_2PO_4$, can be applied as the nucleating agent and flux at the same time [20]. On the other hand, MgF_2 may decrease the connectivity of the silicate glass network by substituting O with F [15]. These three factors can all contribute to our success in fabricating α-cordierite glass-ceramic through volume crystallization at 1050 °C without using TiO_2, the commonly applied nucleating agent of much higher value than P_2O_5.

Noteworthy is that no trace of μ-cordierite could detected in this sample. Peer study has shown μ-cordierite may deteriorate the mechanical properties of cordierite based material through creating internal micro-fractures during the process of its crystalline structure transforming into that of α cordierite [21]. Therefore, our result indicates the formation of this harmful secondary phase has been completely suppressed in the context of as a designed composition of raw materials. This could be a contributing factor to the observed good mechanical property of the as investigated material.

5. Conclusions

Bayan Obo tailing and fly ash, together with a small fraction of (NH_4) H_2PO_4 and MgF_2 were used to fabricate α-cordierite glass-ceramic using the volume crystallization process characterized by a 2.5-hour nucleating process at 760 °C followed by a crystallizing treatment at 1050 °C for 2.5 hours. The glass transition temperature (T_g) and the crystallization temperature (T_c) of α-cordierite were determined at 720 and 1000.4 °C. Block-alike α-cordierite grains were observed to be randomly distributed within the bulk of the material. The density, bending strength and Vickers hardness of this solid wastes derived material were determined as 2.56 g cm^{-3}, 113.5±15.87 MPa, and 7.04±0.26 GPa respectively. Meanwhile, the dielectric constant and loss within the 1-10 MHz frequency range were found out to be at around 7.3 and 10^{-2} respectively, and the average coefficient of thermal expansion over the 100-600 °C temperature range was at around 1.86×10^{-6} °C^{-1}.

Acknowledgments

This study was funded by the 973 project (2012CB722802). This study was also funded by the projects (0901051701 and 0406091701) supported by "Inner Mongolia Science & Technology Plan." Sponsorship from the "Inner Mongolia Autonomous Region Science and Technology Major project: “Fundamental and key technology research for the integrated exploitation of Bayan Obo Mine with high added value" and "Inner Mongolia autonomous region scientific innovation team of integrated exploitation of Bayan Obo mine multi-metal resource" is also fully acknowledged by all authors of the current study. This study was also supported by “The IMUST Innovation Fund (No. 2017QDL-S03).

References

[1] Kim Y and Kang S 2013 *Ceram. Int.* **39** S619-22

[2] Holand W A and Beall G C 2002 *Glass Ceramics Technology*, 1st Edition ed. 735 Ceramic Place, Westerville.
[3] Holand W A and Beall G C. 2002 *J Am. Ceram. Soc.*
[4] Liu F, Huang X, Qu J, Yuan C, Chen G and Ma R, 2018 *J Non-Cryst. Solids.* **481** 329-34
[5] Karimi Y, Zakeri M, Reza M and Faeghinia A 2015 *Int. J. Mater. Res.* **2** 184-7
[6] Kuscer D, Bantan I, Hrovat M and Malič B 2017 *J. Eur. Ceram. Soc.*; **37** (2) 739-46.
[7] Varghese J, Vahera T, Ohsato H, Iwata M and Jantunen H 2017 *Jpn. J. Appl. Phys. S.* **56** (10S) 10.
[8] Zdaniewski W 1975 *J Am. Ceram. Soc.* **58** (5-6) 163-9.
[9] Wang W, Shi Z, Wang X and Fan W 2016 *Ceram. Int.* **42** (3) 4477-85.
[10] Banjuraizah J, Mohamad H and Ahmad Z A 2009 *J. Alloy. Compd.* **482** (1-2) 429-36.
[11] He Y, Cheng W and Cai H 2005 *J. Hazard. Mater.* **120**(1-3) 265-9.
[12] Shao H, Liang K, Peng F, Zhou F and Hu A 2005 *Miner. Eng.* **18** (6) 635-7.
[13] Liu C, Liu L, Tan K, Zhang L, Tang K and Shi X 2016 *Ceram. Int.* **42** (1) 734-42.
[14] Torres F J and Alarcón J 2003 *J Eur. Ceram. Soc.* **23** (6) 817-26.
[15] Kim B H and Lee K H 1994 *J. Mater. Sci.* **29** (24):6592-8.
[16] Volf M B, *Chemical Approach to Glass.* Amsterdam-Oxford-New York-Tokyo, Elsevier, 1984.
[17] Shao H, Liang K, Peng F, Zhou F and Hu A 2005, *Miner. Eng.* **18** 635-7.
[18] Torres F J and Alarcón J 2005 *Ceram. Int.* **31** (5) 683-90.
[19] Kirat G and Aksan M A 2014 *Phys. B: Cond. Mat.* **454** 131-4.
[20] Song L, Wu J, Li Z, Hao X and Yu Y 2015 *J Non-Cryst. Solids.* **419** 16-26.
[21] Wu J and Hwang S 2000 *J. Am. Ceram. Soc.* **83** (5) 1259-65.
[22] Miyashiro A, Iiyama T, Miyashiro T and Yamasaki M 1955 *Am. J. Sci.* **253** 185-208.

IOP Conf. Series: Materials Science and Engineering **474** (2019) 012050 doi:10.1088/1757-899X/474/1/012050

Simulation analysis the effect of heat sink fins on the high-power LED

D S Peng*, K L Liu, M X Zhang, C C Tan

Key Laboratory of Optoelectronic Devices and Systems of Ministry of Education and Guangdong Province, Shenzhen University, Shenzhen 518060, China
*E-mail: sbpengds@szu.edu.cn

Abstract: In this study, the heat dissipation of high-power LED with different heat sink fin thickness and space values is analyzed in detail, in order to tackle the problem of high-power LED overheating. It is shown that an increase in the heat sink fins' thickness initially improves the high-power LED cooling effect, but this improvement is not obvious until the above thickness reaches 3 mm. At the same time, the heat sink fins' space can also enhance the cooling effect, but when it attains 5mm, the cooling ability is decreased instead. Through the comparative analysis of the thermal performance of high-power LED with different heat sink fins thickness and space values, it is proved that the junction temperature is minimum, when the heat sink fin thickness and space are 3 and 5mm, respectively.

1. Introduction

Light-emitting diodes (LED), as the fourth-generation light sources, attract more and more attention and are increasingly applied in many fields, including displays, illumination lamps, and so on. Moreover, they have many advantages, including good monochromatic properties, long lifetime and high efficiency. Although LED has high luminous efficiency, 70% of the energy in the device is consumed in the form of heat [1-2]. We should reduce the LED chip junction temperature, in order to broaden the application fields, especially in the automotive headlamp industry. Now, the chip junction temperature is very high, if the power of the LED exceeds 1W, even up to 5W, or its power density reaches 100 W/cm^2 [3]. This may lead to dominant wavelength drift, decrease of photoelectric conversion efficiency and reduction of lifetime [4]. Moreover, if the heat is not released in time, some problems can occur, including the decline in stability, reduction of photoelectric conversion efficiency, damage to printed circuit boards, and declining life span. Urgent measures are required to minimize the heat generated in the LED chips. Among the effective solutions, the heat sink is a good candidate for reducing the temperature in the LED chip and improving the photoelectric conversion efficiency [5]. Several alternative LED chip temperature-reducing methods have been proposed by researchers [6-10]. In this paper, the comparative analysis of the affection of heat sink fin thickness and space on the heat radiation for high power LED.

2. Theoretical model

The equations of continuity, momentum and energy with buoyance are given below as equations (1), (2) and (3), respectively, to calculate the heat sink transfer capability of heat.

$$\frac{\partial \rho}{\partial t} + \nabla \cdot (\rho U) = 0 \tag{1}$$

$$\frac{\partial (\rho U)}{\partial t} + \nabla \cdot (\rho U U^T) = -\nabla p + \nabla \tau + S_V \tag{2}$$

IOP Conf. Series: Materials Science and Engineering **474** (2019) 012050 doi:10.1088/1757-899X/474/1/012050

$$\frac{\partial(\rho h_{tot})}{\partial t} - \frac{\partial(p)}{\partial t} + \nabla \cdot (\rho U h_{tot}) = \nabla \cdot (\lambda \nabla T) + \nabla \cdot (U \cdot \tau) + U \cdot S_V \quad (3)$$

Here S_V is the buoyance in the equation of momentum, as shown in equation (2), and the radiation in the equation of energy is also neglected in the model [11], as shown in equation (3). The essence of heat conduction is that the molecular thermal motions of a large number of materials collide with each other. The heat can be transferred from the part of high temperature to that of low temperature, or from the high temperature objects to the low temperature objects. The heat conduction equation is shown as equation (4).

$$Q = \frac{kS(T_1 - T_2)}{L} \quad (4)$$

where Q is the thermal flux, k is the heat conductivity coefficient, S is the heat conduction area, L is the heat conduction thickness, T_1 and T_2 are the contact surface temperature, respectively. It can be concluded that the heat conduction is directly proportional to the thermal conductivity, the heat conduction area and the temperature difference, and inversely proportional to thickness, as shown in equation (4).

The heat transfer inside the solid can only be carried out in a conductive manner. Whereas, the heat transfer process between the fluid and the substrate and radiator wall usually includes convection and conduction, there is thermal radiation for high temperature solids or fluids [12]. Thermal radiation is also a way of heat transfer; it is different with heat conduction and convection. It can transfer heat directly from one system to another by medium. If the object temperature is above 0K, it can radiate heat outside because of the personal temperature, it is called thermal radiation. The radiation capability of black body is highest in the same temperature, compared others, and the heat flux of radiation can be calculated by equation (5).

$$E = \sigma T_b^4 \quad (5)$$

where T_b is the black body surface thermodynamic temperature, σ is the black body radiation constant.

Heat produced by chip ration, convection by liquid and air are loaded in the simulation. In generally, the photoelectric transformation efficiency of chip is 20%. The changed heat between wall and liquid is shown as equation (6).

$$Q = h \times A \times \Delta T \quad (6)$$

where h, A, and ΔT are surface-delivered heat coefficient, delivered heat area and temperature difference between wall and liquid, respectively. The convection coefficient is $10 w/m^2 {}^\circ C$, the room temperature is 25°C.

Table 1. The key parameters of R, G, B LED chips

Parameter	Red LED chip	Green LED chip	Blue LED chip
Peak wavelength(nm)	620	525	470
Full width half maximum (nm)	18	40	24
Forward current(mA)	350	350	350
Forward voltage(V)	2.7	3.6	3.6
Luminous efficiency(lm/W)	38	40	11

The heat radiation is simulated by the finite element method using the ANSYS commercial software. The array of chip on the base is 12×12, the power of all chips is 166W. The chips key parameter is shown as table 1. Because the part of model is symmetric, the 1/4 model is studied in the paper. The three dimensional eight panel point unit of SOLID70 is selected in the course of heat analysis. The R, G, B based chip is reduced to a unit in the simplified figure of chip model, as shown in figure 1. The size of chip is 1.5×2×0.01.

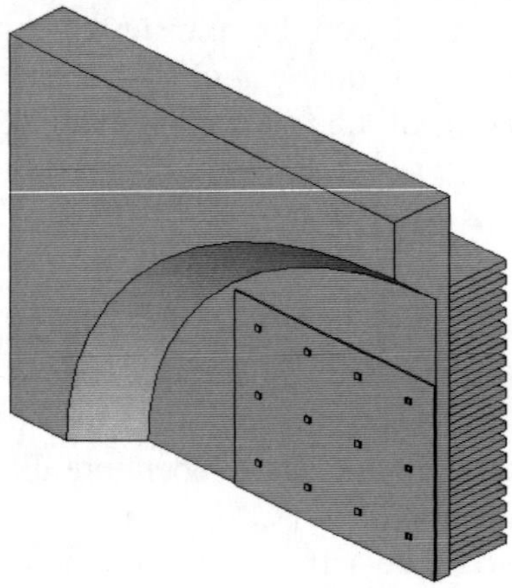

Figure 1. Schematics of chip array on the base in 1/4 model

3. Results and Discussion

The effect of heat sink fins thickness on the LED junction temperature is studied. The LED temperature distribution based on different heat sink fins thickness is shown in figure 2. When the heat sink fins thickness is low, the surface area of radiation is increased with the heat sink fins' thickness. It can improve the heat dissipation. However, the space between fins gets narrower with the increase in the heat sink fins' thickness. The heat dissipation becomes slower, when the heat sink fins thickness is above 3 mm.

The LED temperature distribution based on different heat sink fins space is shown in figure 3. When the heat sink fins space is narrow, the interspace between the fins is small; the heat dissipation ability can be improved with the increase in the heat sink fins' space. However, the heat sink fins are far from chip center with the heat sink fins' space increase. So the heat dissipation becomes worse instead, when the heat sink fins space is above 5 mm.

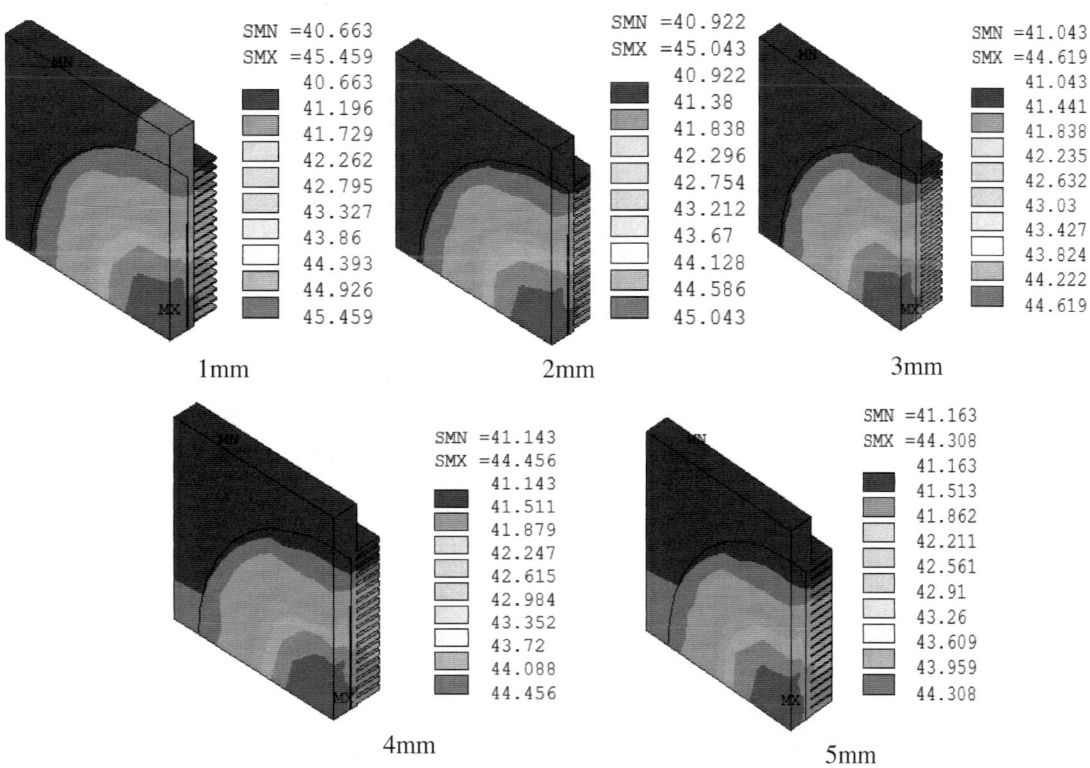

Figure 2. The LED temperature distribution with different heat sink fins

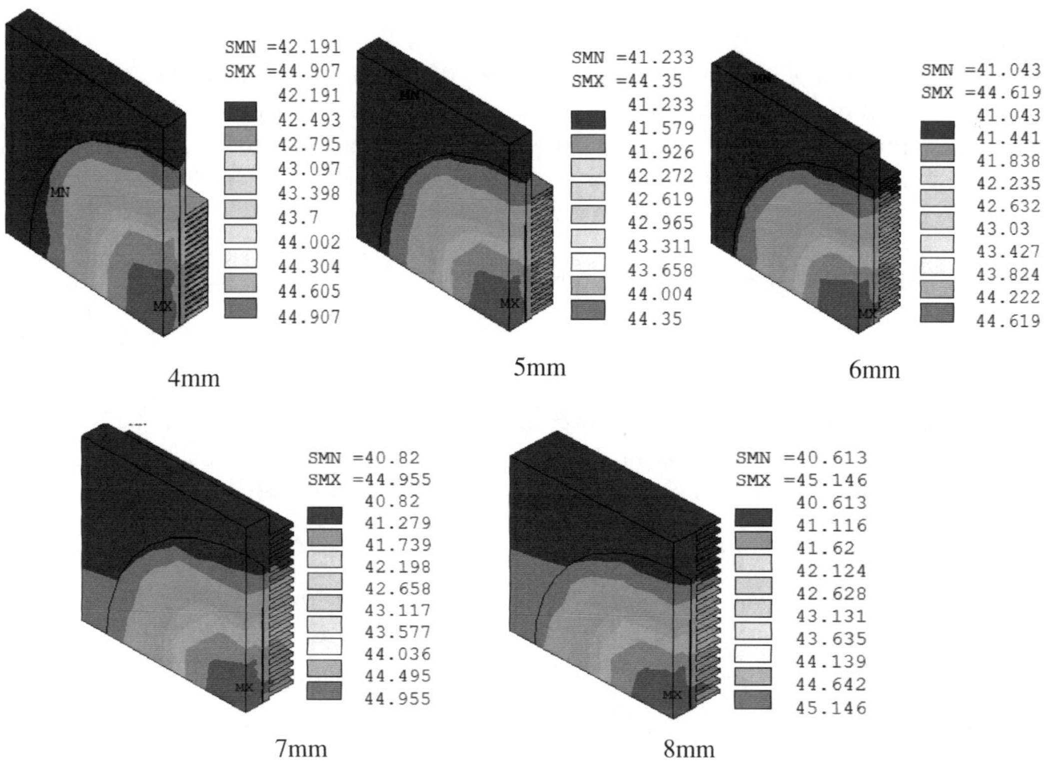

Figure 3. The LED temperature distribution with different heat sink fins

Based on figures 2 and 3, the relationship between the heat sink fins thickness, space and the LED junction temperature are gained, respectively, as is shown in figure 4.

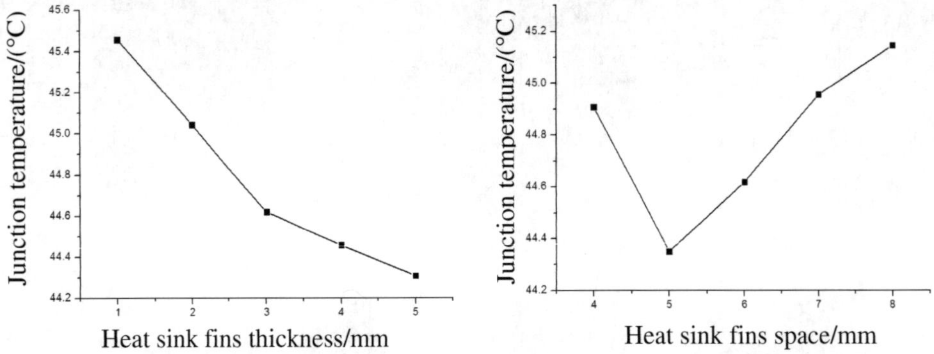

Figure 4. The LED junction temperature curve with different heat sink fins thickness and space

It is shown that the heat dissipation is more significant with the added of heat sink fins thickness, but when the thickness is more 3mm, the space between fins is narrower and narrower, the flow air gets worse, the heat dissipation becomes slower and slower, the decline of LED junction temperature is not obvious, as shown in figure 2 and 4. The LED junction temperature decreases with the added of heat sink fins space, but when the space is beyond 5mm, the heat sink fins is far from chip center when the size of heat sink is the same, the temperature rises instead, as shown in figures 3 and 4.

4. Conclusions

The heat dissipation of high-power LED at different heat sink fins' thickness and space values was analyzed in detail. It is shown that an increase in heat sink fins' thickness improves the heat dissipation, which effect is most pronounced when the thickness exceeds 3 mm. The airflow can be affected, when the space between the heat sink fins is too small, then the heat cannot be effectively dispersed. On the other hand, if the space is too large, the chip center is far from the heat sink fins and the heat in chip center cannot be dispersed. The high-power LED junction temperature is minimum, when the heat sink fin thickness and space are 3 and 5 mm, respectively.

Acknowledgments

The authors would like to acknowledge the funding support by the National Natural Science Foundation of China (Grant No. 60806017 and 61136001), the Rising Industry Development Foundation of Shenzhen, China (Grant No. JCYJ20140418095735627).

References

[1] Maaspuro M, Tuominen A 2013 *Microelectron. Reliab.* **53** 428–34
[2] Zhou J, Huang J, Wang Y and Zhou Z 2016 *Appl. Therm. Eng.* **93** 122–30
[3] Yung K C, Liem H, Choy H S and Cai Z X 2014 *Appl. Therm. Eng.* **63**:105–18
[4] Arik M, Petroski J and Weaver S 2002 *In: Proc. IEEE Iintersociety Conf. Thermal Phenomena, Hawaii.* **A**:113–20

[5] Minseok H 2012 *Microelectron. Reliab.* **32**:836–44
[6] Feng S, Shi M, Yan H, Sun S, Li F and Lu T J 2018 *Appl. Therm. Eng.* **132**:30–7
[7] Bladimir 2013 *Appl. Therm. Eng.* **59**:648–59
[8] Zhao X J, Cai Y X and Wang J 2015 *Appl. Therm. Eng.* **75**:248–58
[9] Li J H, Ma B K, Wang R S and Han L 2011 *Microelectron. Reliab.* **51**:2210–15
[10] Jeng T M, Tzeng S C and Lin Y C 2015 *Int. J. Heat Mass Transf.* **66**:233–9
[11] Shin D H, Baek S H and Ko H S 2016 *Int. J. Heat Mass Transf.* **93**:516–28
[12] Tang Y, Lin L, Zhang S, Zeng J, Tang K, Chen G and Yuan W 2017 *Energy Convers. Manage.* **151**:1-10

Simulation of dental resin monomer mixtures

S B Hong[1,2], H B Liu[1], L Wang[1], X Zhang[1,2], G Chen[1,2], H X Liu[2,3], S Y Huang[1] and B Li[1,2,4,*]

[1] School of Materials Science and Engineering, Wuhan University of Technology, Wuhan, China, 430070

[2] Research Center for Materials Genome Engineering, Wuhan University of Technology, Wuhan, China, 430070

[3] International School of Materials Science and Engineering, Wuhan University of Technology, Wuhan, China, 430070

[4] State Key Laboratory of Materials Processing and Die & Mould Technology, Huazhong University of Science and Technology, Wuhan, 430074, China

*E-mail: libei@whut.edu.cn.

Abstract. In this work, molecular dynamics simulations of Bis-GMA/TEGDMA resin systems were carried out to investigate the rheological and dynamic properties of dental materials. The effect of the resin diluent monomer TEGDMA content r_d on the viscosity and diffusion behaviors of the resin monomer mixtures was studied as well. The viscosity of the resin monomer mixture was calculated using the Green-Kubo method. It was shown that the mixture viscosity decreases apparently with r_d, showing good agreements with experiments. Moreover, the diffusion coefficient of the Bis-GMA/TEGDMA monomer mixtures was evaluated, and it rises from 0.273×10^{-11} m/s to 9.710×10^{-11} m/s as r_d increases from 0 to 100 wt%. In particular, as r_d is < 50 wt%, the diffusion dynamics of the resin monomer system was significantly suppressed due to the rigid benzene rings and hydroxyl groups in Bis-GMA. On the other hand, the diffusion coefficient of the materials increases rapidly with r_d at $r_d > 50$ wt%. The findings in this work are believed to improve the atomic/molecular scale understanding of the rheological and dynamic performances of Bis-GMA/TEGDMA dental materials.

1. Introduction

In the state-of-the-art restorative dentistry, resin composites are becoming increasingly important and prevalent as compared to silver amalgams ascribing to their superior mechanical properties, excellent esthetics, and good biocompatibility [1-2]. Dental resin composites usually consist of a continuous polymeric matrix, a photoinitiator system, and nano/micron dispersed inorganic fillers. The widely used matrix contains the bisphenol-α-glycidyl methacrylate (Bis-GMA) and triethylene glycol dimethacrylate (TEGDMA) monomers. In the resin matrix, Bis-GMA monomer reduces the volume shrinkage, while TEGDMA monomer has considerable mobility and improves degree of conversion [3-4].

Dental resin composites are usually incorporated with inorganic fillers by varying filler compositions, content and size. The fillers could improve mechanical strength, and weaken water sorption and thermal expansion. However, the filler loading is limited by the dental rheological properties [5]. For example, the resin matrix viscosity, as a measure of the resistance opposed by the material itself to an imposed flow motion, influences the maximum filler content. It is reported that a lower viscosity matrix can accept more fillers at constant plasticity [5]. Moreover, the matrix viscosity also affects the handling characteristics of the dental resin composite, i.e., its flowability, ease of placement and shaping on the restored site, and adherence to the tooth tissues [6]. Besides, it is well established that the dental resin matrix could absorb liquids from saliva and food, and this in return affects clinical performance [7]. Therefore, it is crucial to explore dynamic features and rheological performance of the resin models to produce restorative materials with better handling and clinical behaviors.

Meanwhile, computational simulations are widely adopted in examining features and behaviors in physics, chemistry, material, and biological science. Among them, molecular dynamics (MD) method characterizes organic-inorganic hybrid materials from an atomic/molecular perspective quite well [8]. As such, we employ an MD simulation method to study the dynamic and rheological features of Bis-GMA/TEGDMA rein models. It is expected to increase our fundamental understanding of the structure and performance of dental materials.

2. Methods

2.1. Molecular Model

In the MD simulations, we created Bis-GMA monomer and TEGDMA monomer via Materials Studio, as shown in Figures 1(a) and (b). The Bis-GMA monomer has a symmetrical structure with two rigid benzene rings and two methacrylate moieties, giving rise to high viscosity and the hydrophilic nature of the resin composites. The diluent TEGDMA is incorporated into different proportions of Bis-GMA to tune the viscosity of the resulting resin monomer blends. Six different mixtures of Bis-GMA and TEGDMA with varying TEGDMA mass ratios rm, that is, 0, 30, 50 60, 80 and 100 wt%, were modeled. According to the content of the diluent agent, the systems were designated as S0, S30, S50, S60, S80, and S100, respectively. The typical equilibrated atomic structure of the mixture (i.e., r_m = 30 wt%) is also shown in figure 1(c).

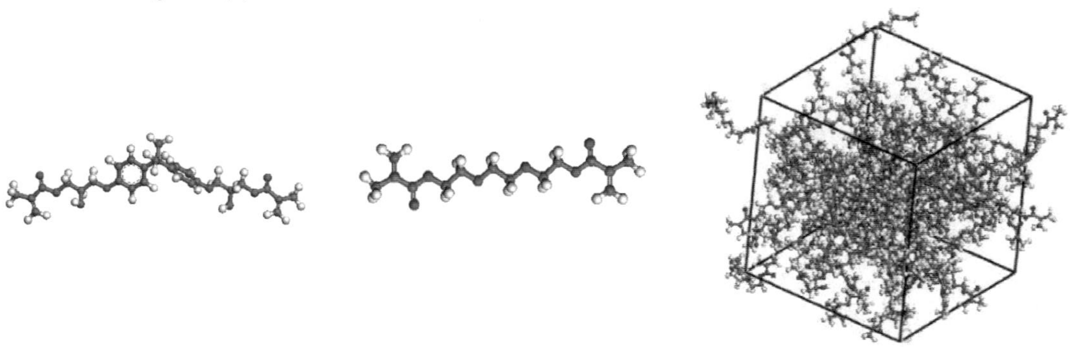

(a) (b) (c)

Figure 1. (a) Bis-GMA monomer model, (b) TEGDMA monomer model, and (c) the final snapshot of the Bis-GMA/TEGDMA monomer mixtures with r_m = 30%.

2.2. MD Simulation Details

We chose COMPASS II force field [9] to describe the interactions between atoms and molecules. The Bis-GMA monomer and TEGDMA monomer were at first optimized using the steepest descent method during energy minimization. Then, Bis-GMA and/or TEGDMA molecules were randomly placed in a cubic domain to represent the monomer mixture with a given r_m. For each mixture, the number of atoms was around 2500 ~ 4000 to improve computational efficiency. The systems were relaxed at 296 K and 1 bar for a period of 1 ns to obtain reasonable densities as compared with experiments, as illustrated in table 1.

Then, the resin blend systems were further equilibrated at 296 K for 10 ns. The system temperature was retained constant using the Nośe-Hoover-Langevin thermostat [10], and the pressure was coupled

IOP Conf. Series: Materials Science and Engineering **474** (2019) 012051 doi:10.1088/1757-899X/474/1/012051

to the Berendsen barostat [11]. The long-range electrostatic interactions were computed with the Ewald summation method [12].

Table 1. Comparison of density (g/cm^3) of the Bis-GMA/TEGDMA mixtures using MD simulations and experiments.

Bis-GMA/TEGDMA monomer mixture	Density	
	Simulation	Experiment [13-14]
S0	1.147	1.151
S30	1.133	1.142
S50	1.117	1.121
S60	1.100	1.113
S80	1.082	1.096
S100	1.072	1.072

3. Results and Discussions

3.1. Viscosity

The shear viscosity of the Bis-GMA/TEGDMA mixtures was evaluated according to the Green-Kubo (GK) expression as [15]

$$\eta = \frac{V}{3K_BT}\int_0^\infty \langle \sum P_{xy}(t)P_{xy}(0)\rangle dt \qquad (1)$$

where η is the shear viscosity, V is the cell volume, T is the ambient temperature, K_B is the Boltzmann constant, and Pxy represents the xy component of the stress, respectively.

It is known that the angle brackets in equation (1) indicates an ensemble average. In the GK computation, it requires deliberate selection of some quantities including the MD simulation time, t_D, the time length over which the stress autocorrelation function (SACF) is calculated, t_W, and the time interval between time windows, t_S. Generally, t_D is in order of a few nanoseconds and t_S is some femtoseconds for systems with particles ranging from 258 to 25,800 [16]. The time window t_W should be large sufficiently to observe the SACF decay, and it may range from a few to hundreds of picoseconds. However, these relatively small values of t_W were chosen mostly when the shear viscosities of the systems of interest are only about a few mPa·s. If target systems with ultrahigh viscosities like Bis-GMA/TEGDMA mixtures are modeled, a much larger t_W is required to obtain the computation accuracy of Equation (1). In this regard, we calculated the shear viscosity of the mixtures using a rather large t_W (t_W = 1 ns) and the results are shown in figure 2.

Figure 2(a) shows the normalized SACF as a function of t_W at various TEGDMA mass ratios r_m. For all the systems, the SACF fluctuates and decays quickly towards zero. More importantly, the SACF curves approach to zero after 10 ps, indicating that the selection of t_W ($t_W \geq$ 10 ps) might be appropriate in cutting off the contribution of SACF at longer integration time in equation (1). However, if we examine the integrals of the shear viscosities with t_W in the insets of figure 2(a), the viscosities continue to increase with t_W, and no acceptable plateau was observed. This implies that the contribution of SACF after 10 ps should not be neglected and a more massive effective t_W is required. In this work, we took 0.20, 0.33, 0.96, and 2 ns as the effective t_W where the plateau of the shear viscosity curves begin for the mixtures with various TEGDMA mass ratios r_m respectively, as seen in figure 2(b). The calculated shear viscosities and the experimental values are also listed in table 2. As expected, the shear viscosity decreases with r_m from both the simulations and experiments. For S80 and S100 samples, the simulated viscosities agree well with the experimental observations. While for S50 and S60, the discrepancy between them increases significantly. This is probably due to two reasons. Firstly, the simulation time t_D is not sufficiently long at high values of t_W (i.e., around 1 ~ 2 ns) to provide enough sampling when calculating SACF. Secondly, the SACF contribution beyond t_W is still non-negligible, and it should be taken into consideration. While in the case of low r_m ($\leq$ 50 wt%), the lack of computational power and longer simulation time make it difficult to verify the detailed viscosity convergence to a reasonable accuracy. However, the results indeed demonstrated the precise quantification of viscosity increment as r_m decreases.

Table 2. Comparison of shear viscosity (mPa·s) of the Bis-GMA/TEGDMA mixtures using the MD simulations and experiments.

Bis-GMA/TEGDMA monomer mixture	Viscosity	
	Simulation	Experiment [6]
S50	139.81	144
S60	54.21	66
S80	27.87	22
S100	11.27	7.7

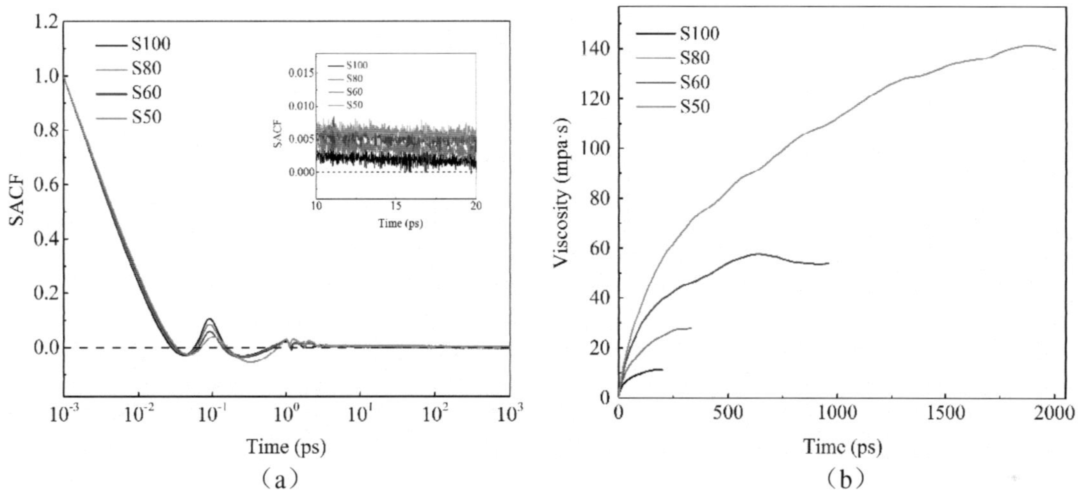

Figure 2. (a) The normalized SACF as a function of t_W at various TEGDMA mass ratios r_m and (b) the effective t_W where the plateau of the shear viscosity curves begin for the mixtures with various TEGDMA mass ratios r_m.

3.2. *Diffusion Coefficient*

In addition to shear viscosity, the mixture dynamics was investigated by calculating diffusion coefficient D of the monomers, and it was computed using the Einstein relationship, as in [17]:

$$\langle \Delta r(t)^2 \rangle = \frac{1}{N}\sum_{i=1}^{N}[r_i(t) - r_i(0)]^2 \quad (2)$$

$$D = \lim_{t\to\infty}\frac{1}{6t}\langle \Delta r(t)^2 \rangle \quad (3)$$

where $\langle \Delta r(t)^2 \rangle$ is the mean square displacement (MSD) of N molecules. The diffusion coefficients of the monomer mixtures were calculated by directly fitting a straight line in the MSD plots from 150 to 400 ps in Figure 3(a). The diffusion coefficients rise from 0.273×10^{-11} m/s of S0 (pure Bis-GMA monomers) to 9.710×10^{-11} m/s of S100 (pure TEGDMA monomers), and more results were shown in Figure 3(b), where the curve might be divided into two straight lines intersected at 50 wt%. When r_m is < 50 wt%, the diffusion coefficients of the mixtures are mainly dominated by the rigid Bis-GMA monomer, and the increment is quite small, due to the two stiff benzene rings and hydroxyl groups in

Bis-GMA. In contrast, the vise verse was an observer at r_m > 50 wt% and the mixtures exhibit much higher mobility as the diluent agent content increases.

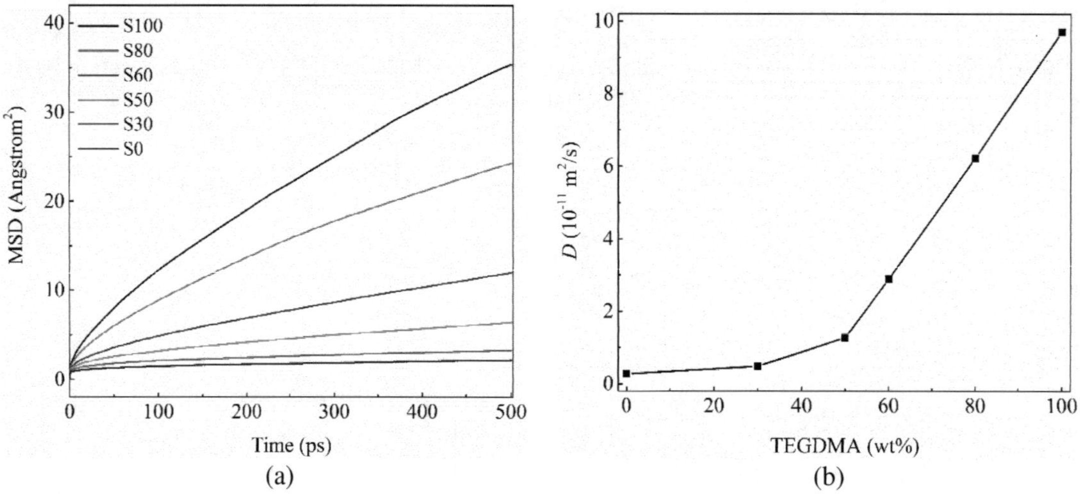

Figure 3. (a) The MSD of the Bis-GMA/TEGDMA mixtures and (b) the diffusion coefficient D at various r_m.

4. Conclusions

In the present work, we carried out MD simulations to study the viscosity and diffusion coefficient of Bis-GMA/TEGDMA monomer mixtures. The viscosity of the mixtures was fundamentally calculated by the GK method. It was shown that the time window t_W plays an essential role in determining the computation accuracy of the viscosities and a large t_W should be adopted for systems of interest with high target viscosities. By choosing appropriate t_W, the simulated viscosities agree well with experimental observations. Moreover, the dynamic property of the monomer mixtures was quantified by calculating the monomer diffusion coefficients. It was found that the two rigid benzene rings and hydroxyl groups in Bis-GMA result in suppressed diffusion dynamics of the mixtures when the TEGDMA mass ratio r_m is < 50 wt%. When r_m is > 50 wt%, the diffusion coefficient of the materials increases rapidly as the diluent agent content increases. The findings in this work could enhance our fundamental understanding of the dynamics and rheological features of resin materials.

Acknowledgments

The authors would like to acknowledge the financial support from the National College Students' Innovation and Entrepreneurship Training Program (WUT: 20181049701047), the Excellent Dissertation Cultivation Funds of Wuhan University of Technology (2017-YS-011), the National Natural Science Foundation of China (51603160), and the Open Fund Project of State Key Laboratory of Materials Processing and Die & Mould Technology, Huazhong University of Science and Technology (No. P2017-011).

References

[1] Khosroshahi M E, Atai M and Nourbakhsh M S 2008 *Laser. Med. Sci* **23** 399-406.
[2] Shi S and Nie J 2008 *Dent. Mater* **24** 530-5.
[3] Lu H, Stansbury J W, Nie J, Berchtold K A and Bowman C N 2005 *Biomaterials* **26** 1329-36.
[4] Tian M, Gao Y, Liu Y, Liao Y, Hedin N E, and Fong H 2008 *Dent. Mater* **24** 235-43.
[5] Taylor D F, Kalachandra S, Sankarapandian M and McGrath J E 1998 *Biomaterials* **19** 197-204.
[6] Lee J H, Lee I B and Um C M 2006 *Dent. Mater* **22** 515-26.
[7] Musanje L and Darvell B W 2003 *Dent. Mater* **19** 414-22.

[8] Li B and Wong C H 2013 *PolymeR* **54** 6008-18.
[9] Sun H, Jin Z, Yang C, Akkermans R L C, Robertson S H, Spenley N A, Miller S and Todd S M 2016 *J.Mol.Model* **22** 1-10.
[10] Samoletov A A, Dettmann C P and Chaplain M A J 2007 *J. Stat. Phys* **128** 1321-36.
[11] Berendsen H J C, Postma J P M, Gunsteren W F V, DiNola A and Haak J R 1984 *J. Chem. Phys* **81** 3684-90.
[12] Ewald P P 1921 *Ann. Phys* **369** 253-87.
[13] Moszner N and Salz U 2001 *Prog. Polym. Sci* **26** 535-76.
[14] Dewaele M, Truffier-Boutry D, Devaux J and Leloup G 2006 *Dent. Mater* **22** 359-65.
[15] Ogawa H, Shiraishi Y, Kawamura K and Yokokawa T 1990 *J. Non-Cryst. Solids* **119** 151-8.
[16] Nevins D and Spera F 2007 *Mol. Simulat* **33** 1261-6.
[17] Dong R Y and Cao B Y 2012 *Int. J. Numer. Meth. Eng* **92** 229-37.

Preparation of submicron boron carbide at a high heating rate

J Zhao*, W H Zhou and S L Hu

Department of Special Materials Engineering, China Institute of Atomic Energy. Beijing 102413, China

*E-mail: zhaojing.mail@qq.com.

Abstract. Organosol gel technology is a research hotspot of low-temperature synthesis of ultrafine boron carbide powders at this stage. Based on this technology, trimethyl borate is directly used as a pyrolysis material, which eliminates the process of forming a gel and simplifies the operating procedures. During the experiment, the reaction temperature and heating/cooling rates played a significant role in developing uniform submicron boron carbide powders. In this paper, uniform submicron-sized boron carbide was prepared at a high temperature of 1600°C and a rapid heating rate of 400°C/min. The ratio of the particle size distribution to 1 μm or less was 91%.

1. Introduction

Boron carbide (B_4C) as a neutron absorbing material has a broad neutron capture energy spectrum, can accommodate helium generated by irradiation, does not form strong secondary radiation, and its nuclear wastes are easy to handle[1]. These advantages over other materials attribute to its wide usage in the nuclear industry. For example, the fast-neutron breeder reactor uses sintered boron carbide pellets as control rods [2]. However, boron carbide is a ceramic material with strong covalent bonds, so its sintering property is reduced. To obtain a higher sintered density, the average particle size of boron carbide powders is less than 1 μm [3].

At present, the preparation of boron carbide mainly includes the carbothermal reduction [4]-[7], self-propagating magnesium reduction [8]-[9], direct synthesis [10]-[12], organosol gel [13]-[15], chemical vapor deposition [16]-[18] and many other components. Among them, the carbothermal reduction has achieved industrialization. However, the obtained boron carbide powders have a large particle size (20-40 μm), ball milling or other pulverization methods are required to prepare boron carbide powders as are necessary for sintering. The manufacturing process is complicated, and the purity of the products is not high. Organosol gel method is the focus of research on the low-temperature synthesis of boron carbide. The essence of the technique is to introduce boron atoms into the organic carbon structure, forming B-O-C bonds to achieve molecular-level mixing of raw materials, thereby lowering the reaction temperature, preparing ultrafine powders. However, boric acid is difficult to form a gel with organics, the yield of this method is low [15],[19], so the technique is still in the laboratory research. In contrast, trimethyl borate itself contains B-O-C bonds, eliminating the need to form a gel process, significantly reducing the operational steps, as shown in figure 1. Therefore, this paper attempts to directly use trimethyl borate as a pyrolysis material to study the optimal conditions for preparing uniform submicron-sized boron carbide.

IOP Conf. Series: Materials Science and Engineering **474** (2019) 012052 doi:10.1088/1757-899X/474/1/012052

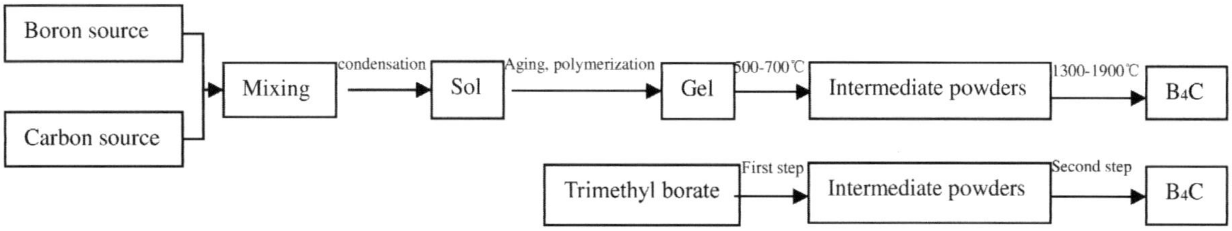

Figure 1. Comparison of organosol-gel technology and trimethyl borate pyrolysis experiment.

2. Experimental

The rate of temperature during the reaction has a significant influence on the formation of a uniform fine boron carbide. Due to the limited heating/cooling capability for the tube furnace used (the maximum heating rate can only reach 20°C/min), this experiment designed a corundum micro-reactor to achieve a high heating rate (about 400°C/min). Figure 2 shows a schematic of the reactor. The device consists of a corundum tube with inner diameter of 6 mm, connects the open end of the corundum tube to a T-shaped form to provide a feed (Ar and trimethyl borate) inlet (through another fine corundum tube with an inner diameter of 1 mm) and the gas products' outlet. Since trimethyl borate is easily decomposed at low temperatures, the experiment used a two-step method: firstly, the micro-reactor was inserted into the furnace tube, and the high-purity Ar was purged. When the tube furnace was preheated to 850°C, the carrier gas Ar and the raw material trimethyl borate were introduced for reaction, and the products were deposited on the bottom of the corundum tube. Secondly, the micro-reactor was removed, and the tube furnace was heated to 1600 °C. Then the micro-reactor was directly pushed into the hot zone to obtain a high-speed heating rate. After the reaction completed, the micro-reactor was quickly withdrawn. The corundum tube was washed with ethanol, and the products were dissolved in ethanol, then dried to obtain the final products.

The pressure changes during the reaction were monitored using an Y-YBS precision digital pressure gauge (by Huai'an Tianyi Instrument Co., Ltd.). The final products were analyzed by X-ray diffraction (Bruker, D8 advance) and Fourier transform infrared spectroscopy (Thermo, Nicolet iS10) for the phase identification. Besides, boric acid content was titrated by automatic potentiometric titrator (Mettler Toledo, T5). After gold spraying, the morphology of the products was analyzed by a cold field emission scanning electron microscope (Hitachi, Japan, S4800). And the particle sizes were characterized using a laser particle size analyzer (HORIBA LA-950).

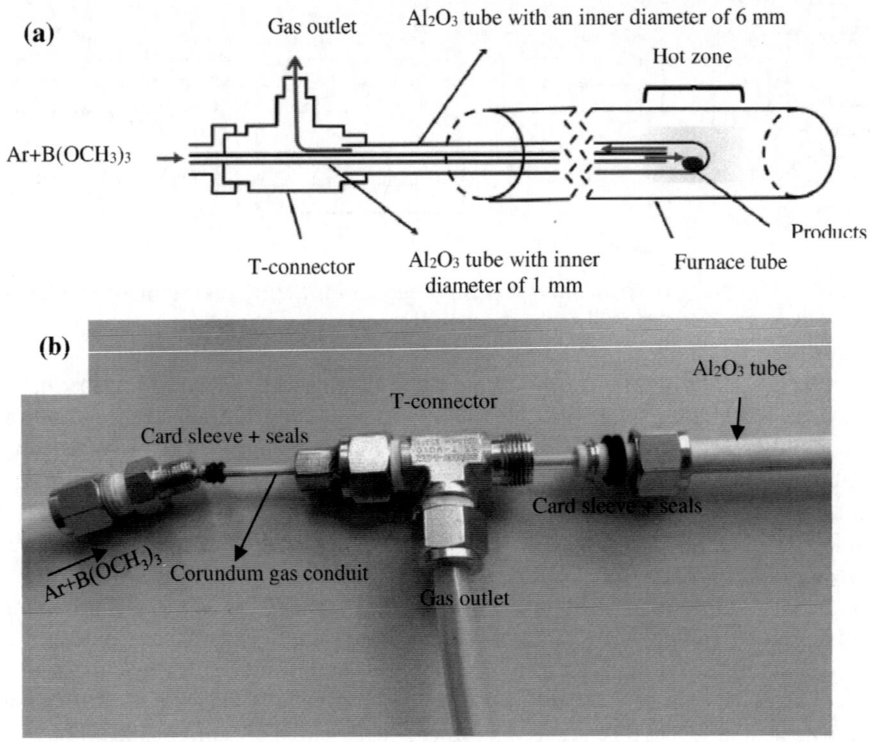

Figure 2. Schematic (a) and photo (b) of the corundum micro-reactor for high heating rate experiment.

3. Results

3.1. Impacts of the first step pyrolysis temperature

It has been documented that trimethyl borate is stable at 470°C [20]-[21]. Based on the limitation of the minimum operating temperature of the high-temperature tube furnace, this study selected 650, 750, 850, and 950°C as the temperatures of the first step of pyrolysis. In the experiment, the flow rate of Ar was controlled at 3-3.5 mL/min, and the feed rate of trimethyl borate was set to 4 mL/h. Exhaust gases were discharged into the air after being absorbed by water. When the temperature reached 650 °C, the number of bubbles in the absorption bottle was small, and the pressure inside the tube had not changed substantially, as shown in figure 3. However, there was a phenomenon of sucking up indicating that the reaction temperature was low and trimethyl borate could not be completely reacted. When the temperature reached 750°C, both the number of bubbles and the pressure increased. But there was a bright white mist at the outlet end instead of depositing at the bottom of the corundum tube. After testing, the substance sticking to the wall of the outlet pipe was boric acid. It was inferred that the reaction temperature was lower, resulting in a slower reaction rate, and the reaction could not be completed entirely within this heating interval. When the temperature reached 850°C, the number of bubbles further increased, so do the pressure inside the tube, but no white mist appeared at the outlet end. The temperature also increased to 950 °C, the pressure inside the tube suddenly went up and down, after 1 h feeding, the pressure rose linearly (not shown in the figure), and the internal corundum gas conduit was blocked. This indicated that the reaction temperature was too high and the products were deposited in the corrugated gas conduit (inner diameter of 1 mm), causing clogging and inability to continue feeding.

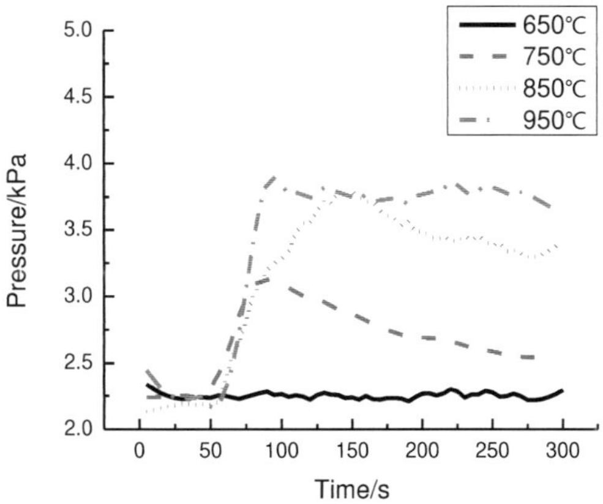

Figure 3. Pressure variation in the tube at different temperatures.

The change in pressure was mainly caused by the gas generated by the reaction. The entire reaction process was not sealed, but open, so the pressure changes were minimal and could only be compared through the overall trend. At the beginning of the reaction, the pressure changed considerably. When the reaction proceeded for some time, the amount of the solid product carbon would affect the feed rate more or less, so the pressure dropped

3.2. Analysis of the first step pyrolysis products

The reaction temperature was set at 850°C, and the first pyrolysis product was mainly concentrated at the bottom of the corundum tube. Depending on the shape of the corundum tube, the first step pyrolysis products were cylindrical, and the products were deposited from the outside to the inside. Compared to the solid phase reaction mechanism of the gel process, the reaction proceeded more thoroughly, the products were more homogeneously mixed and was more suitable for industrial production. Figure 4 shows the XRD analysis of the first-step pyrolysis products with a broad peak; the main reason is that the products are mostly amorphous and the amorphous form of the reaction is also generally larger than the crystal of the same substance. There is a sharp diffraction peak consistent with the position of the boron oxide standard card PDF#06-0297 at 28°; and no distinct carbon diffraction peak appears in figure 4, but the products are black, so it is inferred that the carbon in the products is mainly amorphous carbon, which is not detected. Figure 5 is a comparison of the Fourier transform infrared spectrum of the products at 850 °C with pure boric acid. As seen in figure 5, the absorption peaks attributed to boric acid, mainly, the intense bands of O-H stretching (3224 cm^{-1}), B-O stretching (1472 cm^{-1}), and B-O-H deformation (1195 cm^{-1}) vibrations, further illustrating that also contains a certain amount of boric acid [22].

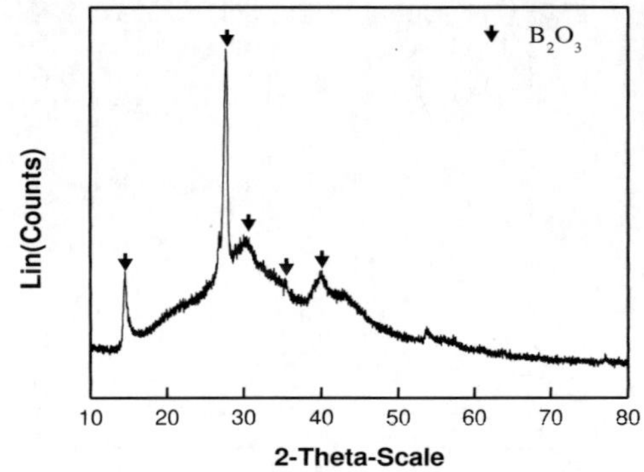

Figure 4. XRD pattern of the first-step pyrolysis products obtained by reaction at 850°C for 4 h.

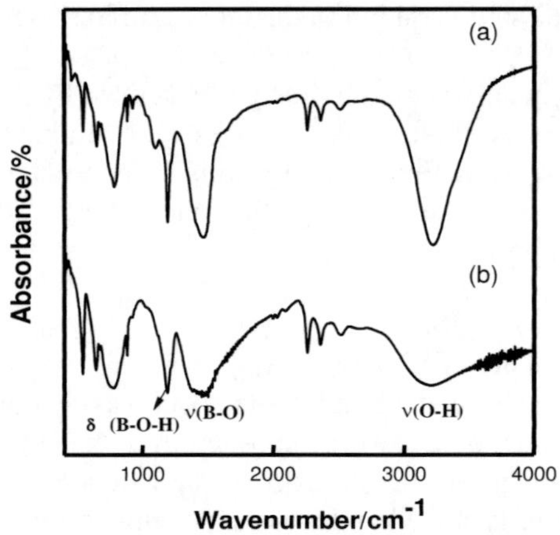

Figure 5. The infrared spectrum of the pyrolysis products at 850 °C: (a) infrared spectrum of pyrolysis products at 850°C; (b) infrared spectrum of pure boric acid.

The first-step pyrolysis product was cylindrical, to facilitate scanning, it was divided into two along the axial direction. Figure 6 presents SEM images of the first-step pyrolysis product. Because boric acid is readily soluble in water, wherein (b) is a micrograph of the same product after washing with hot water for 40 seconds. Time period should not be too long, because the carbon produced will be dispersed into small particles in water. In figure 6(b), one can see that boric acid is scattered over the carbon base. This form is more closely mixed than direct mixing of boric acid and carbon powder in industrial production. Moreover, the product is produced layer by layer by gas phase cracking, and it is expected to achieve a more uniform distribution of boric acid and carbon by the reaction ratio in the transverse and longitudinal directions.

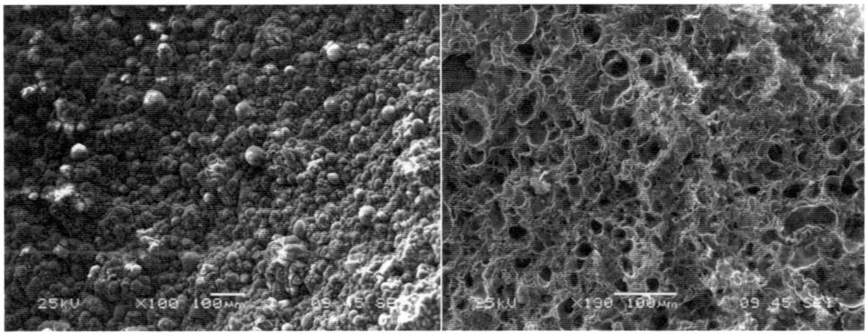

Figure 6. SEM images of 850 °C pyrolysis products:
(a) untreated; (b) treated with hot water.

3.3. Impacts of the second step reaction temperature and time

The experiment used the same batch of raw materials to react at 1400 °C and 1600 °C respectively. The flow rate of Ar was controlled at 1 ml/min. During the process, the pressure changes at different reaction temperatures were recorded with a precision digital pressure gauge. It can be seen from figure 7 that the higher the reaction temperature, the faster the pressure changes, and it can quickly return to the background pressure. During the reaction, the heat and total reaction time have a leading role in the formation of products morphology (including uniformity). The higher the reaction temperature, the faster the reaction rate. However, at high temperature, the longer the reaction time, the more severe the agglomeration of the products. Therefore, the experiment accurately controlled the reaction time by pressure changes. It is apparent from figure 7 that the reaction was completed when the reaction temperature of 1600°C was maintained for 20 minutes.

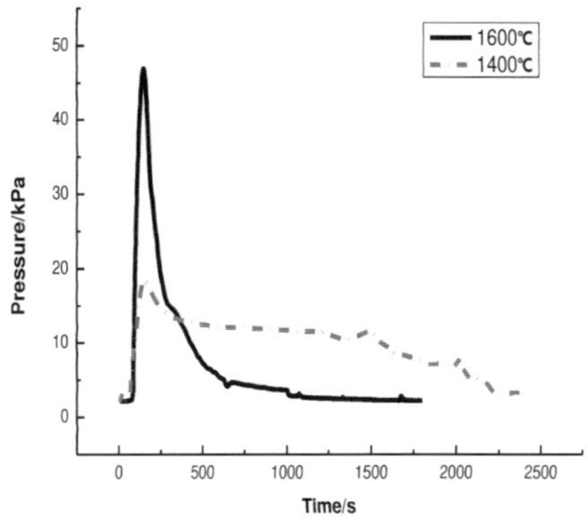

Figure 7. Pressure changes at different reaction temperatures and reaction times.

The experiment was carried out at 1600°C for 20 min, and products were analyzed by XRD. Figure 8 shows that the products coincide with the position of the diffraction peak of the boron carbide standard card PDF #35-0798, indicating that the products are boron carbide.

IOP Conf. Series: Materials Science and Engineering **474** (2019) 012052 doi:10.1088/1757-899X/474/1/012052

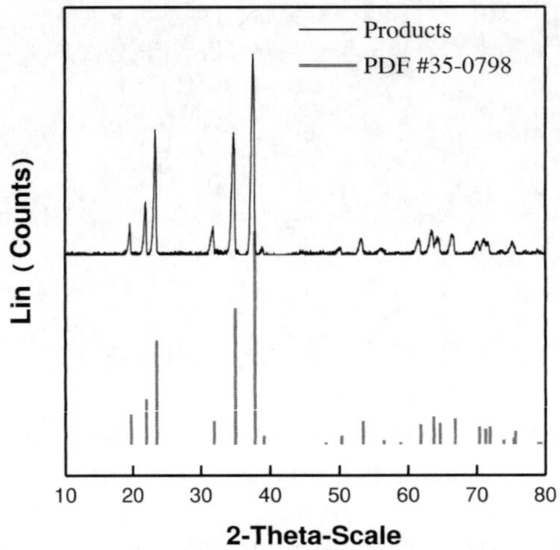

Figure 8. XRD of the products obtained by controlling the temperature heating/cooling rates at 400 °C·min^{-1}.

The particle size distribution of boron carbide powders formed in this study is shown in figure 9. As seen in this plot, boron carbide particles exhibited a bimodal size distribution with 91 % of the particles (D90) in the size range of 0.15-1.00 μm. About 60.87% of powders had particles with a size range from 455 to 877 nm.

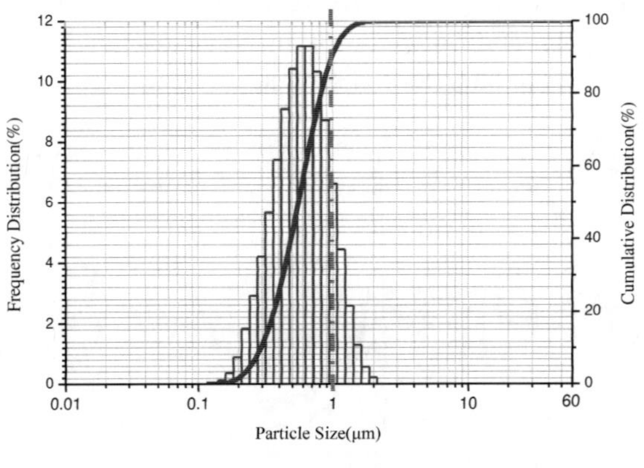

Figure 9. Laser particle size analysis of products obtained by controlling the heating/cooling rates at 400°C·min^{-1}.

4. Conclusions

In this paper, boron carbide was prepared by using trimethyl borate as a pyrolysis feedstock. Since the raw material itself contained the B-O-C bond, it was no longer necessary to prepare it separately. In the experiment, the pyrolysis products of trimethyl borate at 850 °C were mainly boron source and carbon. The reaction temperature and total reaction time in the subsequent carbothermal reduction process were found to have a dominant effect on the formation of product morphology (including the uniformity). The higher the reaction temperature, the faster the reaction rate. The process is dominated by the nucleation rate, whereas boron carbide particles of uniform size are readily formed. And at the same reaction temperature, the longer the reaction time, the more likely it is to cause the agglomeration of product particles. In this paper, a uniform submicron-sized boron carbide was obtained at 1600°C at high heating/cooling rate of 400 °C/min rapid, and the proportion of the product size below 1 μm reached 91%.

IOP Conf. Series: Materials Science and Engineering **474** (2019) 012052 doi:10.1088/1757-899X/474/1/012052

References

[1] Zheng X J 2015 *The Nuclear Industry with Boride* (Beijing: Chemical Industry Press) pp 109-19.
[2] Su X P, Du A B and Han X H 2015 *Shandong Ceram.* **38**(4) 3-5.
[3] Zheng X J 2014 *Boron Carbon and Boron Compounds* (Beijing: Chemical Industry Press) p 17.
[4] Rao M P L N, Gupta G S and Manjunath P 2009 *Thermochim. Acta.* **482**(1/2) 66-71.
[5] Rao M P L N, Gupta G S and Manjunath P 2009 *Int. J. Refract. Met. Hard Mater.* **27**(3) 621-8.
[6] Cao Z W 2006 *Mech. Eng.* **6**:57.
[7] Jia B R, Qin M L and Li H 2010 *Mater. Rev.* **24**(5) 32-8.
[8] Farzaneh F, Golestanifard F and Sheikhaleslami M S 2015 *Ceram. Int.* **41**(10) 13658-62.
[9] Singh P, Singh B and Kumar M 2014 *Ceram. Int.* **40**(9) 15331-4.
[10] Anselmi T U, Munir Z A and Kodera Y 2005 *J. Am. Ceram. Soc.* **88**(6) 1382-7.
[11] Sonber J K, Murthy T S R C and Subramanian C 2013 *Trans. Indian Ceram. Soc.* **72**(2) 100-7.
[12] Fu B, Li S R and Wang Y L 1997 *At. Energy Sci. Technol.* **31**(5) 400-7.
[13] Sinha A, Mahata T and Sharma B P 2002 *J. Nucl. Mater.* **301** 165-9.
[14] Kakiage M, Tahara N and Yanase I 2011 *Mater. Lett.* **65**(12) 1839-41.
[15] Ling S 2010 *Low-Temperature Synthesis of Precursor Cracking Route for Boron Carbide.* (Harbin： Harbin Engineering University).
[16] Berjonneau J, Chollon G and Langlais F 2006 *Proc. Electrochem. Soc.* **153**(12) C795-C800.
[17] Karaman M 2010 *AIChE J.* **52**(12) 4161-6.
[18] Karaman M 2010 *AIChE J.* **8**(3) 2914-9.
[19] Pilladi T R, Ananthansivan K and Anthonysamy S 2013 *Powder Technol.* **246** 247-51.
[20] Makishima S, Yoneda Y and Tajima T 1957 *J. Phys. Chem.* **61**(12) 1618-9.
[21] Zheng X J 2010 *New Boron-Containing Materials* (Beijing: Chemical Industry Press) p 171.
[22] Parsons J L and Milberg M E 1960 *J. Am. Ceram. Soc.* **43** 326-30.

IOP Conf. Series: Materials Science and Engineering **474** (2019) 012053 doi:10.1088/1757-899X/474/1/012053

Study on fracture behavior of NiTi Alloy by the digital image correlation method

L Y Ma, Y J Li* and G Han

School of Civil Engineering and Architecture, University of Jinan, Jinan 250022, China

*E-mail: cea_liyj@ujn.edu.cn

Abstract. Due to its unique super-elasticity and shape memory effect, NiTi alloy has been widely used in aerospace, civil engineering, and other fields. As it is well known, the defect of the material itself will affect its mechanical properties and service life, especially the fracture properties. To study the influence of the defect on the fracture properties of NiTi alloy, in this paper, the relationships between the stress intensity factor (SIF), stress values, and the number of fitting terms of two different NiTi alloys are investigated within the framework of fracture mechanics approach. The results obtained strongly indicate that SIF values of two specimens with different defects gradually increase with the stress; when the number of fitting terms reaches 8, the error is stable within 2%. Moreover, the SIF in both specimens tends to be stable with the increase in the number of fitting terms.

1. Introduction

Due to the needs of structural design, material processing, and other factors, the structure always contains a specific defect. The presence of defects will affect the mechanical properties of the material, and the change of the mechanical properties of the material varies with the geometrical shape of the defect. Therefore, it is necessary to study the fracture mechanics of defective components. In fracture mechanics, the stress intensity factor is an important parameter to characterize the fracture properties of materials. Dally [1] used the resistance strain gauge method to test the stress intensity factor (SIF) theory of aluminum plates. Xi [2] used the photoelastic method to study the stress intensity factor of the I-type crack tip of polyester photoelastic materials. At present, the digital image correlation (DIC) method is most widely in the SIF measurement.

NiTi alloy [3] has a unique and excellent hyperelastic and shape memory effect, and its stress-strain curve [4] has visible hysteresis characteristics (caused by the internal phase transition process of the material), does not cause damage to the material, and the NiTi alloy also has excellent corrosion resistance [5] and fatigue resistance [6]. Due to its unique effect, NiTi alloy has been widely used in various fields, such as aerospace [7], civil engineering [8], and other fields. The defects in the material and scratches formed during its processing are the factors that cause the SIF variation. Therefore, it is of theoretical and engineering significance to study the SIF of NiTi alloy defect specimen for structural design and safety reliability evaluation. Given this, the tensile tests of NiTi alloys with different defects have been carried out, and the variation trend of SIF with the change of stress and the number of fitting terms has been obtained. To get more accurate measurement data, this paper optimizes the DIC carrier, speckle, and obtains the production parameters of higher quality speckle.

2. Extraction of SIF based on DIC

2.1. Principle of DIC

It is the basic principle of DIC to compare the gray feature of speckle image before and after deformation. The tested surface needs to have enough recognizable mark points, generally by spraying black and white paint spots on the specimen surface as a speckle mark point, and assume that the deformation of these speckles is the deformation of the specimen itself.

According to the study [9], the displacement expressions of the area to be tested are as follows:

$$u_j = \frac{1}{2u}\sum_{n=1}^{\infty} a_n r_j^{\frac{n}{2}}\left[\left(k + \frac{n}{2} + (-1)^n\right)\cos\frac{n}{2}\theta_j - \frac{n}{2}\cos(\frac{n}{2} - 2)\theta_j\right] + u^* - \alpha r_j sin\theta_j \quad (1)$$

$$v_j = \frac{1}{2u}\sum_{n=1}^{\infty} a_n r_j^{\frac{n}{2}}\left[\left(k - \frac{n}{2} - (-1)^n\right)\sin\frac{n}{2}\theta_j + \frac{n}{2}\sin(\frac{n}{2} - 2)\theta_j\right] + v^* + \alpha r_j cos\theta_j \quad (2)$$

where r and θ are the polar coordinates of any point of the area to be tested, K is the elastic constant; the superscript j represents the j_{th} data point of the n data points, the coordinate is (x_j, y_j); $u^*$and v^* are the rigid body displacement along x-direction and y respectively, α is the in-plane angle.

Expression of SIF obtained from the fracture mechanics is as follows:

$$K_I = \sqrt{2\pi}a_1 \quad (3)$$

Based on formulas (1) and (2), a_1 and other coefficients can be fitted by the displacement obtained by DIC, while SIF (K_I) is calculated by formula (3).

According to works of literature, DIC has the following advantages over other methods of measuring defects:

1) Compared with the resistance strain gauge method [1], it has the advantage of full-field measurement, and sufficient deformation information can be obtained;

2) Compared with the photoelastic method [10], the experimental equipment is simple, and the environmental factors have little influence. It is considered to develop outdoor applications of this measurement method;

3) Compared with the moiré interferometry [11], the equipment is simple, the required magnification is small, and the requirements for light are low;

4) High-speed cameras can be used for dynamic measurements [12] to compensate for the shortcomings of static measurements.

2.2. Speckle optimization

Speckle is the crucial carrier of DIC deformation information. Making high quality and easy to transfer speckle is one of the critical steps to improve the accuracy of DIC measurement. There are mainly two kinds of speckle: natural randomly distributed grayscale [13] and artificially produced spot [14]. In scientific studies, artificial speckles are mostly used, and carbon powder together with paint is mainly used as speckles. However, the average size of the particles is about 5μm, which is too large and easy to be distorted under the depth of field microscope. Therefore, the graphite powder with an average size of about 1.5μm is selected to optimize the fabrication of speckles. In this study, Huaixi Wang's [15] research was used to streamline the production process of speckle.

The speckle production process [16] is shown in figure 1:

1) Determining the ratio of graphite powder to epoxy curing agent, weigh graphite powder, and prepare epoxy curing agent;

2) Mixing the curing agent with the graphite powder, and centrifuging with a centrifuge to obtain a mixture solution;

3) Pouring the solution onto the acrylic plate in the silicone machine, and apply the glue;

4) Stewing release acrylic plate for 24h, curing;

5) Observing the resulting speckle under an optical microscope.

The quality of speckle is influenced by the concentration of epoxy resin and graphite powder, radial velocity and solution curing time, the speed of glue-throwing and the time of glue-throwing. In this

IOP Conf. Series: Materials Science and Engineering **474** (2019) 012053 doi:10.1088/1757-899X/474/1/012053

paper, the above factors are optimized to obtain the high-quality speckle, as shown in figure 2, the production parameters, in turn, are 3ml: 0.1g, 2000r/min, 10min, 3500r/min, 50s.

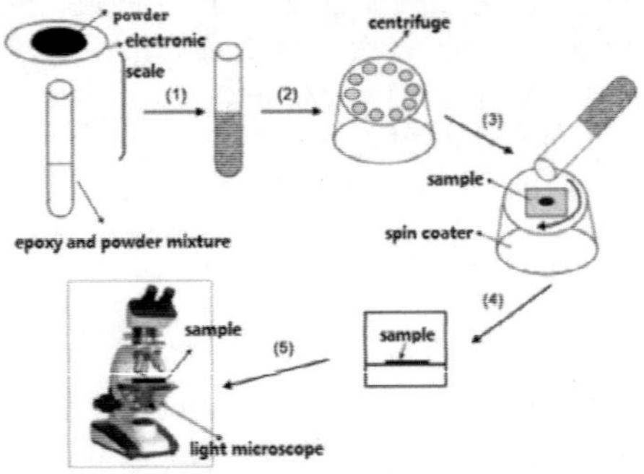

Figure 1. Flowchart of the speckle.

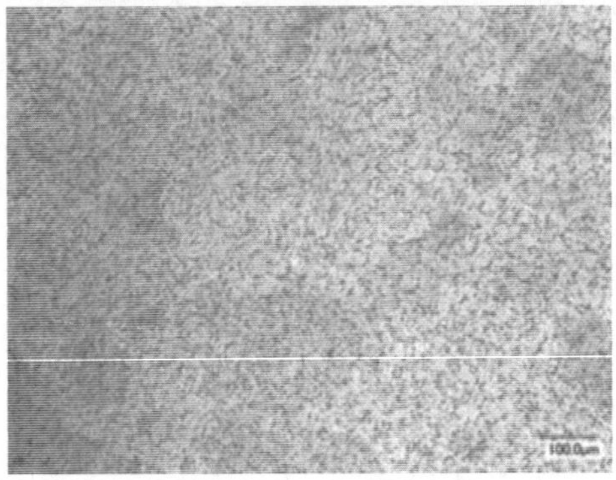

Figure 2. Speckle patterns.

3. Experimental

3.1. Experimental conditions

In this study, two specimens (figure 3) with groove and curved defects were produced, each having a thickness of 1 mm. The specimens with defects were subjected to pre-treatment such as grinding, polishing, etching, etc., and then underwent the tensile tests. The polishing temperature was below 15°C.

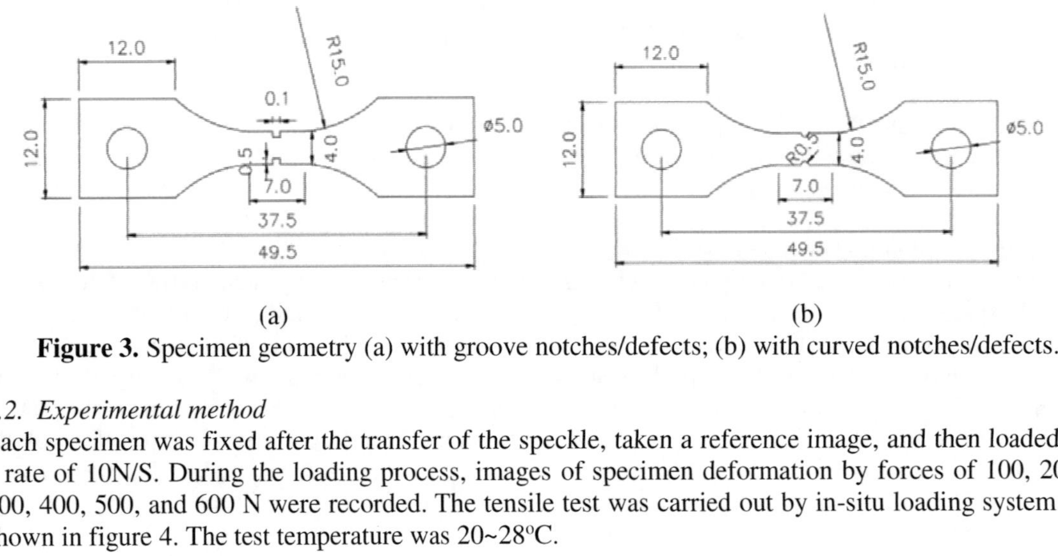

(a) (b)

Figure 3. Specimen geometry (a) with groove notches/defects; (b) with curved notches/defects.

3.2. Experimental method

Each specimen was fixed after the transfer of the speckle, taken a reference image, and then loaded at a rate of 10N/S. During the loading process, images of specimen deformation by forces of 100, 200, 300, 400, 500, and 600 N were recorded. The tensile test was carried out by in-situ loading system as shown in figure 4. The test temperature was 20~28°C.

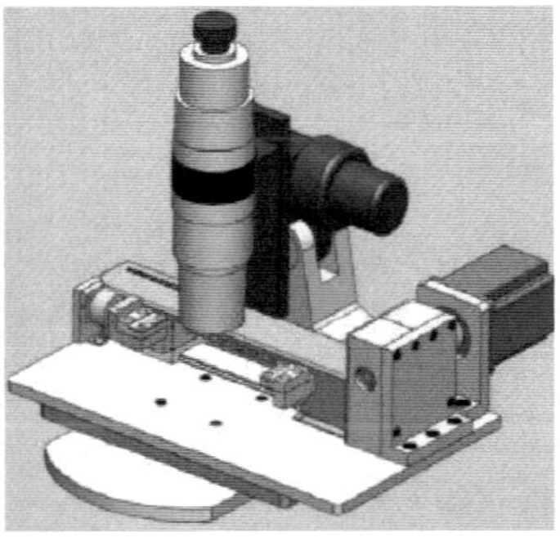

Figure 4. In-situ loading system.

4. Analysis and discussion of experimental results

4.1. Experimental results

The deformed images were processed by DIC to obtain the displacement fields under different stresses. Figure 5 illustrates the displacement fields of two defected specimens at the stress value of 25MPa.

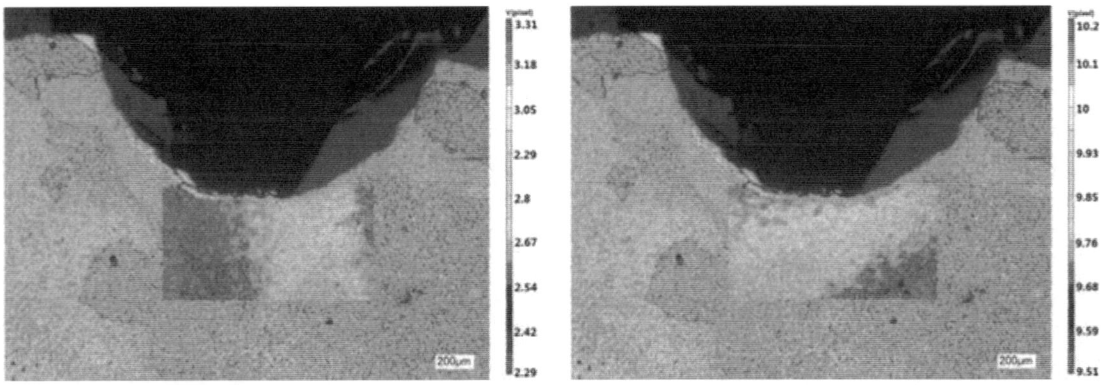

U field V field

(a) Curve-defect displacement field.

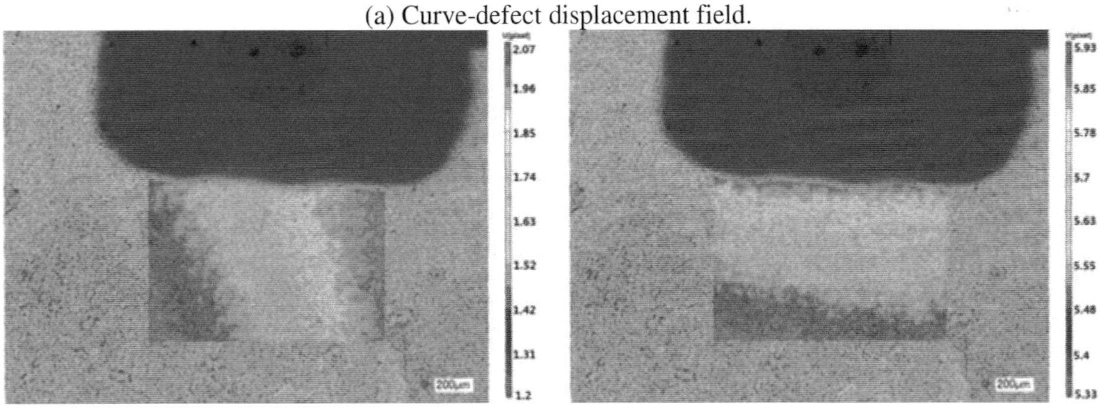

U field V field

(b) Groove-defect displacement field.

Figure 5. The displacement fields of two defected specimens at the stress of 25MPa.

4.2. SIF variation laws

The calculated results are shown in figure 6, where the specimen with curved defects is denoted as specimen 1 and that with grooved defects as specimen 2.

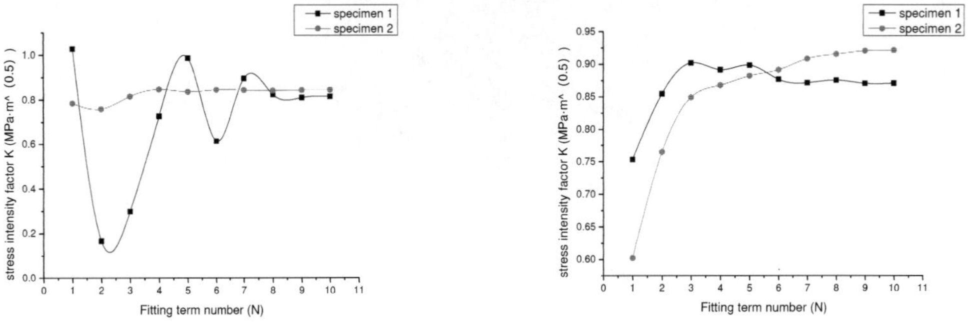

25MPa stress 150MPa stress

(a) SIF values versus the number of fitting terms.

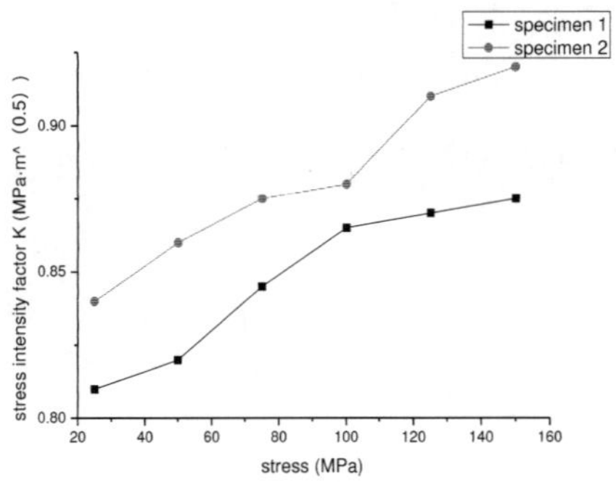

(b) SIF values versus stress.

Figure 6. SIF variation in two specimens.

Table 1. SIF value versus the number of fitting terms at the stress level of 25MPa.

Fitting term number	1	2	3	4	5	8	10	Stable value
Specimen 1	1.028	0.1664	0.299	0.7262	0.9864	0.8232	0.8149	0.81
Specimen 2	0.7843	0.7588	0.8153	0.8466	0.8363	0.8411	0.8441	0.84
Error 1(%)	26.91	79.46	63.09	10.35	21.78	1.63	0.61	/
Error 2(%)	6.63	9.67	2.94	0.79	0.44	0.13	0.49	/

Table 2. SIF value versus the number of fitting terms at the stress level of 150MPa.

Fitting term number	1	2	3	4	5	8	10	Stable value
Specimen 1	0.7532	0.8546	0.9021	0.8916	0.8987	0.8755	0.871	0.875
Specimen 2	0.6023	0.7649	0.8491	0.8679	0.8823	0.9155	0.9216	0.92
Error 1(%)	13.92	2.33	3.10	1.90	2.71	0.06	0.46	/
Error 2(%)	34.53	16.86	7.71	5.66	4.10	0.49	0.17	/

Table 3. SIF values versus stress level

Stress (MPa)	25	50	75	100	125	150
Specimen 1	0.81	0.82	0.845	0.865	0.87	0.875
Specimen 2	0.84	0.86	0.875	0.88	0.91	0.92

4.3. Analysis of SIF variation law

According to the study [17], the NiTi alloy was tested at 28 °C, and the hyperelastic performance was not apparent before the stress reached 300 MPa. The maximum stress in the experiment was 150 MPa, and the material was still in the elastic range, meeting the underlying assumptions of elastic mechanics.

It is found from figure 6(a) that as the number of fitting terms increases, the fluctuation of SIF value is getting smaller, and tends to be stable after the number of fitting terms reaches 8. As shown in Figure 6(b), the SIF value of the two kinds of defective specimens increases with the increase of the stress, while the SIF value of the groove-shaped specimens is larger.

4.4. Discrete analysis of SIF value

Although the general trend of SIF values is easily identified, the discrete points cannot be ignored.

1) When the stress is 25 MPa, the data input is provided from table 1.

When a single a point is selected from the curved-defect specimen is selected to calculate the SIF value, the error is 26.91%. When two items are selected for the fitting calculation, the error reaches 79.46%. As the number of fitting terms increases, the SIF value gradually approaches a fixed value. When the number of fitting terms reaches 8, the error is controlled within 2%. The variation of the SIF value fitting of the groove-defect specimen is lower during the process of fitting, and is stable within 10%.

2) When the stress is 150 MPa, the data input is provided from table 2.

When the number of fitting terms is less than 5, the SIF value of the curved-defect specimen is larger than that of the groove-defect one, and as the number of fitting items increases, the opposite trend is gradually observed. Finally, the SIF values in both specimens tend to be stable. The error is controlled within 1%.

3) As shown in table 3 and figure 6(b), the SIF value of the two kinds of defective specimens increases with the stress, while the SIF value of the groove-shaped specimens is larger. The process of increasing the value of K has a certain undulation, and when the stress is 100 MPa, the crack growth rate in the groove-defect specimen is slightly reduced, then it continues to increase, while that in the curved-defect one slightly rises first, and then slowly grows.

5. Conclusion

In this paper, the mechanical tests and microstructure analysis of two kinds of defective specimens are carried out. From the results of this study, the following conclusions can be drawn:

- ✓ The calculation of the SIF of NiTi alloy material is controlled by the number of fitting terms. As the number of fitting terms increases, the accuracy of SIF measurements is improved.
- ✓ When assessing the SIF value, the selection of the number of calculation points will affect the result.
- ✓ As the stress increases, the SIF values will rise, and the SIF variation trend in specimens with different defects will become inconsistent. Finally, the SIF values in both specimens tend to be stable and the error is controlled within 1%.

Acknowledgment

The authors are grateful for the financial support of the Project of Shandong Province Higher Educational Science and Technology Program (Grant No. J16LJ12)

References

[1] Dally J W and Sanford R J, 1987. *K I. Exp Mech.*, **27**(4): p. 381-8.
[2] Zhao X, et al., 2017. *Chinese Journal of Engineering*, **39**: 1288-94.
[3] Thierry B, et al., 2000. *J. Biomed. Mater. Res.*, **51**(4): p. 685-93.
[4] Machado G, et al., 2015. *Mater. Des.*, **65**: p. 212-20.
[5] Jamesh M I, et al., 2015. *Corros Sci.*, **97**: p. 126-38.
[6] Pirani C, et al., 2016. *Int Endod J.*, **49**(5): p. 483-93.
[7] Li J F, et al., 2014. *Materials Review*, **28**: 104-8.
[8] Zhuang P, et al., 2015. *Journal of Architecture and Civil Engineering*, **32**: 96-103.
[9] Hao W F, et al., 2015. *Journal of Aeronautical Materials*, **35**: 90-5.
[10] Wang W C and Hwang C H, 1998. *Int J Fracture*, **91**(4): p. 311-21.
[11] Zhang C Y, et al., 2016. *Applied Laser*, **4**: 434-9.
[12] Fan Y F, et al., 2015. *Journal of Experimental Mechanics*, **30**(5): p. 590-8.
[13] Zhao P L, et al., 2017. *China Measurement & Test*, **43**(9): 8-12.
[14] Yu Q, et al., 2017. *Hot Working Technology*, **3**: 1-5.
[15] Wang H, et al., 2012. *Mea Sci Technol.*, **23**(3): p. 035402.
[16] He G L, 2016. *Research on Microscale Digital Image Correlation Methods and Techniques*, Jiangsu University.
[17] Wang X M, et al., 2009. *Superelastic Mechanical Properties and Applications of NiTi Alloys*, Science Press.

Study on innovative modified silicate composite grounding electrode

R Liu[1,*], Z G Fang[1], X H Wang[1] and Z Zhang[2]

[1]China Electric Power Research Institute, Beijing, 100192, China
[2]State Grid AC Engineering Construction Company, Beijing,100052, China

*Email: liurui@epri.sgcc.com.cn

Abstract: In view of the current corrosion of metal grounding materials and the low degree of soil adhesion, this paper proposes a modified silicate composite grounding body, which is coated with metal materials by modified silicate materials mainly adding carbon fibers and made by high pressure extrusion, the modified silicate material was analyzed for its micro-morphology and characteristics, and further introduces its structural design and calculation method of ground resistance. Given the typical box-ray grounding device for 220kV transmission line power and its ground resistance requirements under different soil resistivity, CDEGS is used for simulation calculation to get the number of grounding electrodes and ground resistance that meet the standard requirements. It was found that the modified silicate composite grounding is effective in grounding system for lightning protection.

1. Introduction

A lightning strike is one of the leading causes for tripping of transmission lines. According to the statistics, in 2016, about 52% of the tripping of transmission lines at 500kV and above was caused by lightning strikes, of which 91.7% was generated by shielding failure and 8.3% by inverse flashover. For many years, researchers and technical personnel have been dedicated to studies on lightning protection for transmission lines and have achieved many relevant results, such as the development of a new generation of lightning positioning system, the introduction of differential lightning protection, and a series of measures including reducing shielding angle of grounding wires, reducing the ground resistance of the tower, enhancing the insulation of transmission lines, using anti-shielding failure lightning rods, and installing lightning arresters on the line. These results are of practical significance to improving the lightning protection of transmission lines. The domestic and foreign operating experience and theoretical analysis show that reducing the ground resistance of the tower is the most effective measure to protect transmission lines against a lightning strike. Therefore, reliable and stable grounding is necessary for the safe and stable operation of transmission lines.

The early power system design techniques used in China learned from the experience of the Soviet Union. Zinc-galvanized flat steel and round steel were the common grounding materials. These materials get corroded easily, especially in soils with acidity, alkalinity, high salinity, and high moisture content. It usually takes 5-7 years to renovate or even replace the tower grounding grid [1]. The grounding materials buried underground increase the workload and cost of transmission line operation and maintenance. To prevent corrosion, copper and copper-plated steel are often used as grounding materials in foreign countries. Due to the high cost of copper materials, domestic researchers once introduced grounding materials such as copper-clad steel, zinc-clad steel, stainless steel, alloy cladding steel, and high-silicon chromium iron [2]. The use of copper-clad steel follows the different production process; zinc-clad steel is developed from the corrosion protection method of sacrificial anode. A thick zinc layer is coated on the surface of the steel in an extrusion cladding process to overcome the disadvantage of a thin coating of hot-dip galvanized steel and thus prevent corrosion. This material is not commonly used in practical engineering; stainless steel has excellent corrosion resistance in the atmosphere, but it is rarely used in grounding projects; the coating of alloy cladding steel is made from various alloy materials, mainly represented by aluminum-magnesium alloy. It is made in a thermal spraying or a hot dip galvanizing process. Now the application of this technology stays in the research stage.

At present, the study on non-metallic grounding materials provides new ideas about anti-corrosion and resistance reduction of grounding materials. Non-metallic grounding materials, predominantly graphite-based and carbon fiber-based materials, are widely recognized by researchers because of their excellent conductivity, chemical stability, resistance to acid and alkali, and excellent oxidation resistance. Hu and Qiu [3] used highly pure crystalline flake graphite as the primary conductive material to make a flexible graphite composite grounding material with a Litz-layered structure. They tested its performance and considered that the flexible graphite composite grounding material has a stable structure, low skin effect and inductance effect, and a high degree of compatibility with soil. These features facilitate transportation and construction and guarantee a wide range of applications. Shao and Le [4] introduced the preparation and performance of JR nanocarbon anti-corrosion conductive coating and its use in grounding grid anti-corrosion. They argued that this coating could meet the ground resistance and thermal stability requirements of grounding grid while adequately protecting the grounding grid from being corroded in the damp soil environment. Zhou et al. [5] carried out the transient response characteristics and frequency-dependent grounding impedance analysis of the carbon grounding module. The results show that the carbon-based module can be effectively applied to the lightning protection grounding system and verify the correctness of the simulation technology. The applicability of the distributed parameter circuit model to the ground electrode.

This paper presents a modified silicate composite grounding electrode and discusses its composition and electrical conductivity principle. It also analyzes its microscopic morphology and characteristics and introduces its structural design and calculation method of ground resistance. The findings of this study can serve as reference and guidance for the application of composite grounding materials of modified silicate in engineering grounding.

2. The composition of modified silicate composite grounding electrode

The modified silicate composite grounding electrode is mainly composed of modified silicate and conductive core. The modified silicate (commonly known as electrical conductive concrete) is made from Portland cement, bentonite, fly ash, silicon powder, polyacrylonitrile-based (PAN-based) carbon fiber, stainless steel fiber, and moisturizer mixed in a certain proportion. The conductive part mainly comes from PAN-based carbon fiber uniformly mixed in the mortar.

Bentonite, known as "universal soil", is a natural clay containing montmorillonite (which takes up 85-90%). It is non-corrosive and stable and has a resistivity of 2.5Ω•m when the moisture content reaches 300%. The concrete that is mixed with soil-like materials such as bentonite has better crack resistance, impermeability and cohesiveness to maintain close contact with the conductive core and the ground layer, thereby eliminating the contact resistance between the grounding electrode and the soil. Thus it has good compatibility with soil. Fly ash can improve the water impermeability, gas impermeability, sulfate resistance and chemical corrosion resistance of concrete, reduce the heat of hydration, improve the high temperature resistance of concrete, reduce particle separation and precipitation, reduce shrinkage and cracking of concrete, suppress stray current corrosion of the metal part of the concrete. Silica powder, whose scientific name is "silica fume", can reduce $Ca(OH)_2$ content, increase concrete compactness, and effectively improve the weak acid corrosion ability. Besides, silica powder concrete is denser, and its pore structure is refined to lower the transmission rate of harmful ion and reduce the formation of soluble $Ca(OH)_2$ and ettringite, making it also resistant to salt corrosion, especially to chloride salt and sulfate. In particular, in harsh environments such as with chloride salt

IOP Conf. Series: Materials Science and Engineering **474** (2019) 012054 doi:10.1088/1757-899X/474/1/012054

pollution, sulfate erosion and high humidity, the durability of concrete can be doubled or increased by several times. Adding an appropriate amount of stainless steel fiber in concrete can significantly improve the tensile, fracture, bending, shear, and splitting resistance and other mechanical properties of the concrete.

Carbon fiber has a variety of carbon properties, the fiber characteristics, excellent conductivity and wear resistance [6]. Compared with graphite, carbon fiber has the following features: conductive fiber is more likely to form a conductive network than a conductive particle. The finer the fiber, the smaller the diameter. With the same volume content, more fibers will have a higher total length, and they are more easily overlapped to form a network of interconnected conductive fibers; the longer the fibers, the easier the overlapping.

Compared with other non-metallic grounding materials, the modified silicate composite grounding material is composed of electrical conductive concrete and stainless steel core which has excellent conductivity. This material has a stable structure and thus shows clear advantages in soil compatibility, mechanical strength and stability. Its application in tower grounding is of practical significance.

3. Microscopic morphology of modified silicate composite grounding electrode

Carbon fiber is a micro-crystalline graphite material produced after carbonization and graphitization of organic fibers. The microscopic morphology of carbon fibers extends along the axial direction of the fibers, as shown in Figure 1. Carbon fiber has good self-lubrication and electrical conductivity, and reinforced concrete mixed with carbon fibers has high compressive strength and tensile strength. Due to its properties, carbon fiber can significantly increase the strength of concrete. Because of its excellent resistance to air oxidation, it has been widely studied in recent years. In general, the higher the carbon fiber content, the better the electrical conductivity of the concrete. With the same carbon fiber content, the better the dispersion of the carbon fiber is, the easier it is to form a complete conductive network. The resistivity of carbon fiber concrete is affected by temperature. The higher the temperature is, the lower the concrete resistivity is, and vice versa.

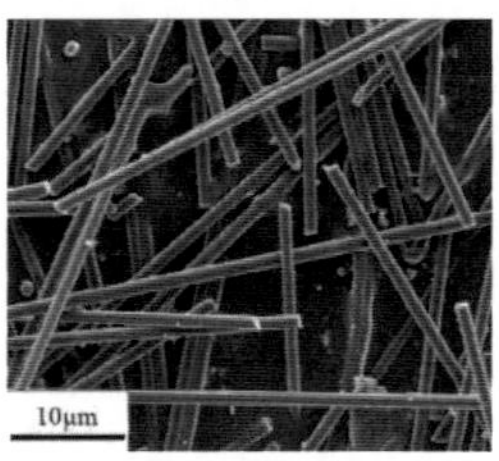

Figure 1. SEM image of carbon fibers

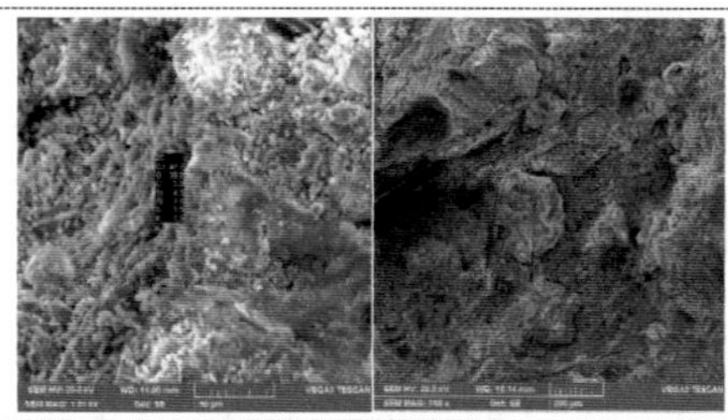

Figure 2. Microscopic morphology of modified silicate concrete after incorporation of carbon fibers

Figure 2 shows that the modified silicate hydrates are tightly bonded with few gaps. After incorporation of carbon fibers, the carbon fibers are evenly distributed, and tight bonding between the carbon fibers and the modified silicate hydration products can be observed clearly. The hydration products can fill into the gaps between the carbon fibers to form a dense microstructure, thereby improving the mechanical properties of the modified silicate hydration products. With higher carbon fiber content, as the external load to destroy the bonding force between the modified silicate hydration products and carbon fibers and to break the fibers increases, the compressive strength of the modified silicate concrete and its cohesion with the inner stainless steel core material will be enhanced.

IOP Conf. Series: Materials Science and Engineering **474** (2019) 012054 doi:10.1088/1757-899X/474/1/012054

4. Grounding characteristics of modified silicate composite grounding electrode

The performance of the modified silicate composite grounding electrode should meet the requirements in GB/T 21698-2008 "Technical specifications of composite grounding device" and DL/T 380-2010 "Technical condition of material for reduced ground resistance".

Grounding electrode test should be carried out under standard test atmospheric conditions [7]: temperature of 10-35°C, relative humidity of 50-80%, and air pressure of 86-106kPa. Three samples were used in each set of tests for the comparison purposes.

(1) Resistance test

Three modified silicate composite grounding electrodes were used, and both sides (outer sides) of each electrode were coated with silver paste in a concentric area of 15mm. The resistance of the grounding rods measured after 1 hour was 6.4-8.5Ω. The resistance test wiring is shown in figure 3.

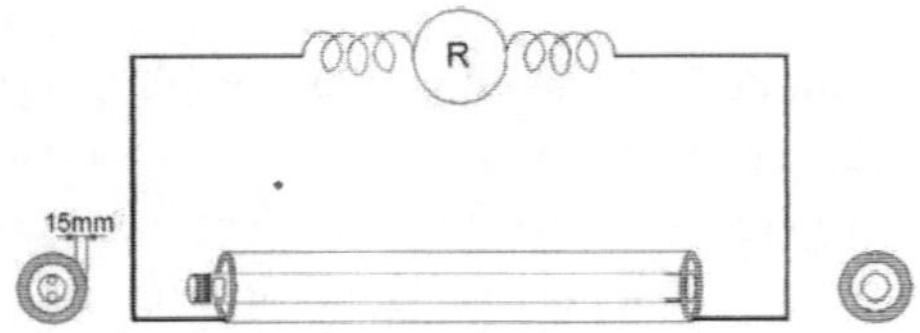

Figure 3. Schematic diagram of resistance test circuit (grounding electrode)

Three samples of modified silicate material were tested. They were coated with silver paste on the entire left and right sides (40mm x 40mm). The resistance of the ground rods measured after 1h was 6.2-7.5Ω. The resistance test wiring was shown in figure 4.

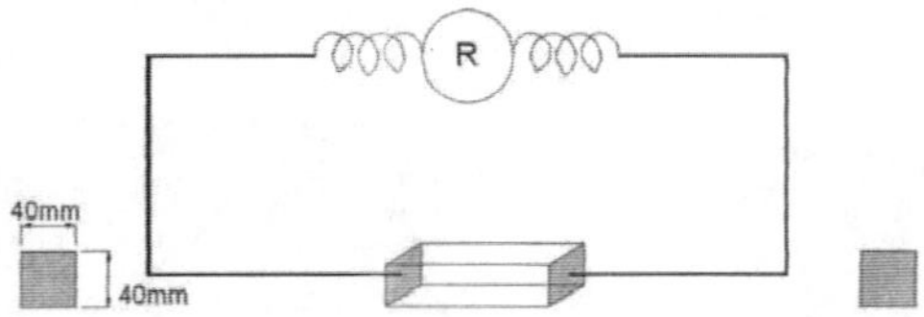

Figure 4. Schematic diagram of resistance test circuit (materials)

(2) Test of resistance-reducing coefficient

Three modified silicate composite grounding electrodes and three metal rods were tested. An open space with soil resistivity of 500Ω·m or more was selected. The metal grounding electrodes with the same size as the grounding electrodes and the modified silicate composite grounding electrodes were buried in the soil. The three-electrode method was used to measure power frequency ground resistance and resistance reducing coefficient of the grounding electrodes and the metal grounding electrodes which was the power frequency ground resistance ratio of the grounding rods and the metal grounding electrodes. The ratio of the resistance-reducing the coefficient of the grounding rods measured after 720h was 0.8-0.85.

(3) Pulse current withstand test

Three modified silicate composite grounding electrodes were buried in the soil inside a metal barrel. Before the pulse current withstands test the measured power frequency current was 1A, and the determined resistance was U/I, according to the wiring diagram, as shown in figure 5.

IOP Conf. Series: Materials Science and Engineering **474** (2019) 012054 doi:10.1088/1757-899X/474/1/012054

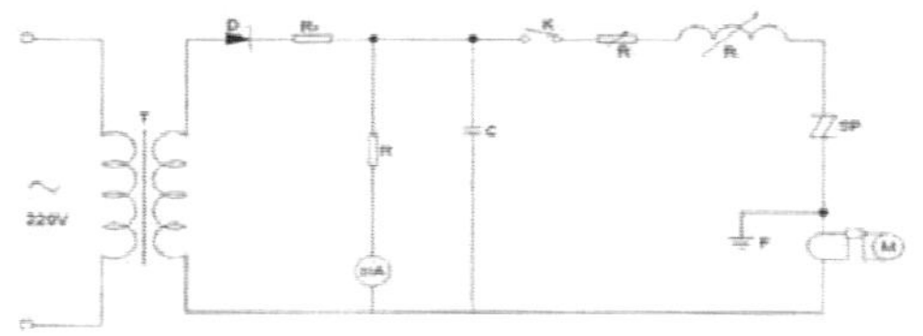

Figure 5. Wiring diagram of pulse current withstand test

Three modified silicate composite grounding electrodes were used and applied with impulse current of 8/20μs and 1kA for 20 times, each with a time interval of 60s. There was a longer interval of 30min between every five times of application. The resistance of samples under power frequency current of 1A was measured again after the test. The resistance of the grounding electrodes varied from -5.4 to 12.9%.

(5) Power frequency current withstand test

Three modified silicate composite grounding electrodes were applied with power frequency current of 10A 5 times, each for 10s. The time interval for every two application was 30min. The wiring diagram of the power frequency current withstand test after the test is shown in figure 6. The resistance-reducing coefficient varied between -34 and -25%.

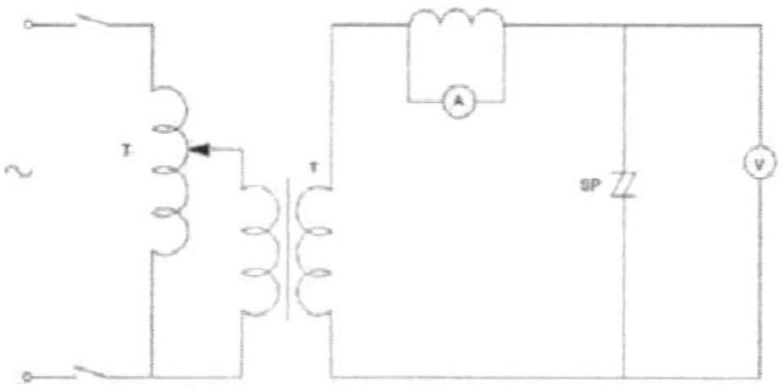

Figure 6. Wiring diagram of power frequency current withstand test

5. Structure and ground resistance calculation of modified silicate composite grounding electrode

The shape of the grounding electrode is a factor that affects the ground resistance reduction of the tower. At present, common grounding module shapes include cylindrical, rectangular, plum-, and REX-shaped. To improve the grounding effect of electrical conductive concrete grounding module, the most common cylindrical and rectangular grounding electrodes were prepared, as shown in figure 7. The cylindrical grounding electrode was buried vertically while the rectangular grounding electrode was buried horizontally. The connecting wire between the grounding electrodes was copper wire-wrapped with an insulating material.

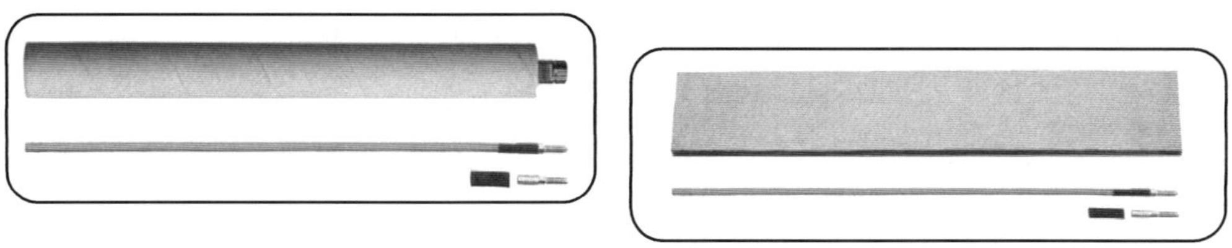

Figure 7. New modified silicate composite grounding electrode

If the ground resistance of any shaped grounding device buried in a single medium was [8]:

$$R_{SM} = F(\rho, S_0, G) \tag{1}$$

$$R_{DM} = F(\rho_C, S_0, G) + F(\rho, S_1, G) - F(\rho_C, S_1, G) \tag{2}$$

where R_{SM} and R_{SM} are resistances of a single and dual dielectric electrodes, respectively (Ω); S_0 is the electrode surface area (m^2); S_1 is the interface area (m^2); G is the geometric coefficient of a given electrode characterized by a special shape.

This equation can be used to analyze the ground resistance of a ground electrode of any shape buried in the soil and wrapped by another material with a different resistivity.

$$R_c = \frac{1}{2\pi L_r}\{\rho_C \ln(D / d) + \rho[\ln(8L_r / D) - 1]\} \tag{3}$$

where ρ and ρ_c are resistivities of soil and concrete, respectively (Ω·m); L_r is the length of the vertical grounding electrode (m); *d* is the grounding conductor diameter (m); *D* is the concrete shell diameter (m).

$$R_R = \frac{\rho}{2\pi L_r}[\ln(8L_r / d) - 1] \cdots\cdots \tag{4}$$

$$R_c = \frac{1}{2\pi L_r}\{(\rho - \rho_c)[\ln(8L_r / D) - 1] + \rho_c[\ln(8L_r / d) - 1]\} \cdots\cdots \tag{5}$$

The above two equations two values of ground resistance: that of the concrete-wrapped ground electrode and that of the concrete vertical grounding electrode with a diameter of *D*, which is buried in the soil.

The ground resistance of the tower grounding device, for which the grounding plate is laid, is calculated similarly to the case of buried strip-shaped horizontal grounding electrode. The ground resistance RC of the horizontal grounding electrode (grounding plate) of length *A* and width *B* can be calculated as follows:

$$R_c = \frac{1}{2\pi A}\{(\rho - \rho_c)\ln(4A / B) + \rho_c \ln(4A / B)\} \tag{6}$$

where ρ and ρ_c are resistivities of soil and the modified silicate material, respectively (Ω·m); *A* and *B* are the length and width, respectively, of the wrapped grounding electrode (m).

Besides, the ground resistance of the concrete wrapped vertical ground electrode can also be calculated by the following equation:

$$R_c = \frac{K_1 K_2 \rho}{2\pi L}\ln\frac{4L}{d} \tag{7}$$

Given the typical box-ray grounding device for 220kV transmission lines in regions with different soil resistivity, rectangular modified silicate composite grounding electrodes with a box side length (foot distance) of 10m were buried along the external extension line at an angle of 45° in the four corners. The rectangular modified silicate composite grounding electrode was 1m-long, 0.2m-wide, and 0.04m-thick. The relative resistivity of the composite grounding electrode (vs. annealed copper) was 45.6, and the relative permeability was 1 (vs. vacuum). The two grounding electrodes had a spacing of 1m and were connected by a special connection line. According to the GB50545-2010 "Design specification for 110-750kv overhead transmission lines," the CDEGS special grounding analysis software was used to simulate the composite grounding electrodes. The required number of grounding electrodes and ground resistance values are listed in table 1.

Table 1. Design of box-ray grounding device for 220kV transmission lines.

Soil resistivity (Ω·m)	Upper limit of ground resistance (Ω)	Number of grounding electrodes (pieces)	Ground resistance (Ω)
100	10	2	7.7
500	15	8	12.2
1000	20	11	18.9
2000	25	18	25
2500	30	19	30

6. Conclusions

This paper introduces a modified silicate composite grounding electrode. It analyzes its microscopic morphology and characteristics and further proposes its structural design and calculation method of ground resistance. Given the typical box-ray grounding device for 220kV transmission line power and its ground resistance requirements under different soil resistivity, CDEGS is used for simulation calculation to get the number of grounding electrodes and ground resistance that meet the standard requirements. The finding of this study will guide the application of composite grounding materials of modified silicate in engineering grounding.

Acknowledgments

This study was funded by the Research Project of the State Grid Corporation of China (GYW17201700241)

References

[1] Xiao W, Hu Y C, et al. 2017 *Transactions of China Electrotechnical Society* **32**(2):85-94.
[2] Zhou P P, Wang S, et al. 2010 *Electric Power Construction* **31**(8):50-54.
[3] Hu Y C, Ruan J J, et al. 2014 *Power System Technology* **38**(1):2851 -2857.
[4] Shao J R, Le Y T, et. al. 2006, *Electric Equipment* **7**(4):70-72.
[5] Zhou Y X, Leng F G, He G X, and Ji X K. 2009, *China Building Materials Science and Technology* **131**(1):42-50.
[6] *Test Report of Grounding Rods Used for Electrical Conductive Concrete*, 2016. Quality Testing Center of Equipment and Instrumentation in the Power Industry.
[7] IEEE Std 80-2013, *IEEE Guide for Safety in AC Substation Grounding*.
[8] Lee K S, Chung Y K, Lee B H. *ICLP*, Shanghai, China

IOP Conf. Series: Materials Science and Engineering **474** (2019) 012055 doi:10.1088/1757-899X/474/1/012055

Improved LayerWise Theory method application to stress analysis for composite tube in pure bending

C Zhang[1,2,*], R Chang[1]

[1]Department of Civil Engineering, Xiamen University Fujian, Xiamen, China
[2]Xiamen Engineering Technology Center for Intelligent Maintenance of Infrastructures, Xiamen, Fujian, China
*E-mail: chzhang@xmu.edu.cn

Abstract. An improved approach using LayerWise theory (LWT) is developed for the stress analysis in pure bending composite tubes. The variables in the circumferential direction are separated from the local displacement, so the eigenvalue techniques are avoided. Then the composite tube is discretized only in the radial direction, and the related displacement in the local displacement is assumed as the simple interpolation polynomial function. Based on the principle of minimum potential energy, the local nodal variables are solved to calculate the strains and stresses for the composite tube. Several composite tubes are analyzed to illustrate the proposed approach feasibility. Besides, the exact solution by unified connected parameter method is employed for comparison. They are in good agreement with each other.

1. Introduction

The composite tubes have proved to be very useful for many structures. As the main load-bearing components, their design requires high-precision stress calculations. However, the classic laminated theory cannot be satisfied due to the large arc curved surface while the numerical simulation with redefined finite element method is quite time-consuming. The methods in elastic theory have been developed to calculate the stressed state of composite tubes in pure bending. Lekhnitskii [1] has provided the solution for the composite tube with only one cylindrically orthotropic layer under bending moment. Jolicoeur and Cardou [2] and Zhang and Hoa [3] generalized this solution to the composite tube with several layers of any winding angles. However, these methods in elastic theory are very complicated. Besides, the particular technique has to be employed to deal with the singular parameters for the layers of some specific winding angles. Hoa et al [4] used the LayerWise theory method (LWT) [5] to calculate the cantilever tube. The global displacement along the axis of the composite tube is theoretical. The composite tube is discretized in the radial direction, and the related variables are assumed as the simple functions. However, the eigenvalue techniques have to be used to obtain the local displacement in the circumferential direction. Thai C H etc. [6] utilized LWT and increase human-made constraints to calculate stress and strain in composite laminates. Xing Y F etc. [7] extended the application of the LWT theory to the calculation of stability problems and dynamic problems of composite laminates. In this paper, a new LWT-based approach is developed to calculate the composite tube of pure bending. The displacement in the circumferential direction is separated from the local displacement. So, the composite tube is discretized only in the radial direction. Based on the principle of minimum potential energy, the local nodal displacements are solved to calculate the strains and stresses for the composite tube.

2. New approach with Layer Wise theory

2.1. Variable separation from the local displacements

It is convenient to consider the composite tube with the cylindrical coordinate system, and the related constitutive equation between strain and stress can be expressed as

$$\boldsymbol{\varepsilon}=\boldsymbol{S\sigma} \tag{1}$$

in which

$$\boldsymbol{\varepsilon}=\begin{Bmatrix}\varepsilon_r\\ \varepsilon_\theta\\ \varepsilon_z\\ \gamma_{z\theta}\\ \gamma_{zr}\\ \gamma_{r\theta}\end{Bmatrix}\quad \boldsymbol{\sigma}=\begin{Bmatrix}\sigma_r\\ \sigma_\theta\\ \sigma_z\\ \tau_{z\theta}\\ \tau_{zr}\\ \tau_{r\theta}\end{Bmatrix}\quad \boldsymbol{S}=\begin{bmatrix}S_{11} & S_{12} & S_{13} & S_{14} & 0 & 0\\ S_{12} & S_{22} & S_{23} & S_{24} & 0 & 0\\ S_{13} & S_{23} & S_{33} & S_{34} & 0 & 0\\ S_{14} & S_{24} & S_{34} & S_{44} & 0 & 0\\ 0 & 0 & 0 & 0 & S_{55} & S_{56}\\ 0 & 0 & 0 & 0 & S_{56} & S_{66}\end{bmatrix} \tag{2}$$

The reduced elastic parameters are also employed as [1]

$$\beta_{ij}=S_{ij}-\frac{S_{i3}S_{j3}}{S_{33}} \tag{3}$$

It is a generalized plane strain problem for the composite tubes of pure bending. Lekhnitskii defined it in[1] as follows:

$$D=S_{13}\sigma_r+S_{23}\sigma_\theta+S_{33}\sigma_z+S_{34}\sigma_{z\theta} \tag{4}$$

where D is a function of r and θ, and used this expression with (1) and (3), to rewrite the six equations of the strain tensor, which were then integrated. In case of moments $M_x \neq 0$ and $M_y = 0$, the following results are obtained for D and respective displacements [2]:

$$D=\kappa_x r\sin\theta \tag{5}$$

and

$$\begin{aligned}u_r&=-\frac{z^2}{2}\kappa_x\sin\theta+\tilde{U}(r,\theta)+u_r'\\ u_\theta&=-\frac{z^2}{2}\kappa_x\cos\theta+\tilde{V}(r,\theta)+u_\theta'\\ w&=\kappa_x zr\sin\theta+\tilde{W}(r,\theta)+w'\end{aligned} \tag{6}$$

In equation (6), $\tilde{U},\tilde{V},\tilde{W}$ are local displacements related to the warping of the cross-section, satisfying [2]

$$\begin{aligned}\frac{\partial\tilde{U}}{\partial r}&=\beta_{11}\sigma_r+\beta_{12}\sigma_\theta+\beta_{14}\tau_{z\theta}+\frac{C_{13}}{C_{33}}D\\ \frac{\partial\tilde{V}}{r\partial\theta}+\frac{\tilde{U}}{r}&=\beta_{12}\sigma_r+\beta_{22}\sigma_\theta+\beta_{24}\tau_{z\theta}+\frac{C_{23}}{C_{33}}D\\ \frac{\partial\tilde{V}}{\partial r}+\frac{\partial\tilde{U}}{r\partial\theta}-\frac{\tilde{V}}{r}&=\beta_{56}\tau_{rz}+\beta_{66}\tau_{r\theta}\\ \frac{\partial\tilde{W}}{\partial r}&=\beta_{55}\tau_{rz}+\beta_{56}\tau_{r\theta}\\ \frac{\partial\tilde{W}}{r\partial\theta}&=\beta_{14}\sigma_r+\beta_{24}\sigma_\theta+\beta_{44}\tau_{z\theta}+\frac{C_{34}}{C_{33}}D\end{aligned} \tag{7}$$

Since we only consider the pure bending of the composite tube, the tensile and torsion stresses are not included in equation (6). Two stress functions are used to express the strain-compatibility equations. According to their non-homogeneous terms, separating variables can be obtained for the two stress functions, and the further stresses can be expressed as the following form:

$$\begin{aligned} &\sigma_r = s_r(r)\sin\theta; \sigma_\theta = s_\theta(r)\sin\theta; \sigma_z = s_z(r)\sin\theta; \\ &\tau_{z\theta} = s_{z\theta}(r)\sin\theta; \tau_{rz} = s_{rz}(r)\cos\theta; \tau_{r\theta} = s_{r\theta}(r)\cos\theta \end{aligned} \tag{8}$$

Using the unified connected parameter method, the stresses in equation (8) can be derived in theory, and the displacements in equation (6) can be obtained finally. However, not only very complicated are the procedures, but also the particular technique should be employed to deal with some singular parameters. The LWT-based approach is a better candidate to calculate the composite tube. In the LWT-based approach, since some variables of the displacement are determined, it is only necessary to consider the other variables in the further discretization. We attempt to separate the variables in the circumferential direction from the local displacement in equation (6) to improve the LWT-based approach. Substituting equation (7) into equation (6) yields [2]:

$$\begin{aligned} &\frac{\partial \tilde{U}}{\partial r} = (\beta_{11}S_r + \beta_{12}S_\theta + \beta_{14}S_{z\theta} + \frac{C_{13}}{C_{33}}\kappa_x r)\sin\theta \\ &\frac{\partial \tilde{V}}{r\partial\theta} + \frac{\tilde{U}}{r} = (\beta_{12}S_r + \beta_{22}S_\theta + \beta_{24}S_{z\theta} + \frac{C_{23}}{C_{33}}\kappa_x r)\sin\theta \\ &\frac{\partial \tilde{V}}{\partial r} + \frac{\partial \tilde{U}}{r\partial\theta} - \frac{\tilde{V}}{r} = (\beta_{56}S_{rz} + \beta_{66}S_{r\theta})\cos\theta \\ &\frac{\partial \tilde{W}}{\partial r} = (\beta_{55}S_{rz} + \beta_{56}S_{r\theta})\cos\theta \\ &\frac{\partial \tilde{W}}{r\partial\theta} = (\beta_{14}S_r + \beta_{24}S_\theta + \beta_{44}S_{z\theta} + \frac{C_{34}}{C_{33}}\kappa_x r)\sin\theta \end{aligned} \tag{9}$$

According to the non-homogeneous terms in equation (9), one can obtain the following

$$\begin{aligned} &\tilde{U} = U(r)\sin\theta \\ &\tilde{V} = V(r)\cos\theta \\ &\tilde{W} = W(r)\cos\theta \end{aligned} \tag{10}$$

So far, the variables of the displacement in the axial direction are determined via equation (6), and those in the circumferential direction are defined via equation (10). Therefore it is necessary to discretize the composite tube only in the radial direction, i.e., versus r value.

2.2. *The numerical solution in the radial direction*

Based on the separating variables for the displacement of pure bending composite tube, the LWT is employed for further analysis. The tube is discretized only in the radial direction into (N-1) elements including N nodes. And the related variables are assumed as the simple interpolation polynomial function as follows:

$$U(r) = \sum_{i=1}^{2} U_i^e \phi_i^e$$
$$V(r) = \sum_{i=1}^{2} V_i^e \phi_i^e \quad (11)$$
$$W(r) = \sum_{i=1}^{2} W_i^e \phi_i^e$$

where $\phi_1^e = \frac{r_2^e - r}{r_2^e - r_1^e}, \phi_2^e = \frac{r - r_1^e}{r_2^e - r_1^e}$ (12)

Upon the substitution of equation (11) into equations (10)and (6), the strains can be calculated as follows

$$\varepsilon_r = e_r(r)\sin\theta; \varepsilon_\theta = e_\theta(r)\sin\theta; \varepsilon_z = e_z(r)\sin\theta$$
$$\gamma_{z\theta} = e_{z\theta}(r)\sin\theta; \gamma_{zr} = e_{zr}(r)\cos\theta; \gamma_{r\theta} = e_{r\theta}(r)\cos\theta \quad (13)$$

where

$$e_r = \sum_{i=1}^{2} U_i^e(\phi_i^e)'; e_\theta = \sum_{i=1}^{2}(\frac{\phi_i^e}{r}U_i^e - \frac{\phi_i^e}{r}V_i^e); e_z = \kappa_x r$$
$$e_{z\theta} = -\sum_{i=1}^{2}\frac{\phi_i^e}{r}W_i^e; e_{zr} = \sum_{i=1}^{2}(\phi_i^e)'W_i^e \quad (14)$$
$$e_{r\theta} = \sum_{i=1}^{2}(\frac{\phi_i^e}{r}U_i^e - \frac{\phi_i^e}{r}V_i^e) + \sum_{i=1}^{2}(\phi_i^e)'V_i^e$$

The relationship between the nodal displacements and strains can be derived from equation (14),

$$\boldsymbol{e} = \boldsymbol{B}_e \boldsymbol{q}_e \quad (15)$$

where

$$\boldsymbol{e} = \begin{Bmatrix} e_r \\ e_\theta \\ e_z \\ e_{z\theta} \\ e_{zr} \\ e_{r\theta} \end{Bmatrix} \quad \boldsymbol{q}_e = \begin{Bmatrix} U_1^e \\ V_1^e \\ W_1^e \\ U_2^e \\ V_2^e \\ W_2^e \\ \kappa_x \end{Bmatrix} \quad \boldsymbol{B}_e = \begin{bmatrix} (\phi_1^e)' & 0 & 0 & (\phi_2^e)' & 0 & 0 & 0 \\ \frac{\phi_1^e}{r} & -\frac{\phi_1^e}{r} & 0 & \frac{\phi_2^e}{r} & -\frac{\phi_2^e}{r} & 0 & 0 \\ 0 & 0 & 0 & 0 & 0 & 0 & r \\ 0 & 0 & -\frac{\phi_1^e}{r} & 0 & 0 & -\frac{\phi_2^e}{r} & 0 \\ 0 & 0 & (\phi_1^e)' & 0 & 0 & (\phi_2^e)' & 0 \\ \frac{\phi_1^e}{r} & (\phi_1^e)' - \frac{\phi_1^e}{r} & 0 & \frac{\phi_2^e}{r} & (\phi_2^e)' - \frac{\phi_2^e}{r} & 0 & 0 \end{bmatrix} \quad (16)$$

By substituting equation (14) in equation (1) and comparing the result with equation (8), we obtain

$$\boldsymbol{s} = \boldsymbol{Ce} = \boldsymbol{CB}_e\boldsymbol{q}_e \quad (17)$$

in which

$$\mathbf{s} = \{s_r \quad s_\theta \quad s_z \quad s_{z\theta} \quad s_{zr} \quad s_{r\theta}\}^T \quad (18)$$

$$\boldsymbol{C} = \boldsymbol{S}^{-1} \quad (19)$$

As is well known, the principle of minimum potential energy can be expressed as:

$$\iiint_{\Omega_0} \delta E d\Omega + \delta P = 0 \quad (20)$$

whereas the strain energy can be calculated as follows

$$\iiint_{\Omega_0} \delta E d\Omega = \pi l \sum_e \int_e \boldsymbol{s}^T \boldsymbol{\delta e}\, rdr \tag{21}$$

and the virtual work can be calculated as follows

$$\delta P = M \frac{\partial}{\partial z}(\delta u_r \sin\theta + \delta u_\theta \cos\theta)\big|_0^l = -Ml\delta\kappa_x \tag{22}$$

Substitution of equations (15) and (17) into equation (20) yields

$$\boldsymbol{Kq} = \boldsymbol{f} \tag{23}$$

where

$$\boldsymbol{f} = \{0 \quad \dots \quad 0 \quad Ml\}^T_{1\times(3N+1)} \tag{24}$$

$$\boldsymbol{q} = \{U_1 \quad V_1 \quad W_1 \quad \dots \quad U_N \quad V_N \quad W_N \quad \kappa_x\}^T_{1\times(3N+1)} \tag{25}$$

while K is assessed via the element stiffness matrix as follows.

$$\boldsymbol{K}_e = \pi l \int_e \boldsymbol{B}_e^T \boldsymbol{C} \boldsymbol{B}_e rdr \tag{26}$$

Now, the composite tube is assembled after its discretization. Next, the nodal displacements in equation (23) can be derived to calculate the stresses and strains in the composite tube.

3. Examples of numerical calculation

The stress analysis of three composite tubes subjected to pure bending with bending moments M_x=0.3kN·m and M_y=0. The material parameters are listed in table 1. The proposed approach is adopted to solve this problem. The exact solution by unified connected parameter method for the same problem is also employed to verify it. The stress distributions are depicted in figures 1-3. The numerical results obtained by the proposed approach and the exact solution by the unified connected parameter method are in good agreement. So the current approach can be used to obtain a satisfactory solution for the pure bending of composite tubes.

Table 1. Composite material (Carbon/Epoxy) properties

E1/GPa	E2/GPa	E3/GPa	G12/GPa	G13/GPa	G23/GPa	v12	v13	v23
155	12.1	12.1	4.4	4.4	3.2	0.248	0.248	0.458

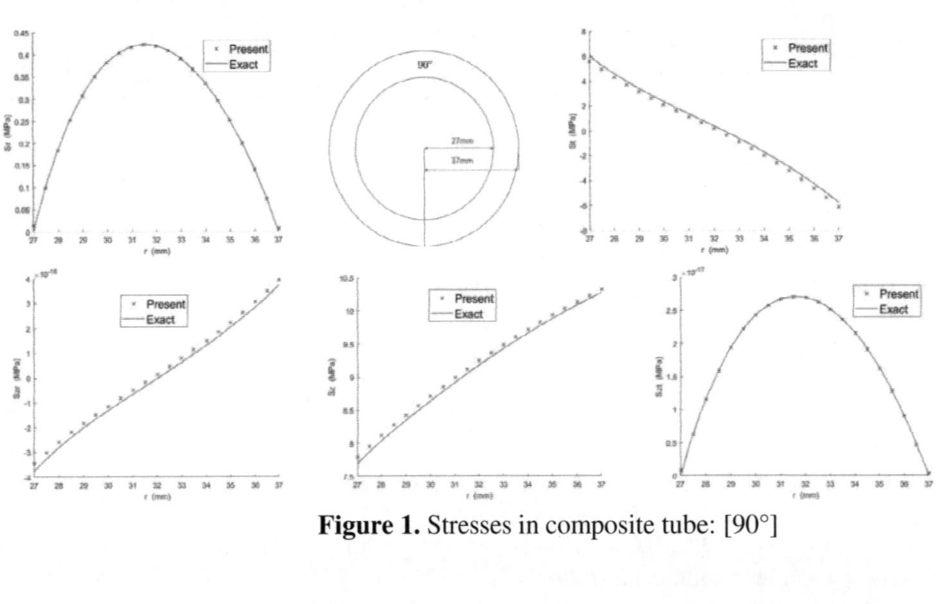

Figure 1. Stresses in composite tube: [90°]

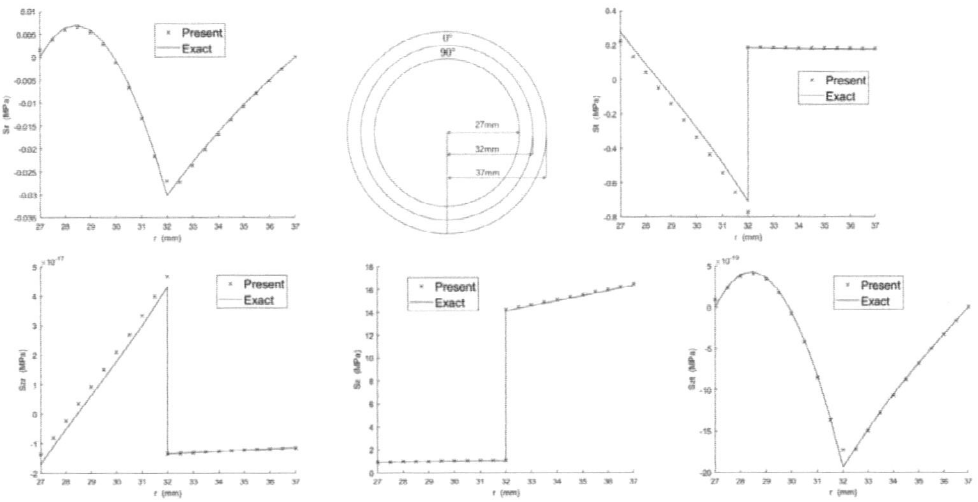

Figure 2. Stresses in composite tube: [90°/0°]

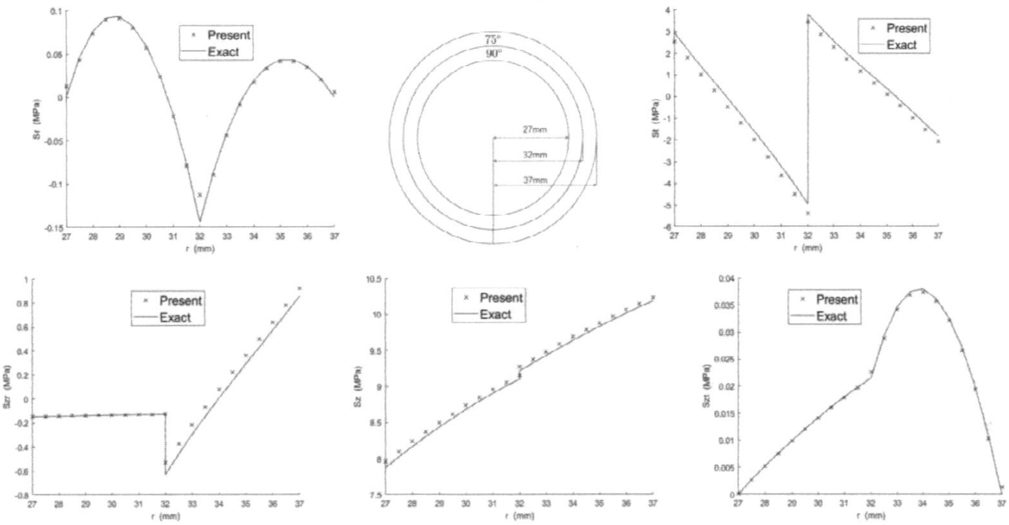

Figure 3. Stresses in composite tube: [90°/75°]

4. Conclusions

An improved LayerWise Theory-based approach is developed to calculate the stresses for composite tubes of pure bending. Firstly, the variables in the circumferential direction are separated from the local displacement. Secondly, the composite tube is only discretized in the radial direction, and the related variables in the local displacement are assumed as the simple interpolation polynomial function. Thirdly, based on the principle of minimum potential energy, the local nodal displacements are solved to calculate the strains and stresses for the composite tube. Several composite tubes are analyzed to illustrate the proposed approach. Besides, the exact solutions by unified connected parameter method are employed for comparison. They are in good agreement with each other.

References

[1] Lekhnitskii S G 1981 *Theory of Elasticity of an Anisotropic Body*, Mir Publishers, Moscow [In Russian].

[2] Jolicoeur C and Cardou A 1994 *Journal of Engineering Mechanics*, **120** (12):2556-2574.

[3] Zhang C, Hoa S V and Liu P 2014 *Compos Struct* **109**(1):57- 67.
[4] Yazdani H, Hoa S V and Hojjati M 2016 *Compos Struct* **140**:433-445.
[5] Robbins D H and Reddy J N. 2010 *Int. J Numer Meth Eng* **36**(4):655-677.
[6] Thai C H, Wahab M A and Nguyen-Xuan H 2018 *Comptes Rendus Mécanique*, **346**(1):57-76.
[7] Xing Y F, Wu Y, Liu B, et al. 2017 *Compos Struct* **170**:158-168.

The composite curing deformation prediction with the account of mold factors

C L Guan, J L He*, X B Yang, J Dai and J Zhang

College of Mechanical and Electrical Engineering, Central South University, Changsha, 410083, China

*E-mail: hejilin@csu.edu.cn

Abstract: In this paper, the effect of interaction on the curing deformation of resin matrix composite materials were investigated through introducing a shear layer to identify the interaction between the mold and the component, and a numerical calculation model was established between the parameter and the maximum deformation. Then, the experimental data in the existing literature was compared with numerical calculation model data to verify the accuracy of the calculation model. The results of numerical model proved that the curing deformation does not have a direct relationship to the shear modulus and the thickness of the shear layer, but only to the ratio of the shear modulus to the thickness. The experimental results show that the calculation model hardly changes with the transformation of the length of the components, but varies with the ply of the laminates. When comparing the curing deformation obtained by calculation model with the deformation measured by experiment, it can be concluded that the model can predict the curing deformation of composite laminates with different length and thickness. Taking the practical significance into consideration, the introducing of ratio parameter can reflect the influence of the mold on the curing deformation of the components more concise. The numerical model between the parameters and the maximum deformation is established based on the shear layer, which can predict the curing deformation of the laminates with different lengths and thicknesses more accurately.

1. Introduction

Since the coefficient of thermal expansion of the metal mold is much larger than that of the composites, there is a mismatching in the process of heat deformation during the curing process of autoclave molding. Hence, under the action of large pressure, the bottom layer of the components is tightly fitted on the upper surface of the mold. Due to the inconsistent amount of heat deformation, shear stress will occur between the mold and the material, the shear force generated can be transmitted to the respective layers through the resin during the viscoelastic phase. However, at this stage, the modulus of the resin matrix is still relatively small, the interaction cannot be well transferred, resulting in a stress gradient in the thickness direction. That stress will remain interiorly after curing cycle, which will be released after demoulding and will lead to the curing deformation. [1-2]

To date, a great number of papers have been published on the effect of the mold during the curing of composites. Twigg *et al.* [3-4] found that the component and the mold are neither completely slipped nor stationary during the viscoelastic stage. Bapanapalli *et al.* [5] introduced a shear layer into the model to study the interaction between mold and component. By changing the parameters of the shear layer, the simulated values close to the experimental values were obtained, so then the shear layer parameters under specific materials, mold and pressures were also determined. This method can accurately predict the solidification deformation caused by the action of the mold, but the shear layer parameters need to be redefined in different processes and structures. Arafath *et al.* [6] introduced a shear layer to express the interaction between the mold and composite layer which shows that the curing deformation of composites depends on the choice of the elastic modulus of the resin matrix. Zeng *et al.* [7] used the friction theory to establish three-dimensional curing model of large aeronautical structural components and the warpage deformation of which was predicted based on commercial finite element software. The accuracy of this three-dimensional model was also verified by experiments.

Yue *et al.* [8-10] studied the mold action in the curing process by embedding fiber grating in the composite components and established a finite element model to predict the deformation based on the shear layer. Yuan *et al.* [11] established a finite element model to predict curing deformation using aof shear layer method. The three-dimensional simulation model of L-type composites was established taking into account the mold factors, which shows that the prediction of curing deformation is more accurate after considering the action of mold.

This paper also introduces a shear layer to identify the interaction between mold and component. Most of the previous studies have obtained specific shear layer performance parameters with the complicated calculation. In this paper, a ratio parameter λ is used to reflect the influence of the mold on the curing deformation of the components, which is more concise. The numerical model between the parameters and the maximum deformation is established based on the shear layer, which is proved to more accurately predict the curing deformation of the laminates with different lengths and thicknesses.

2. A calculation Model between warpage deformation and Shear layer

The shear layer, as the medium of force transfer, is similar to the black box identification method. It is assumed that the introduced shear layer is located on the upper surface of the mold and the bottom layer of the composite component. When the temperature rises, the interaction between the mold and the component is produced through the shear layer. The relationship between the mold and the shear layers and the components is shown in figure 1.

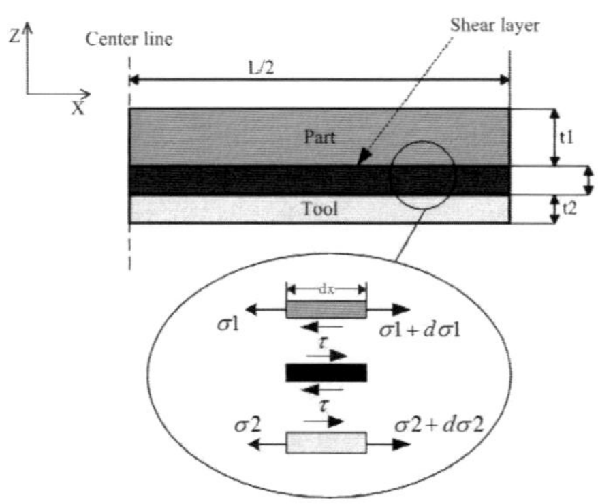

Figure 1. Sketch of X coordinate direction.

Due to the mismatch of the thermal expansion coefficient between the mold and the material during the curing process, the interface between themis interacts. Assuming this interaction isa shear stress, which can be determined as:

$$\tau=\frac{\mathrm{d}\sigma_1}{\mathrm{d}x}\cdot t1=-\frac{\mathrm{d}\sigma_2}{\mathrm{d}x}\cdot t2 \tag{1}$$

where σ_1 and σ_2 are the stresses of the component and the mold respectively; $t1$ and $t2$ are the thickness of the component and the mold.

According to Hooke's law:

$$\frac{\mathrm{d}u_1}{\mathrm{d}x}=\frac{\sigma_1}{E_1}+\alpha_{1,j}\cdot\Delta T \tag{2}$$

$$\frac{\mathrm{d}u_2}{\mathrm{d}x}=\frac{\sigma_2}{E_T}+\alpha_T\cdot\Delta T \tag{3}$$

where u_1 and u_2 are the displacements of the component and the mold in the X direction respectively during the curing process; $\alpha_{1,j}$ is the coefficient of thermal expansion of the component in the X direction; α_T is the coefficient of thermal expansion of the mold.

For the shear layer, the shear force can be described as:

$$\tau=G\cdot\gamma=G\cdot\frac{u_1-u_2}{t} \tag{4}$$

where G is the shear modulus of the shear layer and t is the thickness of the shear layer.

From equations (1) - (4), we get:

$$\frac{d^2\tau}{dx^2} = \tau \cdot \eta \tag{5}$$

where $\eta = \dfrac{G(E_1 \cdot t1 + E_T \cdot t2)}{t \cdot E_1 \cdot E_T \cdot t1 \cdot t2}$.

From figure 1, the boundary conditions can be expressed as:

$$\tau(0) = 0; \sigma_1(L/2) = \sigma_2(L/2) = 0 \tag{6}$$

where L is the length of the component.

From the above equations:

$$\tau = k \cdot [\exp(\eta^{\frac{1}{2}} x) - \exp(-\eta^{\frac{1}{2}} x)] \tag{7}$$

Where $k = \dfrac{G \cdot (\alpha_{1,j} - \alpha_T) \cdot \Delta T}{t \cdot \eta^{\frac{1}{2}} \cdot [\exp(\eta^{\frac{1}{2}} \cdot L/2) + \exp(-\eta^{\frac{1}{2}} \cdot L/2)]}$.

From equation (1):

$$\sigma_1 = \int_0^{L/2} \frac{\tau}{t1} dx \tag{8}$$

The bending moment caused by the stress obtained by equation (8) can be expressed as:

$$M = \int_0^{t1} \sigma_1 \cdot b \cdot (z - \frac{t_0}{2}) dz \tag{9}$$

where b is the width of the laminate and t_0 is the thickness of the single layer of the material. According to the knowledge of material mechanics, the maximum deformation of the laminate which under the action of bending moment M can be expressed as:

$$v_{\max} = \frac{M}{2E_1 I}(L/2)^2 \tag{10}$$

where a cross-sectional moment of inertia $I = (b \cdot t1^3)/12$. It can be calculated from equation (7)-(10) that:

$$v_{\max} = k_1 \cdot \frac{3(\alpha_{1,j} - \alpha_T) \cdot \Delta T \cdot (t1 - t_0) \cdot L^2}{2 \cdot t1^3 \cdot E_1} \tag{11}$$

where $k_1 = \dfrac{G \cdot [\exp(\eta^{\frac{1}{2}} \cdot L/2) + \exp(-\eta^{\frac{1}{2}} \cdot L/2) - 2]}{t \cdot \eta \cdot [\exp(\eta^{\frac{1}{2}} \cdot L/2) + \exp(-\eta^{\frac{1}{2}} \cdot L/2)]}$ is a fixed value.

The shear layer parameter λ and the parameter λ_1 which is related to the thickness and the modulus of elasticity of the component and the mold are introduced. Let $\lambda = \frac{G}{t}$, $\lambda_1 = \frac{E_1 \cdot t1 + E_T \cdot t2}{E_1 \cdot E_T \cdot t1 \cdot t2}$, and λ_1 is a fixed value, then k_1 can be expressed as:

$$k_1 = \frac{1}{\lambda_1}(1 - \frac{2}{\exp(L \cdot \lambda^{\frac{1}{2}} \cdot \lambda_1^{\frac{1}{2}} / 2) + \exp(-L \cdot \lambda^{\frac{1}{2}} \cdot \lambda_1^{\frac{1}{2}} / 2)}$$

It can be seen that if λ is known, the maximum deformation in the X direction can be calculated and predicted.

3. Numerical examples

3.1. Determination of the shear layer parameter λ

In this paper, by adjusting the coefficient of shear modulus to a thickness of the shear layer λ, the curing deformation of the component is consistent with the experimental results, then the parameters of the shear layer under specific process conditions are obtained. The example was taken from the literature [12], using prepreg T800HB/3900-2 as the laminate material which monolayer thickness is 0.16mm. The mould material is 6061-T6 aluminum and invar with each thickness of 10 mm. The mold parameters are shown in table 1, and the material parameters are shown in table 2.

Table 1. Properties of mold.

Mould material	t	α_T
Aluminum	10mm	23.1με/°C
Invar	10mm	2.9με/°C

Table 2. Properties of T800HB/3900-2.

Properties	E_1 /GPa	E_2, E_3 /Gpa	α_1/°C^{-1}	α_2,α_3/°C^{-1}	$\upsilon_{12}, \upsilon_{13}$	υ_{23}	G_{12},G_{13} /GPa	G_{23} /GPa
Value	169	8.62	-0.001×10^{-6}	8.95×10^{-6}	0.355	0.41	5	1.22

According to [6], the component is laminate with four layers, which $L \times b = 254mm \times 25mm$, the material of the mold is 6061-T6 aluminum. Using the method described in this paper, the relationship between the maximum deformation of composite laminates and coefficient λ is shown in figure 2.

It can be seen from figure 2 that when the coefficient λ is about 0.19, the calculation are most consistent with the experimental results. The results of the comparison are shown in table 3:

IOP Conf. Series: Materials Science and Engineering **474** (2019) 012056 doi:10.1088/1757-899X/474/1/012056

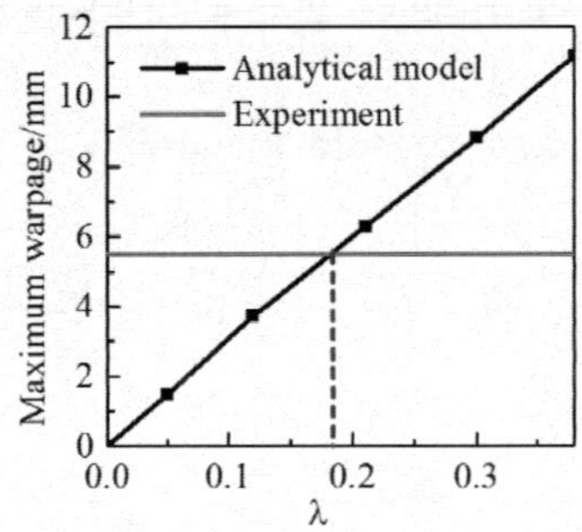

Figure 2. The relationship between the maximum deformation of the composite component and λ.

Table 3. Deformation comparison of T800HB/3900-2 composite.

Experiment	Analytical model	Error
5.498	5.60	1.8%

3.2. *Effects of characteristic parameters on coefficients* λ

Mezeix *et al.* [12] carried out curing deformation experiments on laminates of different length and thickness using molds made of steel. The effect of the interaction between the mold and the composites on the curing deformation was studied. The calculation model of this paper is used to calculate the curing deformation of laminates of different sizes. The results are shown in table 4 and the corresponding coefficients λ are shown in table 5.

Table 4. Comparison between experimental and calculated deformation of laminates of different thickness and length.

Ply	L(mm)	Experiment (mm)	Analytical model (mm)	Simulation (mm)	The analytical model error	The simulation model error
4	300	5.2	5.16	5.10	0.8%	2%
	400	9.19	9.17	8.0	0.2%	13%
	500	23.3	14.32	22.5	38.5%	5%
8	300	1.45	1.47	1.48	1.4%	2%
	400	2.37	2.6	2.96	9.7%	25%
	500	4.5	4.07	5.0	9.6%	11%
16	300	0.32	0.323	0.31	0.9%	3%
	400	0.6	0.57	0.71	5.0%	18%
	500	1.47	0.9	2.65	38.7%	80%

Table 5. λ of different plies

Ply	4	8	16
λ	0.72	0.68	0.55

From the calculation results in Table 4, it can be seen that the model in this paper can predict the curing deformation of the laminates well. At the same time, the coefficient λ hardly transforms with the change of the length, but with the change of the thickness of the laminates, the coefficient must be recalculated to predict the components of any length under the thickness. It can be seen from Table 5 that during the curing process of T800HB/3900-2 composites, the relationship between the coefficient λ and the number of layers can be obtained by the least squares method under the action of the Invar mold:

$$\lambda = -0.01446t + 0.785 \tag{12}$$

where t represents the number of laminates layers. The fitted curve between coefficient λ and laminate ply is shown in figure 3.

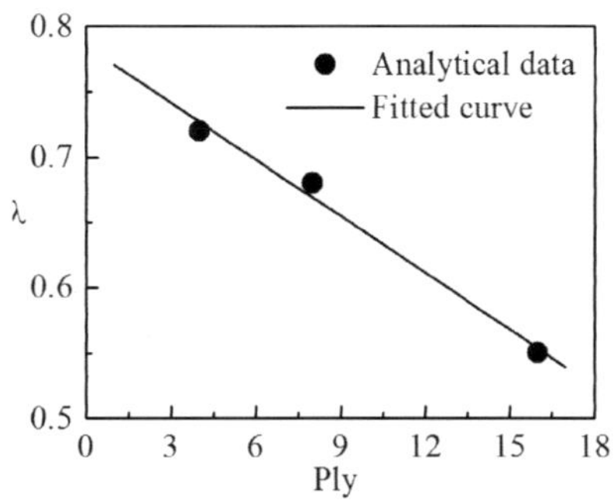

Figure 3. The fitted curve between coefficient λ and laminate ply.

3.3. Verification of numerical calculation model and shear layer parameters

To verify the accuracy of the numerical calculation model and the shear layer parameters λ, the experimental data in the literature [12] is used for validation. The component is a laminate with 6 layers, which size is $300mm \times 300mm$, the coefficient of that thickness can be calculated to be 0.698 by using the fitted curve. The calculations are shown in table 6:

Table 6. Verification of fitting accuracy of λ.

Analytical model	Experiment	Error
2.52	2.38	5.9%

It can be seen from table 6 that the coefficient obtained by the fitted curve can predict the maximum curing deformation of laminates accurately, thus proving the correctness of this fitting.

4. Conclusions

In this paper, a shear layer is introduced for the mold-component interaction during the curing process of the composites, that is, the interaction between the mold and the component is transmitted through the shear layer. The numerical model of curing deformation of composite laminates is established by introducing a shear layer parameter λ. The main conclusions are:

- ✓ It is proved by this model that the curing deformation is not directly related to the shear modulus and the thickness of the shear layer, but only to the ratio of the shear modulus to the thickness λ.
- ✓ Combining the experimental Several LED chip temperature-reducing methods have been proposed by researchers [6-10]. results with the calculation model, it can be concluded that λ hardly changes with the transformation of the length of the components, but varies with the ply of the laminates.
- ✓ The curing deformation obtained by using the calculation model of this paper is compared with the deformation measured by experiment. The results show that the model can predict the curing deformation of composite laminates with different length and thickness.

References

[1] Yue G Q 2010 *Harbin Institute of Technology*, 2010.

[2] Wang Y X. 2015 *Dalian University of Technology*, 2015.

[3] Twigg G, Poursartip A and Fernlund G 2004 *Composites Part A: Applied Sci. Manufact.*, **35**(1):121-133.

[4] Twigg G, Poursartip A and Fernlund G 2003 *Composites Science & Technology*, **63**(13):1985-2002.

[5] Bapanapalli S K and Smith L V 2005 *Composites Part A Applied Science & Manufacturing*, **36**(12):1666-74.

[6] Arafath A R A, Vaziri R and Poursartip A 2009 *Composites Part A Applied Science & Manufacturing*, **40**(10): 1545-57.

[7] Zeng X and Raghavan J 2010 *Composites Part A Applied Science & Manufacturing*, **41**(9):1174-83.

[8] Yue G Q, Zhang B M, Dai H F and Du S Y 2010*Acta Materiae Compositae Sinica*, **27**(6):167-71.

[9] Yue G Q, Zhang B M, Du S Y and Dai F H. 2010 *Fiber Reinforced Plastics/Composites*, **5**: 62-5.

[10] Yue G Q, Zhang J Z and Zhang B M 2013 *Acta Materiae Compositae Sinica*, **30**(4):206-10.

[11] Yuan Z Y, Yuan Y J, Wang J B and Wei S M 2016 *Acta Materiae Compositae Sinica*, **33**(4): 902-9.

[12] Mezeix L, Seman A, Nasir M N M, Aminanda Y, Rivai A, Castanié B, Olivier P and Ali K M 2015 *Composite Structures*, **124**:196-205

The morphology-controlled deposition of ZnO nanocrystalline

D Zhou, L N Wang*, J T Liu, Y Yu, Z Q Luan, Z H Gao, G Y Jin, Y T Chen and W Z Huang

Key Laboratory of Marine Biophysics of Liaoning Province, Dalian Ocean University, Dalian, China

*E-mail: l.n.w@163.com

Abstract. The ZnO nanocrystalline was deposited on quartz substrate by a CVD method. Our investigation proves that the morphology-controlled deposition of nanocrystalline can be realized by adjusting the substrate temperature. The nanocrystalline consisted of nanorods, nanobulks and nanowires. It turns out that higher temperatures are more favorable for the foration of nanorods and nanowires, while lower ones promote the formation of nanobulks. The SEM and XRD methods were used to characterize the microstructural properties of the sample. The produced nanorods were about 6.25-28.25μm in length with great uniformity. An intense 002 peak was observed, which indicated that nanorods had a strongly pronounced hexagonal wurtzite structure, while the lattice parameter was very close to its theoretical value.

1. Introduction

Zinc oxide (ZnO) is one of II-VI group semiconductor materials, which are widely used in gas sensors [1-3], light-emitting diodes [4-6], optoelectronic devices, nanogenerators and transparent conducting electrodes [7-10]. The properties of optoelectronic devices are greatly influenced by the morphology of ZnO nanostructures. There are many nanostructures have been reported, such as nanorods, nanowires, nanoflakes and nanoflowers [11-14]. Several kinds of technique have been reported to synthesis ZnO nanostructures, for instance the ZnO nanorods has been made by electron beam lithography and hydrothermal method [15-16], the epitaxial electrodeposition and wet chemical deposition also be used to synthesis ZnO nanopillars [17-18]. But all these methods need much more complicated preparation or introduce expensive mental catalysts such as Au, Ag. For that, the more catalyst-free method and much cheaper deposition methods are needed to fabricate ZnO nanostructures. The chemical vapour deposition (CVD) method is one of the most effective way to realize stable and repeatable ZnO nanostructures' synthesis.

Among many metal oxides, which can be used to make optoelectronic devices, the ZnO is veru promising one. Recently, numerous publications on the usage of ZnO nanowires or nanorods in gas sensors have appeared [19-21]. There are several factors, which can influence the sensitivity of the devices, such as the surface-to-volume and the growth orientation. The high-quality ZnO nanostructures are critical to improve the sensitivity of ZnO gas sensor devices. The growth mechanism of ZnO nanostructures have many difficulties to overcome, such as strong dependence on the substrate, the sensitivity to the growth environment, length, and diameter, which hinder the controlled deposition. In this paper, we synthesized ZnO nanorods on the quartz tube surface by a simple CVD method. The nanorods have a high crystal quality, which allows one to use them in gas sensors.

2. Experimental

The ZnO nanorods were synthesized on the quartz tube surface by the CVD method with two different positions. The experimental system consisted of three parts: a digital control equipment, the quartz tube furnace and double oxygen mix nitrogen supply. The quartz tube is about 120cm in length and 8cm in diameter. A small quartz tube with 30cm in length and 4cm in diameter was put in the middle of the CVD system. The small quartz tube was rinsed with alcohol and deionized water step by step, then dried by clean nitrogen. The ZnO powder 8g and graphite powder 2g were mixed together which play a part of reaction source at one end of the small quartz tube. The diagram of the experiment was shown in figure 1. During the whole deposition process, the nitrogen plays a role of protection gas, the experiment can be described as follows: first, the nitrogen was introduced into the quartz tube with a flow rate of 30 sccm (standard-state cubic centimeter per minute), when the impurity gases were all exhaust, the whole system become to be heated. Second, when the temperature up to 800°C, the oxygen was introduced into the quartz tube which act as reaction gas with a flow rate of 20 sccm. The whole system was maintained at 1000°C for 60 minutes, then cool down to the room temperature. At last two different type of ZnO nanorods were found at the other end of the quartz tube. on the nozzle one marked as sample *a* and sample *b* was 4cm away from sample *a*.

The morphology and structure properties of the sample were carried out by scanning electron microscope (SEM) and X-ray diffraction (XRD) respectively. The SEM system was from Hitachi SU1510, which operated at 15 KV, and the XRD was RigakuD/max3B, which performed in the θ-2θ configuration.

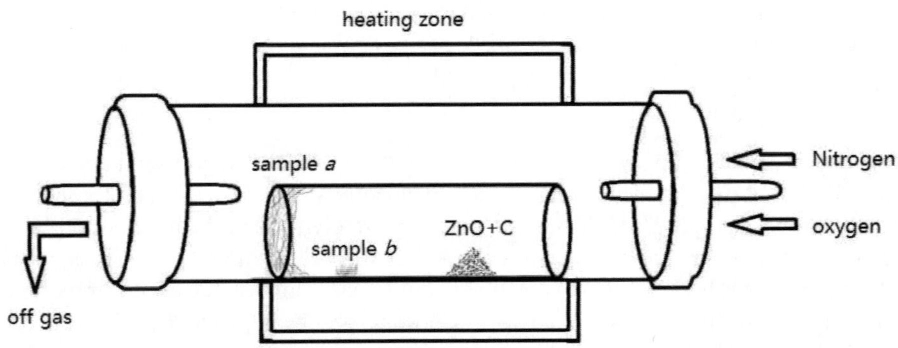

Figure 1. The schematic of experimental system.

3. Results and Discussion

After deposition, the two samples were characterized by SEM. Both samples included multiple structures. Sample A consisted of nanorods and nanobulks with various lengths, according to figure 2(a), the smallest one was about 6.25μm, the longest one was about 28.25μm, the diameters of the nanorods were also different. Numerous nanobulks also emerged around the nanorods and exhibited a non-uniform distribution. Some small holes were also observed in the sample. Sample B also consisted of nanorods and nanobulks, but there were some differences between the two samples. In sample B, nanorods were more uniform and nanobulks were more sparse, while some nanowires also emerged on the top of the nanorods. The two samples were deposited under the same conditions, the only difference between the two samples being the deposition location. From figure 1, one can ee that the reaction source was located at the heating zone center with the temperature of 1000°C. Since the marginal point of the CVD system was close to room temperature, sample B was close to the heating zone center, which made its deposition temperature higher than that of sample A. Moreover, different substrate temperatures led to different morphologies: the higher substrate temperature was more

IOP Conf. Series: Materials Science and Engineering **474** (2019) 012057 doi:10.1088/1757-899X/474/1/012057

favorable to fabricate nanorods and nanowires, while the lower one was more likely to form nanobulks. It can be deduced that the morphology-controlled deposition of ZnO nanostructures can be realized by adjusting the substrate temperature. During the whole deposition process, the mix powder of ZnO and graphite participated in the redox reactions as follows:

$$ZnO + C = Zn + CO \qquad (1)$$

$$2CO + O_2 = 2CO_2 \qquad (2)$$

$$2Zn + O_2 = 2ZnO \qquad (3)$$

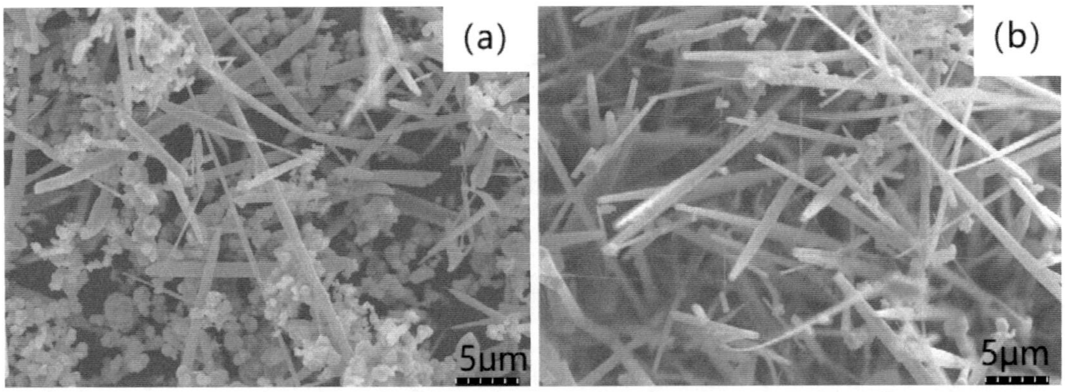

Figure 2. SEM images of ZnO nanorods.

The microstructural properties of ZnO was characterized by XRD, as shown in figure 3. The two samples were mixed together to do the test. Here eight peaks can be observed, the most intense one is the 002 peak, which is located at 34.55°. This peak can give much information about the crystal structure: a highly pronounced (intense) one implies that nanorods are more likely to grow along the c-axis and have the hexagonal wurtzite structure. The 002 peak should be found at 34.42° in theory, the shift of 002 peak would mean the presence of some defects such as vacancy and interstitial defects. The other peaks were located at 31.91°, 36.56°, 47.88°, 56.86°, 63.07°, 68.34° and 69.89°. The lattice constants a and c , which describe the crystal structure, can be calculated as follows:

$$a = \frac{\lambda}{2\sin\theta}\left(\frac{4}{3}\left(h^2 + hk + k^2\right) + \left(\frac{al}{c}\right)^2\right)^{\frac{1}{2}} \qquad (4)$$

$$c = \frac{\lambda}{2\sin\theta}\left(\frac{4}{3}\frac{\left(h^2 + hk + k^2\right)c^2}{a^2} + l^2\right)^{\frac{1}{2}} \qquad (5)$$

In the latter formulas, h, k, and l represent the Miller indices (for example the 002 peak corresponds to h=0, k=0, l=2). The parameter λ=0.15418nm is the X-ray wavelength, θ is the diffraction angle. According to equation (5), the constant of (002) peak can be calculated as c=5.191Å. Based on the literature (JCPDS NO.36-451), the lattice constant is c=5.200Å when the ZnO nanocrystalline exhibits a hexagonal wurtzite structure. It can be deduced from the obtained constant value(c=5.191Å) that the obtained structure is hexagonal wurtzite one.

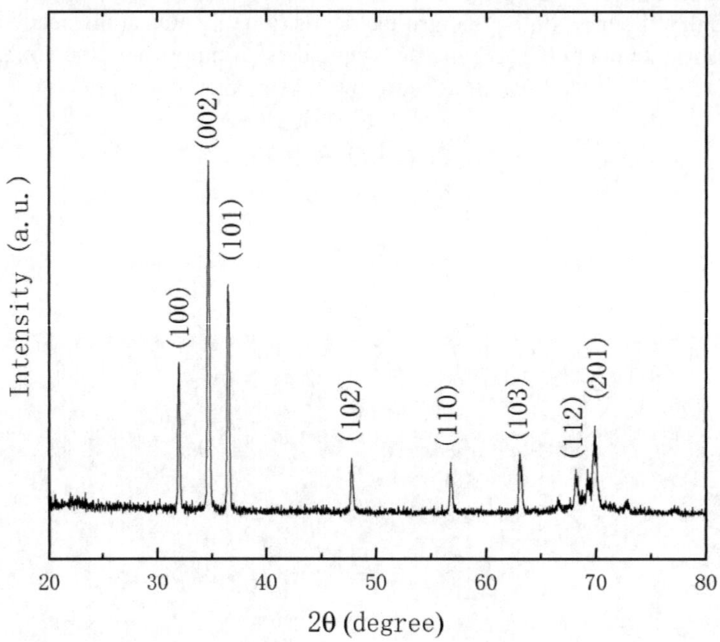

Figure 3. XRD pattern of ZnO nanocrystalline.

4. Conclusions

In this paper, ZnO nanorods and nanobulks were synthetized over the quartz substrate by the CVD method. The two samples were prepared with different deposition locations. Uniform nanorods and nanowires were found to be produced near the heating zone, where nanobulks were rarely observed. Near the edge of the quartz tube, the nonuniform nanorods and nanobulks were found. This difference may be attributed to different substrate temperatures. Higher substrate temperatures are more favorable for producing nanorods and nanowires. From the obtained SEM images, it can be seen that the two samples have different morphology, which implies that ZnO nanocrystalline can be realized via the controlled deposition by adjusting the substrate temperature. The XRD pattern also characterizes the obtained crystal structure: an intense 002 peak signifies that nanorods have a hexagonal wurtzite structure, which theoretical value is also consistent with the calculated lattice parameter c=5.191Å.

Acknowledgments

This work was financially supported by the Science and Technology Funds from Liaoning Education Department (No. QL201719); Nature Science Foundation of Liaoning Province (No.20170540109 and No.201601285); Nature Science Foundation of China (No.11647104); College Students' innovation and Entrepreneurship Project of Dalian Ocean University (No.201710158000033).

References

[1] Chen Y, Xu P C, Xu T, Zheng D and Li X X 2017 *Sens. Actuators B* **240** 264-7
[2] Zou A L, Qiu Y, Yu J J, Yin B, Cao G Y, Zhang H Q and Hu L Z 2016 *Sens. Actuators B* **227** 65-72
[3] Zou A L, Hu L Z, Qiu Y, Cao G Y, Yu J J, Wang L N, Zhang H Q, Yin B and Xu L L 2015 *J. Mater. Sci: Mater Electron* **26** 4908-12
[4] Liang S, Sheng H, Liu Y, Huo Z, Lu Y and Shen H 2001 *J. Cryst. Growth* **225** 110-3

[5] Tsukazaki A, Ohtomo A, Onuma T, Ohtani M, Makino T, Sumiya M, Ohtani K, Chichibu S, Fuke S, Segawa Y, Ohno H, Koinuma H and Kawasaki M 2004 *Nat. Mater.* **4** 42-6
[6] Tsukazaki A, Kubota M, Ohtomo A, Onuma T, Ohtani K, Ohno H, Chichibu S.F and Kawasaki M. 2005 *Japan J. Appl. Phys., Part 2* **44** 643-5
[7] Mishra Y K, Kaps S, Schuchardt A, Paulowicz I, Jin X, Gedamu D, Freitag S, Claus M., Wille S, Kovalev A, Gorb S and Adelung R 2013 *Part. Part. Syst. Charact.* **30** 731-731.
[8] Singh A, Singh A, Singh S, Tandon P, Yadav B C, Yadav R R 2015 *J. Alloy. Compd.* **618** 475–83.
[9] Yin B, Qiu Y, Zhang H, Ji J and Hu L 2014 *Cryst. Eng. Comm* **16** 6831-5
[10] Yin B, Qiu Y, Zhang H, Ji J and Lei J. 2015 *J. Mater Sci: Mater Electron* **26** 742-6
[11] Mosquera E, Bernal J, Zarate R, Mendoza F, Katiyar R and Morell G 2013 *Mater. Lett.* **93** 326-9
[12] Zhang S, Yan C L, Zhang H X and Lu G Y 2015 *Mater. Lett.* **148** 1-4
[13] Liu C, Wang B Q, Liu T, Sun P, Gao Y, Liu F M and Lu G Y 2016 *Sens. Actuators B* **235** 294-7
[14] Aladpoosh R and Montazer M 2016 *Carbohydr. Polym.* **141** 116-25
[15] Behera B and Chandra S 2016 *Sens. Actuators B* **229** 414-24
[16] Kim Y J, Lee C H, Hong Y J, Yi G C 2006 *Appl. Phys. Lett.* **89** 163128
[17] Zou C W, Liang F, Xue S W 2015 *Appl. Surf. Sci.* **353** 1061-1069
[18] Liu R, Vertegel A A, Bohannan E W, Sorenson T A, Switzer J A, 2001 *Chem. Mater.***13** 508.
[19] Matsuu M, Shimada S, Masuya K, Hirano S, Kuwabara M 2006 *Adv. Mater.* **18** 1617.
[20] Chen H, Ma S Y, Jiao H Y, Yang G J, Xu X L, Wang T T, Jiang X H, Zhang Z Y 2016 *J. Alloy. Compd.* **687** 342-351.
[21] Zou C W, Wang J, Xie W 2016 *J. Colloid. Interf. Sci.* **478** 22-28

Fatigue life prediction of flexible bearings based on ABAQUS and nCode-DesignLife

D S Zhao[1], Y Z Wang[1,*] and K Zhao[2]

[1]School of Mechatronic, Engineering and Automation Shanghai University Shanghai 200072, China
[2]Technology Centre of Cixing Bearing GROUP CO. LTD, Ningbo, 315301, China
*E-mail: meyzwang@shu.edu.cn

Abstract. Flexible bearings are critical components of harmonic drives. In service, the corresponding positions of the long and short axes of the inner and outer rings of a flexible bearing regularly change. This leads to a fatigue fracture of the outer ring because of the cyclic alternating load. In this study, a multi-body contact model and a dynamic model were established using ABAQUS to analyze the load spectrum of a flexible bearing and a wave generator. The dynamic analysis results were employed to simulate the fatigue life of the inner and outer rings using nCode-DesignLife. The results show that the inner ring underwent significant fatigue damage at the long-axis position, and the corresponding fatigue life was low. As the damage on the outer ring was due to continuous alternating deformation, the inner and outer surfaces of the outer ring were significantly damaged. Besides, the outer ring underwent fatigue damage prior to the inner ring under the same working conditions. The results provide theoretical references and practical guidance for the design and life estimation of flexible thin-wall bearings.

1. Introduction

Flexible bearing, which are intermediate deformation components used in harmonic drives, are key elements affecting the work performance, load carrying capacity, and service life of harmonic drives [1-2].

Current studies on harmonic drives have reduced the flexible bearing and wave generator to a rigid cam, ignoring the complicated internal contact state and kinematic characteristics. The failure of the flexible bearing is one of the critical factors leading to the failure of the harmonic drive [3]. The accuracy of the results of most studies that predicted the service life of flexible bearings using the Lundberg–Palmgren theory, which requires that the outer ring of the bearing be installed on a rigid bearing chock and with the load remaining unchanged [4-5], is questionable.

Considering the particularity of the pre-deformation and loading conditions, a dynamic model of a flexible bearing is established to analyze the stress distribution and load spectrum. Based on the dynamic analysis results, a fatigue analysis software was used to simulate the fatigue life of the flexible bearing. The research results have strong engineering application value for the design and optimization of flexible thin-wall bearings.

2. A dynamic model of flexible bearing

The explicit dynamics method was used to simulate the rotation of the flexible bearing under loading conditions. First, the loading model of the flexible bearing is analyzed, and an explicit dynamic analysis model of the flexible bearing is then established.

2.1. Analysis of loading model of flexible bearing

According to the principle of harmonic drives, the flexible bearing is associated with the flexspline, and the loading is relatively complicated. The rigid and flexible wheels are meshed in the long-axis region and are distributed in a mostly symmetrical manner [6-7]. Figure 1 shows the distribution of the loading.

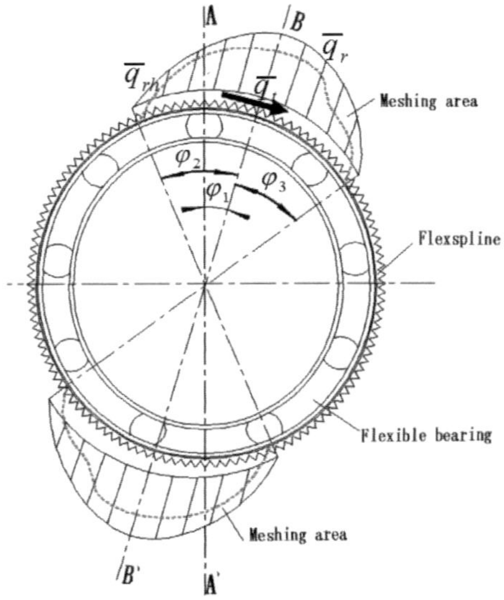

Figure 1. Load distribution of flexspline and flexible bearing.

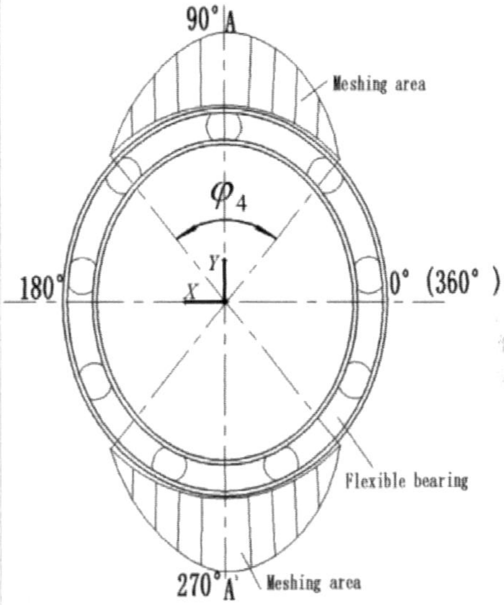

Figure 2. Approximate loading model of the flexible bearing.

In figure 1, the load distribution of the flexible bearing is approx. $\varphi_2 = \varphi_3$, and in the region where φ_2 is located, the meshing force between the flexspline and the rigid wheel can be calculated as follows.

$$\begin{cases} \bar{q}_t = \bar{q}_{t\max} \cos[\pi(\varphi - \varphi_1)/2\varphi_2] \\ \bar{q}_r = \bar{q}_t \tan\alpha \end{cases} \quad (1)$$

where $\bar{q}_t$ and $\bar{q}_r$ are circumferential and radial components, respectively, of the meshing force; α is the meshing angle of the flexible wheel.

In a harmonic reducer, if the torque experienced by the flexible wheel is *T*, then *T* and $\bar{q}_{t\max}$ have the following relationship.

$$T = 4\int_{\varphi_1}^{\varphi_1+\varphi_2} b_R \left(\frac{d_g}{2}\right)^2 \bar{q}_{t\max} \cos[\pi(\varphi-\varphi_1)/2\varphi_2]d\varphi \quad (2)$$

$$\bar{q}_{t\max} = \pi T / (2\varphi_2 d_g^{\ 2} b_R) \quad (3)$$

According to the thin-walled ring theory, the deformation of the ring due to the tangential load is the same that due to the corresponding radial load. Hence, the relationship between the radial load $\bar{q}_{rt}$ and the tangential load $\bar{q}_t$ can be given as follows.

$$\bar{q}_{rt} = \int \bar{q}_t d\varphi \tag{4}$$

The following relationship can be obtained by performing the Fourier series expansion on $\bar{q}_t$ in the above equation.

$$\begin{aligned}\bar{q}_t &= \bar{q}_{t\max} \cos[\pi(\varphi-\varphi_1)/2\varphi_2] \\ &= \frac{4\varphi_2}{\pi^2}\bar{q}_{t\max} + 8\varphi_2 \bar{q}_{t\max} \sum_{k=2}^{\infty} \frac{\cos k\varphi_2 \cos k(\varphi-\varphi_1)}{\pi^2 - 4k^2\varphi_2^2} \\ &= \bar{q}_{t0} + \sum_{k=2}^{\infty} \bar{q}_{tk} \cos k(\varphi-\varphi_1)\end{aligned} \tag{5}$$

Here, $\bar{q}_{t0} = \frac{4\varphi_2}{\pi^2}\bar{q}_{t\max}$, $\bar{q}_{tk} = \frac{8\varphi_2 \bar{q}_{t\max} \cos k\varphi_2}{\pi^2 - 4k^2\varphi_2^2}$, and k takes 2, 4, 6...

The constant *K* is ajusted by the torque *T* of the output shaft. Therefore, only the variation component in the integral equation is required.

$$\bar{q}_{rt} = \int \bar{q}_t d\varphi = \int \sum_{2,4,6,\cdots} \bar{q}_{tk} \cos k(\varphi-\varphi_1) d\varphi \tag{6}$$

$$\bar{q}_{rt} = \sum_{2,4,6,\cdots} \bar{q}_{tk} \sin k(\varphi-\varphi_1) d\varphi + C \tag{7}$$

Because the wave generator only bears the positive pressure transmitted from the flexible wheel, the integral constant *C* can be determined using the one-way contact condition.

$$C = -\sum_{2,4,6,\cdots} \frac{\bar{q}_{tk}}{k} \sin k(\gamma-\varphi_1) d\varphi \tag{8}$$

The total radial load on the outer ring of the bearing can be obtained as the sum of the equivalent radial load and the radial load of the meshing force, as follows.

$$\bar{q}_{rh} = \bar{q}_r + \bar{q}_{rt} = \bar{q}_r + \sum_{2,4,6,\cdots} \bar{q}_{tk} \sin k(\varphi-\varphi_1) d\varphi + C \tag{9}$$

where $\bar{q}_{rh}$ and $\bar{q}_{rt}$ are total and equivalent radial loads, respectively; $\bar{q}_r$ is the radial load of the meshing force.

The distribution of the actual load acting on the flexible bearing is indicated using the red dotted line shown in figure 1. The effect of partial loading is ignored, and the load is simplified in this study, as shown in figure 2. The load distribution is as follows:

$$\bar{q}_{rh} = \bar{q}_{r\max} \cos(\pi\varphi/\varphi_4) \tag{10}$$

2.2. A dynamic model of flexible bearing

The flexible bearing model CSF_25_80 was established. Table 1 lists its geometric parameters.

Table 1. Bearing parameters.

Main parameters	Value
Inner diameter of bearing [mm]	45.1
Thickness of inner ring [mm]	3
Width of the inner ring [mm]	6.35
Outer diameter of bearing [mm]	58.3
Thickness of outer ring [mm]	3
Width of the outer ring [mm]	9
Diameter of the ball [mm]	5.556
Long-axis radius of standard cam [mm]	22.95
Short-axis radius of standard cam [mm]	22.15
Number of balls	23

The model is subdivided into full hexahedral meshes via scanning and mapping, and the contact area is refined to improve the analysis accuracy and solution speed.

It is assumed that the inner ring rotates with the camshaft, the outer ring undergoes only radial elastic deformation, and the applied load rotates with the shaft. The applied radial load is 1000 N, and the span angle (φ_4) is 120°. The rotation speed of the camshaft is 1000 rev/min.

2.3 Analysis of dynamic simulation results of flexible bearing

The groove bottom circles of the inner and outer raceways are selected as the analysis paths (the start and end points of the paths are both at the 180° azimuth, as shown in figure 2; the paths are selected in the clockwise direction). figure 3 shows the displacement curve (in the Y-axis direction, shown in figure 2) of the paths when the assembly and loading are complete.

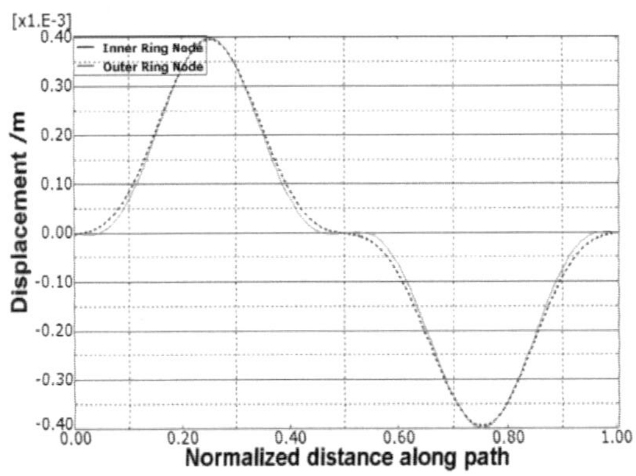

Figure 3. Displacement distribution on the bottom circle of the inner and outer rings.

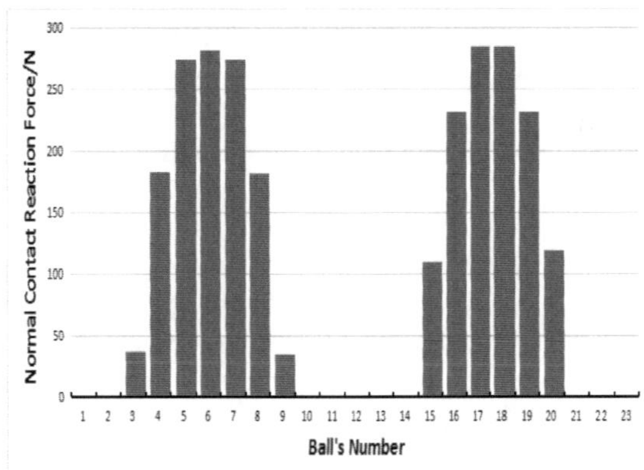

Figure 4. The contact reaction force of each rolling element of the flexible bearing.

Figure 3 indicates that the inner ring undergoes the most considerable radial deformation along the long axis and the smallest radial deformation along the short axis; the radial deformation of the outer ring is similarly distributed. As the inner ring is directly assembled on the elliptical shaft and the outer ring is forced to deform when contacted with the rolling elements, the radial deformation of the inner ring is slightly higher than that of the outer ring at the proximal short-axis position.

Figure 4 shows the distribution of the contact force of each rolling element when the assembly and loading are complete. The analysis results show that the rolling elements in the vicinity of the long axis are subjected to a large load, whereas those in the vicinity of the short axis are not subjected to any loading. Among the twenty-three rolling elements, thirteen are loaded, and the load distribution is approximately symmetric concerning the short axis of the elliptical axis.

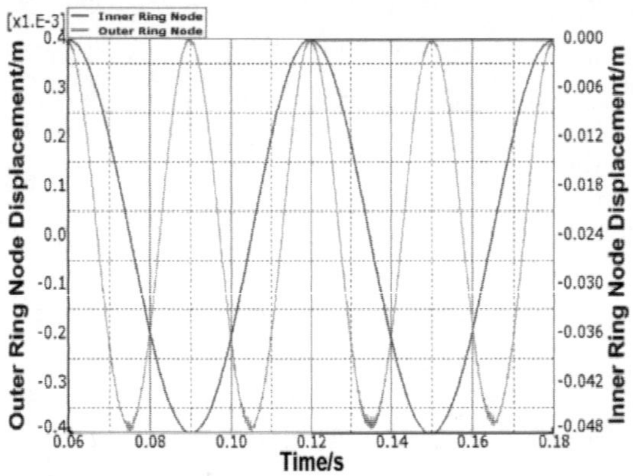

Figure 5. Displacement evolution curve of the outer and inner ring nodes.

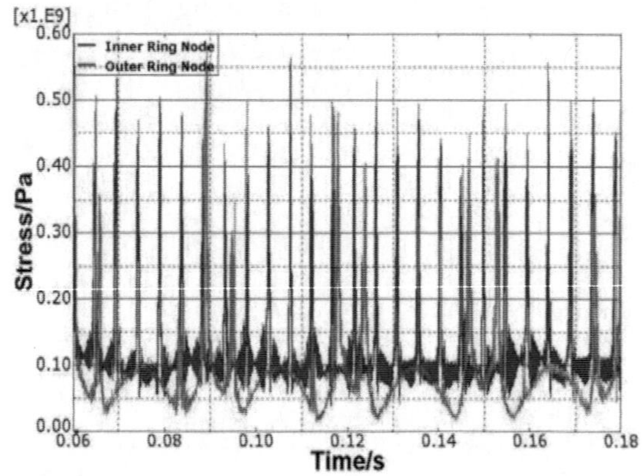

Figure 6. Stress evolution curve of the inner and outer ring nodes.

The nodes in the 90° azimuth of the inner and outer ring raceway grooves, shown in figure 2, are selected as the research nodes. Figures 5 and 6 show the displacement curves (along the Y-axis direction) and stress curves of the chosen nodes during operation, respectively.

Figure 5 shows that the spatial displacement of the inner ring node reciprocates once from 0.06 s to 0.12 s (elliptical axis rotates one week), the displacement of the outer ring node undergoes four cyclic alternations between −0.4 mm and 0.4 mm, and the change in the outer ring node displacement is in the form of a regular cosine curve over time.

At 0.09, 0.12, 0.15, and 0.18 s, when the long axis passes through the selected node of the outer ring, the stress increases sharply because of the combined action of deformation and loading. As the inner ring is assembled with the camshaft, the shape of the deformation remains unchanged during the movement. When the rolling ball passes through the position of the node, the contact stress mutates.

3. Simulation analysis of fatigue life

The failure of a flexible bearing can be due to many reasons. The contact stress and cyclic bending stress in service are two main factors causing fatigue failure [8].

3.1. Determination of fatigue analysis parameters for flexible bearing

According to the fatigue analysis principle of nCode-DesignLife [9-10] and based on the dynamic analysis results, the loading spectra of the inner and outer rings can be obtained. Their material properties were defined according to reference [11]. The S-N curve equation for GCr15 steel under an experimental condition with 99% reliability can be written as follows.

$$\lg(\lambda) = 43.3934 - 12.4999\lg(S) \tag{11}$$

where S is cyclic stress and λ is the time to fracture under cyclic loading.

3.2. Results and analysis

Figures 7 and 8 show the fatigue damage and life cloud diagrams of the inner and outer rings, respectively.

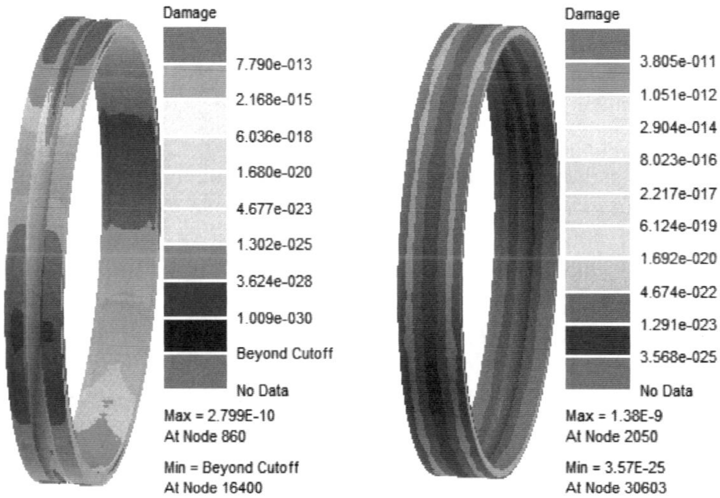

Figure7. Inner ring fatigue damage cloud diagram (a)—L and outer ring fatigue damage cloud diagram (b)—R.

Figure 7(a) shows that the region with a higher fatigue damage on the inner ring is mainly concentrated on the surface near the long and short axis of the cam outline, and the damage is more severe in the long-axis raceway than that in other areas. The damage on the inner ring is maximum at node N861 with a damage size of 2.799×10^{-10}. As shown in figure 7(b), the fatigue damage on the outer ring is mainly concentrated on the surface area, and the damage on the outer and inner surfaces near the raceway is more significant. The damage on the outer ring is maximum at node N1613 with a damage size of 1.38×10^{-9}.

The fatigue life of the components obtained using the nCode-DesignLife software is defined as the number of cycles corresponding to the load history. A working cycle is equal to 0.12 s.

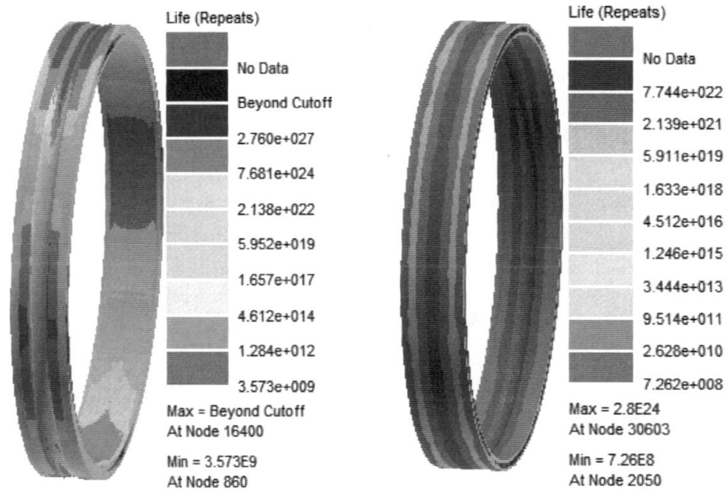

Figure 8. Inner race fatigue life cloud diagram (a)—L and outer race fatigue life cloud diagram (b)—R.

Figure 8(a) shows that the lowest fatigue life position of the inner and outer rings corresponds to the highest fatigue damage position. The minimum fatigue cycle number for the inner ring is 3.573×10^{9}.

Figure 8(b) shows that the minimum fatigue cycle number for the outer ring is 7.262×10^8. The minimum fatigue life of the inner and outer rings under the same working conditions is 24206.67 h.

4. Conclusions

The conclusions drawn from this study are as follows.

- ✓ When the flexible bearing was installed on the camshaft, the inner ring deformed only once and then rotated synchronously with the camshaft. The relative position of any node on the outer ring regularly changed in service. As the change in the displacement could be represented in the form of a regular cosine curve over time, the outer ring was always subjected to alternating bending.
- ✓ The fatigue damage on the inner ring was mainly due to the contact stress, whereas the outer ring simultaneously underwent contact fatigue failure and bending fatigue failure. In service, the outer ring underwent fatigue failure before the inner ring. Through simulation, the maximum fatigue life of the selected flexible bearing was found to be 24206.67 h under rated conditions.

Acknowledgements

This study was supported by the National Project No. 863 (Grant No.2015AA043005), Ningbo Science and Technology Innovation Team Project (Grant No.2015B11012) and Ningbo Science and Technology Research Projects (Grant No.2014B1006) of China.

References

[1] Zhang D L and Liu F 2015 *J. Journal of Mechanical Transmission.* **39**(05) 50-53
[2] Huang B, Wang Y Z, Zhao K, et al. 2017. *Int. Conf. on Mechanical Design. Springer, Singapore.* 227-237
[3] Wang Y Z, Jiang Y, et al. 2017 *J. Lubr. Eng.* **42**(03) 58-54
[4] Tedric A H and Michael N K 2009 *Rolling Bearing Analysis Fifth Edition: Essential Concepts of Bearing* (Beijing: CRC Press) p 167-173
[5] Fu J F, Dong H J and Shen Y W 2007 *J. China Mechanical Engineering.* **18**(18) 2210-4
[6] Kayayasi O and Erzincanli F 2007 *J. Mater. Des.* **28**(2) 441-7
[7] Jiang Y, Wang Y Z, et al. 2017 *J. Machine Design and Research.* **33**(03) 91-4
[8] Feng D Y, Chen X Y, et al. 2017 *J.Bearing.* **6** 60-4
[9] Lin W Y and Tao S 2016 *J. Machinery and Electronics.* **34**(05) 21-4
[10] Lu L M, Yu Y Y and Zeng G W 2017 *J. Journal of Mechanical Transmission.* **41**(03) 114-9
[11] Li W 2007 Fatigue Behavior of GCr15 Steel in Gigacycle Regime *D. Southwest Jiaotong University.* 24-30

Study on engaging characteristics of wet friction pair with copper-based powder metallurgical friction material

Y Z Wang, S Y Ren*, Y Li, W Zhang and X M Chu

School of Mechanical Engineering & Automation, Beihang University, Beijing 100191, China.
*E-mail: 1585777318@qq.com

Abstract: Copper-based powder metallurgy (Cu-PM) has become quite common as the frictional material for clutches in transmission system. However the traditional research on friction pair is still limited to the static, qualitative and semi empirical studies. Based on the study of the physical properties of Cu-PM materials, the mathematical model of the wet friction pair is established. Through a simulation example, the transmission torque and the oil film pressure are obtained. Furthermore, a test bench has been set up to test the wet friction pair with Cu-PM friction materials. The correctness of the mathematical model was proved by comparing the experimental data and simulation data.

1. Introduction

The friction pair is one of the key parts of the wet clutch in the transmission. It consist of friction discs and steel discs [1-2]. Under the action of pressure, the friction disc joins the steel disc to transfer the torque of the transmission system. The performance of the frictional pair directly determines the design level and performance of the transmission system.

The structure of the friction pair is shown in figure 1, and the friction discs and steel discs are alternately arranged. In the process of wet clutch engagement, axial pressure makes the friction sheet and dual steel sheet stick together. Torque transfer from input shaft to output shaft by friction force on the contact surface between the discs and viscous shear force of the lubricating oil film.

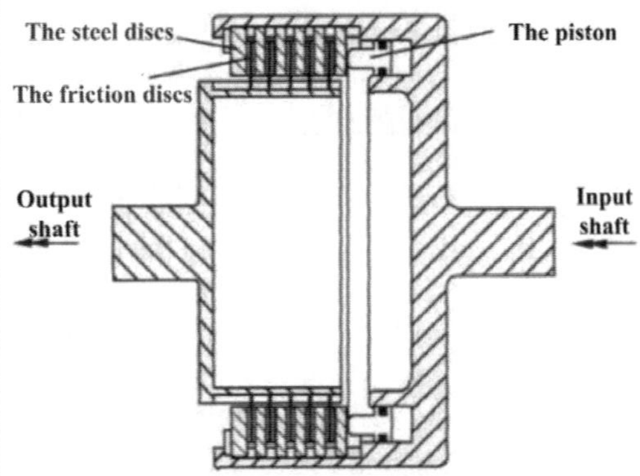

Figure1. The structure of the friction pairs

The mechanism of torque transfer in the dynamic process of engagement friction pair has been studied step by step. Burton [3-5] pointed out that the friction torque is produced by the shear torque of the lubricating oil when the oil film thickness is thicker, so the friction material and its lubrication characteristics have a decisive influence on the friction performance of the friction pair. Matsumoto [6-7] studied the influence of elastic deformation and permeability of friction materials on the engagement performance. Komvopoulos [8] found that rough surface friction is the main mechanism of boundary lubrication. Razzaque [9-10] introduced the influence of the groove shape of the friction plate into the study of the ring extrusion film. Liu [11] simplified the friction pair into an axisymmetric model, neglecting the groove and the centrifugal action of oil. Zhu and Wang [12-13] established the finite element thermal analysis model. But the effect of lubricating oil and the groove on the friction disc are neglected.

In previous studies, the permeability of materials has been neglected, such as references [9-10], and the shape of grooves is seldom involved, such as references [11-13]. In this paper, the mathematical model of Cu-PM material friction pair with radial grooves was established and the boundary conditions considering the permeability of materials are also established. The research in this paper has guiding significance for the design and manufacture of friction pairs of Cu-PM materials.

2. Physical properties of Cu-PM friction materials and their effect on friction pair performance

2.1Physical properties of Cu-PM friction materials

The friction material used in this study is Cu-PM friction material (figure 2). It is made of copper or copper alloy with strengthening elements, lubrication and friction elements. The strengthening elements are iron, nickel, aluminum, tin, zinc and manganese, the friction elements are silicon dioxide, silicon monoxide and zirconia, and the lubricating elements are graphite and molybdenum disulfide. The specific mass fraction is shown in table 1.

Figure 2. Copper based powder metallurgy friction disc

Table 1. Composition and physical properties of Cu based powder metallurgy friction materials

Component						Characteristic			
Cu	Sn	Pb	Fe	C	SiO_2	Density, $kg \cdot m^{-3}$	Hardness, HB	Thermal conductivity, $W \cdot m^{-1} \cdot K^{-1}$	Ultimate strength, MPa
68~76	8~10	7~9	3~5	6~8	2~4	6.0~6.2	25~28	15.5	24.0

The Cu-PM friction material has the advantages of good thermal conductivity and stable friction coefficient. The sliding friction coefficients f_c at different relative sliding velocities v_{rel} are obtained by the sliding experiment of friction discs and steel discs, and then the relationship between the sliding friction coefficient f_c and the relative sliding velocity v_{rel} is established as below.

$$f_c = 0.13 - 0.008\log_{10}\left(v_{rel}\right) \quad (1)$$

2.2 Influence of Cu-PM friction material on contact surface slip

The microstructure of the Cu-PM friction material is shown in figure 3[14]. The observation scale is 100 μm. It can be seen from the figure 3 that the internal pores of the Cu-PM friction material are criss-crossed and interpenetrated, and the porosity is high. Under the pressure, the lubricating oil can discharge the gap between the friction disc and the steel disc from the pores.

At the contact interface between the lubricating oil film and the friction material, the velocity u of the lubricating oil in the radial direction is no longer the same as the radial velocity of the friction plate, that is, $u \neq 0$, and the sliding boundary condition of the lubricating oil can be expressed as:

$$u = u_B \qquad v = r \cdot w_1 \qquad w = w'$$
$$\frac{\partial u}{\partial z} = \frac{\alpha}{\sqrt{\Phi}}\left(u_B - \frac{\Phi}{\mu}\frac{\partial P}{\partial r}\right) \tag{2}$$

In formula (2), α=0.2 is the Beavers-Joseph slip factor, Φ is the permeability of the porous friction material, and u_B is the velocity boundary condition of the lubricating oil in the radial direction of the contact interface with the porous friction material, as shown in figure 4 .

Figure 3. Microstructure of copper based powder metallurgy materials (a) parallel direction (b) vertical direction

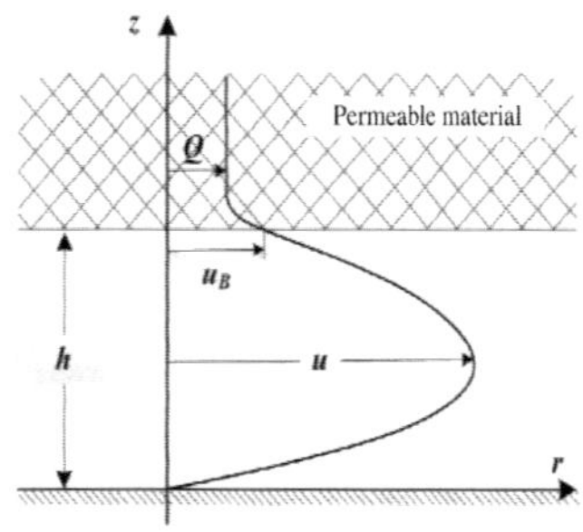

Figure 4. Interface slip boundary condition

2.3 Influence of Cu-PM friction material surface roughness on friction torque contact surface slip
The friction torque transmitted by the friction pair is composed of the viscous shear torque of the lubricating oil and the friction torque. The friction torque is calculated by the rough surface model (figure 5) and the friction coefficient model.

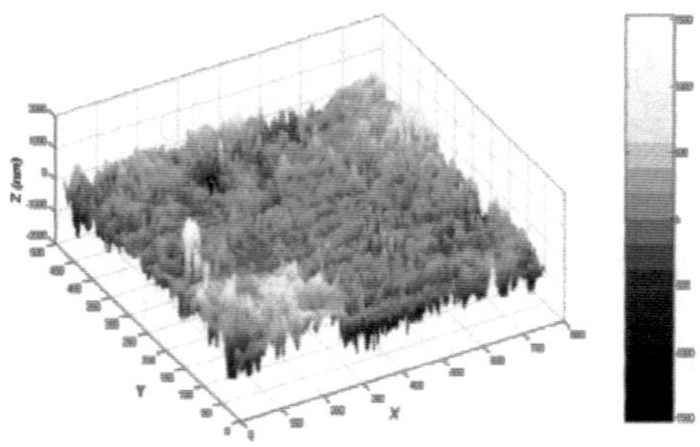
Figure 5. Surface topography of Cu-PM friction material obtained from the acquired data

According to the Hertz theory and statistics, the expression of the real contact area A_c and the contact pressure F_c is:

$$A_c = \pi\eta A_N \beta\sigma F_1\left(h/\sigma\right) \tag{3}$$

$$F_c = \frac{4}{3}\eta A_N E'\beta^{1/2}\sigma^{3/2}F_{3/2}\left(h/\sigma\right) \tag{4}$$

where η is the distribution density of the micro-convex peaks on the rough surface, σ is the standard deviation of the height-probability distribution, and A_N is the rough surface and the smooth surface, E' is the equivalent elastic modulus.

3. Dynamic simulation of friction pairs of Cu-PM friction materials.

3.1 Dynamic model of friction pair engagement process

In the process of friction pair engagement, under the action of axial loading pressure (figure 6), the friction disc is gradually bonded with the steel disc, and the torque is transmitted by friction torque and viscous shear torque.

The oil film between the friction discs and the steel discs is a ring film. The axial load during the engaging process is less than 5MPa. The differential equations of motion of viscous incompressible fluid are formulated in cylindrical coordinates as follows.

$$\begin{cases} \rho\left(\frac{Du}{Dt}-\frac{v^2}{r}\right)=\rho\cdot f_r-\frac{\partial p}{\partial r}+\mu\left(\nabla^2 u-\frac{u}{r^2}-\frac{2}{r^2}\frac{\partial v}{\partial \theta}\right) \\ \rho\left(\frac{Dv}{Dt}-\frac{u\cdot v}{r}\right)=\rho\cdot f_\theta-\frac{1}{r}\frac{\partial p}{\partial \theta}+\mu\left(\nabla^2 v-\frac{v}{r^2}+\frac{2}{r^2}\frac{\partial u}{\partial \theta}\right) \\ \rho\frac{Dw}{Dt}=\rho\cdot f_z-\frac{\partial p}{\partial z}+\mu\nabla^2 w \end{cases} \quad (5)$$

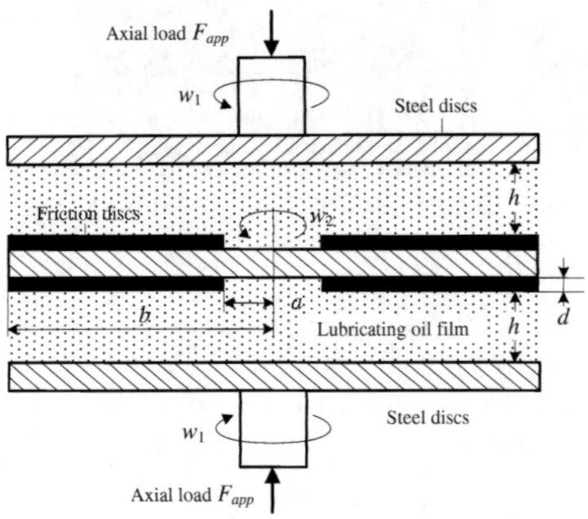

Figure 6. Structure of friction pair dynamic model

In formula (5), u, v and w are the velocity components along the direction of r, θ and z, ρ is the density of lubricating oil, μ is the dynamic viscosity of lubricating oil, f_r, f_θ and f_z are the components of unit mass of lubricating oil in the direction of r, θ and z, respectively.

$$\begin{aligned} \frac{D}{Dt}&=\frac{\partial}{\partial t}+u\frac{\partial}{\partial r}+\frac{v}{r}\frac{\partial}{\partial \theta}+w\frac{\partial}{\partial z} \\ \nabla^2&=\frac{\partial^2}{\partial r^2}+\frac{1}{r}\frac{\partial}{\partial r}+\frac{1}{r^2}\frac{\partial^2}{\partial \theta^2}+\frac{\partial^2}{\partial z^2} \end{aligned} \quad (6)$$

3.2 Method for solving dynamic model

In this study, the multi-grid method is used to solve the difference equation. The solution of the modified Reynolds equation difference form at each level is solved by the line relaxation method. The expressions (5) and (6) can be reduced to the following form:

$$\begin{aligned} A_0 p_{i,j}^{n+1}+A_2 p_{i,j+1}^{n+1}+A_4 p_{i,j-1}^{n+1}&=b-A_1 p_{i+1,j}^{n}-A_3 p_{i-1,j}^{n} \\ &-A_5 p_{i+1,j+1}^{n}-A_6 p_{i-1,j+1}^{n}-A_7 p_{i-1,j-1}^{n}-A_8 p_{i+1,j-1}^{n} \end{aligned} \quad (7)$$

Given $p_{i,j}{}^{n}$ is pressure of each node in the niteration step, the pressure $p_{i,j}{}^{n+1}$ of each node in the n+1 iteration step can be solved according to equation (7).

3.3 Example

The dynamic characteristics of friction pair during the engagement process are obtained by solving the numerical model of wet friction pair with radial groove. The input parameters are shown in table 2.

Through numerical solution, the dynamic load characteristics of the friction pair are obtained. The transmission torques are shown in figure 7. At the beginning of the engaging process (t=0s, n_1=1500rev/min), the process is in the stage of hydrodynamic lubrication, and the load pressure of oil

IOP Conf. Series: Materials Science and Engineering **474** (2019) 012059 doi:10.1088/1757-899X/474/1/012059

film is higher; with the increase of axial load, the thickness of oil film decreases rapidly, and the process enters the stage of boundary lubrication (t=0.25s). When the viscous shear torque T_h of oil film reaches its peak value, it begins to decrease, and the friction torque T_c increases continuously. When the relative angular velocity is 0(t=1.4s), the viscous shear torque T_h of oil film decreases to 0, and the friction torque T_c increases to the maximum value. The wet friction pair engagement process is completed.

Table 2. Input parameters in the experiment

Parameter		Value
Friction disc inner diameter	a	0.051 m
Friction disc outer diameter	b	0.0685 m
Thickness of friction material	d	5.0×10^{-4}m
Friction disc speed	n_1	1500r/min
Loading pressure	F_{app}	1.0 MPa
Moment of inertia	I	0.5 kg·m^2
Initial lubricant film thickness	h_0	2.54×10^{-5}m

The pressure distribution of the lubricating oil film is also obtained by the numerical solution. Figure 8 shows the pressure distribution of the oil film with radial groove at t=0.25s. It is found that the pressure of the oil film is 0 in the inner and outer diameters, and the pressure of the oil film is very small except around the positions of the groove, and it is periodically distributed in the circumference. The maximum value of oil film pressure appears in the middle of groove. This is because the thickness of oil film changes in the groove position. According to the hydrodynamic pressure effect, the oil film pressure in the groove position increases and the peak value appears.

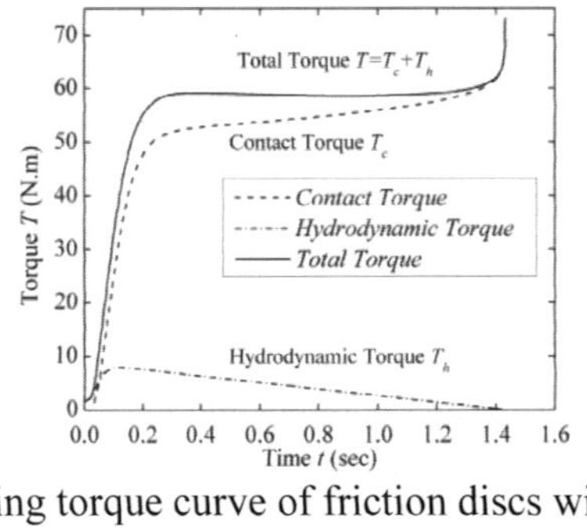

Figure 7. Engaging torque curve of friction discs with radial groove

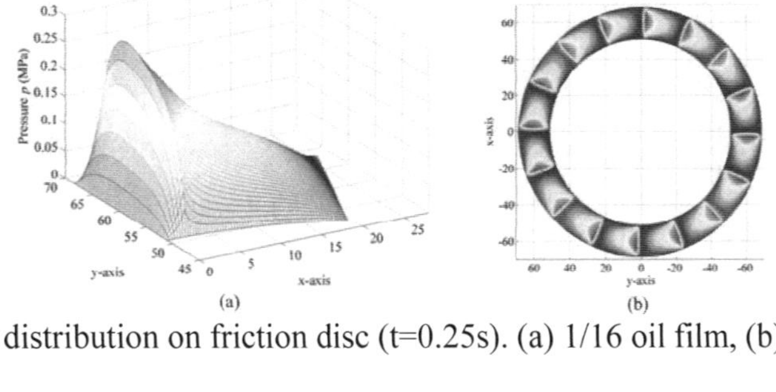

Figure 8. Pressure distribution on friction disc (t=0.25s). (a) 1/16 oil film, (b) complete oil film

IOP Conf. Series: Materials Science and Engineering **474** (2019) 012059 doi:10.1088/1757-899X/474/1/012059

4. Test bench and experimental results analysis

A test bench (figure 9) was built to test the wet friction pair with Cu-PM friction materials. It includes motor, flywheel, torque sensor, speed sensor and test box. During the test, the starting and stopping of the motor, the speed of the motor, the pressure of the hydraulic system and the oil supply of the lubrication system are controlled by computer. The maximum loading pressure is 10MPa, the maximum fuel flow is 10L/min, and the acquisition card model is MCC USB-2408 which has 12-bit resolution and 500 kS/s sampling rate.

Figure 9. Dynamic performance test bench for friction pair engaging process

The experimental data of relative rotational speed and transfer torque during the engaging process are compared with the simulation results as shown in figure 10.

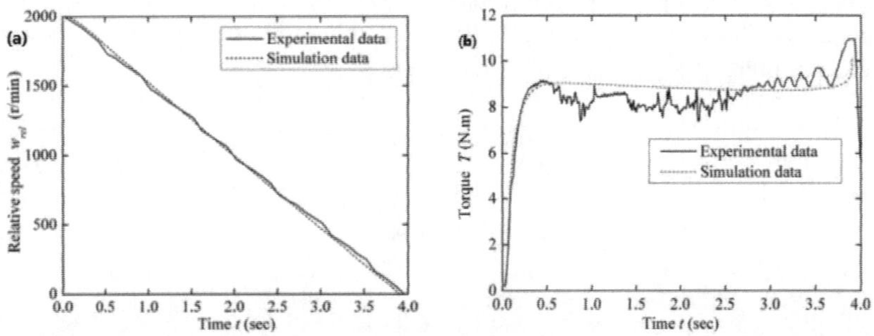

Figure 10. Comparison of experimental and simulation results (a) relative speed; (b) torque.

Figure 10 (a) shows that the experimental engaging time is t_{exp}=3.923s, which is 4.84% larger than the simulation result t_{simu} =3.903s. This is probably because of the pressure system of the test bench has some delay. The experimental peak value of the transmission torque is about T_{max_exp}=9.01N.m, and the simulation result is T_{max_simu}=9.05N.m (figure 10 (b)). The experimental results of torque and relative speed are consistent with the simulation results.

5. Conclusions

In this paper, the physical properties of Cu-PM friction material were studied, and its influence on the friction pair was analyzed. The mathematical model of engaging process of the friction pair with Cu-PM materials was established. Experimental data prove the correctness of this model.

During the engaging process, the viscous shear torque T_h gradually increases and reaches its peak value at t=0.07s, then begins to decrease gradually. Meanwhile, the friction torque T_c increases rapidly at the beginning, slows down at t=0.25s and reaches its peak value at t=1.4s, when T_h becomes 0.

The pressure of the oil film is zero in the inner and outer diameters, and the pressure of the oil film is very small except around the positions of the groove. The shape of groove plays a significant role in the oil film thickness variation. Therefore, the influence of different groove shapes, such as spiral groove and square groove, on the friction pair performance of Cu-PM is a promising direction for further research.

References

[1] Matsumoto T, Ohuma M and Kuse T 1998 *Drive System Technique* **2** p16 -23
[2] Matsumoto T 1994 *SAE Transactions: Journal of Passengers Cars* **103** p1445 -56
[3] Burton R A and Gainesburton R. *Tribology Transactions* **35**(4) p756-760
[4] Burton R A 1960 *Tribology Transactions* **3** (1) p1-10
[5] Burton R A 1980 *Wear* **59** p501-508
[6] Matsumoto T 1993 *SAE Paper 932924*
[7] Miyazaki T and Matsumoto T 1998 *Tribology* **120** (4) p393-398
[8] Komvopoulos K and Suh N P 1985 *ASME*
[9] Razzaque M M and Takahisa Kato 2001 *Tribology Transactions* **44** (1) p97-103
[10] Razzaque M M and Kato T 1999 *Journal of Tribology* **121** (1) p56-61
[11] Liu X C, Zhang Z G and Shi X H 2015 *J Chongqing (Natural Science Edition)*
[12] Zhu H Q 2012 *Sliding Friction Characteristics and Heat Load Characteristics of Wet Clutch Zhejiang University (Doctoral dissertation)*
[13] Wang H W, Zhang X Q and Zhang J L 2013 *Journal of Beijing University of Technology* **33** (1) p47-51
[14] Wang X F, Huang Q Z and Yin X L 2008 *Journal of Central South University (Natural Science Edition)* **03** p517-521

IOP Conf. Series: Materials Science and Engineering **474** (2019) 012060 doi:10.1088/1757-899X/474/1/012060

Chitosan blend iron oxide nanostructure-based biosensor for healthy & malignant tissue glucose/urea detection

M W Akram[1, 2, *], M F Alam[1, 3, 4,], H N Ji [1,2,***] A Mahmood[3], Tariq Munir[3], M Z Iqbal[4], M R Saleem[3], N. Amin[3] and A G Wu[4]**

[1] Institute of Fundamental and Frontier Sciences, University of Electronic Science and Technology, Chengdu, Sichuan province, China

[2] School of Materials and Energy, University of Electronic Science and Technology of China, Chengdu 610054, China

[3] Department of Physics, GC University, 38000, Faisalabad, Pakistan

[4] Key Laboratory of Magnetic Materials and Devices & Division of Functional Materials and Nanodevices, Ningbo Institute of Materials Technology and Engineering, Chinese Academy of Sciences, No. 1219 Zhongguan West Road, Ningbo, China

*E-mail: waseem.physicist@gmail.com; **E-mail: fakharphy@gmail.com;

***E-mail: hainingji@163.com (H N Ji);

Abstract. A urease trapped chitosan-Fe_3O_4 based nano-composite biosensor of controlled morphology is successfully fabricated by applying the aqueous chemical growth technique. Final product of nanodevice was characterized by applying various techniques. Log of the concentration of various solutions (1 to 80 mM) ranged between 0.0 to 2.5 mM for urea and glucose sample and an output signal of ~ 42 mV per decade was detected at room temperature. It disclosed the presence of chitosan composite with Fe_3O_4 results enhanced active surface area of sensing along with up to the mark precession, as compared to traditional biosensor for immobilization of enzymes. The current developed form of chitosan-Fe_3O_4 nano-composite biosensor illustrated a significant stable potentiometric response in the minimal time period. Besides, nafion/polypyrrole addition to the top surface of CH-Fe_3O_4 nanocomposite not only protected it from degradation but also upgraded the characteristic of sensor regarding rapid electron transfer enhancement and higher shelf life of this hybrid form of nanomaterial. The advanced form of this voltammetry implied that Ur-GLDH/CH-Fe_3O_4 bioelectrode is found to be very sensitive in the range of 4-120 mg/dl urea/glucose concentration and can detect as low amount as 0.4mg/dl.

1. Introduction

Initially, the glucose-based biosensors were mostly immobilized and used electrochemically to detect the glucose in blood and food products. Later on, the concept of glucose enzyme-based electrode was proposed by Clark and Lyons in 1962 [1]. In connection to this, the biosensor immobilized-glucose on an electrode provided higher simplicity and sensitivity. However, coated magnetic nanoparticles (NPs) are becoming more prominent due to their significant role in biomedical applications, for instance, hyperthermia therapy, drug delivery, biosensors and magnetic separation [2, 3]. As an example, chitosan coated magnetic nanoparticles can provide efficient solid-state nanomaterial with versatile biosensing applications. It is reported that chitosan-coated Fe_3O_4 NPs were biocompatible and exhibited strong super-paramagnetic properties hence offered low toxicity [4]. These particles could potentially improve the delivery, and recovery of biomolecules for various biomedical applications. Both chitosan and iron oxide are low-cost material which enabled an extensive use in many practical biomedical applications [5].

Chitosan-based NPs are essential and excellent stabilizing agents, film-forming abilities, mechanical strength, biocompatibility, and non-toxicity, higher permeability towards the water, surface reactivity, and cost-effectiveness among various NPs, used in this context [6]. The amino groups of chitosan provide a hydrophilic environment and thereby enhance the biocompatibility. It is reported that the pH of chitosan-coated iron oxide nanoparticles should be above 6, which promotes biomedical compatibility [7]. Nanoparticle-based biosensors are being preferred due to their large surface to volume ratio, high absorption power, and high surface reactivity nowadays [8]. These immobilized nanoparticles can transfer the electrons between electrodes via active sites of the enzyme. A preferred method of synthesizing Fe_3O_4 NPs is the co-precipitation method because it is simple and carried out under mild conditions without the use of harmful solvents [9]. A key challenge in the synthesis is to achieve optimal size and shape of the nanoparticles by controlling the pH, temperature, ionic strength and salt properties of the pH during incorporation of the surface coating agent. NPs typically aggregate due to high surface energy and magnetization.

Currently, desired, stable and well-shaped nano-biosensor fabrication is a challenging task. Anyhow, the primary focus of this study is to develop most reliable, facile, up to the mark, novel ultra-small nanoparticles (individual/hybrid) with multiple future applications.

2. Materials and method

2.1. Reagents and preparation of solutions

Iron (II) Chloride Tetrahydrate $FeCl_2.4H_2O$ (0.274M), Iron (III) Chloride Hexahydrate $FeCl_3.6H_2O$ (0.213M), Iron Acetylacetonate, Oleic Acid & Oleylamine, Chitosan, and triethylamine has been purchased from Aladdin China Mainland. Indium tin oxide (ITO) coated glass plates has been attained from the Blazers, UK. All relevant chemicals used are of molecular biology grade. The substances were prepared by deionized water.

2.2. Preparation of iron oxide nanoparticles

Combine all above chemicals into a suitable capacity of the beaker. For homogenous perform sonication of 3-4 minutes and allow it for 15 minutes of vigorously magnetic stirring. Now the solution is ready for the next step. For the further pursuit of the desired nanomaterial, place an updated form of a chemical solution in hydrothermal reactor Teflon and the internal space of autoclave around Teflon filled with 10 ml ethanol and 2.5 ml of deionized water. Proceed the reaction for approximately 04 hours at 300°C till the few nanometers of iron oxide nanoparticles were developed. On the next day put off the cover and wait for the oven cool down to room temperature. After thorough examination, the dark brown/blackish precipitate can be visualized via naked eye. Wash this prepared form of iron nanoparticles with ethanol thrice gently with the help of centrifuge having 11500 rpm for 10 min at room temperature. Next, add 97% of n-hexane C_6H_{14} for perfect dispersion and a suitable amount of cyclohexane for non-polar chain/mixing mode. Finally, store it at room temperature the nanomaterial should be dark brown/black color.

2.3. Preparation of hybrid form of chitosan with iron NPs (CH-Fe_3O_4 NPs)

Take 0.5% chitosan (CH) solution and dissolve CH (50 mg) in 100 ml of acetate buffer (0.05 M; pH 4.4). The defined amount of Fe_3O_4 nanoparticles was spread in the CH solution at 25 °C by continuous stirring to get a viscid solution of CH with homogeneously dispersed Fe_3O_4 nanoparticles. The resultant form of CH-Fe_3O_4 nano-composite was dried 60°C for 5 h and characterized.

3. Results and Discussion

3.1. Characterization

To confirm the morphological and structural analysis of iron nanoparticles (Fe_3O_4) and CH-doped iron nanoparticles (CH-doped iron NPs), the scanning electron microscopy (SEM) was employed for the structural analysis. A globular aggregated form of nanoparticles morphology was observed, the particle sizes of Fe_3O_4 and CH-Fe3O4 composite were investigated to be from 5 nm to 30 nm (indicated by marking some nanoparticles).

Figure 1 shows the scanning electron microscope (SEM) analysis of Fe_3O_4 NPs based working electrodes. A series of experiments and various synthesis processes were performed for the perfect form of fabricated biosensing working electrodes within a range of 5nm to 30 nm.

Figure 1. Scanning Electron Microscopy (SEM) Analysis of Iron Oxide Nanoparticles

3.2. XRD Analysis

Calculation: Structural parameters such as lattice constant "a", unit cell volume and crystallite size "D" were calculated from the XRD data using equations (1- 3) described below.

$$a = \frac{\lambda}{2sin\theta} = \sqrt{h^2 + k^2 + l^2} \quad ....(1), \quad V_{cell} = a^3 \; ....(2), \quad D = \frac{k\lambda}{\beta_{hkl}cos\theta} \; ....(3)$$

Here *h*, *k*, and *l* are the Miller indices, V is the unit cell volume, Z represents eight molecules per unit cell of the spinel structure, "N_A" is the Avg.'s number and "*M*" is the molecular weight of the sample. Here "k" is the shape factor, λ is the wavelength of the X-rays, θ is the diffraction angle, and β_{hkl} represents the full-width half maxima.

Analysis: figure 2 shows the XRD pattern of Fe_3O_4 ferrite synthesized by chemical method. A bi-phase structure was observed. The observed patterns have prominent diffraction peaks with these reflection planes (220), (311), (222), (400), (440) and (511), which is indexed by computer software Jade 5 [10]. The lattice constant "*a*" was measured by the equation. Corresponding to the heat treatment temperature, the crystallite size was 94 nm. Crystal sizes of less than 50 nm have been described in this study, and it has been reported that crystalline sizes below 50 nm can be used to provide a suitable signal to noise ratio in high-density recording media received [11]. Many researchers have previously reported similar results for polycrystalline spinel ferrites synthesized by several methods [12-13].

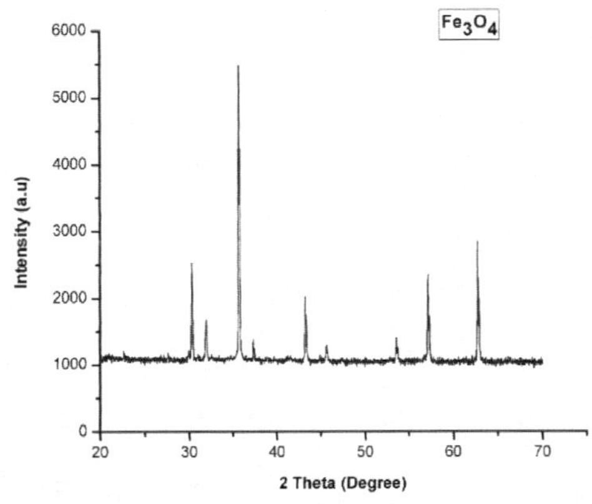

Figure 2. XRD pattern of Fe_3O_4 ferrites.

Table 1. Lattice constant (a), crystallite size (nm), unit cell volume of Fe_3O_4.

Crystallite size (nm)	Lattice constant (Å)	Volume (Å)^3
94	8.4146	595.799

3.3. Sensitivity measurements

As a result of hydrolysis unease enzyme presented onto the surface of Iron Oxide NPs along with polypyrrole composite mounted through a glass substrate via Ti and Au thin film deposition. The potentiometric measurement was conceded out at different values of pH ranging from 3 to 11 as depicted by figure 3. Physiological investigation response for calibration of stability and performance Fe_3O_4 NPs based biosensor were inspected using a pH range of 3 to 11, but the neutral pH, which was 7, was selected as the best pH range for stability, performance, and sensitivity parameter. Besides, the selectivity of the fabricated nano-biosensing electrode or were tested by the accumulation of various interfaces, e.g., glucose and uric acid to test the solution containing urea.

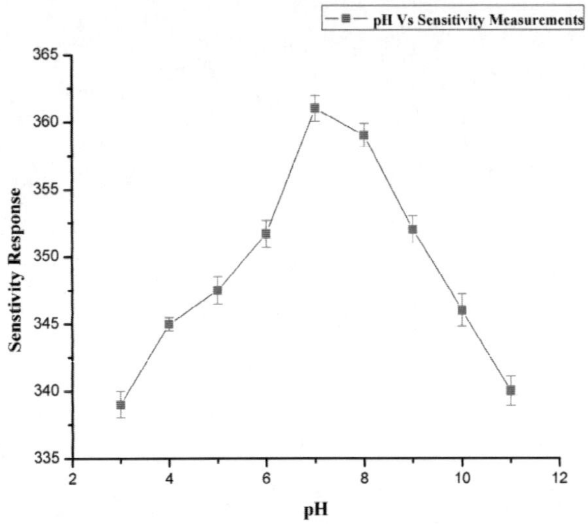

Figure 3. Sensitivity response of biosensor based on Fe_3O_4 at pH values range of pH 3-11.

The measured potential difference values (EMF values) the from the working electrode of biosensor were found influence free from a volume of the test solution and the area of the dipped tip of the working electrode in the test solution. A series of experiments were performed for precise measurements and fabricated working electrode of good reproducibility and repeatability. As sensitivity response of nano-based biosensor was shown in graph depicted EMF (mV) vs. log of concentration (range 1 to 80 mM) 0.00 to 2.5 mM. As the described graph show that EMF concentration rises as increasing of solution (glucose and urea concentration). Fe_3O_4 NPS based biosensor shows the linear curve trend between EMF and sample concentration as shown in figure 4. It is evident from the figure as long as the biological sample became concentrated due to stroma's or cancerous tissue EMF (mV) polarity signal shows an increasing trend too, similar data were already published in many articles [14].

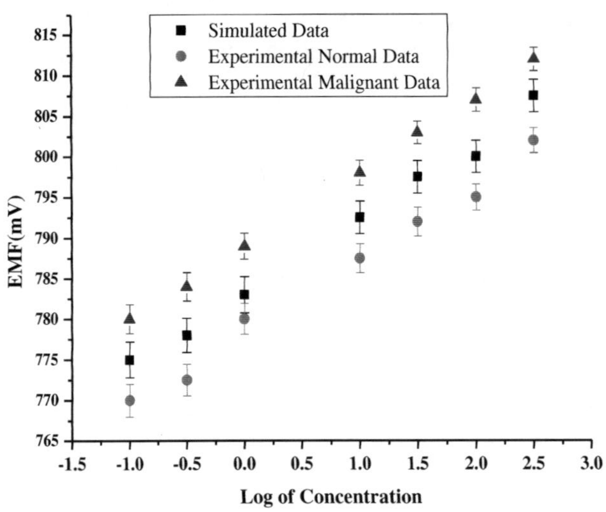

Figure 4. Sensitivity response curve of Fe_3O_4 NPs at pH=7.

3.4. Capsulation of chitosan coating in iron oxide NPS layer

Take a suitable amount of pluronic acid and iron oxide nanoparticles allow for homogeneity till 30 minutes and incorporate dropwise form of Ccitosan solution and allow rotating by using magnetic stirrer until 4 hours at room temperature for reaction process after reaction completion washed by using suitable quantity of ethanol thrice and stored final product after filtration.

4. Conclusion

A urease-trapped CH-Fe_3O_4 nano-biocomposite-based urea biosensor of desired morphology has been successfully fabricated by applying the aqueous chemical growth technique and characterized by various methods. Log of the concentration range of multiple solutions (1mM to 80 mM) 0.00 to 2.5 mM for urea and glucose and an output signal (≈42 mV per decade) was detected at room temperature.

The current developed form of CH-Fe_3O_4 nanocomposite biosensor illustrates significant stable potentiometric response in very minimal time overall performance like reproducibility, selectivity, stability, and sensitivity of the biosensor is very up to the mark and was excellent due to a versatile and unique characteristic magnetic based biosensor valuable for medical application and other related areas like to decide non-hygienic form of food. Additionally, CH-Fe_3O_4 based sensor is also achievable for clinical diagnosis as well. However, the said biosensor showed slight influence once inspected through the addition of a variety of interferers such as glucose and uric acid of healthy and abnormal tissues/cells, which reflects that the sensor is helpful for urea detection along with such interferes.

Acknowledgments

This work was supported by the National Natural Science Foundation of China 51302030, 51272038, 61474015, 61474014, and the National Program on Key Basic Research Project (973 Program) 2013CB933301, 2018YFA0306100.

References

[1] Yang L, Ren X, Tang F and Zhang L 2009. *Biosensors and Bioelectronics*, **25**(4), 889-5.
[2] Sophie L, Delphine F; Marc P; Alain R; Caroline R; Luce V E and Robert N M 2008 *Chemical Reviews* **108** (6), 2064-110.
[3] Harivardhan Reddy L; José L A; Julien N and Patrick C 2012 *Chemical Reviews* **112** (11), 5818-78.
[4] Nehra P, Chauhan R P, Garg N and Verma K 2018. *British Journal of Biomedical Science*, **75**(1), 13-8
[5] Miao Y and Tan S N 2000. *Analyst*, **125**(9), 1591-4
[6] Aziz T, Masum S M, Qadir M R, et al. 2016 *Int Res J Pure Appl Chem.* **11**(1):1–9.
[7] Kaushik A, Solanki P R, Ansari A A, Ahmad S and Malhotra B D 2008. *Electrochemistry Communications*, **10**(9), 1364-8.
[8] Zhang F 2015. *Photon Upconversion Nanomaterials* Springer Berlin Heidelberg 255-84
[9] Arakha M, Pal S, Samantarrai D, Panigrahi T K, Mallick B C, Pramanik K and Jha S 2015. *Scientific Reports*, **5**, 14813.
[10] Hankare P P, Jadhav S D, Sankpal U B, Chavan S S, Waghmare K J and Chougule B K 2009. *J. Alloys and Compounds*, **475**(1-2), 926-9.
[11] Ashiq M N, Ehsan M F, Iqbal M J and Gul I H 2011. *J. Alloys and Compounds*, **509**(16), 5119-5126.
[12] Panda R N, Shih J C and Chin T S 2003. *J.Magnetism and Magnetic Materials*, **257**(1), 79-86.
[13] Tahar L B, Artus M, Ammar S, Smiri L S, Herbst F, Vaulay M J and Fievet F 2008. *J. Magnetism and Magnetic Materials*, **320**(23), 3242-50.
[14] Zhu C Z, Yang G H, Li H, Du D and Lin Y H 2014. *Analytical Chemistry* **87** (1) 230-49.

Effect of austenitizing temperature on microstructure formation of quenched X65 micro-alloyed pipeline steel

Z X Qiao[1,*], H X Mi[1], J Huo[2], Y B Wan[1] and X Chen[1]

[1]School of Mechanical Engineering, Tianjin University of Commerce, Tianjin 300134, China
[2] Tianjin Heavy Industries Research & Development Co., Ltd., Tianjin 300457, China

*Email: qzhxia@tjcu.edu.cn

Abstract. The effect of austenitizing temperature on the microstructure formation in micro-alloyed X65 pipeline steel was investigated using an optical microscope (OM), transmission electron microscope (TEM) and microhardness measurement. The results show that the austenite grains coarsening temperature of X65 steel reaches 1100°C. When austenitization is carried out below 1000°C, a large number of fine microalloying carbides precipitate during quenching within some α ferrite laths, with a uniform orientation. With austenitizing temperature increasing, the amount of precipitated carbides increases. Such fine carbide precipitates play significant strengthening for the steel. When the austenitizing temperature exceeds 1000°C, no carbide precipitation occurs in α ferrite laths, because microalloying solute atoms have too low mobility to precipitate under such conditions. This results in that X65 steel exhibits the maximum microhardness when quenched from 1000°C.

1. Introduction

Low carbon micro-alloyed high-strength pipeline steel has been recognized as a preferable material and widely used for transporting oil and gas in the deep-water environment, due to its high strength-toughness combination [1-3]. Its excellent mechanical performance results mainly from the grain refinement and precipitation strengthening through the formation of fine and highly stable particles of microalloying carbides. Grain refinement is principally achieved from the fine carbide particles strain induced in austenite during hot deformation, which can inhibit austenite grain growth during reheating and retard the recrystallization of austenite. The particles responsible for precipitation strengthening mainly precipitate during or after $\gamma \rightarrow \alpha$ phase transformation on the γ/α interface or in the matrix of ferrite [4-8].

At present, quenching-tempering treatment after rolling is commonly applied in the production of micro-alloyed pipeline steel to obtain optimum comprehensive mechanical properties by grain refining and precipitation hardening. Thus, it is necessary and quite essential to design optimal parameters for the technology of quenching-tempering treatment (QTT). Reheating is the first critical step for QTT because the initial state of prior austenite obtained in this process can influence the development of final microstructure significantly. Many previous studies have been done on the relationship between reheating temperature and the final microstructures after quenching to room temperature, and it was shown that the reheating temperature influences the microstructures and properties of steels mainly through its effect on the size of prior austenite grains [9-11]. For steels with additions of microalloying elements like niobium, vanadium, and titanium, fine sparingly soluble precipitates can efficiently inhibit austenite grain growth at relatively low temperatures. As heating temperature increases to some level, microalloying carbide particles completely dissolve into austenite, and afterward, sudden abnormal grain growth occurs. Therefore, aiming to obtain fine austenite grains, the high heating temperature is usually avoided during QTT. In previous studies, few researchers paid much attention to the influence of austenitization temperature on the behavior of $\gamma \rightarrow \alpha$ phase transformations during the cooling process to room temperature from the austenite region. However, for micro-alloyed steels, the reheating temperature is a significant factor responsible for the dissolution extent of microalloying atoms into austenite and thus would affect the precipitation behavior of carbide particles as phase transformation occurs in the continuous cooling process.

In the present work, the investigation was carried out on how austenitizing temperature influence the produced microstructure after quenching to room temperature from austenite, using X65 steel, typical micro-alloyed pipeline steel. TEM observations were performed on the obtained microstructures, and furtherly the results were confirmed by microhardness measurement. It is expected to provide a helpful theoretical basis for optimizing the technological parameters for quenching-tempering treatment.

2. Experimental materials and procedures

The nominal composition of the investigated X65 pipeline steel is (in weight percent): 0.09C, 1.29Mn, 0.29Si, 0.15Ni, 0.18Mo, 0.06Cr, 0.034Al, 0.13Cu, 0.05V, 0.03Nb and 0.001Ti. Heat treatments were conducted in a Muffle furnace with a thermocouple for detecting the temperature. The samples with a dimension of 10mm×10 mm × 10mm were mechanically cut from an X65 hot-rolled steel pipe. To determine the effect of austenitizing temperature on the microstructure, the samples were heated to eight different austenitization temperatures respectively (850, 900, 950, 1000, 1050, 1100, 1150, and 1200°C), held for 10 min, and then quenched into the water at room temperature.

Heat-treated samples were subjected to microstructural examination via an OLYMPUS optical microscope and JEM-100CX II transmission electron microscope. Prior austenite grain boundaries were displayed by etching metallographic specimens with a saturated picric acid solution after mechanical polishing. Austenite grains sizes were determined by mean linear intercept method. Foil specimens for TEM were prepared by mechanical grinding to about 100μm in thickness, followed by dual jet electrolytic polishing using an electrolyte of 10% perchloric acid and 90% ethanol. To determine the hardness of the quenched samples, microhardness measurements were performed on a Vickers microhardness tester, using a load of 200g.

3. Results and Discussion

3.1. Austenite grain size

Figure 1 presents the typical optical micrographs of X65 pipeline steel specimens quenched from different austenitizing temperatures, revealing prior austenite grain boundaries. With the austenitizing temperature increased from 900°C to 1200°C, the size of austenite grains grows markedly. The average sizes of austenite grains of the specimens were measured, and the results are shown in figure 2, as a function of austenitizing temperature. As can be seen, austenite grains grow slightly (from 10 to 20μm) in relatively low-temperature range (850~1050°C), and then a sudden grain coarsening occurs (above 1100°C), up to 110μm as the austenitizing temperature increased to 1200°C. Thus the critical grain coarsening temperature of this steel is about 1050°C. The weight percent of Nb dissolved in austenite at various austenitizing temperatures can be calculated according to the following empirical equation (1), and then the undissolved Nb weight percent were also obtained [12]. The calculation results were plotted in figure 2, together with the austenite grain size values:

$$\log_{10}[\mathrm{Nb}][\mathrm{C}+\frac{12}{14}\mathrm{N}]=-\frac{6770}{T}+2.26 \qquad (1)$$

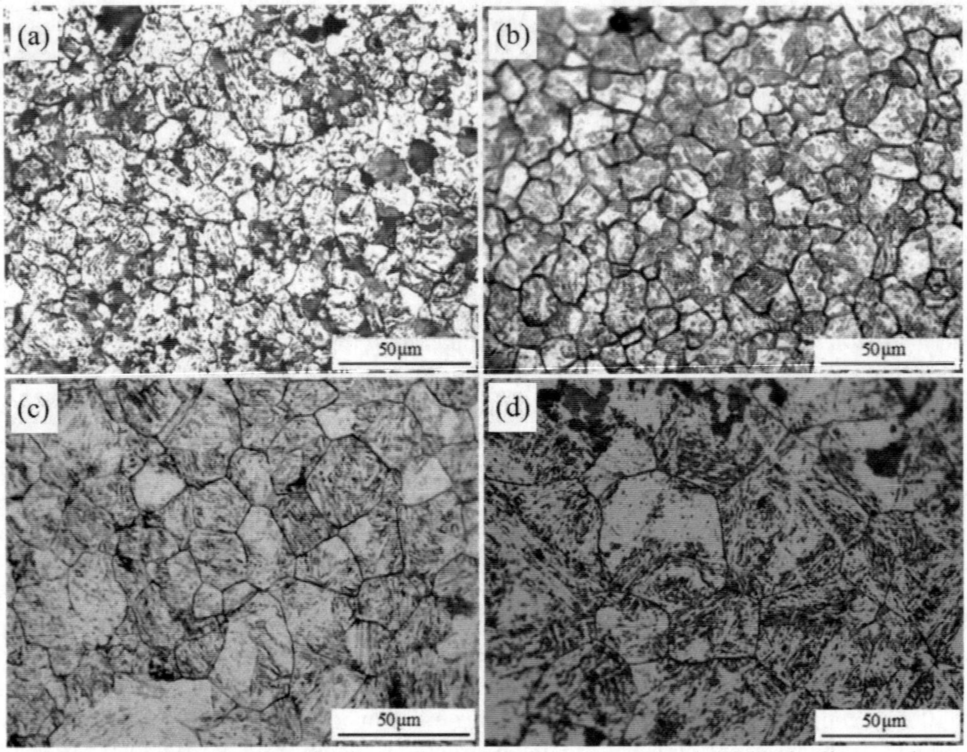

Figure 1. Typical optical micrographs displaying prior austenite grains in X65 pipeline steel specimens austenitized at different temperatures for 10min, (a) 900°C, (b) 1000°C, (c) 1100°C, (d) 1200°C

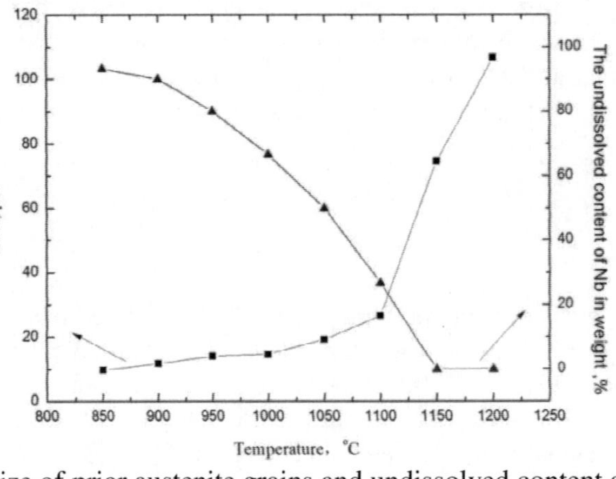

Figure 2. The average size of prior austenite grains and undissolved content of Nb in X65 pipeline steel specimens, as a function of austenitization temperature.

From figure 2, the undissolved Nb content decreases monotonously with increasing austenitizing temperature. It is noted that the decrease ratio below 1050 °C is lower than that above 1050 °C. At 1150°C, all added Nb dissolved into austenite. At temperatures below 1050°C, more than half of the added Nb keeps out of solution and exists in the form of carbide particles, which can contribute to restraining austenite growth. Therefore, austenite grains keep slow growth as the austenitizing temperature does not exceed 1050°C. This indicates that the percent of undissolved micro-alloy carbide particles is a critical factor for prior austenite grains size.

TEM microstructural examinations were carried out on the specimens quenched from different austenitizing temperatures, as shown in Figure 3. It was found that the produced microstructure consists of aggregates of α ferrite laths, parallel to each other. A high density of dislocations exists within the α ferrite laths. Some adjacent α ferrite clusters angled each other at about 60°, as shown in figure 3b. The widths of α ferrite laths are not uniform, in the range of 0.1~1μm. Detailed observations indicate that a lot of fine needle-like particles are distributed in relatively wide α ferrite laths. It is also noted that such wide laths with the distribution of fine particles can only be observed in the specimens austenitized below 1000°C. No fine precipitates were found within the α ferrite laths in the specimens austenitized at 1100 and 1200°C. Figure 3c is a micrograph with higher magnification, in which the particles are shown more clearly. The needle-like particles are about 100 nm in length, and most of them are distributed in a common direction. This indicates that the carbide particles are precipitated coherently from the parent austenite, but not like the auto-tempering process of martensite. Precipitates in auto-tempered martensite are usually round in shape and without preferred orientation [13]. Previous studies also proved these oriented carbide particles to be M_3C phase [14]. The formation of coherent interfaces between M_3C precipitates and austenite matrix phase can reduce the interfacial nucleation energy. These fine carbide particles contribute much to the strengthening of the steel, named precipitation hardening. It should be noted that such fine precipitates were not observed in the specimens austenitized at higher temperatures (1100 and 1200°C), as shown in figure 3d.

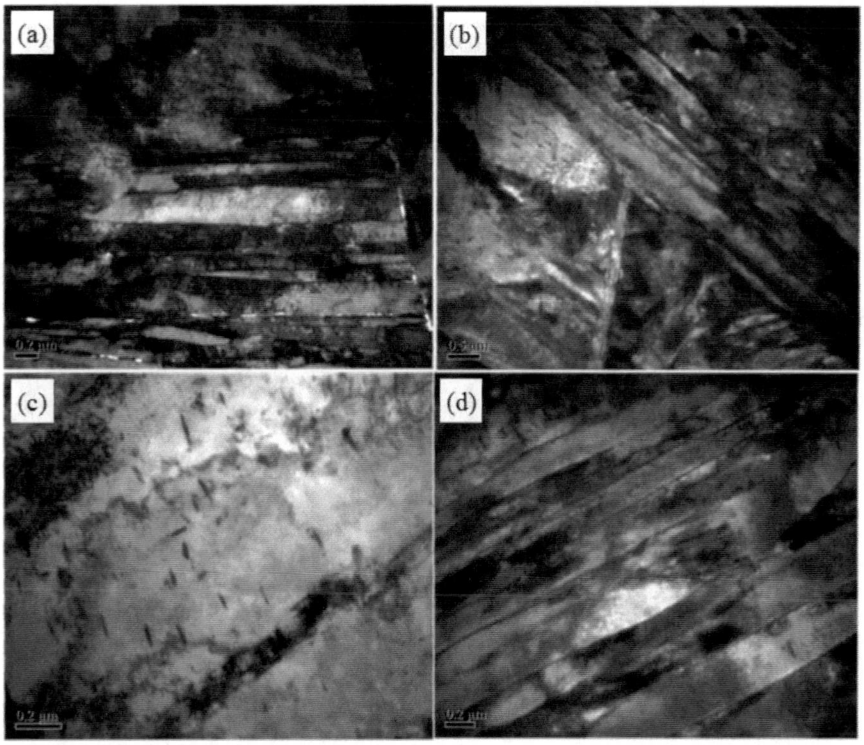

Figure 3. TEM images of water-quenched X65 pipeline steel specimens austenitized at different temperatures for 10min, (a) 900°C, (b) 950°C, (c) 1000°C, (d) 1100°C

3.2. *Microhardness*

Figure 4 shows the Vickers microhardness results of the quenched specimens from different austenitized temperatures. It was shown that the microhardness value (HV) increases rapidly with austenitizing temperature increasing from 900 to 1000°C, and then decreases significantly. The specimen austenitized at 1000°C exhibits the maximum microhardness. It was known from figure 2 that austenite grains size increases monotonously with increasing austenitizing temperature. The width of α ferrite laths formed within coarse austenite grains would be relatively wide and long, and the steel with such microstructure will exhibit low hardness. Thus, from the point of grains size, microhardness value should decrease monotonously with increasing austenitizing temperature. The significant decrease of microhardness as the austenitizing temperature exceeds 1000°C agrees with this case, as shown in figure 4. But the increasing trend of hardness value in a temperature range below 1000°C is in consistent with this and should be attributed to some other factor. With the TEM examination results being fully considered, the unusual increase below 1000°C should be attributed to the precipitation of fine carbide particles within α ferrite laths. In the process of quench cooling, micro-alloying solute atoms in prior austenite precipitate finely as carbides, which are uniformly distributed within lath-like α ferrite. These fine carbides supply significant precipitation strengthening for the steel. With the austenitizing temperature increasing, more micro-alloying carbides decompose, and more solutes go into austenite, and thus more fine needle-like precipitates form within α ferrite laths. Therefore, higher austenitizing temperature leads to higher microhardness. At the early stage that micro-alloying atoms enter the austenite matrix, they aggregate mainly at sites near dislocations. When precipitation occurs, the micro-alloying carbides nucleates preferentially at the dislocations. Once austenitizing temperature increases to a high enough extent, the micro-alloying atoms, such as Nb, Mo, have sufficient energy to leave dislocations and dissolve into austenite matrix stably. More time and energy are needed for the micro-alloying atoms stably existing in austenite matrix to diffuse and nucleate as carbides, and thus the solute atoms tend to keep within austenite after the specimen is quenched into water. This can explain why precipitated carbides were not observed in specimens austenitized at 1100 °C and above.

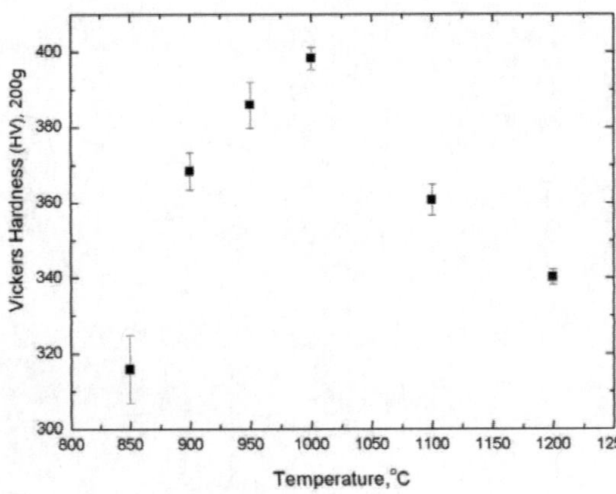

Figure 4. Vickers hardness values of water-quenched X65 pipeline steel specimens austenitized at different temperatures for 10min, with a load of 200g.

From the above analysis, it can be concluded that there exists an optimal quenching temperature for microalloyed steels, at which a lot of fine carbide particles can form within lathlike α ferrite and provide high strength for steels. The optimum austenitizing temperature for the present X65 pipeline steel is about 1000°C.

4. Conclusions

Austenitizing temperature has a considerable effect on the precipitation behavior of microalloying carbides. The investigated X65 microalloyed pipeline steel has a critical coarsening temperature of austenite grains of 1050°C. The quenched specimens subjected to austenitization at about 1000°C exhibit the maximum hardness, which is attributed to the significant precipitation of fine carbides at this temperature. If austenitized at lower temperatures, there would be fewer microalloying solute atoms into austenite, and thus fewer precipitates form during the subsequent cooling process, leading to less precipitation strengthening. On the contrary, if austenitization proceeds at too high temperatures, microalloying solute atoms would be dissolved stably into austenite matrix, and it would be more difficult for carbide precipitates to nucleate.

Acknowledgments.

This work was supported by the Key Technologies R & D Program of Tianjin (Grant No. 18YFZCGX00050) and the National Natural Science Foundation of China (Grant Nos. 51204121).

References

[1] Reip C P, Shanmugam S and Misra R D K 2006 *Mater. Sci. Eng. A***424** 307-17
[2] Joseph O O, Loto C, Sivaprasad S, et al. 2018 *Engineering Fracture Mechanics* **189** 1-12
[3] Lan L Y, Qiu C L, Zhao D W, et al. 2011 *J. Iron Steel Res. Int.* **18** 57-63
[4] Rakhshkhorshid M, Zadeh H M and Hashemi S H 2015 *Int. J Adv. Manuf. Tech* **79** 1623-31
[5] Shen W F, Zhang C, Zhang L W et al. 2018 *J. Mater. Eng. Perform.* **27** 4616-24
[6] Qi J J and Huang Y H 2006 *Microalloyed Steels* (Beijing, Metallurgical Industry Press) p 1-5
[7] Qiao G Y, Cao Y B, Liao B et al. 2018 *T. Indian I. Metals* **71** 627-37
[8] Shanmugama S, Ramisettib N K, Misra R D K et al. 2007 *Mater. Sci. Eng. A* **460-461** 335-43
[9] Seikh A H, Soliman M S, Al Majid A et al. 2014 *Adv. Mater. Sci. Eng* **2014** 1-8
[10] Maropoulos S, Karagiannis S and Ridley N 2008 *Mater. Sci. Eng. A* **483-484** 735-9
[11] Lee S J, LeeY K 2005 *Materials Science Forum* **475-479** 3169-72
[12] Porter L F, Repas P E 1982 *J Met.* **34** 14-21
[13] Bhadeshia H K D H 2001 *Bainite in Steels Transformations, Microstructure and Properties, 2nd ed.* (UK, IOM Communications, Ltd.) p 67
[14] Liu C X, Zhang D T, Liu Y C et al. 2011 *Nucl. Eng. Des* **241** 2411-5

3D processing maps and a modified constitutive model of 16Cr-5Ni-Mo supermartensitic stainless steel

Y W Sun*, Y P Zhong, L S Wang and F X Fan
Luoyang Ship Material Research Institute, Luoyang 471000, China
*E-mail: weiyong09@163.com

Abstract. The hot deformation behavior of 16Cr-5Ni-Mo supermartensitic stainless steel was investigated using a Gleeble-1500D thermo-mechanical simulator in the temperature range of 850 to 1150℃ and strain rate range of 0.01 to 10 s^{-1}. The hot deformation apparent activation energy of the steel was obtained as 424.173 kJ/mol. Arrhenius type constitutive equation was established according to flow stress data to describe the flow behavior. Meanwhile, the 3D maps of power dissipation and processing based on dynamic materials mode (DMM) were developed for the typical strains of 0.2, 0.4, 0.6 and 0.8, respectively. The results show that with the high deformation temperature (1000-1150℃) and low strain rate (0.01—0.1 s^{-1}), it is dominated by the dynamic recrystallization (DRX) for the softening mechanism. And the critical strains for the onset of DRX at different conditions are obtained. The optimized processing parameters for hot working of 16Cr-5Ni-Mo supermartensitic stainless steel are in the temperature range of 980—1150℃ and strain rate range of 0.02—0.36 s^{-1} at strains of 0.6 and 0.8.

1. Introduction

Supermartensitic stainless steels are widely used as structural components in applications such as impeller blade [1], propeller blade bolts [2], power plant [3], high strength fasteners served in the marine environment, and so on. These steels have excellent mechanical properties, good corrosion resistance, good fatigue property, and excellent weldability [4-5]. And they can be used as bars, sheets, and forged parts. 16Cr-5Ni-Mo steel is a low-carbon supermartensitic stainless steel containing less than 0.05 wt.% C. The steel is strengthened by the solution treatment and lath martensite transformation. Recently, some high strength 16Cr-5Ni-Mo steel fasteners used as 12.9 grade are serving in safety in marine engineering equipment. Due to the excellent corrosion resistance, high strength, and toughness, the need of 16Cr-5Ni-Mo is huge in the future market. However, some defects often occur during the forging of products resulting in the waste of time and economic cost. Hot deformation behavior research of steels that usually conducted in the stability range corresponding to the single austenite phase plays an essential role in hot working for the production of parts [6], and it is necessary to study the constitutive models and processing maps of the steel.

Recently, the investigations of hot deformation are focused on the aluminum alloys, magnesium alloys, copper alloys, titanium alloys, alloy steel, superalloy, and duplex stainless steel, etc. [7-10]. The hot deformation of supermartensitic stainless steels has been fewer reported in recent years, especially on the aspect of the onset of DRX. And it is vital that these steels don't have perfect hot workability due to the harmful phase under the high temperature. The investigation of 16Cr-5Ni-Mo supermartensitic stainless steel should be paid attention to the hot deformation behavior which is primarily used as an essential engineering material. In this study, the effects of deformation strain, temperature, and strain rate on the hot deformation behavior of 16Cr-5Ni-Mo steel are investigated while the typical Arrhenius model constitutive equation is also established. It is noted that in this work, three-dimensional (3D) maps are developed to instruct the hot deformation process and provide the optimum hot working parameters. Therefore, the present work is helpful in solving the problems of 16Cr-5Ni-Mo steel during practical production.

2. Materials and Experimental Procedures

The chemical compositions of the 16Cr-5Ni-Mo supermartensitic stainless steel used in this work are shown in table 1. The steel was supplied with a diameter of 95 mm bar, which was prepared by vacuum induction melting and electroslag remelting. Then, the ingots were forged to obtain the homogeneous microstructure and reduce the defects.

Hot compression tests of the 16Cr-5Ni-Mo steel were carried out on a Gleeble-1500D thermo-mechanical simulator at the temperature of 850, 900, 1000, 1100 and 1150 ℃ with strain rates of 0.01, 0.1, 1 and 10 s^{-1}, respectively. Before each test, the sample was maintained at the solution temperature of 1150 ℃ to achieve a stable microstructure. Every sample was hot compressed to a total true strain of 0.9. And all the samples were quenched in water quickly after hot deformation to preserve the deformed microstructure.

The compressed samples were sectioned parallel to the compression axis and polished according to the standard procedure and then etched with a solution of 5g $FeCl_3$, 10 ml HCl and 50 ml H_2O. And then the microstructures were observed using ZEISS Observer. Zim optical microscopy.

Table 1.Chemical compositions of 16Cr-5Ni-Mosupermartensitic stainless steel used in this study (wt%)

C	Si	Mn	S	P	H	N	O	Al	Cr	Ni	Mo
0.04	0.50	0.80	0.0016	0.0076	0.00017	0.028	0.0023	<0.010	16.08	4.43	1.34

3. Results and Discussion

3.1. Constitutive equation of 16Cr-5Ni-Mo steel

Constitutive equations are often used to predict the resistance to deformation of materials during forging, rolling and so on. In this study, the class Arrhenius constitutive equations [11] can be summarized with the deformation parameters of flow stress σ, deformation temperature T and strain rate $\dot{\varepsilon}$:

$$\dot{\varepsilon} = A_1\sigma^{n'} \exp(-Q/RT) \qquad (\alpha\sigma<0.8, \text{ low stress}) \tag{1}$$

$$\dot{\varepsilon} = A_2 \exp(\beta\sigma)\exp(-Q/RT) \qquad (\alpha\sigma>1.2, \text{ high stress}) \tag{2}$$

$$\dot{\varepsilon} = A[\sinh(\alpha\sigma)]^n \exp(-Q/RT) \qquad (\text{for all stress}) \tag{3}$$

where A_1, A_2, A, α, β, n, n', and β represent material constants; Q is the apparent activation energy, which is described as the degree of difficulty for deformation; the value of R is 8.314 $J\cdot mol^{-1}\cdot K^{-1}$; T is Kelvin temperature. And the value of α is calculated by [12]: $\alpha = \beta/n'$.

Furthermore, the influence of strain rate and temperature on the hot deformation behaviors can be described by the Zener-Hollomon parameter Z [13], which is shown in equation (4). And then substituting equation (3) into equation (4) gives equation (5).

$$Z = \dot{\varepsilon} \exp[\frac{Q}{RT}] \tag{4}$$

$$Z = A[\sinh(\alpha\sigma)]^n \tag{5}$$

With the natural logarithms based on Equations. (1), (2) and (5), the material constants and apparent activation energy are obtained as follows:

$$\ln \dot{\varepsilon} = \ln A_1 - \frac{Q}{RT} + n' \ln \sigma \tag{6}$$

$$\ln \dot{\varepsilon} = \ln A_2 - \frac{Q}{RT} + \beta\sigma \tag{7}$$

$$\ln Z = \ln A + n \ln[\sinh(\alpha\sigma)] \tag{8}$$

In this research, the true stress at the total strain of 0.9 is used for the above linear regression analysis to determine the relevant parameters of the 16Cr-5Ni-Mo steel. The relationship among $\dot{\varepsilon}$, σ and T of the steel is illustrated in figure1. The value of n' can be obtained from the slope of ln(strain rate)-lnσ at the temperature of 1150℃, while β is determined from the slope of ln(strain rate)-σ at the temperature of 850℃. For the certain temperature and strain rate, the apparent activation energy Q is expressed as:

$$Q = R\frac{\partial \ln[\sinh(\alpha\sigma)]}{\partial(1/T)}\bigg|_{\dot{\varepsilon}} \frac{\partial \ln \dot{\varepsilon}}{\partial \ln[\sinh(\alpha\sigma)]}\bigg|_T = RS_1S_2 \tag{9}$$

From figure1(c) and (d), the slopes of ln[sinh($\alpha\sigma$)]-1/T and ln(strain rate)-ln[sinh($\alpha\sigma$)] is obtained on average. And the values of S_1 and S_2 are 5.028 and 10.147, respectively. The values of the 16Cr-5Ni-Mo steel constants during hot deformation can be calculated as follows: n'=6.2343, β=0.0677, α=0.011, n=4.985, ln$A$=33.510, Q=424.173 kJ/mol. Therefore, based on the above experimental data and model, the Arrhenius constitutive equation of the steel can be expressed as follows:

$$\dot{\varepsilon} = 3.57 \times 10^{14}[\sinh(0.011\sigma)]^{4.985} \exp(-\frac{424173}{RT}) \tag{10}$$

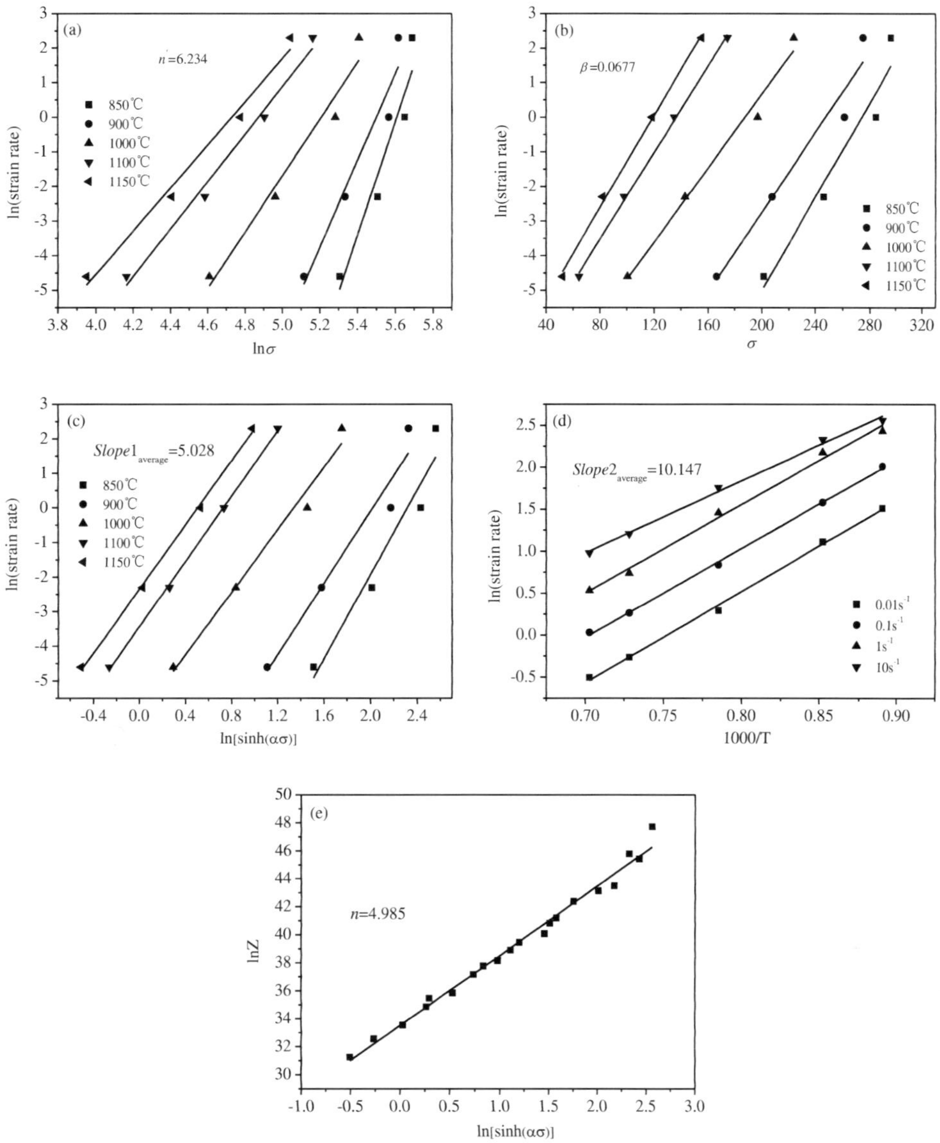

Figure 1 Relationships between temperature, peak stress and strain rate of 16Cr-5Mo-Ni steel: (a) ln(strain rate)-lnσ,(b) ln(strain rate)-σ; (c) ln(strain rate)-ln[sinh($\alpha\sigma$)]; (d) ln[sinh($\alpha\sigma$)]-$1/T$; (e) ln[sinh($\alpha\sigma$)]-ln$Z$

3.2. Constitutive models considering dynamic recrystallization (DRX) of 16Cr-5Ni-Mo steel

In above investigations, using of Arrhenius mode constitutive equation has been the common modeling approach for predicting the deformation parameters during hot deformation. However, the model consists of a few material constants, which can derive from the hot compression tests. And it is important that the influence of dislocation mechanism is ignored in the above equation. During the deformation, the dislocation mechanism plays an important role on the DRX. Hence, it is necessary to establish a model considering the DRX of 16Cr-5Ni-Mo steel by the Sellars model. The Sellars models about the critical conditions for DRX are shown as follows [14]:

$$\varepsilon_c = k\varepsilon_p \quad (11)$$

$$\varepsilon_c = aZ^b \quad (12)$$

where k, a and b are material constants; ε_c is the critical strain; ε_p is the peak strain, which can be obtained from the true stress-strain curves.

Based on the DRX kinetic theory, the onset of DRX is dependent on the work hardening rate θ, and the lnθ-ε curves are fitted and smoothed using a third-order polynomial to obtain the critical strain for DRX[15]. Figure 2 shows the relationship between $-\partial(\ln\theta)/\partial\varepsilon$ and ε at different conditions. From Figure 2, the critical strain for DRX at evaluated temperatures and strains can be calculated.

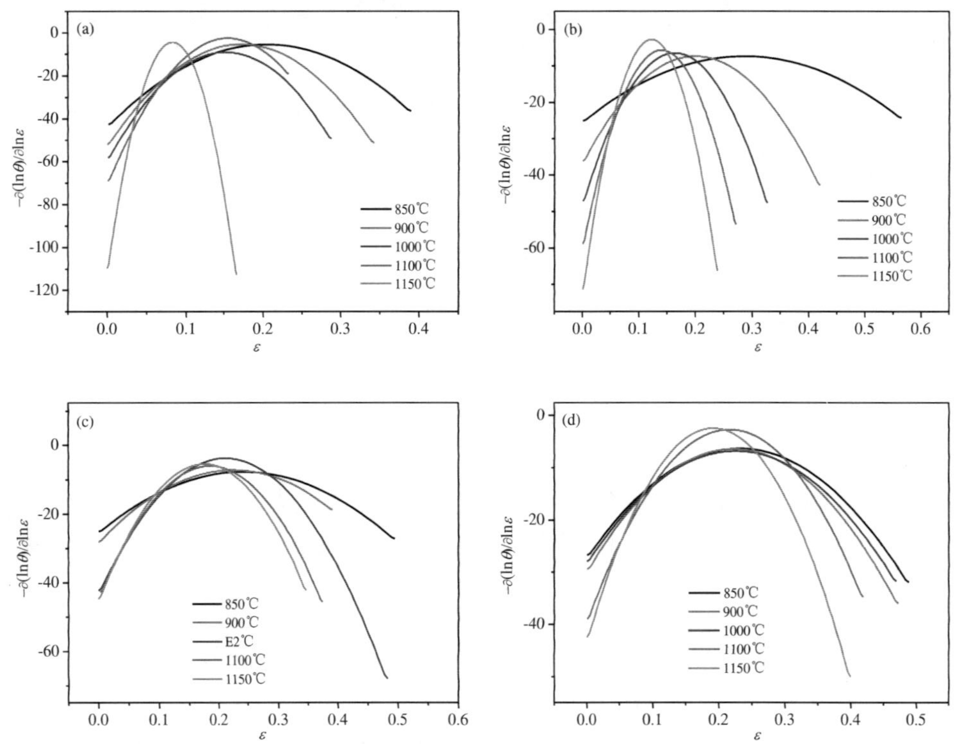

Figure 2. The relationship between $-\partial(\ln\theta)/\partial\varepsilon$ and εat different strain rate:

(a) 0.01 s^{-1}; (b) 0.1 s^{-1}; (c) 1 s^{-1}; (d) 10 s^{-1}

Meanwhile, by plotting the values of ln(ε_c) concerning ln(Z) and linear regression analysis of 16Cr-5Ni-Mo steel as shown in Figure 3, the dependence of ε_c on Z can be described as:

$$\ln\varepsilon_c = 0.051\ln Z - 3.726 \quad (13)$$

$$\varepsilon_c = 0.024Z^{0.051} \quad (14)$$

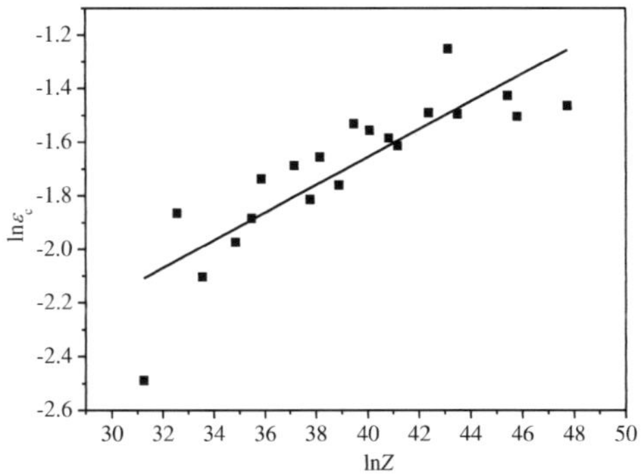

Figure 3. Relationship between ln(Z) and ln(ε_c)

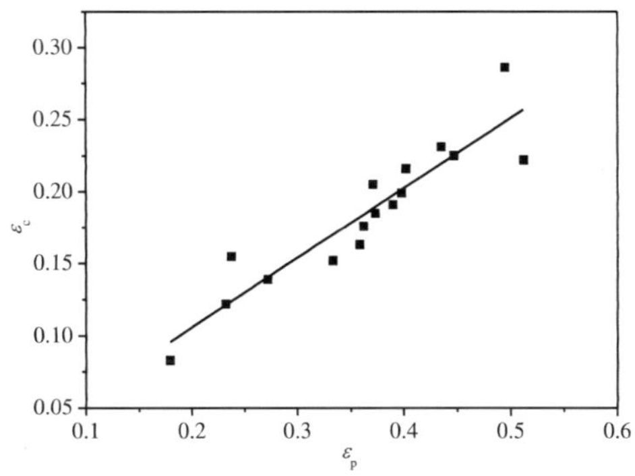

Figure 4. The relationship between the critical strain (ε_c) and peak strain (ε_p)

Based on the values of ε_c and ε_p, the relationship between the critical strain (ε_c) and peak strain (ε_p) can be plotted as shown in Figure 4. And the slope of the fitting line gives the equation in the following:

$$\varepsilon_c = 0.483\varepsilon_p + 0.009 \tag{15}$$

Therefore, the constitutive model for the 16Cr-5Ni-Mo steel considering the DRX can be summarized as follows:

$$\begin{cases} \dot{\varepsilon} = 3.57\times10^{14}[\sinh(0.011\sigma)]^{4.985}\exp(-\dfrac{424173}{RT}) \\ \varepsilon_c = 0.024Z^{0.051} \\ \varepsilon_c = 0.483\varepsilon_p + 0.009 \end{cases}$$

3.3. Processing maps of 16Ni-5Mo-Ni steel

To avoid the hot deformation defects and select the optimal hot working parameters, the method of dynamic material mode (DMM) is often used to establish the processing map. Prasad has developed the DMM which is successfully applied to the research for hundreds of materials [16]. In the DMM, the forging under hot deformation conditions is considered to be a dissipator of power. When the material has enough power to dissipate, the processing conditions are considered to be safe or stable. The total dissipation power P includes G and J, which can be expressed as [17]:

$$p = \sigma\dot{\varepsilon} = G + J = \int_0^{\dot{\varepsilon}} \sigma d\dot{\varepsilon} + \int_0^{\sigma} \dot{\varepsilon} d\sigma \tag{16}$$

where G content is defined as the power dissipated by plastic work; the J content is related to the power dissipated in microstructure evolution.

Based on data of the flow stress, the strain rate sensitivity m can be given in the following equation[18-19]:

$$m = [\frac{\partial(\ln \sigma)}{\partial(\ln \dot{\varepsilon})}]_{\varepsilon T} \tag{17}$$

And the efficiency of power dissipation η given regarding m parameter can be described as follows[20-21]:

$$\eta = \frac{J}{J_{max}} = \frac{2m}{m+1} \tag{18}$$

The efficiency of power dissipation η is a vital material parameter referring to the power dissipated by metallurgical mechanisms such as the DRX, DRV, wedge cracking, and so on. Consequently, the higher the value of the ηparameter, the better the material dissipates the flow of energy through microstructure evolution. Meanwhile, it means that the material deformed with high ηvalue shows high workability. To display the power dissipation parameters of the steel visually, the three-dimensional (3D) diagrams of the power dissipation efficiency changing with deformation temperature and strain rate are established in Figure. 5 at strains of 0.6 and 0.8, respectively.

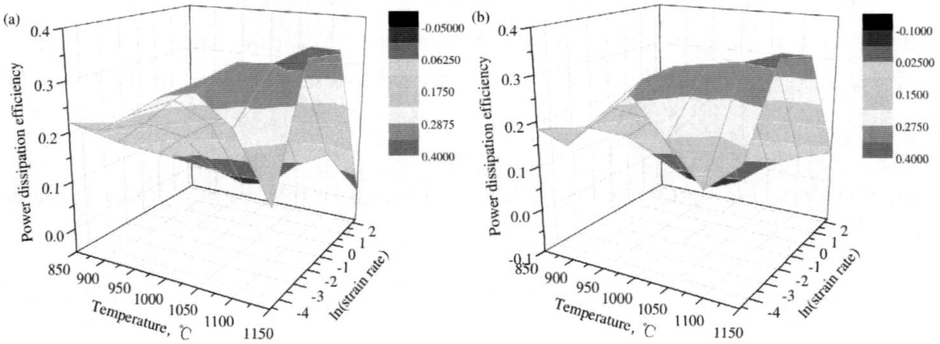

Figure 5. 3D diagrams of power dissipation efficiency of 16Cr-5Ni-Mo steel at strains: (a) 0.6; (b) 0.8

In Figure 5, the power dissipation efficiency is associated with the microstructure evolution, for example, DRX and DRV. Generally, the regions in which the value of power dissipation efficiency is more than 0.3 will be often considered as safe workability. And the conditions of DRX and DRV need a high value of power dissipation. In Figure 5(a), the 3D map displays one region with power dissipation efficiency range of 0.34—0.38 at temperatures of 1050—1150 ℃ and strain rates of 0.02—0.6 s^{-1}. In Figure 5(b), the 3D map displays one region with power dissipation efficiency range of 0.31—0.34 at temperatures of 980—1150 ℃ and strain rates of 0.02—1 s^{-1}. Above regions exhibit high power dissipation which is usually responding to good workability. However, other defects during hot deformation can also lead to high ηvalue, such as wedge cracking, void formation and so on. Therefore, the ηparameter cannot provide comprehensive advice for hot working. And the parameter ξrepresenting the instability has been considered to determine the workability region for the material. Factor ξcan be expressed as [22-23]:

$$\xi(\dot{\varepsilon}) = \frac{\partial \ln(\frac{m}{m+1})}{\partial \ln(\dot{\varepsilon})} + m < 0 \quad (19)$$

where $\xi(\dot{\varepsilon})$ represents the instability parameter. When the value of $\xi(\dot{\varepsilon})$ is negative, the hot deformation behavior is regarded as instability.

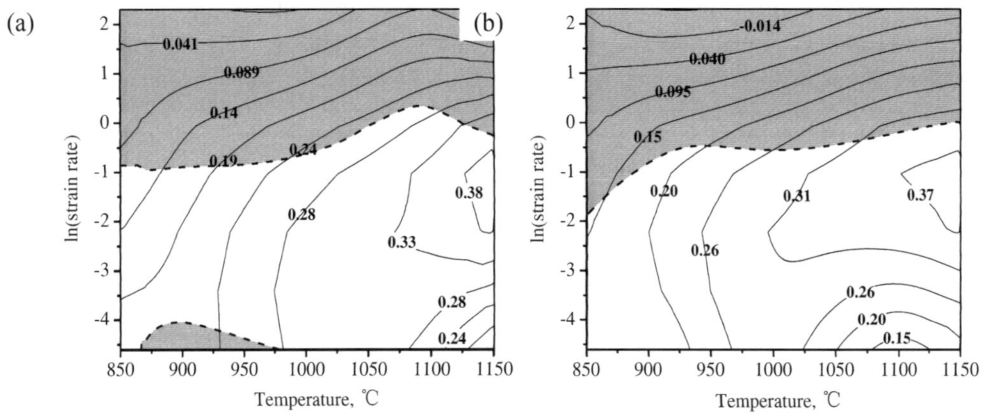

Figure 6. Process maps of 16Cr-5Mo-Ni steel under different temperatures and true strain at strains: (a) 0.6; (b) 0.8

To predict the workability of the 16Cr-5Ni-Mo steel precisely, the processing maps of the steel, composed of power dissipation map and instability map, are drawn in Figure 6. Figures 6(a), (b) show the processing maps at strains of 0.6 and 0.8, respectively. In maps, the contour numbers represent the power dissipation efficiency, and the color regions indicate the flow instability. It can be seen that with increasing of strain, the proportion of flow instability increases. It is pointed out that the variation trend of instability is nearly the same for all the conditions. Due to the considerable deformation strain in the practical production, the strain of over 0.6 is often used which is discussed deliberately in this work. From Figure 6(a) and (b), it can be found that the flow instability occurs at the high deformation strain rate. It is because that at high strain rate, there is not enough time for dislocation slipping and moving, resulting the stress concentration in localized defects and formation of cracking finally. Moreover, at low deformation temperature, such as 850—950 ℃, it also exists the instability region with the low strain rate. For the 16Cr-5Ni-Cr supermartensitic stainless steel, the δ-ferrite appears inevitably. Especially in the deformation, cracking generates along the δ-ferrite as brittle phase easily. Therefore, it should avoid harmful regions during hot working. Based on the processing maps and microstructure evolution, the optimized processing parameters for hot working of 16Cr-5Ni-Mo supermartensitic stainless steel are in the temperature range of 980—1150 ℃ and strain rate range of 0.02—0.36 s^{-1} at strains of 0.6 and 0.8.

4. Conclusion

(1) For 16Cr-5Ni-Mo steel, the following rate equation by class Arrhenius mode is determined:

$$\dot{\varepsilon} = 3.57 \times 10^{14}[\sinh(0.011\sigma)]^{4.985}\exp(-\frac{424173}{RT})$$

(2) The critical strain for the DRX of 16Cr-5Ni-Mo steel is expressed as: $\varepsilon_c = 0.024Z^{0.051}$ and $\varepsilon_c = 0.483\varepsilon_p + 0.009$, and the equations are used to predict the kinetics of DRX.

(3) Based on DMM, the processing maps of 16Cr-5Ni-Mo steel were established by considering power dissipation maps and instability maps. The optimum domain with high values of power dissipation and without flow instability is in the temperature range of 980—1150 ℃ and strain rate range of 0.02—0.36 s^{-1} at strains of 0.6 and 0.8.

References

[1] Wu G Z, Wang M Q, WANG C F, et al, 2012. *Trans Mater Heat Treat*, **33**(1): 136-40.
[2] Zhong Z Q, Tian Z L, Yang C, et al, 2015. *Eng Fail Analy,* **57**: 129-36.
[3] Ahgaie K M, ADHAMI F, 2010. *Mater. Sci. Eng. A*, **527**: 1052-7.
[4] Zou D N, HAN Y, Yan D N, et al., 2011. *Mater Des*, **32**: 4443-8.
[5] Yuan W H, Gong X H, Sun Y Q, et al., 2016. *J Iron Steel Res Int*, **23**(4): 401-8.
[6] Mirzadeh H, Cabrera J M, Abbas N, 2011. *Acta Mater*, **59**: 6441-8.
[7] Tan Y B, Ma Y H, Zhao F, 2018. *J Alloys Comp*, **741**: 85-96.
[8] Amir H, Mary A W, 2017. *J Mag Alloys*, **5**: 369-87.
[9] Gong B, Duan X W, Liu J S, et al., 2018. *Vacu*, **155**: 345-57.
[10] Meng Q G, Bai C G, Xu D S, 2018. *J Mater Sci Technol*, **34**: 679-88.
[11] Sellars C M, Mctegart W J, 1966. *Acta Metall*, **14:** 1136-8.
[12] Mcqueen H J, Ryan N D, 2002. *Mater Sci Eng A*, **322**(1-2): 43-63.
[13] Zener C, Hollomon J H, 1944. *J Appl Phys*, **15**: 22-32.
[14] Sellars C M, Whiteman J A, 1979. *Metal Sci*, **13**: 187-94.
[15] Najafizadeh A, Jonas J J, 2006. *ISIJ Int*, **46**: 1679-84.
[16] Prasad Y V R K, Gegel H L, Doraivelu S M, et al., 1984. *Metall Mater Trans A*, **15**(10): 1883-92.
[17] Prasad Y V R K, 1990. *India J Technol*, **28**: 435-51.
[18] Gegel H L, Malas J C, Doraivelu S M. *Processing Modeling of P/M Extrusion*. In: G. Bruggeman, V. Weiss (Eds.), Innovations in Materials Processing, Plenum Press, New York, 1985: 137-159.
[19] Liu X G, Ji H P, Guo H, et al., 2013. *Mater Sci Technol*, **29**(1): 24-9.
[20] Venugopal S, Mannan S L, Prasad Y V R K, 1993. *Mater Sci Technol*, **9**: 899-906.
[21] Wan Z P, Hu L X, Sun Y, et al., 2018. *J Alloys Comp*, **769**: 367-75.
[22] Murty S V S N, Rao B N, Kashyap B P, 2000. *Int Mater Rev*, **45**(1): 15-26.
[23] Ren F C, Chen F, Chen J, et al., 2018. *J Manuf Proc*, **31**: 640-9.

Institute of Physics
Dirac House, Temple Back
Bristol BS1 6BE UK

ISSN: 1757-8981
ISBN 978-1-5108-8117-4

CURRAN ASSOCIATES INC.
proceedings.com

XXI Winter School on Continuous Media Mechanics 2019

IOP Conference Series: Materials Science and Engineering
Volume 581

Perm, Russia
18 – 22 February 2019

IOP
Institute of Physics